The Fast Track to Determining Transfer Functions of Linear Circuits

The Student Guide

$v_{in}(t)$

$T = 2\pi r_c C_1$

$v_{out}(t)$

V_{in} ——o———$\bigwedge\!\!\bigwedge\!\!\bigwedge$———o V_{out}

R_1 r_C

C_1

Transfer function:

$$\frac{V_{out}(s)}{V_{in}(s)} = \frac{1 + s r_C C_1}{1 + s(r_C + R_1)C_1} = \frac{1 + \dfrac{s}{\omega_z}}{1 + \dfrac{s}{\omega_p}}$$

zero

pole

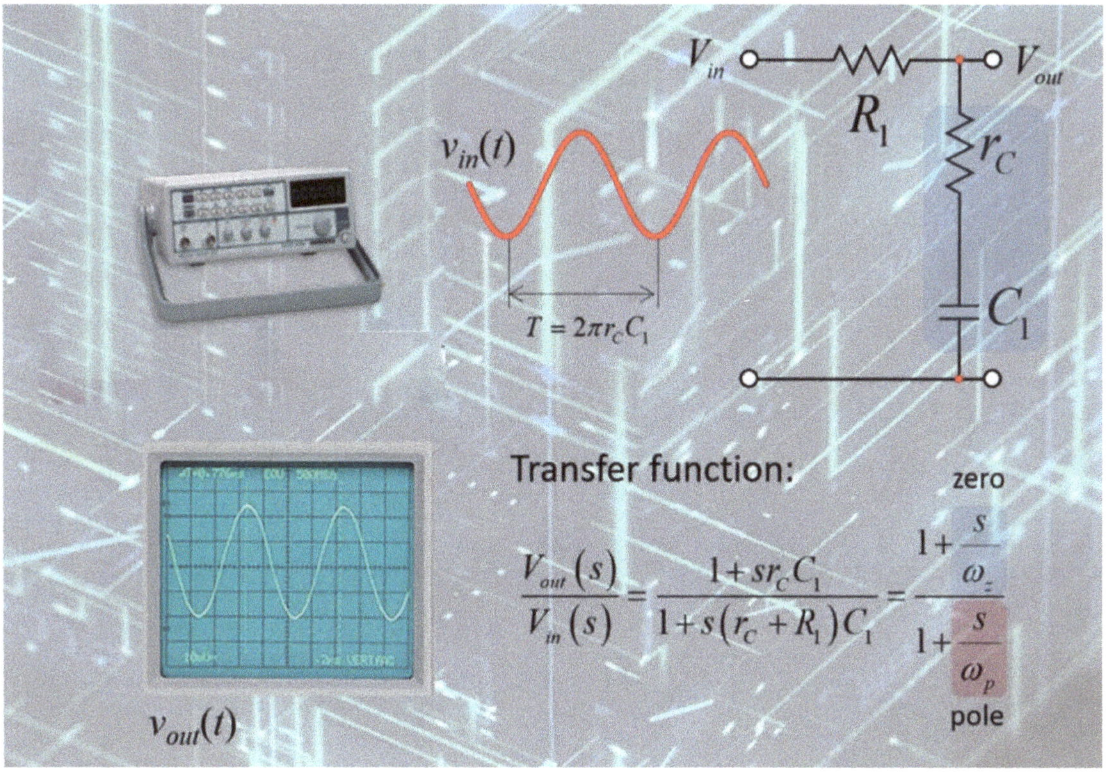

Christophe Basso

The Fast Track to Determining Transfer Functions of Linear Circuits: The Student Guide

Other books by Christophe Basso

Transfer Functions of Switching Converters: Fast Analytical Techniques at Work with Small-Signal Analysis (2021)
Linear Circuit Transfer Functions: An Introduction to Fast Analytical Techniques (2016)
Switch-Mode Power Supplies, Second Edition: SPICE Simulations and Practical Designs (2014)
Designing Control Loops for Linear and Switching Power Supplies: A Tutorial Guide (2012)
Switch-Mode Power Supplies: Spice Simulations and Practical Designs (2008)
Switch-Mode Power Supply SPICE Cookbook (1996)

ISBN 978-1-960405-19-7

Cover Design by Guy D. Corp, www.GrafixCorp.com
Index by Zurain Shahzad https://www.linkedin.com/in/zurain-shahzad-3a367a204/

Faraday Press
1000 West Apache Trail—Suite 126
Apache Junction, AZ 85120 USA

Acknowledgements

THIS NEW BOOK has been reviewed by many friends and colleagues I was fortunate to interact with. Going through the details of so many expressions and figures requires tenacity and patience, thank you all! My former *onsemi* colleagues kindly reviewed parts or entire chapters and I want to thank Stéphanie Cannenterre (France), Joël Turchi (France) and Alain Laprade (US) for their efforts. In the east, Dmitriy (Russia) reviewed and commented many pages while he was applying the FACTs to some of his projects. David Morrison (US) from the How2Power online newsletter commented on the original book structure and made useful formatting recommendations. Riccardo Collura (Italy) from Future Electronics looked at some of the documents and sent me comments on the content. Andrew Krill (US) reviewed part of the work and made interesting suggestions for some sections. Professor Katherine Kim from National Taiwan University and her students—Adrian Keil, Chi-Yuan Huang and Wen-Yen Li—did provide precious feedback after reviewing and exercising parts of the examples.

Finally, as a loyal reviewer, Tomáš Gubek (Czech Republic) thoroughly reviewed all the materials, going into equations and figure details.

I cannot forget my sweet wife, Anne, who has encouraged me to pursue the writing, kindly pausing her piano sessions while I struggled on complicated circuits analyses.

Last but not least, I would like to thank Ken Coffman at Stairway Press for giving me the opportunity to publish my work.

Preface

HAVE YOU EVER felt powerless before an electrical circuit assembling a few insignificant passive elements for which you had to determine the transfer function? I remember panicking many years ago in university when the exam was featuring a simple *LC* network with a few resistors around: "determine the output voltage after shaping the transfer function in a normalized form" they said, while cascading questions based on the correct first answer. Of course, after reaching algebraic paralysis, I was giving up, feverishly jumping to the next problem. I recall the solution was usually simple but I often missed the spark that could have illuminated the path. Those days are long gone of course, but reading posts on the excellent community site *electronics.stackexchange.com* I can sometimes see lost students and engineers, lacking a fast and radical method to get straight to the point, without losing time in fixing equations or factoring expressions.

When analyzing an electrical diagram to determine one of its transfer functions, you can choose among a variety of tools to get you there. The most common one uses the so-called *brute-force* method which consists of writing down Kirchhoff's current and voltage laws (KCL and KVL), trying to give the algebraic magma a meaningful shape—read a factored polynomial expression. If this works fine when a numerical result is wanted, for instance via the implementation of matrixes in a solver, it can quickly skid in a symbolic disaster where you trample and cannot move further. The arm-long equation you have found is a dead end and perhaps even flawed. The solution? Start all over again and try to find the point where you made the sign error or discarded a term in one of the equations. Needless to relate you how frustrating it is to end up this way because I know you have been there!

On the opposite, the method I am proposing in this book builds on reducing the mathematical overhead and, in many cases, completely eliminate it. Yes, you have well read: no equations! This approach uses the so-called *fast analytical circuits techniques* abbreviated FACTs. The FACTs promote a simple approach: if a circuit is too complicated, split it in several simpler subcircuits that you solve individually. Then assemble all intermediate results to form the final one you want. In case the result deviates from what you were expecting, identify the guilty sketch and fix it without restarting all over from scratch. Believe me, this is invaluable when working on difficult circuits with many components, active or passive. Oh yes, I forgot to mention but FACTs work on *RLC* networks but also on active circuits featuring operational amplifiers (op-amps) or transistors.

The FACTs do not date from today and go back to the foundations of control engineering in the 40s, later revisited and improved to take on the formalism I use in this book. To that respect, it is important to emphasize the work carried by Dr. Middlebrook of Caltech in the 80s who really took the technique to an upper level for a practical purpose described as *design-oriented analysis* or D-OA [1]. The term implies that the formula you determine must serve a goal which is to let you design a circuit featuring specific characteristics purposely made visible in your expression. In other words, the formula you came up with unveils its salient points such as gain, multiple poles or zeroes, resonance and so on. These are the characteristics of a *low entropy* expression with well ordered terms, offering insight on what its frequency response could look like and what elements affect it. The FACTs naturally pave the way to this type of well-organized formulas while the brute-force approach delivers *high-entropy* formulas, requiring further energy to rearrange the whole thing into a polynomial form, if you can. More recently, Dr. Vorpérian from JPL, published a book [2] demonstrating how FACTs could efficiently be used for solving structures previously requiring many lines of algebra. Vorpérian promoted his technique with a course I was fortunate to attend in the 2000s. My own book on the topic came out later [3] and focuses on more practical applications starting with basic circuits. It can be seen as a stepping stone to let you acquire the skill step by step and tackle more complex readings afterwards.

This new book differs from the previous ones. It is aimed to be truly practical, leading you to solve your immediate problem by applying recipes and a step-by-step approach. In many cases, the only required accessories are a sheet of paper and a pen. I have used Mathcad® and SPICE for illustrating purposes but any other solver will do. Chapters one to four are truly a crash course on the FACTs, assuming you already know how to manipulate simple Laplace expressions and are familiar with the concept of poles and zeroes described in [3]. The following chapters detail how to determine transfer functions of classical networks from the first to third order. After going through the proposed examples at your own pace, you should be able to master the technique and show your friends how to write the transfer function of a second-order *RLC* network in less than a minute: prepare for glory!

As usual, feel free to send me your comments or any typos you may find at cbasso@orange.fr. I will maintain an errata list in my personal web page [4] as I did with the previous books. Thank you and happy reading to you all.

For teachers who could use PDFs of the illustrations in their classes, I will make them available on my website. Professors: for support of your classes, feel free to contact me via email.

—Christophe Basso, November 2023

References

1. R. D. Middlebrook, *Methods of Design-Oriented Analysis: Low-Entropy Expressions*, Frontiers in Education Conference, Twenty-First Annual conference, Santa-Barbara, 1992.
2. V. Vorpérian, *Fast Analytical Techniques for Electrical and Electronic Circuits*, Cambridge 2002, 0-521-62442-8.
3. C. Basso, *Linear Circuit Transfer Functions – An Introduction to Fast Analytical Techniques*, Wiley, 2016.
4. http://powersimtof.com/Spice.htm

About the Author

CHRISTOPHE BASSO WORKS as a business development manager for Future Electronics in France. In this role, he provides technical assistance to customers developing power switching converters in Europe. Before this position, he was a Technical Fellow with *onsemi* in Toulouse, France. He led an application team dedicated to developing new offline PWM controller's specifications. Christophe has originated numerous integrated circuits among which the NCP120X series has set new standards for low standby power converters.

Christophe has released several books on power conversion and simulation. His last title was published with Faraday Press and is entitled *Transfer Functions of Switching Converters*. In this work, he applied the fast analytical circuits techniques for determining the four transfer functions of many switching converters.

Christophe has over 25 years of power supply industry experience. He holds 25 patents on power conversion and often participates in conferences and trade magazines including How2Power. Prior to joining *onsemi* in 1999, Christophe was an application engineer at Motorola Semiconductor in Toulouse. Before 1997, he worked at the European Synchrotron Radiation Facility in Grenoble, France, for 10 years. He holds a *diplôme universitaire de technologie* from Montpellier University (France) and a MSEE from the Institut National Polytechnique of Toulouse (France). He is an IEEE Senior member.

When he is not writing, Christophe enjoys mountain hiking in the Pyrenees.

Table of Contents

Chapter 1: Transfer Functions

THE UPCOMING CHAPTERS teach how to determine transfer functions in a swift and efficient manner with the *Fast Analytical Circuits Techniques* known as *FACTs*.

In the author's opinion, they are unbeatable in terms of simplicity and ease of application in most cases. Applying the *divide-and-conquer* technique—split a complicated schematic into small individual sketches that are independently solved—the FACTs naturally lead to so-called *low-entropy* expressions implying a factored form in which gains, poles, zeroes or resonance—if any—are immediately distinguished. Should an error be spotted when examining the final result, there's no need to restart from scratch as with the classical approach: just identify the guilty sketch, fix it, update the coefficients at play—while leaving the rest intact—and *voilà*.

The term *low-entropy* was forged by Dr. Middlebrook in his founding papers [1], [2] where he made an analogy between a raw polynomial expression and a disorganized system described by thermodynamics as *entropic*. Applying *brute-force* analysis to high-order circuits usually leads to complicated expressions, leaving an algebraic paralysis with *high-entropy* formulas that can't further be used because they are simply intractable.

By applying FACTs, execution speed increases and the final result appears in a well-ordered polynomial form, often without the need for extra factoring efforts. FACTs do not require acquisition of new knowledge; they build on what was already learned in the university and extend the range to drastically simplify analyses.

Finally, and this is an essential point for students and engineers, FACTs are not just another convenient tool for quickly determining a transfer function: they shape the result to fulfill a design goal.

As such, by adopting the adequate mathematical structure, you will naturally highlight the gain you want for this bandpass filter or the quality factor for this resonating network you have to tune.

This approach is described by the term *Design-Oriented Analysis*, or D-OA for which Dr. Middlebrook always insisted on the hyphen linking the two words: you write your final expression for design purposes, not for being the first to find this 4th-order polynomial formula your school mates still sweat at unveiling.

These introductory chapters can be seen as a crash course on FACTs with subsequent chapters documenting numerous examples for quickly exercising your newly acquired skill.

I wrote an introductory part on transfer functions to detail the adopted formalism around time constants. As with any summary, it is incomplete and assumes you are at ease with Laplace transforms and poles-zeroes expressions.

For those wanting to dig into the subject further, beside the classical text books on the subject, I encourage reading references [3] and [4] which cover the topic of transfer functions and FACTs with solid theoretical foundations.

After this quick introduction, it's time to start with a recapitulation about what a transfer function is and how to determine it.

1.1 Transfer Functions

A Transfer Function, often abbreviated TF, is a mathematical relationship linking a *stimulus* to a *response*. This is the formal definition found in the literature.

Practically speaking, inject a signal or a *stimulus* in the considered input and observe its propagation in the circuit to form a *response* observed at the output terminals.

The stimulus or *excitation* signal can be of any shape, but usually, for the so-called *harmonic analysis*, it is a sinewave of sufficiently low amplitude to keep the system linear, but strong enough to distinguish it from the noise.

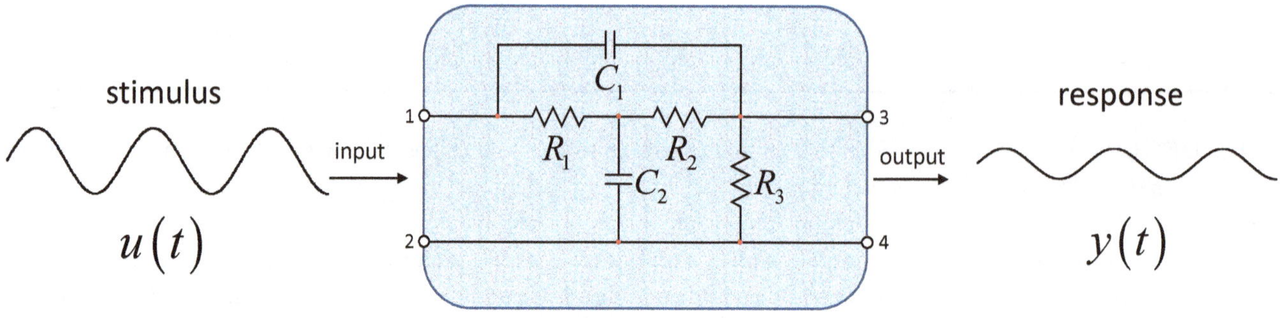

Inject a Sinusoidal Waveform into the Box and Study its Course Throughout

Figure 1.1

In the Laplace-domain, perform analyses with the complex number frequency parameter *s* defined as $s = \sigma + j\omega$ in which σ and ω are real numbers. When performing harmonic analysis, set σ to zero thus having $s = j\omega$ in which ω represents the angular frequency at which the network is excited. Note that the *s* in Laplace notation is italicized and lowercase versus a normal s, also lowercase but not italicized, which designates time in seconds.

Figure 1.1 shows a simple illustration of the measurement principle.

In this picture, the input signal *u*(*t*) is applied at the *input port* between terminals 1 and 2 while the *time-domain* response *y*(*t*) is observed at the *output port* between connections 3 and 4. A *port*, in our example, represents a pair of connecting terminals—making the circuit shown in the figure known as a *two-port* network.

Note that *u* and *y* are lowercase because they represent instantaneous or *time-dependent* waveforms as opposed to a continuous value like I_1 which, for instance, could represent the current entering terminal 1 under a constant bias. Such variable would always be uppercase.

The periodic waveforms as illustrated in Figure 1.1 are characterized by an amplitude and a phase. The input signal *u* travels through the network and may undergo amplification, attenuation and/or delays before forming the response *y*.

The amplitude and phase of the response will vary with the excitation frequency *f* and we will store, at each frequency point *f*, the ratio of the response amplitude $Y(f)$ to the excitation amplitude $U(f)$. As we will soon see, the amplitude can be expressed in volts or amperes depending on the desired transfer function. At each frequency *f*, we save the phase information linking the input and output waveforms.

As *U* and *Y* are *complex* variables affected by a magnitude and a phase, we can write:

$$H(s) = \frac{Y(s)}{U(s)} \tag{1.1}$$

H designates a *transfer function*, a mathematical relationship linking a response signal *Y* to an excitation signal *U*. I called the transfer function *H* in this example, but it can take on any name like *Z* for an impedance, *Y* for an admittance, *G* for a compensation filter and so on.

Please note in (1.1) that the excitation signal *U* resides in the transfer function denominator while the response *Y* sits in the numerator. It will always be this way in this book.

The transfer function is characterized by a magnitude noted $|H(f)|$ and an argument or a phase, $\angle H(f)$ also noted $\arg H(f)$. The stored ratios $\frac{Y(f)}{U(f)}$ correspond to the transfer function magnitude (also called *modulus*) observed at a frequency *f*.

The phase difference between *Y* and *U* represents the transfer function argument or phase at the considered frequency.

As an example, Figure 1.2 shows how to extract this data with an oscilloscope.

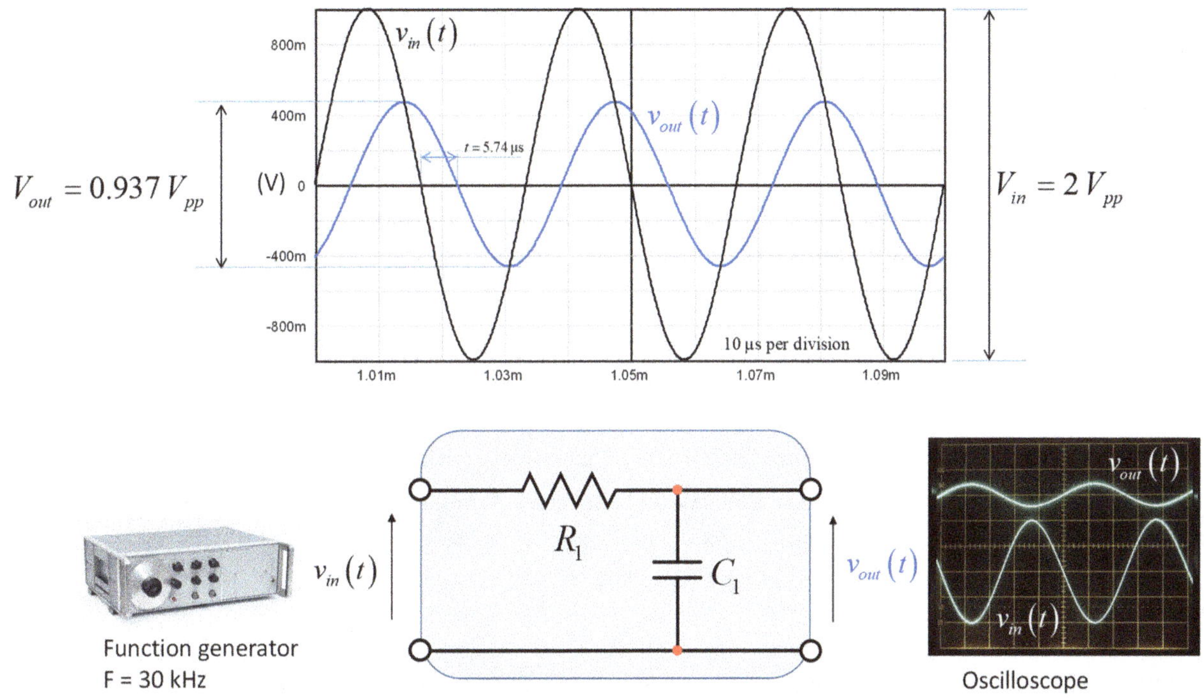

Transfer Function Response Determined by Measuring Amplitude and Phase at Selected Frequencies

Figure 1.2

In this example, the capacitor is 10 nF while the resistance is 1 kΩ.

Inject a 30-kHz, 2-V peak-to-peak sinusoidal voltage with a function generator—the stimulus—and observe the output—the response—on an oscilloscope. Using cursors or evaluating the height of the waveform on the screen, measure an output voltage of 937 mV peak-to-peak.

The magnitude of the transfer function at 30 kHz is the ratio of the two measurements computed as:

$$\left|H(30\ \text{kHz})\right| = \frac{0.937}{2} = 0.468 \tag{1.2}$$

Log-compress it in decibels (dB) as follows:

$$20\log_{10}(0.468) = -6.6\ \text{dB} \tag{1.3}$$

We say this filter provides a 6.6-dB attenuation at 30 kHz or the magnitude is -6.6 dB at that frequency.

The phase is inferred from the time difference between the signals. If we take V_{in} as the reference, we see that V_{out} is delayed or *lags* V_{in} by 5.74 µs implying a negative phase shift in this example. If a 33.33-µs period corresponds to 2π radians or $360°$, then a 5.74-μs time shift implies a phase of:

$$\angle H\left(30\,\text{kHz}\right) = -\frac{5.74\mu}{33.33\mu}\times 360° \approx -62°$$

(1.4)

We can write the filter is affected by a $62°$ phase delay/lag or the argument/phase at 30 kHz is $-62°$.

The transfer function magnitude dimension depends on the observed variables.

In this example, because we observe volts for both input and output variables, the transfer function magnitude is *dimensionless* or *unitless*. For an impedance, a stimulus in amperes and a response in volts lead to a transfer function Z expressed in ohms.

Later in this chapter, you may see a subscripted notation like in H_v, indicating that this is a voltage gain or H_i for a current gain for instance. The subscript can also be a zero as in H_0 which designates a dc gain determined for $s = 0$ or H_∞ determined when s approaches infinity.

The modulus or magnitude of H can only be greater than or equal to zero. It is what makes the difference between an amplitude which can take on any value, positive, zero or negative and a magnitude which can only be zero or positive.

If it is zero, there is no output signal. If $|H|$ is less than 1, *attenuation* has occurred. If $|H|$ is greater than 1, *gain* has occurred. If the magnitude can only be a null or positive number, what about a gain of -2 then? It simply characterizes a stage offering an amplification factor of 2, lagging or leading the excitation signal phase by $180°$.

During your experiment, you could begin analyzing the circuit at 100 Hz and increment by 1 or 10 Hz for the next points and carry on until you reach 10 MHz.

This implies a tremendous number of points and a tedious measurement session.

One good thing is to limit the number of points within a decade like what is done in a SPICE simulator. Usually, 50-100 points per decade are good enough, but peaky responses may require more points per decade for an adequate granularity.

To obtain the log-compressed frequency points—as in a simulator running an ac analysis—a quick routine can be programmed, such as the Mathcad® sheet shown in Figure 1.3:

$\text{Start} := 10^2$ **Start frequency**

$\text{Stop} := 10^7$ **Stop frequency**

$\text{Points} := 100$ **Number of points per decade**

$\text{Number\_of\_decades} := (\log(\text{Stop}) - \log(\text{Start}))$

$\text{Number\_of\_points} := \text{Number\_of\_decades} \cdot \text{Points} + 1$

$k := 0 .. \text{Number\_of\_points}$

$f_k := 10^{\log(\text{Start})+k\cdot\frac{\text{Number\_of\_decades}}{\text{Number\_of\_points}}} \cdot \text{Hz}$

$\omega_k := 2\cdot\pi\cdot f_k$

$f_0 = 100\,\text{Hz}$ $\quad\text{Number\_of\_decades} = 5$ $\quad\text{Number\_of\_points} = 501$

$f_1 = 102.325\,\text{Hz}$

$f_2 = 104.703\,\text{Hz}$

$f_{10} = 125.835\,\text{Hz}$

$f_{30} = 199.251\,\text{Hz}$

$f_{40} = 250.727\,\text{Hz}$

$f_{50} = 315.502\,\text{Hz}$

$f_{100} = 995.415\,\text{Hz}$

...

To Limit the Number of Collected Points, Choose a Smaller Number of Points per Decade

Figure 1.3

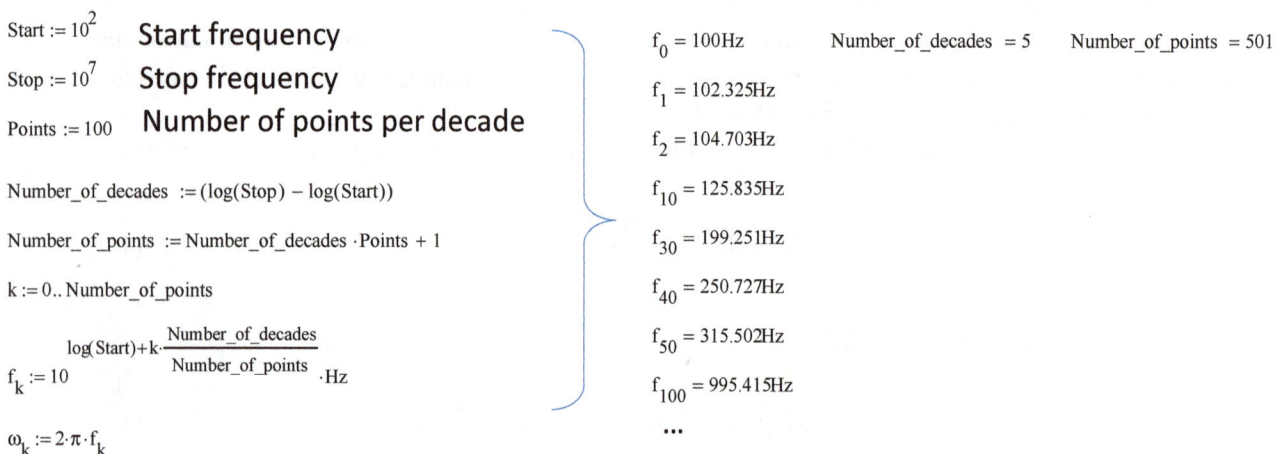

When magnitude and phase points in an array are collected, for instance between 100 Hz and 10 MHz, the response can be graphed in a so-called Bode plot.

The frequency axis is log-compressed with the computed points and the vertical axes displaying the magnitude in decibels and the phase in degrees. Plots like Figure 1.4 can be obtained.

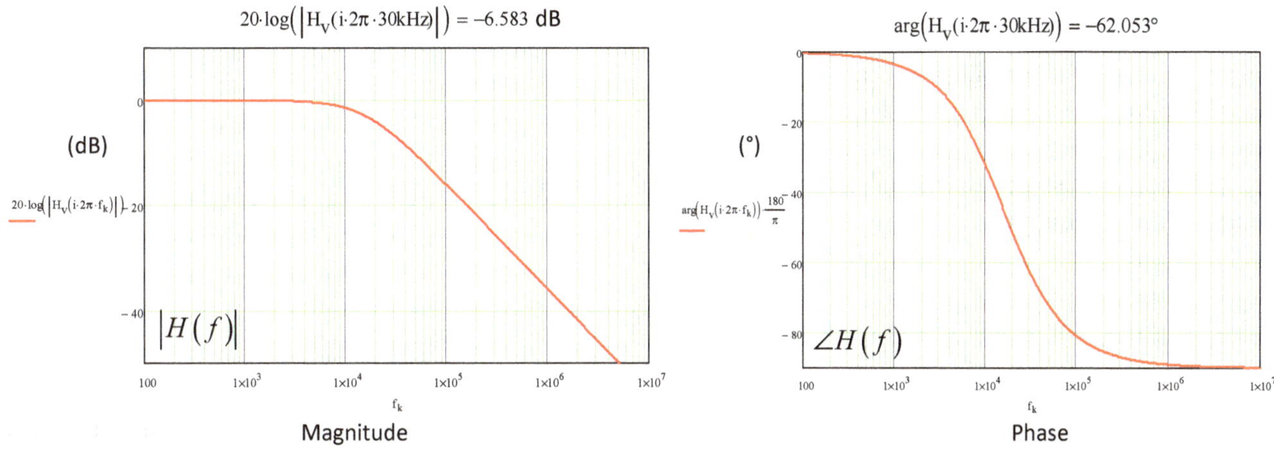

Create a Bode Plot by Collecting and Displaying Magnitude and Phase Data Points

Figure 1.4

Bode plots are widely used when illustrating transfer function responses and are particularly well suited in control engineering for stability analyses. Other graphs such as Nyquist or Nichols also exist and serve different illustration purposes; they won't be used in this book.

1.1.1 Different Types of Transfer Functions

In the previous example, a transfer function linking a voltage response with a voltage stimulus is shown. But, we could also think of a current stimulus bringing a current response or any combinations of the like as represented in Figures 1.5 and 1.6. In these examples, the stimulus and the response are respectively applied and observed at different ports.

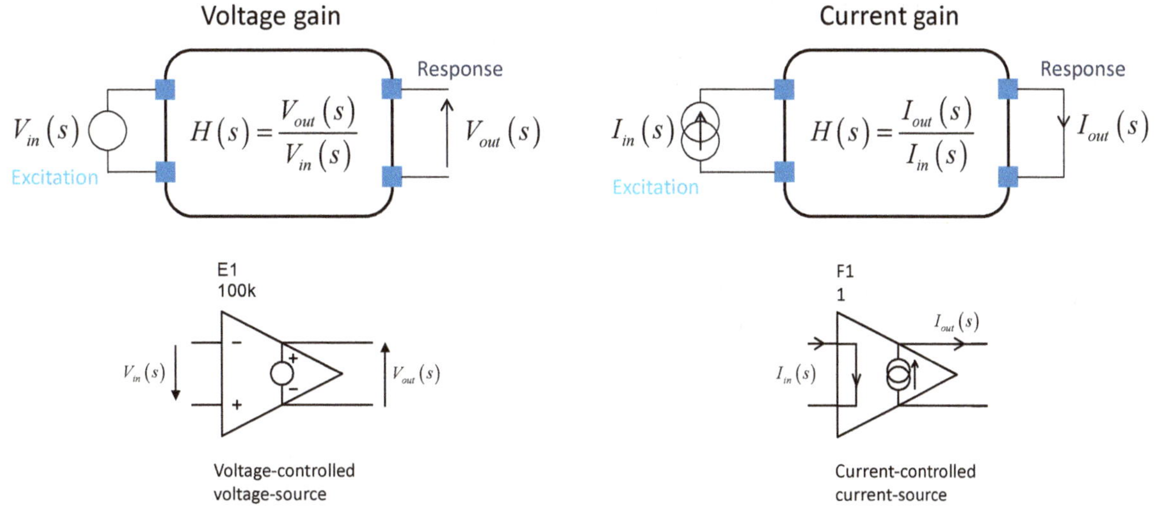

In these Transfer Functions, Stimulus and Responses are Observed at Different Ports

Figure 1.5

The first transfer function is a *voltage gain* where input and output signals are voltage variables:

$$H_v(s) = \frac{V_{out}(s)}{V_{in}(s)} \tag{1.5}$$

A typical case is an amplifier driving loudspeakers: a source like a CD player injects an audio signal which is amplified by a power circuit and drives a loudspeaker enclosure. It can be modeled as voltage-controlled voltage-source or an *E*-source in a SPICE simulator. H_v is unitless, sometimes expressed in [V]/[V].

The second transfer function deals with currents only. A first current is injected in the input port and a second current is observed at the output port:

$$H_i(s) = \frac{I_{out}(s)}{I_{in}(s)} \tag{1.6}$$

It is a unitless *current gain* sometimes expressed in [A]/[A]. A current-controlled current-source is modeled by an *F* primitive in SPICE.

In Figure 1.6, the *transfer admittance*—often abbreviated as *transadmittance*—designates a circuit excited by a voltage source whose response is a current:

$$Y(s) = \frac{I_{out}(s)}{V_{in}(s)} \tag{1.7}$$

The dimension of this transfer function is [A]/[V] or siemens [S].

Amplifiers implementing this type of control—they are modeled as a voltage-controlled current-source (the primitive is *G* in a SPICE simulator)—are often designated in the literature or in components datasheets, as *transconductance* operational amplifiers or OTAs.

Their transconductance g_m is expressed in siemens also, for instance 100 µS for a typical example. *mhos* and even Ω^{-1} are sometimes used, but are not part the *international system of units* (*système international d'unité* or SI).

The fourth transfer function is a *transfer impedance* also known as *transimpedance*.

The excitation is a current source while the observed response is a voltage:

$$Z(s) = \frac{V_{out}(s)}{I_{in}(s)} \tag{1.8}$$

As an example, a transimpedance amplifier is typically used to amplify a photodiode current.

Z is expressed in [V]/[A] and can be modeled in SPICE by an *H* primitive.

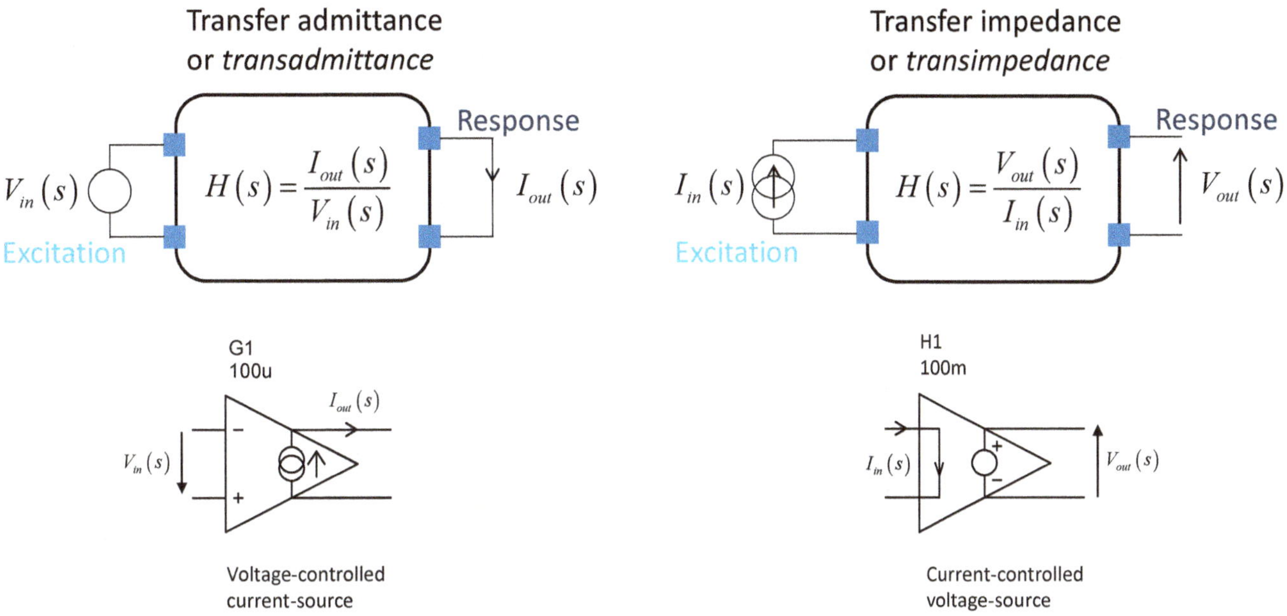

The Stimulus and the Response Shown as Different Dimensions—Still Observed at Different Ports

Figure 1.6

For the two last transfer functions describing an impedance Z and an admittance Y, the stimulus and the response are observed at a common port.

It is therefore important to specify how the excitation is created—and what is considered a response.

For a so-called *driving point impedance* or DPI, the excitation is a current source while the voltage across its terminals represents the response.

On the other hand, if we want to determine an admittance, the excitation will be a voltage source applied at the considered port while the generated current becomes the response. These notions are summarized in Figure 1.7.

By the way, this is a convention and one could argue that applying a voltage stimulus to determine an impedance works equally well. This is true.

However, if you pay attention, in all previous transfer functions definitions, the stimulus variable lies in the denominator while the response appears in the numerator. To maintain this convention for the impedance and admittance definitions, we adopt a similar principle.

The practical application is immediate if to determine the impedance offered by the connecting terminals—as shown in Figure 1.8.

In this illustration, the arrow preceded by the symbol *Z?* or *R?* implies that we need to find the impedance or the resistance (real part of the impedance) offered by the connecting port at which the arrow points. To do so, install a current test generator I_T which generates a voltage V_T across its terminals.

Then, the impedance is simply:

$$Z(s)=\frac{V_T(s)}{I_T(s)} \tag{1.9}$$

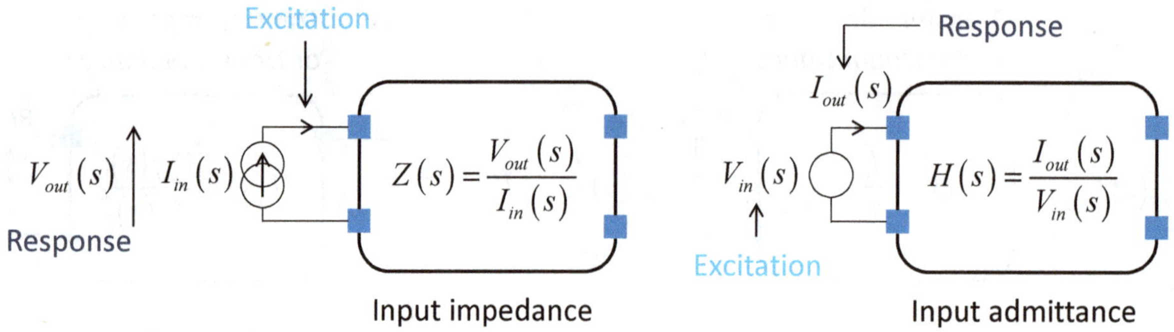
Excitation
Response
$V_{out}(s)$
$I_{in}(s)$
$Z(s)=\frac{V_{out}(s)}{I_{in}(s)}$
Input impedance
$I_{out}(s)$
$V_{in}(s)$
$H(s)=\frac{I_{out}(s)}{V_{in}(s)}$
Input admittance

In these Transfer Functions, Stimulus and Response are Evaluated at a Common Injection Port

Figure 1.7

To determine a resistance R, inject a dc current I_T and measure the collected voltage V_T:

$$R=\frac{V_T}{I_T} \tag{1.10}$$

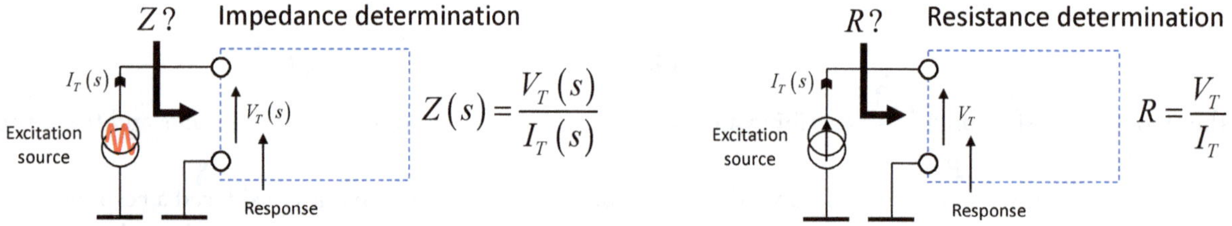
$Z?$ Impedance determination
$I_T(s)$
Excitation source
$V_T(s)$
Response
$Z(s)=\frac{V_T(s)}{I_T(s)}$
$R?$ Resistance determination
$I_T(s)$
Excitation source
V_T
Response
$R=\frac{V_T}{I_T}$

For this Transfer Function, inject a Test Current I_T and Collect Response V_T across the Current Generator Terminals

Figure 1.8

Apply the same technique to determine an admittance as described in Figure 1.9. This time, a voltage source is connected across the input terminals and represents the stimulus. The absorbed current is the response created by the voltage source. Then the admittance is simply:

$$Y(s)=\frac{I_T(s)}{V_T(s)} \tag{1.11}$$

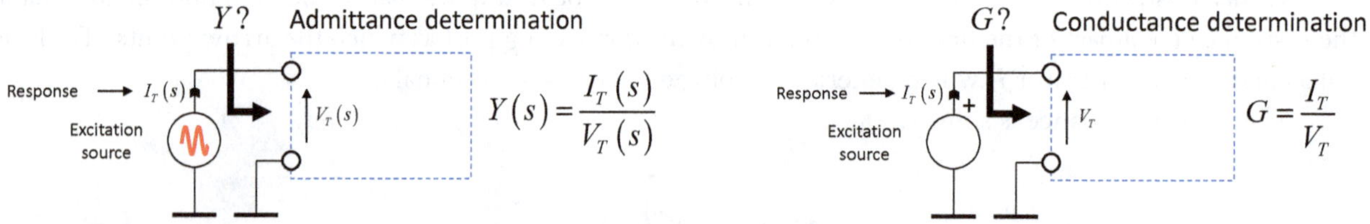
$Y?$ Admittance determination
Response
$I_T(s)$
Excitation source
$V_T(s)$
$Y(s)=\frac{I_T(s)}{V_T(s)}$
$G?$ Conductance determination
Response
$I_T(s)$
Excitation source
V_T
$G=\frac{I_T}{V_T}$

In this Transfer Function, a Voltage Source Biases the Input Terminals and the Absorbed Current Represents the Response

Figure 1.9

If a conductance G (the real part of the admittance) is needed, bias the circuit with a dc voltage source V_T and measure the absorbed current I_T:

$$G(s)=\frac{I_T(s)}{V_T(s)} \tag{1.12}$$

It is important to understand how the test generators are arranged to perform the various operations—like, for instance, determining a gain or impedance. These sketches will often be referred to throughout this book.

In particular, the core exercise of the FACTs consists of determining a resistance R "seen" from the capacitor or the inductance connecting terminals when these elements have been temporarily disconnected.

Sometimes just *inspecting* the circuit to infer this resistance will be enough. Otherwise, use (1.10) to find the analytic expression of R.

1.1.2 Time Constants, Poles and Zeroes of a Transfer Function

The transfer function of a linear time-invariant (LTI) system without pure delays—think of a delay line for instance—can be defined by the ratio of two polynomials: a numerator designated as $N(s)$ with $D(s)$ in the denominator:

$$H(s)=\frac{N(s)}{D(s)} \tag{1.13}$$

The denominator D hosts the *poles* of the transfer function whereas *zeroes* reside in the numerator N.

As we will see in the next section, poles represent roots of the denominator while zeroes designate roots of the numerator.

The number of poles determines the *degree* of the denominator and the *order* of the transfer function. If there is only one pole, it is a 1st-order transfer function.

Should there be two poles, then there is a 2nd-order expression—and so on. In the Laplace-domain, when the stimulus frequency is tuned to a pole position, the denominator cancels and the magnitude of H in (1.13) becomes infinite.

Let's start with a 1st-order denominator:

$$D(s)=1+s\tau \tag{1.14}$$

In this expression, the Greek letter τ (tau) represents a *time constant* whose unit is time [s].

A time constant associates an energy-storing element—a capacitor C or an inductor L—with a resistance R. For example, when the capacitor or inductor are labeled C_1 and L_2, the associated time constants adopt the same label for clarity.

The time constant defined for a capacitor C_1 will then be:

$$\tau_1 = RC_1 \tag{1.15}$$

For an inductor L_2 it is:

$$\tau_2 = \frac{L_2}{R} \tag{1.16}$$

A time constant represents the time needed by a system to change from one *state* to the other. In Figure 1.10, to change the voltage across capacitor C_1 when driven by a resistance, C_1 takes a certain amount of time to charge.

Correspondingly, the current in inductor L_2 also driven by a resistance won't change instantaneously and will take some time to reach its final value.

In both cases, the 10-µs time constant defines the time duration for the variable to start from its initial value and reach 63% of the *forced* value. If you remove the excitation or the force, the circuit is left with initial conditions (for example, 1 V across the capacitor or 1 mA in the inductor) before returning to its initial state. As shown on the right side of Figure 1.10, this *natural* response lets you quantify the time constant.

To express the time constant of a linear electrical circuit, look at the circuit freed from its excitation source—which has been turned off for this exercise.

Temporarily disconnect the capacitor or the inductor from its connecting terminals and "look" through the connections to infer the resistance R seen. This simple process lays the foundations of the FACTs and is illustrated in Figure 1.11. Without writing a single line of algebra, *inspect* the circuit and determine the resistance by mentally evaluating the electrical connections. This is the simplest approach and works well with passive circuits.

With circuits featuring controlled sources, inspection can be impractical and the circuit electrical properties cannot be inferred. Then, use the technique described in Figure 1.8 by injecting a test current I_T and determining the voltage V_T developed across the injection source.

The resistance R can be obtained by identifying one or the linear combination of a group of resistances in a passive circuit, but can often be supplemented with coefficients such as gain or transconductance when active elements (transistors, operational amplifiers or op-amps) are involved. Then, verify the homogeneity of the final result to guarantee that R is correctly expressed in ohms.

The time constants of a circuit are independent from the input stimulus and *solely* depend on the architecture: how R, C and L are connected in the circuit when the excitation source or stimulus is turned off. The number of natural frequencies in the system—and consequently its order—is linked to the number of independent *state variables* in that circuit.

A state variable is associated with an energy-storing element (for instance, the voltage across a capacitor and the current in an inductor) and defines the *state* of the circuit at a given time. A state variable is independent if not *uniquely* defined by other state variables. Counting the number of energy-storing elements in a circuit to determine its order thus needs to be supplemented with a check on the state variables to account for a degenerate case if any. For instance, two capacitors C_1 and C_2 in parallel share the same state variable and form a 1st-order system.

Add a small resistance in series with C_2 to create two independent state variables for a second-order circuit. More details and examples are found in [3].

FACTs are mostly about determining the time constants of the system when set in two different conditions:

a) when the excitation source is turned off and;
b) when the ac response is *nulled* despite the presence of a stimulus signal.

The term *null* implies that despite the presence of a stimulus injected in the network, a response is not observed at the output.

This statement might sound weird right now but bear with me as we will come back to this important point later in the text.

It is important to emphasize that (1.14) is unitless, implying that s shall be multiplied by a single time constant or the sum of several time constants (which is still time) when s is part of higher-order expressions. When determining time constants, it is important to identify them clearly as equations are written involving energy-storing elements, especially when there are many in a circuit.

That way, each time constant is always well identified during the analysis.

As shown in (1.15) and (1.16), I recommend *always* associating the time constant label with the involved energy-storing element.

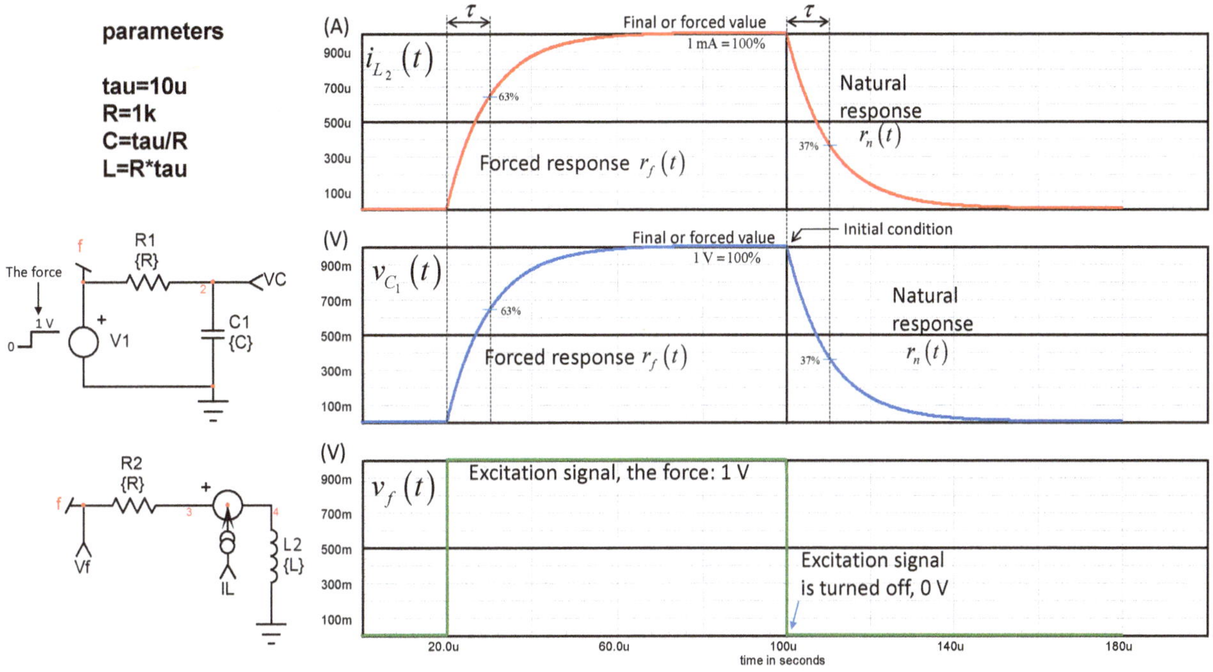

The Time Constant of a System Defines the Time it Takes to Change from One State to the Other

Figure 1.10

Starting with (1.14) and a time constant labeled τ_1, determine the root s_p the denominator cancels:

$$1 + s_p \tau_1 = 0 \tag{1.17}$$

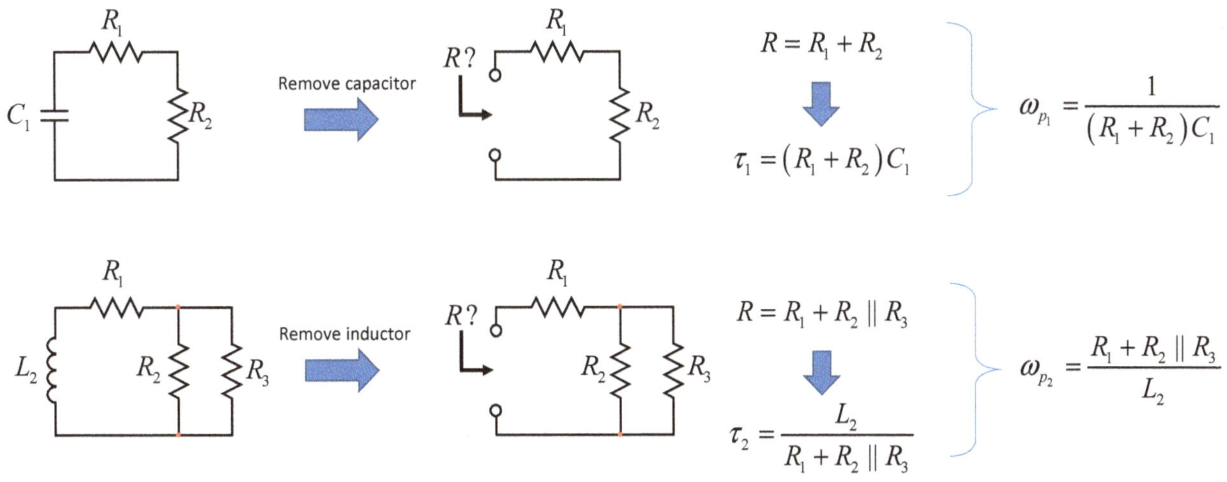

Disconnect the Energy-Storing Component and Look Through its Connections to Determine *R*. Once *R* is Determined, the Pole is Obvious

Figure 1.11

In this simple case, it is immediate and equal to:

$$s_p = -\frac{1}{\tau_1} \tag{1.18}$$

This is a negative root and leads to a decaying time-domain response. Roots can be placed in the *s*-plane, a two-dimensional chart with imaginary and real axes represented in Figure 1.12. The pole s_p is located on the left side and represented by a red cross.

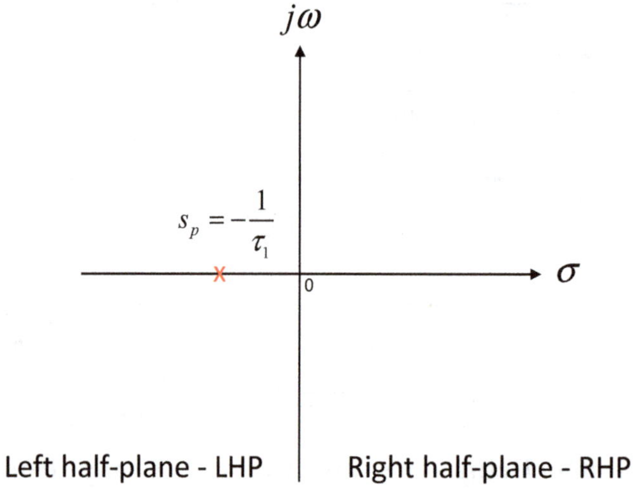

The *s*-plane features Two Axes and Illustrates the Position of Poles and Zeroes

Figure 1.12

In control theory, this is called a *left-half-plane pole* or LHPP also sometimes designated as a *stable* pole since its time-domain response is a decaying exponential waveform. A positive root, instead, is known as a *right-half-plane pole* or RHPP and called an *unstable* pole since it would lead to a diverging or growing time-domain output response. This is why it is interesting to place the poles of a transfer function on the chart and observe how they move in relationship with components values—and keep away from the right half-plane. In a nutshell, there is no difference in magnitude comparing the ac response of a 1st-order network including a LHPP and a RHPP located at the same frequency. However, the phase response will *lead* for the second case rather than lagging as a LHPP normally implies.

A pole is usually written with a subscripted *p* as in ω_p for example. (1.18) root is real and leads to a pole defined by the magnitude of s_p:

$$\omega_p = |s_p| = |\sigma + j\omega| = \left|-\frac{1}{\tau_1} + 0j\right| = \sqrt{\left(-\frac{1}{\tau_1}\right)^2 + 0^2} = \frac{1}{\tau_1} \tag{1.19}$$

In this simple first-order expression, the pole is the inverse of the *natural* time constant. Time constants are always determined by the circuit structure *alone*, without the contribution of the input stimulus. I drew two simple electrical structures in Figure 1.11, where there is no stimulus, just the so-called *natural* structure of the network.

To determine the time constant in these cases, simply temporarily disconnect the considered energy-storing element and "look" through the connecting terminals to determine the resistance. It is easy to identify series and parallel connections in your mind and infer the value of the resistance R by *inspection*. The poles of the two examples are then found instantly.

The denominator of a 1st-order transfer function can be advantageously rewritten in a *low-entropy* form starting with a 1:

$$D(s)=1+\frac{s}{\omega_p} \tag{1.20}$$

The "+" sign tells you this is a LHPP while a "–" such as in $\left(1-\frac{s}{\omega_p}\right)$ immediately points to a RHPP.

In higher-order circuits with multiple energy-storing elements, the denominator can host multiple poles which may not readily show up when you obtain the raw expression from the first analysis pass.

The normalized form for a denominator of degree n follows the format below:

$$D(s)=1+b_1s+b_2s^2+b_3s^3+...+b_ns^n \tag{1.21}$$

The nice thing owing to the FACTs is that you will swiftly determine each individual value for the b coefficients but you will have to rearrange the expression to unveil the poles. The best is to combine them in a factored form such as:

$$D(s)=(1+s\tau_1)(1+s\tau_2)...(1+s\tau_n)=\left(1+\frac{s}{\omega_{p_1}}\right)\left(1+\frac{s}{\omega_{p_2}}\right)...\left(1+\frac{s}{\omega_{p_n}}\right) \tag{1.22}$$

In the above equation, the pole positions appear clearly in the expression and no further manipulations are necessary to meet the wanted *low-entropy* form. This is because poles are real and spread well apart from each other, for example, 100 Hz, 1 kHz and 50 kHz in a 3-pole circuit.

Sometimes, because of the equation complexity, it is not possible to find or organize the denominator with separate poles (or zeroes because this is true for the numerator as well) and you may want to find a normalized polynomial or an expression approaching it. A good read for this exercise is [5] which details how analyzing time constants (to see which ones dominate or can be grouped) leads to approximating polynomials of any order. This is not always an obvious exercise, but you will need to be acquainted with the method to find the best possible format for your transfer function.

By the way, why do we need to shape this final result since we can easily plot its response with any modern program? Because this final transfer function is likely to serve a design purpose: for instance, to design a filter, you will need to link the component values to the poles and zeroes you want. Or adjust some values to match a gain you expect at a given frequency. In a control system, you want to identify the terms affecting the loop gain and potentially jeopardizing margins during the operating lifetime where temperature and age inexorably affect these parameters.

Knowing the origins of these changes lets you implement a robust compensation strategy, neutralizing all these contributors and ensuring long-term stability. These objectives illustrate the principle behind *design-oriented* analysis or D-OA. Without rearranging (1.21), it is impossible to infer any of these results at first sight, whereas, (1.22) unveils the poles and allows the determination of adequate component values.

Consequently, the exercise of determining a transfer function does not stop with a raw expression but requires more work to shape the final expression in a useful form which will lead to design guidelines.

This is the ultimate goal.

Having (1.21) start with a 1 naturally makes the numerator unitless. We will stick to this notation in the rest of the book. Thus, if (1.21) describes a third-order expression (n = 3), it implies that the factor b_1 of s has a dimension of time [s] and combines one of several time constants added together.

On the other hand, the coefficient b_2 of s^2 must display a unit of time square [s^2] and multiplies two time constants or sums products of two time constants. Finally, the coefficient b_3 in front of s^3 has a dimension of time cube [s^3] and involve the product of three time constants or added products of three time constants. This concept can be generalized up to the order n.

In case $D(s)$ does not start with a 1, it simply means that you have one or several poles at the origin: when s = 0, the denominator cancels and the transfer function magnitude goes infinite. It is an integrator. For instance, consider the expression below without a 1 which can be easily rewritten in a factored form:

$$H(s)=\frac{N(s)}{b_1s+b_2s^2}=\frac{N(s)}{b_1s\left(1+\frac{b_2}{b_1}s\right)}=\frac{N(s)}{s\tau_1\left(1+s\tau_2\right)}=\frac{N(s)}{\frac{s}{\omega_{po}}\left(1+\frac{s}{\omega_{p_1}}\right)} \quad (1.23)$$

When s = 0 or if you bias the circuit at dc, the magnitude goes infinite. In reality, the gain of the active element, an op-amp for instance, limits this infinite gain to its open-loop value. Should you have s^2 in the denominator, then there is a double pole at the origin. In this expression, the term ω_{po} is called the 0-dB crossover pole (the magnitude of the integrator has dropped to 1 at this frequency).

The numerator N hosts the *zeroes* of the transfer function. Zeroes are the roots of the numerator.

The angular frequency of a zero is usually written with a subscripted z as in ω_z for example. When the excitation frequency is tuned to a zero position, the magnitude of H is zero.

For instance, assume the below 1st-order numerator:

$$N(s)=1+s\tau_1 \quad (1.24)$$

To determine the zero in (1.24), simply find the root s_z for which $N(s_z)$ = 0:

$$s_z=-\frac{1}{\tau_1} \quad (1.25)$$

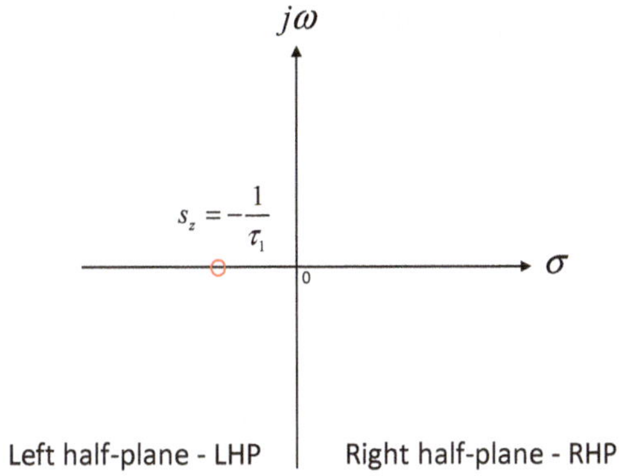

The Zero can be placed in the *s*-plane, here a Negative Real Value

Figure 1.13

This root is real—it does not have an imaginary part—and leads to a zero defined by the magnitude of s_z:

$$\omega_z = |s_z| = |\sigma + j\omega| = \left|-\frac{1}{\tau_1} + 0j\right| = \sqrt{\left(-\frac{1}{\tau_1}\right)^2 + 0^2} = \frac{1}{\tau_1} \tag{1.26}$$

It can also be placed in the *s*-plane, but this time represented by a small circle in Figure 1.13.

This is a negative root and is called a *left-half-plane zero* or LHPZ.

A positive root instead is known as a *right-half-plane zero* or RHPZ.

When present, a RHPZ affects the time-domain response of an input step: rather than immediately increasing, the output would first dip before rising.

For the sake of illustration, three time-domain responses are plotted in Figure 1.14:

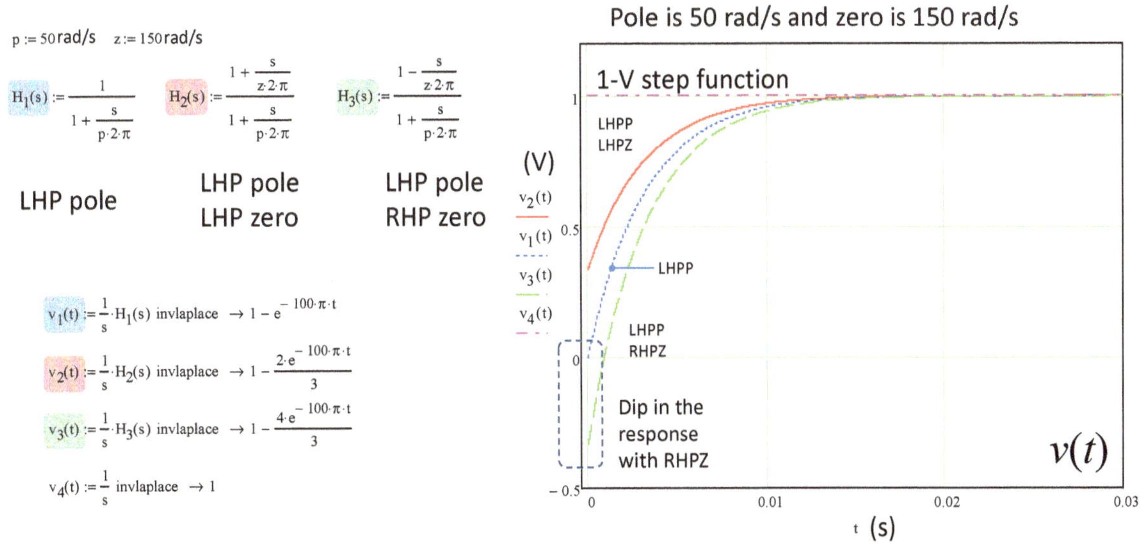

If the Zero is in the Left- or Right-Half Plane, the Time-Domain Responses of an Input Step are Different

Figure 1.14

The blue dashed line shows the classical exponential voltage response of $H_1(s)$ to the 1-V input step.

The time constant depends on the inverse of the left-half plane pole as in (1.19).

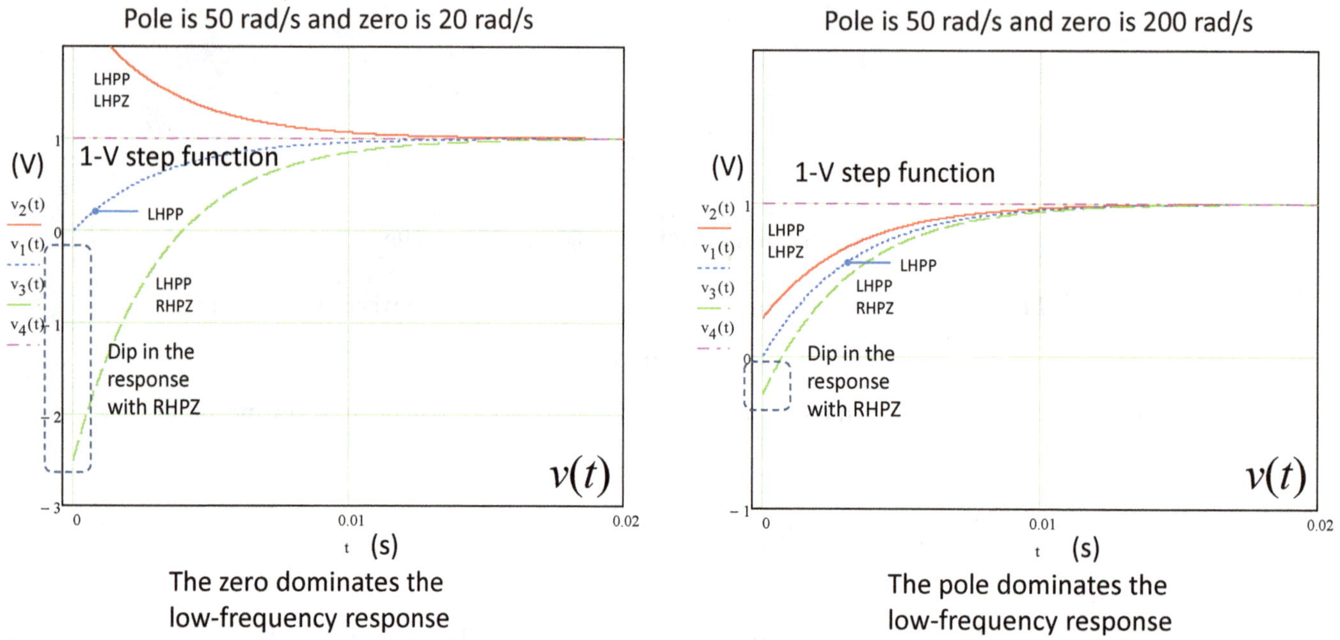

When the LHP Zero Dominates the Low-Frequency Response, Overshoot Appears. It Disappears when the Pole is Placed Lower than the Zero

Figure 1.15

Add a left-half-plane zero as in $H_2(s)$ and you see a differentiating effect pushing the voltage up in the beginning until the pole takes over.

Now, bring the zero in the right-half plane in $H_3(s)$—notice the negative sign in the numerator—and the voltage first dips before rising again.

This is the typical behavior of a system featuring a RHPZ—like a switching converter such as the boost or the buck-boost which include a positive zero in their control-to-output transfer function.

In Figure 1.15, if the zero dominates the response, the differentiating effect brings a clear overshoot (LHPZ) or an undershoot (RHPZ) in the response which eventually settles down to 1 V.

Move the zero to a higher frequency and see how the integrating effect now dominates the response in the right-side graph. You can observe this type of response—differentiating or integrating—when you trim a 10:1 probe on an oscilloscope. Without the RHP zero dip however.

To close this section on the RHP roots, there is no magnitude difference comparing the ac response of a 1st-order network including a RHPP and a RHPZ located at the same frequency. However, the phase response will lag for the second case rather than leading as a LHPZ would normally imply.

With this definition on hand, rewrite (1.24) in a more formal way that we will adopt all along this book:

$$N(s)=1+\frac{s}{\omega_z} \tag{1.27}$$

The "+" sign tells you this is a LHPZ while a "–" sign such as in $\left(1-\frac{s}{\omega_z}\right)$ immediately points to a RHPZ. Note that this expression also starts with a 1 and is dimensionless as already pointed out for the denominator expressions.

The normalized form of the numerator follows what we have already encountered in (1.21) with coefficients usually adopting the letter a:

$$N(s)=1+a_1s+a_2s^2+a_3s^3+...+a_ns^n \tag{1.28}$$

All remarks pertaining to the denominator are valid for the numerator—which remains a unitless expression in our notation.

It means that a_1 is expressed in seconds, a_2 unit is squared seconds and a_3 is in cubic seconds.

As implied by D-OA, factor the expression so that zeroes show up and helps with circuit design. All details given in [5] apply equally well to the numerator and serve a factorization goal such as:

$$N(s)=(1+s\tau_1)(1+s\tau_2)...(1+s\tau_n)=\left(1+\frac{s}{\omega_{z_1}}\right)\left(1+\frac{s}{\omega_{z_2}}\right)...\left(1+\frac{s}{\omega_{z_n}}\right) \tag{1.29}$$

As underlined in the document, in some cases, trying to find an exact factored solution to (1.21) or (1.28) is hopeless.

What you want is an approximation good enough to serve the design purpose. Approximations are often obtained by neglecting parasitic elements or comparing values to ignore one or several particular contributors.

In the end, experience and engineering judgment will let you appreciate if the deviation in magnitude and phase of the approximated solution is acceptable—compared to what the raw format gives.

In case the numerator does not start with a 1, then there is a zero at the origin meaning that when $s = 0$, the magnitude is zero. You have a differentiator which blocks dc.

A zero at the origin means that s factors the entire numerator as in the below example where you have a standard zero with another one at the origin:

$$H(s)=\frac{a_1s+a_2s^2}{D(s)}=\frac{a_1s\left(1+\frac{a_2}{a_1}s\right)}{D(s)}=\frac{s\tau_1(1+s\tau_2)}{D(s)}=\frac{\frac{s}{\omega_{zo}}\left(1+\frac{s}{\omega_{z_1}}\right)}{D(s)} \tag{1.30}$$

Should you have s^2 factoring the whole expression then you have a double zero at the origin.

In this expression, the term ω_{zo} called the 0-dB crossover zero (the magnitude of the differentiator has grown to 1 at this frequency).

Figure 1.11 shows how to determine the resistance driving the capacitor or the inductor in absence of stimulus. We will come back on this technique showing how to turn voltage and current sources off to return the network in its natural structure.

For the zeroes, the exercise also consists of determining a resistance R driving a capacitor or an inductor. However, if the stimulus was turned off for the pole exercise, keep the injection stimulus active while determining the resistance R seen from the capacitor or inductor terminals.

The principle of this technique is shown in Figure 1.16 and is called a *null double injection* or NDI [2].

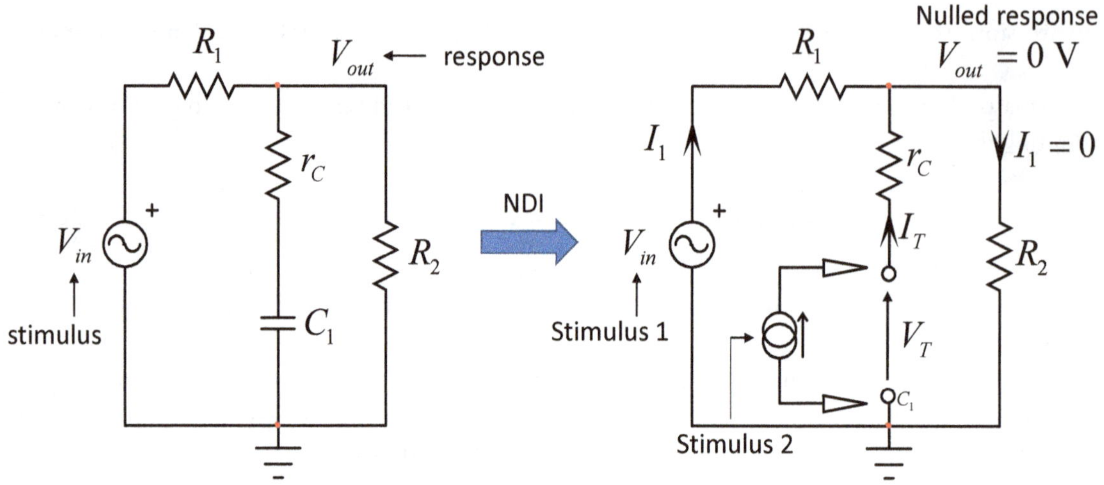

Despite the Presence of Stimulus 1, Consider a 0-V Response – $\hat{v}_{out} = 0$ V – to Determine the Resistance *R*

Figure 1.16

As the name NDI implies, there are two injection stimuli: the input source V_{in} plus a test generator I_T. This current source will be adjusted so that it cancels the current flowing into resistance R_2, naturally zeroing the output.

Note: this is *not* a short circuit applied to the output but resembles the virtual ground found in operational amplifier circuits. When this condition is met, calculating V_T over I_T gives the resistance *R* driving the capacitor. It is important to note that amplitude of stimulus 1 is irrelevant for expressing the value of resistance *R*.

One might wonder how to cancel the output voltage while still having a voltage V_{in} biasing the circuit. A simple SPICE simulation template introduced in [3] easily confirms this result. In this circuit, a high-gain voltage-controlled current-source G_1 is added which injects current until the output is 0 V.

This is realized when the low-side of R_2 is biased to -60 mV.

The ratio of the voltage V_T across the injection terminals and the injected current I_T gives the resistance.

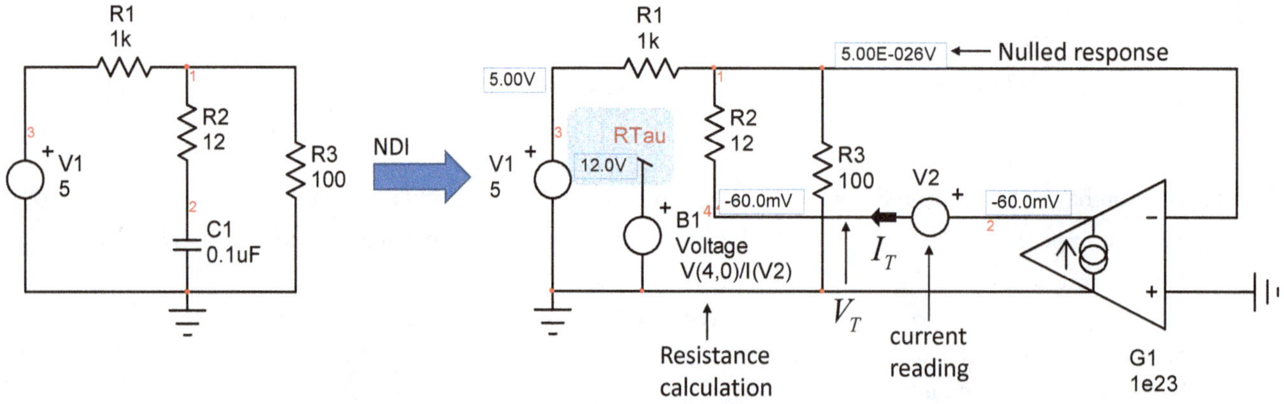

Absorbed Current I_T is exactly that as Injected by V_{in}, Effectively Nulling the Output Node

Figure 1.17

A simple bias point calculation confirms that it works well. The resistance computed by the in-line source B_1 shows that during this NDI, the capacitor is associated with a 12-Ω resistance which is r_C in Figure 1.16. Change V_1 to a different value, 3 V for instance, the value computed by B_1 remains the same.

Now compute the time constant:

$$\tau_1 = r_C C_1 \tag{1.31}$$

...and determine the zero position instantly:

$$\omega_z = \frac{1}{\tau_1} = \frac{1}{r_C C_1} \tag{1.32}$$

During this exercise, we actually considered that the stimulus tuned at the zero frequency propagates in the circuit but does not make it to the output. It's lost somewhere in the electrical structure. After all, we defined the zero as a specific frequency canceling the numerator and making the transfer function magnitude equal to zero. We have simply determined the time constant in this condition emulated by the NDI setup.

Could inspection work also for determining zeroes? Of course. As we'll see later, this leads to the simplest expressions.

1.1.3 Writing the Transfer Functions the Right Way

As underlined in the text above, it is important to properly express a transfer function in *low-entropy* form where poles, zeroes and specific gain or attenuation show up at first sight. We start with the classic example of a transfer function written the same way as in textbooks:

$$H(s) = \frac{N(s)}{D(s)} = \frac{s+4}{(s+0.8)\left[(s+2.5)^2+4\right]} \tag{1.33}$$

First-off, you immediately see that this equation does not satisfy the notation we have encouraged so far with a numerator and denominator starting with a 1. Simple factorization will help find a more suitable expression:

$$H(s) = \frac{N(s)}{D(s)} = \frac{4}{0.8} \cdot \frac{1+\frac{s}{4}}{\left(1+\frac{s}{0.8}\right)\left(s^2+5s+10.25\right)} = \frac{4}{8.2} \cdot \frac{1+\frac{s}{4}}{\left(1+\frac{s}{0.8}\right)\left[1+\frac{s}{2.049}+\left(\frac{s}{\sqrt{10.25}}\right)^2\right]} \tag{1.34}$$

The numerator $N(s)$ features one root obtained when solving:

$$N(s) = 1+\frac{s}{4} = 0 \tag{1.35}$$

Leading to:

$$s_z = -4 \tag{1.36}$$

The zero is thus:

$$\omega_z = 4\,\frac{\text{rad}}{s} \tag{1.37}$$

The denominator features three poles—it is thus a 3rd-order denominator or a 3rd-order transfer function—and the first root appears immediately:

$$1 + \frac{s}{0.8} = 0 \tag{1.38}$$

Which gives:

$$s_{p_1} = -0.8 \tag{1.39}$$

...and a pole located at:

$$\omega_{p_1} = 0.8\,\frac{\text{rad}}{s} \tag{1.40}$$

The second term in the denominator shows a second-order polynomial whose roots can be found by solving:

$$s^2 + 5s + 10.25 = 0 \tag{1.41}$$

It leads to complex conjugate roots:

$$s_{p_2} = -2.5 + 2j \tag{1.42}$$

$$s_{p_3} = -2.5 - 2j \tag{1.43}$$

...and a complex poles pair located at:

$$\omega_{p_2} = \omega_{p_3} = \sqrt{(-2.5)^2 + (\pm 2)^2} = \sqrt{10.25}\,\frac{\text{rad}}{s} \tag{1.44}$$

If we rewrite (1.34) in a symbolic form, we can see that the above results are naturally unveiled by the adopted format:

$$H(s)=H_0\frac{1+\frac{s}{\omega_z}}{\left(1+\frac{s}{\omega_{p_1}}\right)\left[1+\frac{s}{\omega_0 Q}+\left(\frac{s}{\omega_0}\right)^2\right]} \tag{1.45}$$

This is the form you want for writing (1.33) the right way. Note that (1.33) and (1.45) are rigorously identical but only the latter brings insight in the expression. For instance, you see a dc gain (actually an attenuation in this case) H_0 obtained for $s = 0$:

$$H_0=\frac{4}{8.2}\approx 0.488 \tag{1.46}$$

...or:

$$20\cdot\log_{10}\left(H_0\right)\approx -6.2\text{ dB} \tag{1.47}$$

Then resonant frequency and quality factor—or damping ratio—are obtained via a simple equation by identifying the terms in (1.45) with that in (1.34):

$$\omega_0=\sqrt{10.25} \tag{1.48}$$

...and:

$$\omega_0 Q=2.049 \tag{1.49}$$

Solving for Q gives:

$$Q=0.64 \tag{1.50}$$

...which is linked to the damping ratio ζ as:

$$\zeta=\frac{1}{2Q}=0.781 \tag{1.51}$$

These values directly relate to component values and let you design your circuit to fulfil specific goals. Following these guidelines, it is possible to extend the concept to the different types of transfer functions such as gains, impedances or admittances. In all cases, it is recommended to write the transfer function with a factored leading term followed by the fraction associating $N(s)$ and $D(s)$.

As these two have no dimension, the leading term (if any) must carry the unit. For instance, if you determine an impedance Z, the transfer function must be written as:

$$Z(s)=R_0\frac{N(s)}{D(s)} \tag{1.52}$$

In this expression, Z is an impedance with a dimension in ohms and R_0 is a resistance also expressed in ohms. It can be the resistance measured for $s = 0$ or when s approaches infinity (in this case it can be written as R_{inf} or R_∞). It can also be any other value you want to highlight, like a plateau, a peak or a dip in the transfer function.

Figure 1.18 illustrates this concept for various transfer functions.

gain $H(s)=H_0\frac{N(s)}{D(s)}$ — $H(s)$: unitless; H_0: unitless; $\frac{N(s)}{D(s)}$: unitless

impedance $Z(s)=R_0\frac{N(s)}{D(s)}$ — $Z(s)$: [Ω]; R_0: [Ω]; $\frac{N(s)}{D(s)}$: unitless

gain $G(s)=G_\infty\frac{N(s)}{D(s)}$ — $G(s)$: [V]/[V]; G_∞: [V]/[V]; $\frac{N(s)}{D(s)}$: unitless

admittance $Y(s)=Y_0\frac{N(s)}{D(s)}$ — $Y(s)$: [S]; Y_0: [S]; $\frac{N(s)}{D(s)}$: unitless

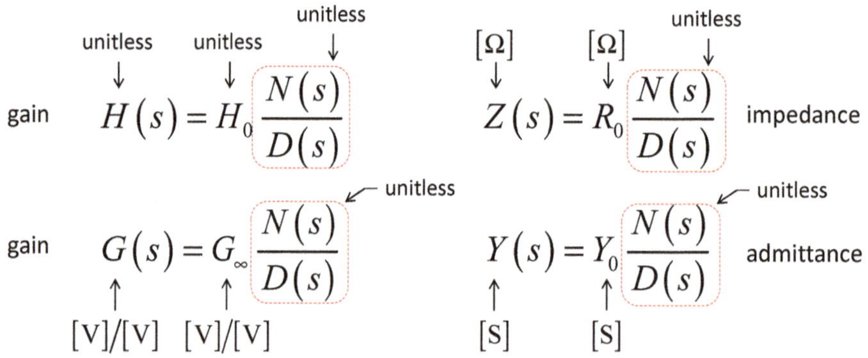

Whenever Possible, Write the Transfer Function so the Leading Term Carries the Unit while the Fraction Remains Unitless

Figure 1.18

Checking the homogeneity of the leading term—making sure R_0 returns a value in ohms for instance—while $N(s)$ and $D(s)$ are unitless is a first step to check the integrity of the derived expression.

1.1.4 Determining an Impedance

Consider the simple example of Figure 1.19. Let's determine the impedance offered by the parallel connection of a resistance and a capacitor affected by its equivalent series resistance (ESR) labeled r_C. The exercise consists of connecting a current source (the stimulus) across the connecting terminals and determining the response V_T collected at the source connecting points. The ratio of the voltage developed across the current source divided by the injected current will give us the transfer function we want. You could immediately see this impedance is:

$$Z(s)=R_1\,||\left(r_C+\frac{1}{sC_1}\right) \tag{1.53}$$

...and you would be right.

However, you must now expand this expression to make it fit the wanted format described in Figure 1.18. Rather than going down that path, let's see how the FACTs help us solve this exercise in a few seconds. In sketch (c), redraw the circuit for $s = 0$: open the capacitor and the dc resistance is immediate:

$$R_0=R_1 \tag{1.54}$$

Now set the stimulus to 0 A (a 0 A current source is equivalent to an open circuit), then, as shown in sketch (d), the resistance "seen" from the connecting terminals of the capacitor leads straight to the circuit time constant:

$$\tau_1 = RC_1 = (r_C + R_1)C_1 \quad (1.55)$$

Which immediately implies a pole set at:

$$\omega_p = \frac{1}{\tau_1} = \frac{1}{(r_C + R_1)C_1} \quad (1.56)$$

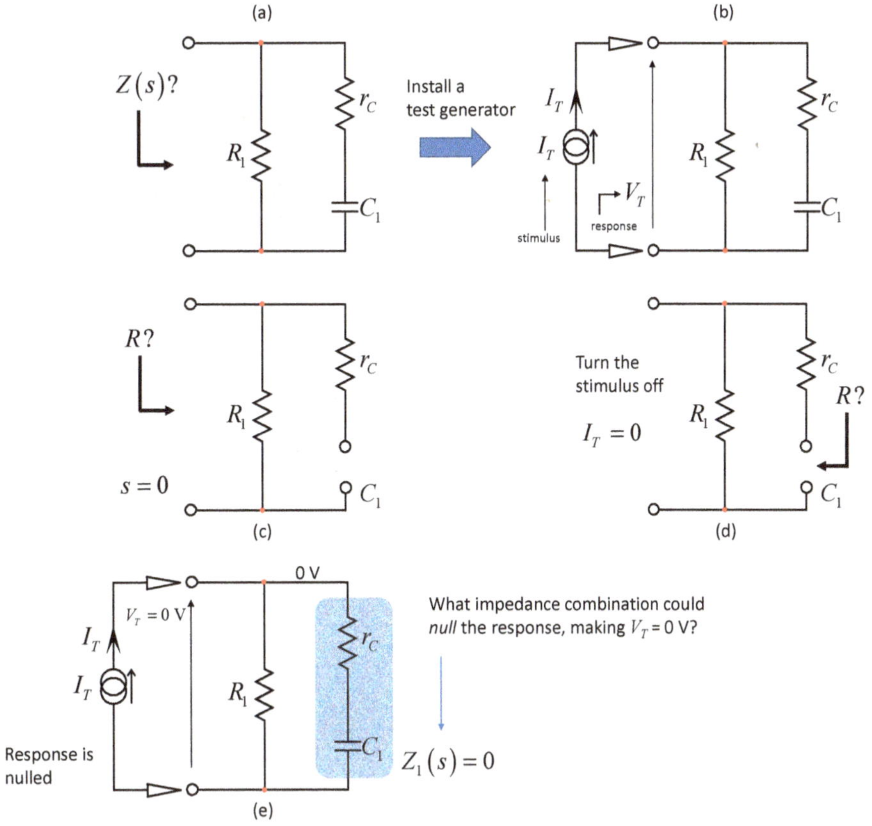

This Impedance offers a Resistive Term at Dc; also as *s* approaches Infinity

Figure 1.19

The zero could be obtained using an NDI, but I said that inspection also works for determining zeroes.

The manifestation of a zero is the zeroed magnitude of the transfer function when the stimulus is tuned at zero frequency.

In the circuit shown in sketch (e), what would consequently bring the response V_T to 0 V? Could R_1 suddenly become a shunt at a certain frequency?

Certainly not, it is a fixed resistance.

What about the series connection of r_C and C_1 labeled $Z_1(s)$: could it become a shunt at some point?

To check it, simply express this impedance and see if a root exists:

$$Z_1(s) = r_C + \frac{1}{sC_1} = \frac{1 + sr_C C_1}{sC_1} = 0 \quad (1.57)$$

The numerator $1+sr_C C_1$ cancels for a root equal to:

$$s_z = -\frac{1}{r_C C_1} \tag{1.58}$$

...leading to a zero located at:

$$\omega_z = \frac{1}{r_C C_1} \tag{1.59}$$

The simple SPICE NDI setup of Figure 1.20 confirms that the resistance driving the capacitor with a nulled response is the resistance r_C whose value is 10 Ω.

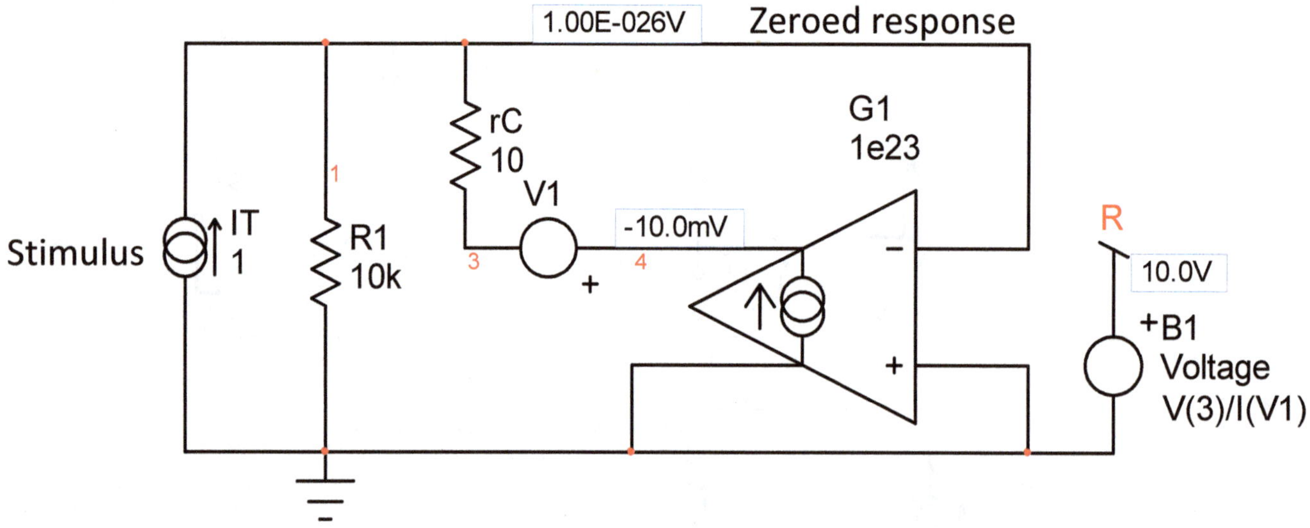

An NDI Confirms the Resistance Value Determining the Zero

Figure 1.20

We can now assemble the transfer function highlighting the resistance determined in these conditions:

$$Z(s) = R_1 \frac{1+sr_C C_1}{1+s(r_C+R_1)C_1} = R_0 \frac{1+\frac{s}{\omega_z}}{1+\frac{s}{\omega_p}} \tag{1.60}$$

In this expression obtained with a few simple electrical diagrams, we highlighted the dc resistance as a leading term.

However, we can observe that at high frequencies, C_1 becomes a short-circuit and the impedance is dominated by a resistive term made of the parallel connection of r_C and R_1.

Could we rearrange our formula differently should we want to highlight the high-frequency response instead?

From (1.60), factor the terms including C_1 in the numerator and the denominator:

$$Z(s) = R_1 \frac{1+sr_C C_1}{1+s(r_C+R_1)C_1} = R_1 \frac{sr_C C_1}{s(r_C+R_1)C_1} \frac{1+\dfrac{1}{sr_C C_1}}{1+\dfrac{1}{s(r_C+R_1)C_1}} \tag{1.61}$$

Simplify the leading term and introduce the pole and zero notation to obtain:

$$Z(s) = R_\infty \frac{1+\dfrac{1}{s\tau_N}}{1+\dfrac{1}{s\tau_D}} = R_\infty \frac{1+\dfrac{\omega_z}{s}}{1+\dfrac{\omega_p}{s}} \tag{1.62}$$

In which:

$$R_\infty = r_C \,||\, R_1 \tag{1.63}$$

Expressions (1.60) and (1.62) are rigorously similar but differently arranged to highlight a particular term that interests you design-wise. The numerator features an *inverted zero* while the denominator hosts an *inverted pole* (see [3] for more details).

To check our expressions, a simple Mathcad® sheet shown in Figure 1.21 will tell us if we correctly determined this transfer function.

When using Mathcad® or any other solver, make sure you always pass units to the assigned values. The solver is capable of checking the homogeneity of the expressions and flags any inconsistency.

This saved me many times when determining transfer functions.

In the graph, the impedance is log-compressed and results in values expressed in dBΩ. However, as the built-in log function can only manipulate unitless numbers, you must divide by 1 Ω before compressing the results.

In this sheet, compare the response of the raw expression given in (1.53) with the two formulas derived in (1.60) and (1.62). As shown in the magnitude and phase plots, all curves exactly superimpose—confirming the derivation is correct.

How do we know if there is not a slight deviation, masked by the thickness of the curves?

The best way to check this is to plot the differences between the magnitude and phase responses you want to compare. If everything is okay, the result cannot be zero but approaches the pico value (the solver resolution) or below as confirmed in Figure 1.22. Any significant deviation in the magnitude or phase differences indicates a flaw in one of the two expressions (or both). When a brute-force expression is difficult to derive, you can simulate the circuit and import data into the solver then superimpose the plots for comparison purposes.

$R_1 := 10k\Omega \quad r_C := 10\Omega \quad C_1 := 0.47\mu F \quad \|(x,y) := \frac{x \cdot y}{x+y}$

$R_0 := R_1 = 1 \times 10^4 \Omega \qquad 20 \cdot \log\left(\frac{R_0}{\Omega}\right) = 80 \text{ dBohms}$

$R_{inf} := r_C \| R_1 = 9.99\Omega \qquad 20 \cdot \log\left(\frac{R_{inf}}{\Omega}\right) = 19.991 \text{ dBohms}$

$\tau_1 := (r_C + R_1) \cdot C_1 = 4.705ms$

$\omega_p := \frac{1}{\tau_1} \qquad f_p := \frac{\omega_p}{2 \cdot \pi} = 33.829Hz$

$\omega_z := \frac{1}{r_C \cdot C_1} \qquad f_z := \frac{\omega_z}{2 \cdot \pi} = 33.863kHz$

$Z_{ref}(s) := R_1 \| \left(r_C + \frac{1}{s \cdot C_1}\right) \qquad Z_1(s) := R_0 \cdot \frac{1 + \frac{s}{\omega_z}}{1 + \frac{s}{\omega_p}} \qquad Z_2(s) := R_{inf} \cdot \frac{1 + \frac{\omega_z}{s}}{1 + \frac{\omega_p}{s}}$

Raw transfer function

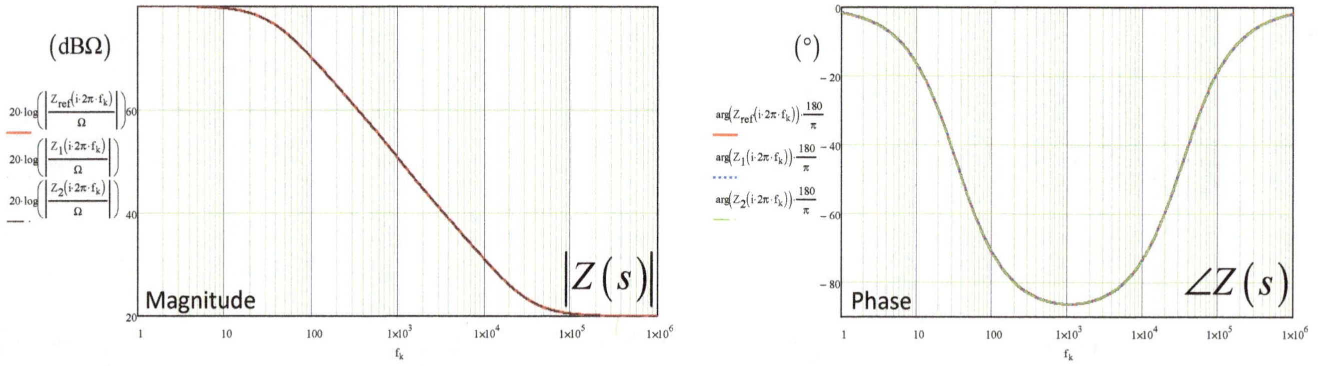

The Magnitude is Flat at 10 kΩ at Dc and Drops to Almost 10 Ω as Frequency Increases

Figure 1.21

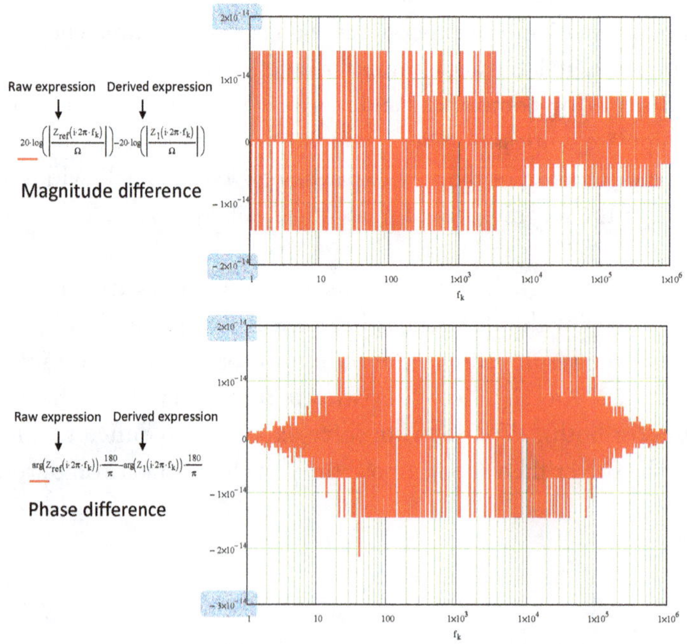

Formulas are Identical when the Magnitude and Phase Differences Hit the Solver Resolution

Figure 1.22

1.2 What Should I Retain from this Chapter?

In this first chapter, we learned key information that is summarized below:

1. When a *stimulus* is applied at the input port of a network, it propagates in the circuit to form a *response* that can be observed at the output port. The mathematical relationship linking the stimulus with the response is called a *transfer function*.

2. The transfer function of a linear circuit can be determined by finding its time constants when the network is analyzed in two situations: with the excitation reduced to zero and when the response is *nulled*. Nulling the response means that despite the presence of a stimulus injecting a signal, the response is equal to zero.

3. A transfer function is made of numerator N and denominator D. Determine the roots of the denominator or the *poles* of the transfer function when the excitation or the stimulus is reduced to zero. On the contrary, the roots of the numerator are the *zeroes* of the transfer function and are obtained when the ac output is nulled with the stimulus back in action.

4. A zeroed excitation means turning off the generator. Practically speaking, a 0 V voltage source is replaced by a wire in the network under study while the 0 A current source is open circuited. Determine the time constants of the circuit in this mode by temporarily disconnecting each energy-storing element and determining the resistance R offered from its connecting terminals.

5. Once the time constants involving each energy-storing element is obtained, assemble them to form the denominator $D(s)$.

6. The zeroes are determined using a null double-injection or *NDI*: the stimulus is back in place but find the time constants of the circuit when the output is nulled. These time constants are assembled to discover the numerator $N(s)$.

1.3 References

1. R. D. Middlebrook, *Methods of Design-Oriented Analysis: Low-Entropy Expressions*, Frontiers in Education Conference, Twenty-First Annual conference, Santa-Barbara, 1992.
2. R. D. Middlebrook, *Null Double Injection and the Extra Element Theorem*, IEEE Transactions on Education, Vol. 32, NO. 3, August 1989 (https://authors.library.caltech.edu/63233/1/00034149.pdf)
3. C. Basso, *Linear Circuit Transfer Functions – An Introduction to Fast Analytical Techniques*, Wiley, 2016.
4. V. Vorpérian, *Fast Analytical Techniques for Electrical and Electronic Circuits*, Cambridge University Press, 2002.
5. R. Erickson, D. Maksimović, *Fundamentals of Power Electronics*, Kluwer Academic Publishers, 2001, pp. 289-293 (https://www.ieee.li/pdf/introduction_to_power_electronics/chapter_08.pdf)

Chapter 2: Fast Analytical Circuits Techniques

THE FIRST CHAPTER defined the concept of a transfer function and how time constants affect the polynomial coefficients. In this upcoming chapter, you will learn that temporarily disconnecting an energy-storing element and "looking" through its terminals in specific conditions let you determine the transfer function swiftly.

2.1 An Introduction to Fast Analytical Circuits Techniques or FACTs

Now that we know how a transfer function should be written and what terms constitute it, it is time to shed light on details we purposely covered quickly in the previous chapters. Start with the stimulus involved in time constant determination. We said that the source must be turned off before considering the time constants of the circuit. The stimulus can be a voltage source or a current source depending on the transfer function to be determined.

Turning off a voltage source—or reducing its output to 0 V—is similar to replacing its symbol by a wire. For a current source, reducing its current to 0 A is similar to physically removing it from the circuit and leaving its connecting terminals unconnected. With these guidelines, per Figure 2.1, redraw the original circuit with the stimulus replaced by a short circuit if it involves a voltage source or open the circuit where the stimulating current source is connected.

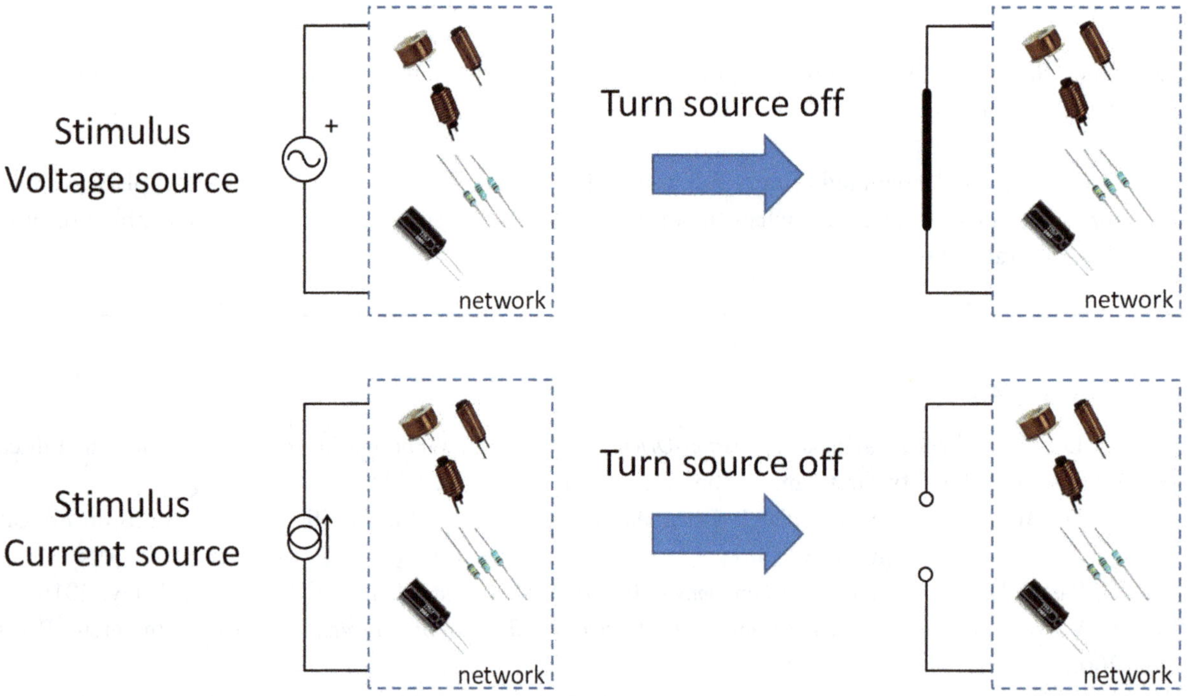

Turn the Stimulus Off and Update the Schematic to Determine the Time Constants

Figure 2.1

Figure 2.2 illustrates this principle in a practical example in which the energy-storing element is temporarily removed to determine the resistance R driving it.

The time constant is immediately determined.

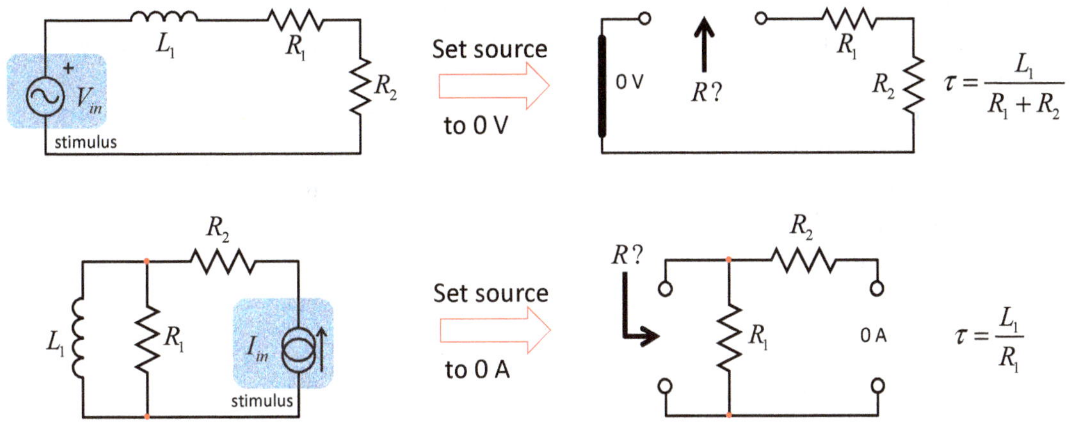

The First Step in Determining the Natural Time Constants is to Turn the Excitation Off

Figure 2.2

This principle also perfectly works with active elements whether you deal with a bipolar transistor or an operational amplifier (op-amp) as revealed by Figure 2.3. In the dc-biased bipolar circuit, replace the transistor by its linear hybrid-π model before starting the analysis as FACTs apply to linear or small-signal circuits only. The input voltage V_{in} is reduced to 0 V—implying that there is no base current i_b and, consequently, no collector current either. The V_{cc} rail is classically grounded considering a 0-V ac component on it as it were decoupled by an infinite capacitance value. Therefore, the only resistance seen from the capacitor connecting terminals is R_c. You immediately see the time constant and the pole of the circuit.

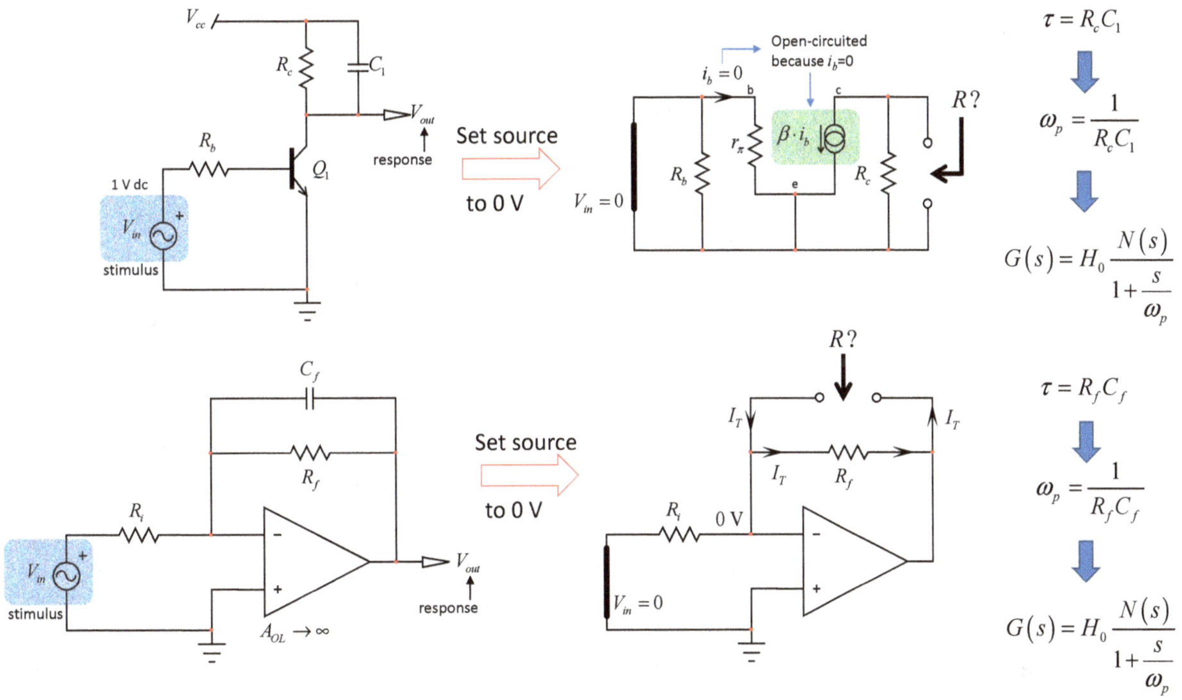

The Principle Remains the Same with Active Components such as a Bipolar Transistor or an Op-Amp when Small-Signal Models are used

Figure 2.3

For the op-amp, as shown in Figure 2.3, use the same approach: reduce the input source to 0 V and look through the capacitor connecting terminals. There is no current flowing through R_i considering 0 V across it so all the test current I_T flows in R_f immediately leading to the time constant and the pole of the circuit. Don't worry, we will come back to these circuits in more details in all the documented examples later.

There is also a so-called *degenerate* case for the current source as illustrated in Figure 2.4. When the voltage across its terminals is zero volt in a given condition, then, for the sake of analyzing the network, replacing the current generator by a wire is the way to proceed. When doing so, the current in the branch remains unchanged and there is still 0 V across the wire.

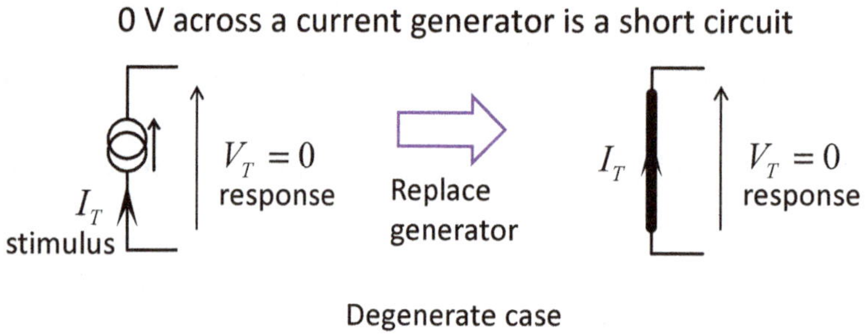

When the Voltage Across a Current Source is 0 V, it can be Replaced by a Wire

Figure 2.4

This observation is particularly useful when working on the input or output impedance of a particular circuit and you want to determine the zeroes by nulling the response.

For instance, in sketch (e) of Figure 1.19 of Chapter 1, we have determined the resistance considering the $r_C C_1$ network shunting the response signal to ground.

An alternative was to consider the degenerate case I just mentioned and look through the capacitor terminals when the current generator was shorted. Inspection immediately reveals that $R = r_C$ in Figure 2.5 since the wire shorts R_1 to ground and leaves r_C alone in the mesh.

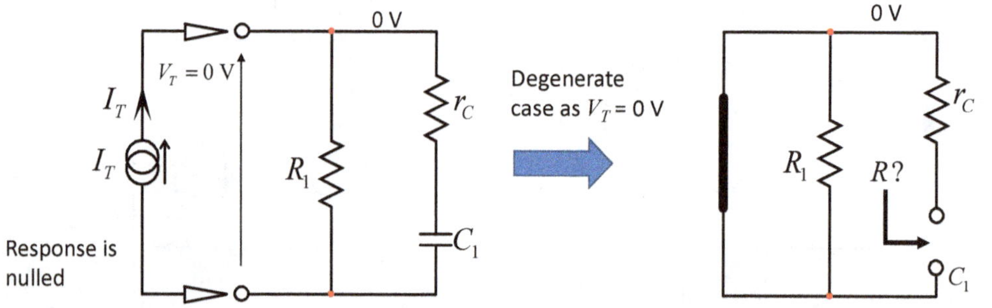

When the Response is Nulled in this Configuration, Analyzing the Circuit is Simple

Figure 2.5

As I wrote before, the natural time constants and the poles are set by the electrical structure of the circuit—the way the elements are connected together—which is revealed when the excitation source is turned off: you place the circuit in its *natural* structure.

Once one or several time constants are determined in this mode, you can write the denominator $D(s)$. If the analysis now consists of expressing several transfer functions around the same electrical circuit, then you will have to apply a stimulus at different places.

For instance, for determining the gain of the circuit, you will excite one port of the network with a voltage source while, for expressing an output impedance, you may connect a current source across the output of the circuit you study.

In all these exercises, there is a stimulus applied to the circuit. If turning off this stimulus brings the circuit back into a common natural structure, then it means the time constants are the same across all exercises. You can therefore immediately reuse the denominator already obtained during the first transfer function analysis and there is no need to rederive it again.

Although probing the response at different points in the circuit can affect the zeroes location as we will see, it *won't* change the denominator expression.

This is true as long as the structure remains the same once the stimulus is turned off. Accounting for this fact will save you a tremendous amount of time as illustrated in Figure 2.6 with a gain and output impedance determination of a simple 1st-order circuit.

In the upper side, the stimulus is turned off and replaced by a wire in the drawing. For the output impedance, considering a perfect input generator V_{in}, we can replace it by a 0-Ω resistance and apply a stimulus to the output via a current source I_T.

In both analyses, you see that the newly formed schematic diagrams are identical meaning that time constants are the same: if you have determined $D(s)$ in the first case, you can reuse it straight away for the second exercise.

Now assume that you want to find the input impedance seen by the input generator V_{in}. In that case, you disconnect V_{in} and apply a current generator I_T across the connecting terminals as documented by Figure 1.8 in Chapter 1.

The injected current I_T will develop a response V_T and combining both variables will lead to the desired impedance.

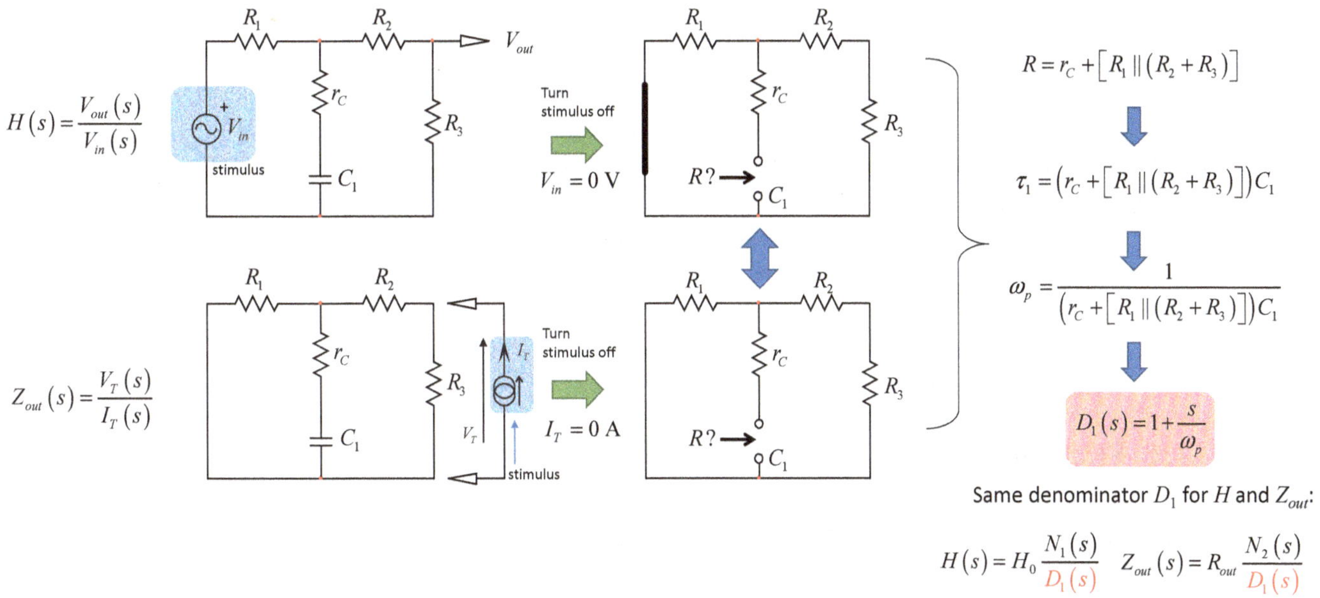

When the Excitation is Turned Off, the Circuit Returns to the Same Natural State. Both Transfer Functions Share a Common Denominator

Figure 2.6

Because we talk about the same circuit, can we reuse the denominator already found in Figure 2.6?

$$Z_{in}(s)?$$

$$Z_{in}(s) = \frac{V_T(s)}{I_T(s)}$$

Install a stimulus

$$I_T$$

$$V_T$$

Turn stimulus off

$$I_T = 0\ A$$

$$R?$$

$$R = r_C + R_2 + R_3$$

$$\tau_1 = (r_C + R_2 + R_3)C_1$$

$$\omega_p = \frac{1}{(r_C + R_2 + R_3)C_1}$$

$$D_2(s) = 1 + \frac{s}{\omega_p}$$

$$H(s) = H_0 \frac{N_1(s)}{D_1(s)}$$

$$Z_{out}(s) = R_{out} \frac{N_2(s)}{D_1(s)}$$

$$Z_{in}(s) = R_{in} \frac{N_3(s)}{D_2(s)}$$

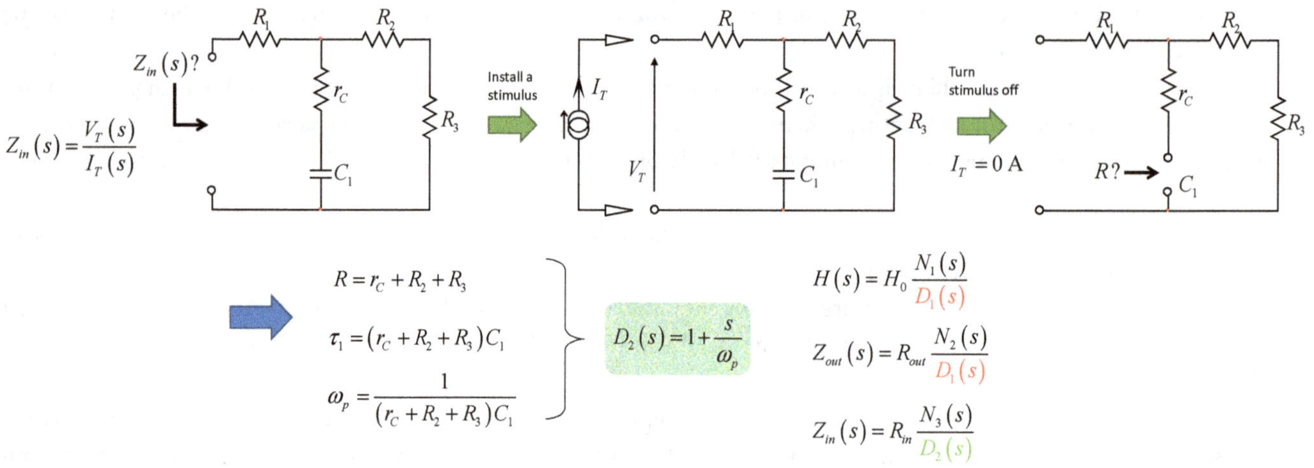

When the Excitation is Turned Off, the Circuit Does Not Return to its Previous Natural State and the Denominator Cannot be Reused

Figure 2.7

No, because opening the current generator when turning it off leaves the left connection of R_1 unconnected clearly revealing a new structure: you cannot reuse the previously determined denominator and need to find the new one. Nothing complicated though as shown in Figure 2.7.

Actually, with some experience in manipulating these concepts, you will discover that in this impedance determination exercise, the denominator obtained from Figure 2.6—when looking at the natural time constants—will become the numerator of your impedance expression.

This is because when the zero is determined by nulling the response across the test current generator, you can actually replace this generator by a short circuit in this very specific case. This is called a *degenerate* case described in [1]. If the current source is shorted, you analyze a circuit that is now similar to that of Figure 2.6 for which you already determined the time constants: the expression found for the denominator in Figure 2.6 becomes the numerator of your impedance expression in Figure 2.7, again saving analysis time.

Determining the pole in a first-order circuit is not complicated when inspection works. Sometimes, especially when you deal with active sources (voltage- or current-controlled sources), inspection is not obvious or simply impossible and you may have to resort to the test generator I_T installed across the connection terminals then invoke Kirchhoff's circuit laws to determine R.

This principle will be extended to higher-order circuits as we will see in upcoming sections.

2.1.1 State of the Passive Element for First-Order Circuits

We have seen that the FACTs consist of studying a linear network when it is placed in different operating conditions such as turning off the stimulus source for the time constant determination. Other typical states are when you look at the circuit for $s = 0$ but also when s approaches infinity.

For these extreme cases, capacitors and inductors take on specific values. In the dc state, the capacitor becomes an open circuit (remove it from the circuit) while the inductor is a short circuit (replace it by a wire in the circuit).

This is what SPICE does when it calculates an operating bias point at the beginning of any simulation: it shorts inductors and open all capacitors to study the circuit in dc conditions and launches a linearization process around that point.

Then, at high frequencies, capacitors become short circuits while inductors open circuit. Figure 2.8 illustrates this fact with simple drawings.

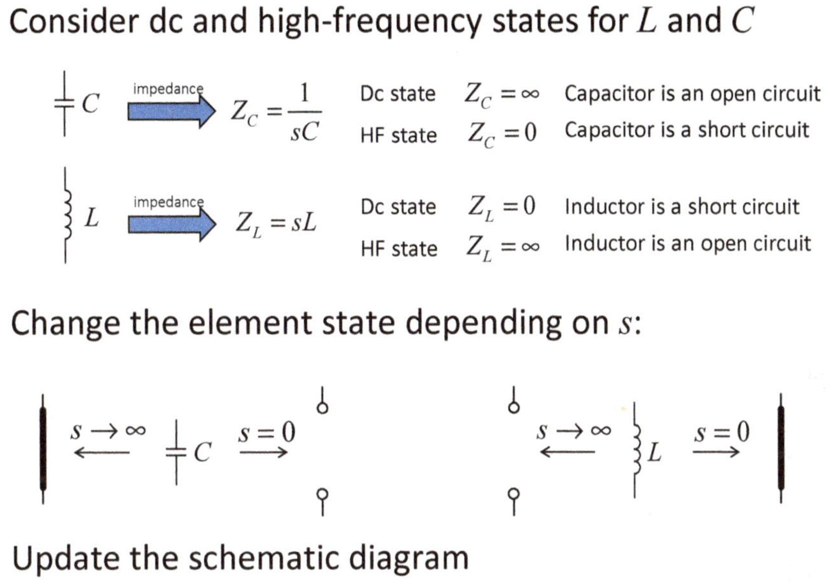
Consider dc and high-frequency states for L and C
C
impedance
$Z_C = \frac{1}{sC}$
Dc state $Z_C = \infty$ Capacitor is an open circuit
HF state $Z_C = 0$ Capacitor is a short circuit
L
impedance
$Z_L = sL$
Dc state $Z_L = 0$ Inductor is a short circuit
HF state $Z_L = \infty$ Inductor is an open circuit
Change the element state depending on s:
$s \to \infty$
C
$s = 0$
$s \to \infty$
L
$s = 0$
Update the schematic diagram

Update the Schematic Diagram to Observe Dc or High-Frequency Conditions

Figure 2.8

From this representation, we can easily determine the gain of a network in the dc condition or at high frequencies simply by drawing intermediate sketches where the energy-storing elements are replaced by their equivalent circuit in the considered condition (dc or high frequency). Consider the simple filter in Figure 2.9.

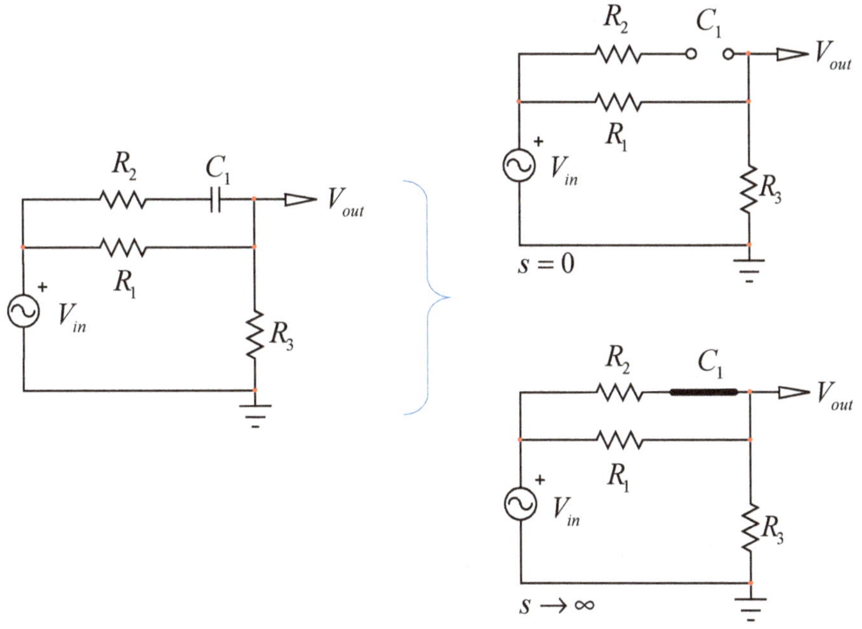

This 1st-Order Filter can be Expressed with a Low-Frequency Gain H_0, but also with a Gain Determined when *s* Approaches Infinity

Figure 2.9

In dc, capacitor C_1 is open and you can determine the gain that we label H_0 to be:

$$H_0 = \frac{R_3}{R_3 + R_1} \tag{2.1}$$

On the contrary, if you now consider a high-frequency observation, short C_1 and determine the gain in this mode:

$$H_\infty = \frac{R_3}{R_3 + R_1 \parallel R_2} \tag{2.2}$$

Because it is a linear circuit, we can use a simple dc operating point calculation using SPICE to check if our results are correct. Just replace the source by a 1-V level and have SPICE compute the output voltage. The displayed value at node 2 is your gain (the stimulus is 1 V) and should exactly match your calculations.

Figure 2.10 shows this technique in action.

$R_1 := 10\text{k}\Omega \quad R_2 := 1\text{k}\Omega \quad R_3 := 2.2\text{k}\Omega \quad \parallel(x,y) := \frac{x \cdot y}{x+y}$

$H_0 := \frac{R_3}{R_1 + R_3} = 0.18 \qquad 20 \cdot \log(H_0) = -14.879 \text{ dB}$

$R_1 := 10\text{k}\Omega \quad R_2 := 1\text{k}\Omega \quad R_3 := 2.2\text{k}\Omega \quad \parallel(x,y) := \frac{x \cdot y}{x+y}$

$H_{\text{inf}} := \frac{R_3}{R_3 + R_1 \parallel R_2} = 0.708 \qquad 20 \cdot \log(H_{\text{inf}}) = -3.004 \text{ dB}$

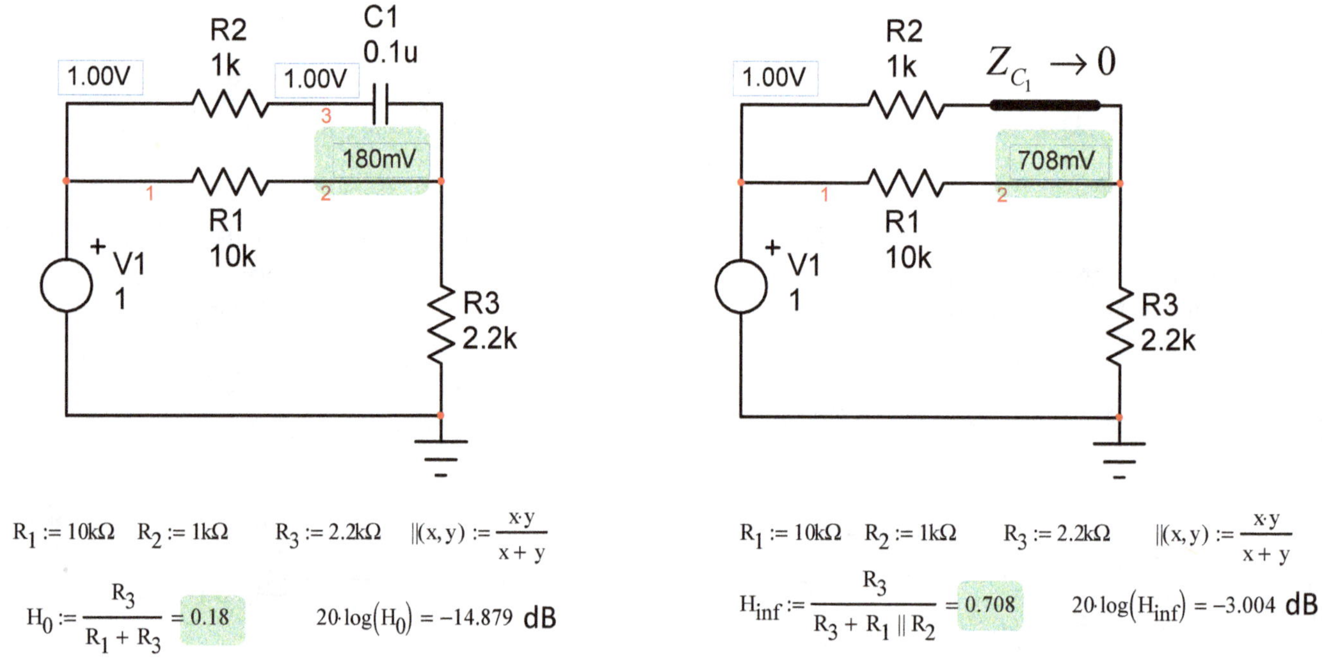

SPICE and its Dc Operating Point Calculation are Extremely Valuable to Check Calculations

Figure 2.10

In this particular case, the various possibilities to express the transfer function are seen, depending on what parameter you want to highlight for design purposes.

If the dc gain is important, you can write:

$$H(s) = H_0 \frac{N_1(s)}{D_1(s)} \tag{2.3}$$

On the contrary, if it's the high-frequency response that you care about, the transfer function will look like:

$$H(s) = H_\infty \frac{N_2(s)}{D_2(s)} \qquad (2.4)$$

In both cases, the time constant is determined as:

$$\tau_1 = C_1\left(R_2 + R_1 \parallel R_3\right) \qquad (2.5)$$

Figure 2.11 confirms this result with another bias point calculation using a 1-A dc source.

Note the implementation of the in-line equation B_1 which extracts the floating voltage available across the current source terminals (nodes 1 and 2) and displays a ground-referenced value. As confirmed by SPICE and Mathcad®, the values are rigorously identical and this certifies a sound process.

A significant deviation between simulated and computed values indicates an error somewhere and has to be corrected before proceeding.

$R_1 := 10k\Omega \quad R_2 := 1k\Omega \quad R_3 := 2.2k\Omega \quad C_1 := 100nF \quad \parallel(x,y) := \frac{x \cdot y}{x+y}$

$R := R_2 + R_1 \parallel R_3 = 2.80328 \times 10^3\,\Omega$ Value in ohms - √

$\tau_1 := R \cdot C_1 = 280.32787\mu s$ Value in seconds - √

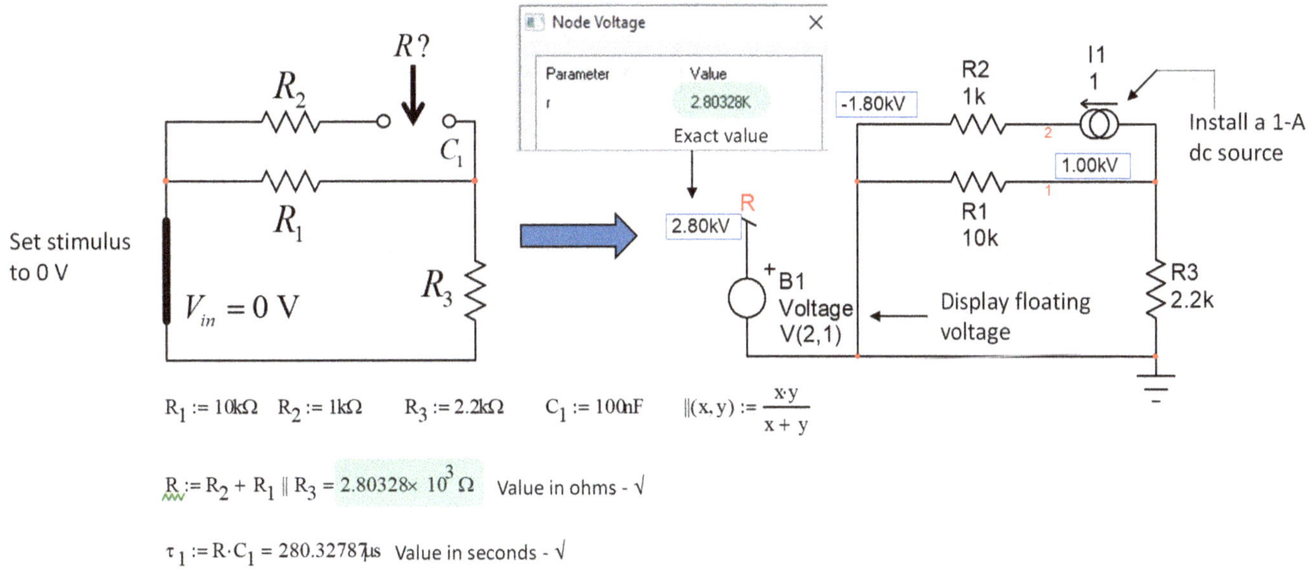

Installing a 1-A Dc Source Leads to the Desired Resistance Value for Comparison with Analytical Results

Figure 2.11

2.1.2 States of Passive Elements in Higher-Order Circuits

When we have one single energy-storing element such as a capacitor or an inductor, the choice is easy: the involved component is either placed in its dc or high-frequency state to determine where we start from.

This is what is called the *reference state* selected for $s = 0$ or $s \to \infty$.

However, when there are more energy-storing elements, we have many possible choices for this state later used for calculating the leading term. Assume the classical 2nd-order filter shown in the right-side of Figure 2.12 where you see a capacitor and an inductor.

To determine the leading term, in which each energy-storing element is set in its reference state, we have four different options described through sketches (a) to (d).

Which one should we pick?

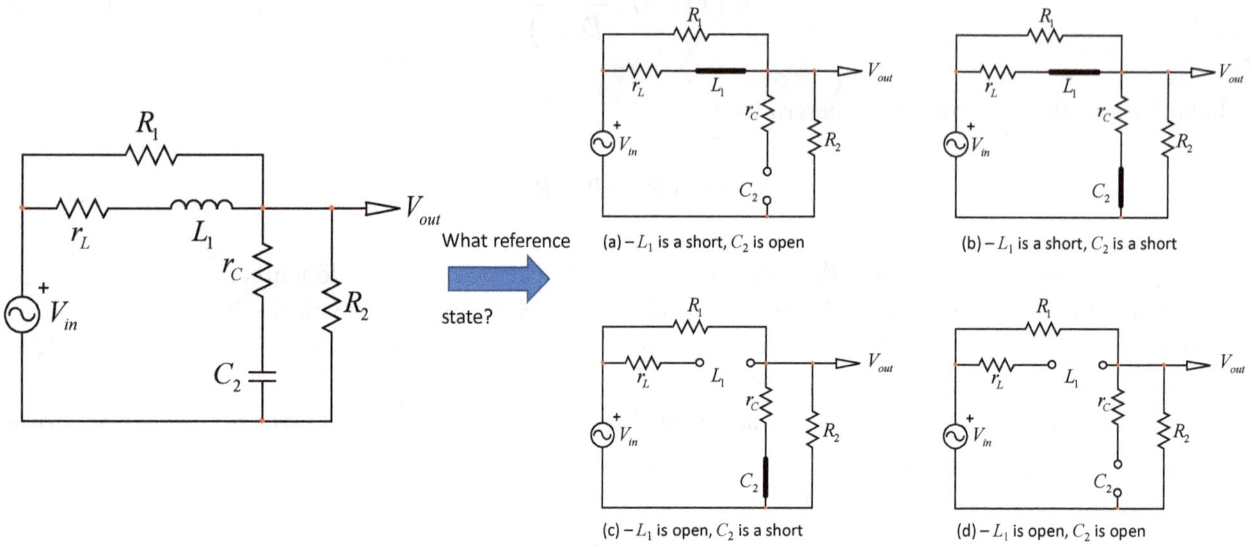

This 2nd-Order Filter Has Two Energy-Storage Elements

Figure 2.12

Actually, we could pick any of the four, provided we process the rest of the analysis accordingly. I have shown this with a worked example in [1]. However, for the sake of a more physical approach, I have selected option (a) which is that of SPICE when $s = 0$: shorted inductors and open capacitors for bias-point determination. This is, in my opinion, the simplest and most straightforward approach as it refers to a dc state familiar to any engineer.

From there, we can immediately determine the dc gain H_0 from sketch (a) which is:

$$H_0 = \frac{R_2}{R_2 + R_1 \| r_L} \tag{2.6}$$

How do we then proceed with the natural time constants determination? We follow the same path except that when we determine the resistance R driving one of the two energy-storing elements, we must define the state of the second or n^{th} one while we run the exercise.

We know that our denominator will follow the below normalized form for a 2nd-order system such as the LC filter we are looking at:

$$D(s) = 1 + b_1 s + b_2 s^2 \tag{2.7}$$

Let's start with b_1 whose unit should be time in seconds considering a unitless denominator. For a second-order network, it is a sum of two time constants:

$$b_1 = \tau_1 + \tau_2 \tag{2.8}$$

To determine any of these terms, temporarily disconnect the considered energy-storing element, C_1 for instance, leave the second one (or the nth ones) in its (their) dc state and determine the resistance R driving C_1. When done, go for the second time constant: temporarily disconnect L_2 and look through its terminals to determine R while C_1 is now set in its dc state or open circuited. The steps are:

1. Turn the stimulus off: short the voltage source or open-circuit the current source
2. Look at the resistance driving C_2 while L_1 is replaced by a short circuit (dc state)
3. Look at the resistance driving L_1 while C_2 is open-circuited (dc state)
4. Sum the two time constants to form b_1

Whether there are 1 or 10 energy-storing elements, it is easy to figure out the sketches for determining b_1: all components not involved in the on-going determination of the resistance R are placed in their dc state during the exercise. In the end, you sum all time constants to determine b_1:

$$b_1 = \tau_1 + \tau_2 + \tau_3 + \ldots + \tau_n \tag{2.9}$$

Look at Figure 2.13 for this first step:

Determine τ_2, L_1 is in it dc state

$$\tau_2 = C_2\left(r_C + R_1 \parallel r_L \parallel R_2\right)$$

Determine τ_1, C_2 is in its dc state

$$\tau_1 = \frac{L_1}{r_L + R_1 \parallel R_2}$$

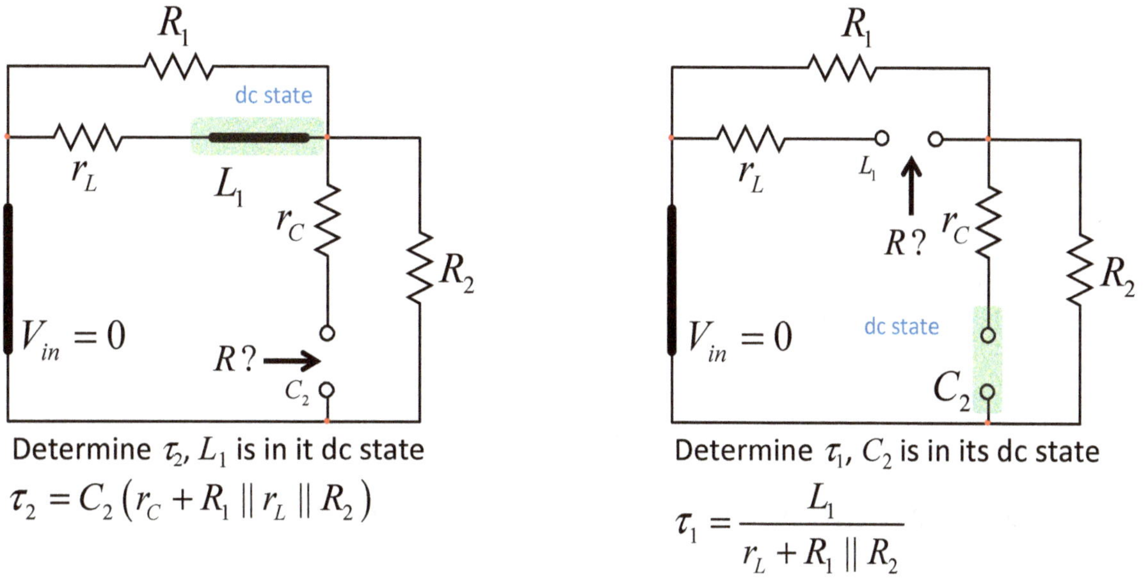

Look Through Energy-Storing Component Terminals while the other Element is Placed in its Dc State

Figure 2.13

From the figure, we can sum τ_1 and τ_2 to determine b_1:

$$b_1 = \tau_1 + \tau_2 = \frac{L_1}{r_L + R_1 \parallel R_2} + C_2\left(r_C + R_1 \parallel r_L \parallel R_2\right) \tag{2.10}$$

In this expression, do not develop the parallel combinations. They are the ones providing insight in the expression.

For instance, in τ_1, because r_L is potentially a small value, it can be neglected if the numerical results show that the parallel combination of R_1 and R_2 is much larger than r_L. In τ_2, assuming r_L is very small, it dominates the parallel arrangement immediately leading to the below simplification:

$$b_1 \approx \frac{L_1}{R_1 \| R_2} + C_2 \left(r_C + r_L \right) \tag{2.11}$$

Should you have developed the parallel arrangements then the insight with r_L may not have been that obvious.

Now, for term b_2, we have to consider a product (or a sum of products in higher-order networks) of two time constants so that D(s) remains unitless. A new notation is thus introduced to illustrate this process:

$$b_2 = \tau_1 \tau_2^1 \tag{2.12}$$

...or:

$$b_2 = \tau_2 \tau_1^2 \tag{2.13}$$

This formalism indicates that the energy-storing element whose label lies in the superscripted (exponent-like) notation is set in its high-frequency state while you determine the resistance R driving the energy-storing element whose label lies in the subscript. In (2.12), we are reusing the time constant τ_1 and multiply it by τ_2^1 : set L_1 in its high-frequency state (an open circuit), determine the resistance R driving C_2 in this mode and finally form the time constant RC_2.

It is shown in [2] that a redundancy exists meaning that if you take the time constant τ_2 and multiply it by τ_1^2 in which C_2 is short-circuited while you determine R driving L_1, then the expression obtained in (2.13) is identical to that of (2.12). Of course, you should always pick the simplest combination between the two options or the one avoiding an indeterminacy.

The complete expression for this 2nd-order denominator is thus:

$$D(s) = 1 + (\tau_1 + \tau_2) s + \tau_1 \tau_2^1 s^2 \tag{2.14}$$

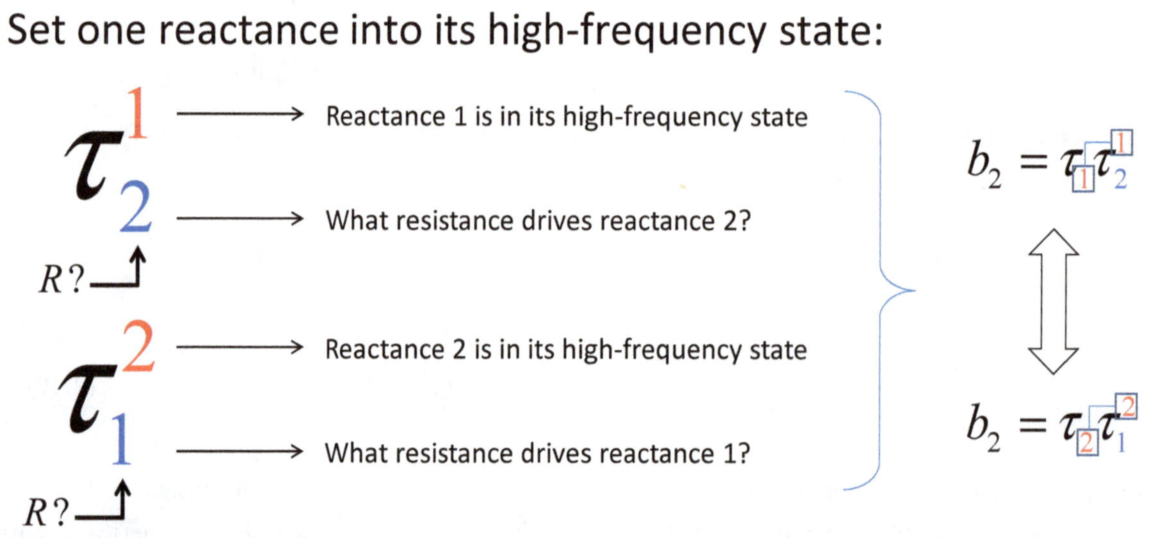
Set one reactance into its high-frequency state:
Reactance 1 is in its high-frequency state
What resistance drives reactance 2?
R?
Reactance 2 is in its high-frequency state
What resistance drives reactance 1?
R?
$b_2 = \tau_1 \tau_2^1$
$b_2 = \tau_2 \tau_1^2$

The Second-Order Time Constants Involve a Different Notation

Figure 2.14

Also equal to:

$$D(s) = 1 + (\tau_1 + \tau_2)s + \tau_2\tau_1^2 s^2 \tag{2.15}$$

Figure 2.14 illustrates the notation and explains that the label sitting in the exponent-like position is set in its high-frequency state while you determine R for the second element. You then multiply the resulting time constant by the already-determined time constant corresponding to the superscripted label position as detailed in the right-side of the picture.

Figure 2.15 shows this theory in action for our LC filter.

We can now form the term b_2—whose unit is s^2—following (2.12) which is calculated as:

$$b_2 = \tau_1\tau_2^1 = \frac{L_1}{r_L + R_1 \| R_2}(r_C + R_1 \| R_2)C_2 \tag{2.16}$$

If we consider r_L and r_C to be very small compared to the other resistors, this expression simplifies to:

$$b_2 \approx L_1 C_2 \tag{2.17}$$

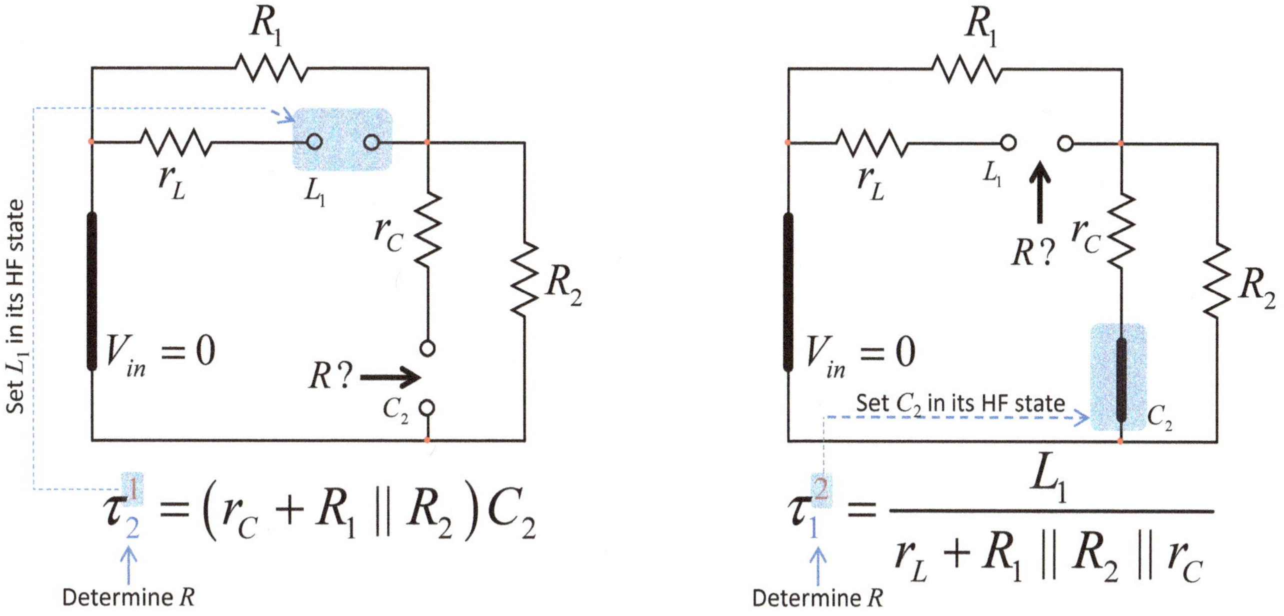

The Second-Order Time Constants are Simple to Determine and Inspection Works Well

Figure 2.15

If we now apply (2.13) to calculate b_2, we find:

$$b_2 = \tau_2\tau_1^2 = C_2\left(r_C + R_1 \| r_L \| R_2\right)\frac{L_1}{r_L + R_1 \| R_2 \| r_C} \tag{2.18}$$

Again, if r_L and r_C are very small and dominate the parallel arrangement, we obtain:

$$b_2 \approx C_2\left(r_C + r_L\right)\frac{L_1}{r_L + r_C} = L_1 C_2 \tag{2.19}$$

This is it; we have determined our denominator without writing a single line of algebra, just drawing simple sketches that can be independently solved by inspection. Following (2.7), we can write:

$$D(s) = 1 + \left[\frac{L_1}{r_L + R_1 \| R_2} + C_2\left(r_C + R_1 \| r_L \| R_2\right)\right]s + \left[\frac{L_1}{r_L + R_1 \| R_2}\left(r_C + R_1 \| R_2\right)C_2\right]s^2 \tag{2.20}$$

Which, considering negligible parasitic contributors with respect to R_1 and R_2 can be approximated as:

$$D(s) \approx 1 + \left[\frac{L_1}{R_1 \| R_2} + C_2\left(r_C + r_L\right)\right]s + L_1 C_2 s^2 \tag{2.21}$$

And we are done, we have determined the denominator in the simplest possible manner—without writing a line of algebra—by inspecting a series of intermediate sketches. Sketches on which you can easily come back if you spot an error in the end: just fix the guilty coefficient while keeping the others untouched and you straighten the whole equation up without restarting from scratch as with other methods.

What is also very interesting is that you can individually test all of your steps with a SPICE simulation and check the bias points match the results you have found. This is illustrated in Figure 2.16 where the Mathcad® screenshot in the lower right corner confirms all values.

If simple schematics like in this example are usually less prone to making mistakes, it is another story with controlled sources as with op-amps or transistors. In this case, SPICE represents an invaluable tool to confirm your analysis is correct and all results are sound. It helped me progress and fix errors in many complicated cases as those illustrated in [1].

Having determined the denominator expression is one thing but we need to rearrange (2.20) in a *low-entropy* form where a quality coefficient Q and a resonant frequency appear. It is not that complicated if you equate the below expressions and determine the coefficients:

$$1 + b_1 s + b_2 s^2 = 1 + \frac{s}{\omega_0 Q} + \left(\frac{s}{\omega_0}\right)^2 = 1 + \frac{2\zeta}{\omega_0}s + \left(\frac{s}{\omega_0}\right)^2 \tag{2.22}$$

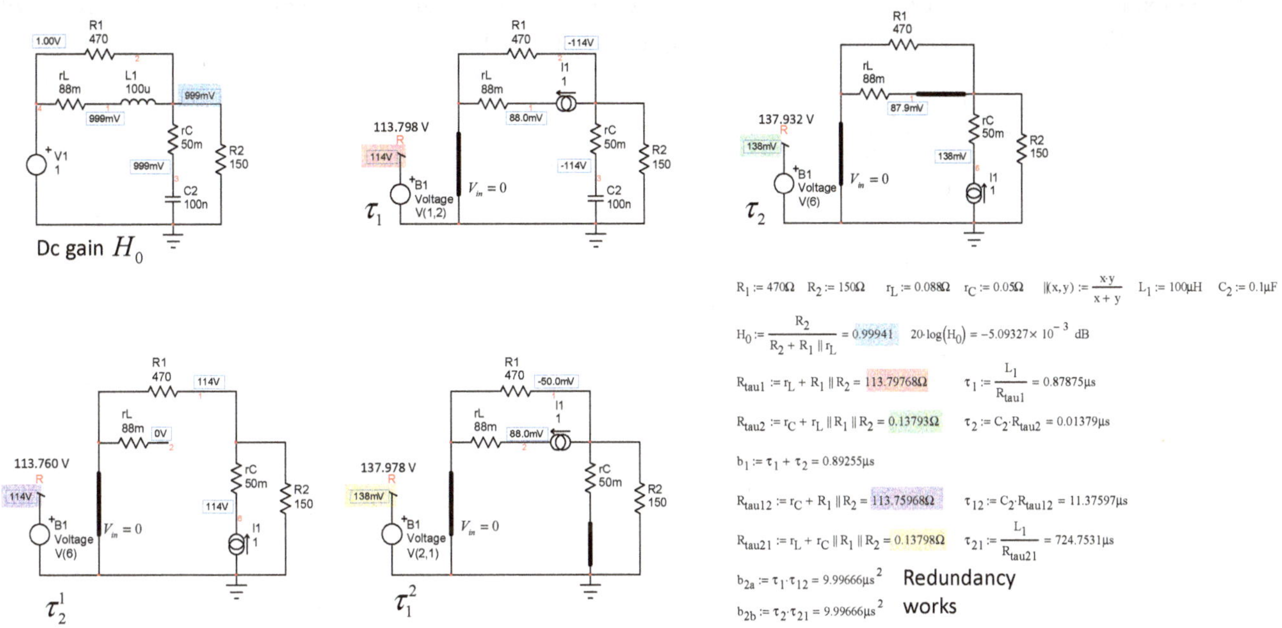

SPICE is an Invaluable Tool to Individually Test all Sketches and Spot Deviations Between Equations and Bias Point Results

Figure 2.16

If you do the math, you will find:

$$Q = \frac{\sqrt{b_2}}{b_1} \tag{2.23}$$

$$\zeta = \frac{1}{2Q} = \frac{b_1}{2\sqrt{b_2}} \tag{2.24}$$

...and:

$$\omega_0 = \frac{1}{\sqrt{b_2}} \tag{2.25}$$

If you substitute the approximate values we found for b_1 and b_2 in (2.23), you have:

$$Q \approx \frac{\sqrt{L_1 C_2}}{\frac{L_1}{R_1 \| R_2} + C_2\left(r_C + r_L\right)} \tag{2.26}$$

If we further consider r_C and r_L to be parasitics and thus of very small values, then this expression can be further simplified as:

$$Q \approx \left(R_1 \| R_2\right)\sqrt{\frac{C_2}{L_1}} \tag{2.27}$$

In this expression, the first term is resistive while the second has the dimension of a conductance (the inverse of a characteristic impedance) so Q is unitless and homogeneity is respected.

$Q := \frac{\sqrt{b_{2a}}}{b_1} = 3.5424$ $\zeta := \frac{1}{2 \cdot Q} = 0.14115$ $\frac{b_1}{2 \cdot \sqrt{b_{2a}}} = 0.14115$ $Q_a := (R_1 \parallel R_2) \cdot \sqrt{\frac{C_2}{L_1}} = 3.59582$ $D_1(s) := 1 + \frac{s}{\omega_0 \cdot Q} + \left(\frac{s}{\omega_0}\right)^2$ $D_2(s) := 1 + \frac{2 \cdot \zeta}{\omega_0} \cdot s + \left(\frac{s}{\omega_0}\right)^2$

$\omega_0 := \frac{1}{\sqrt{b_{2a}}}$ $f_0 := \frac{\omega_0}{2 \cdot \pi} = 50.33762 \text{kHz}$

$D_3(s) := 1 + s \cdot \left[\frac{L_1}{r_L + R_1 \parallel R_2} + C_2 \cdot (r_C + r_L \parallel R_1 \parallel R_2)\right] + \left[\frac{L_1}{r_L + R_1 \parallel R_2} \cdot (r_C + R_1 \parallel R_2) \cdot C_2\right] \cdot s^2$

$D_4(s) := 1 + \left[\frac{L_1}{R_1 \parallel R_2} + C_2 \cdot (r_L + r_C)\right] \cdot s + L_1 \cdot C_2 \cdot s^2$

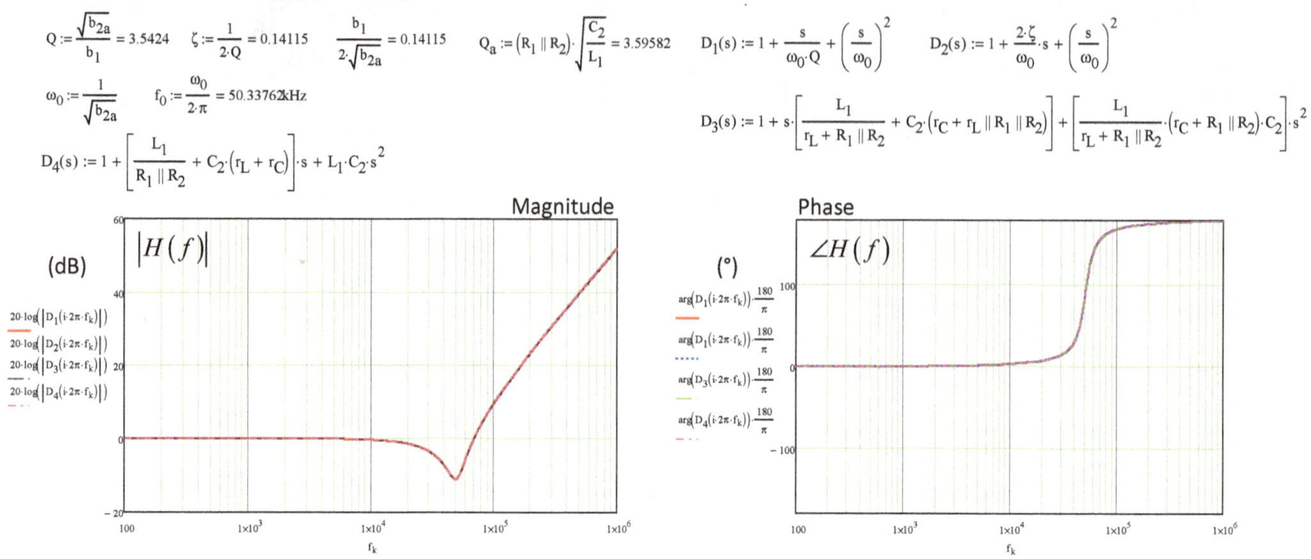

The Denominator Must be Expressed in a Normalized Form in Which a Quality Factor and Resonant Frequency Appears

Figure 2.17

This is what you must learn to do with the FACTs: reduce, rearrange and simplify as much as you can the raw expression to form a compact formula whose practical design application is easy. This obviously requires some engineering judgement to avoid over-simplifying and corrupting the result. Mathcad® or any other solver is a good tool to assess the deviation in magnitude and phase between the full-blown expression and its simplified version. Keep in mind that it is useless to calculate resistance or capacitor values with 3 digits after the decimal point considering the natural tolerance of these elements in the laboratory. Simple and compact formulas are thus the ultimate goal motivating this exercise.

The Mathcad® sheet of Figure 2.17 tells us how all these expressions compare and confirms results are correct as all curves superimpose perfectly.

For a higher-order denominator, you would have more time constants products expressions summed in b_2 as shown in (2.9). If we take the example of a 3rd-order denominator, the denominator would follow the below normalized expression:

$$D(s) = 1 + b_1 s + b_2 s^2 + b_3 s^3 \qquad (2.28)$$

Owing to previous explanations, we would have:

$$b_1 = \tau_1 + \tau_2 + \tau_3 \qquad (2.29)$$

Then:

$$b_2 = \tau_1 \tau_2^1 + \tau_1 \tau_3^1 + \tau_2 \tau_3^2 \qquad (2.30)$$

…with all the terms determined as illustrated by Figure 2.14.

Finally, the 3rd-order coefficient b_3—whose is unit is s^3—would be written as:

$$b_3 = \tau_1 \tau_2^1 \tau_3^{12} \tag{2.31}$$

This is what Figure 2.18 details: two terms are now sitting in the exponent-like position and placed in a high-frequency state.

You then determine the resistance R driving the third term to form the time constant. This notation and principle can be extended to n^{th}-order system where terms set in their high-frequency state gather in the exponent position while you determine the resistance driving the subscripted label.

Now, as with coefficients b_2 defined by (2.12) and (2.13), redundancy or possibility to *reshuffle* the terms exists. And this is a great option in some cases, particularly when indeterminacies occur or the circuit when analyzed with one combination is too complicated. Indeterminacies may potentially appear when you multiply terms together: make sure the product leads to a finite number.

For instance, if during a resistance determination exercise, you find a zero-ohm resistance then $\tau_1 = 0\ \Omega \cdot 1\ \mu\text{F} = 0\ \text{s}$ and it is acceptable. If you multiply this term by $\tau_2^1 = 5\ \mu\text{s}$, then you have $\tau_1 \tau_2^1 = 0 \cdot 5u = 0$ and this is also fine.

If you now mix inductors and capacitors, you can stumble on one of the following issues when associating time constants (beside $\frac{0}{\infty}$ which returns zero):

$$\frac{0}{0}$$
$$\infty \cdot 0$$
$$\frac{\infty}{0}$$
$$\frac{\infty}{\infty} \tag{2.32}$$

In that case, reshuffle the expression as shown in the figure to build coefficient b_3 and find the combination where the indeterminacy is gone.

If this indeterminacy is persistent, then the solution consists of involving a dummy resistance added to the circuit either in series to fight a 0-Ω path (a small value R_s) or in parallel with the involved energy-storing element to limit the resistance of an open circuit to a finite resistance value (a high value R_{INF}).

When the transfer function is determined, then you can rework the result through factorization and have R_s go to zero and R_{INF} approach infinity to simplify and shape the final result. I have successfully applied this method in the many examples I solved in [1].

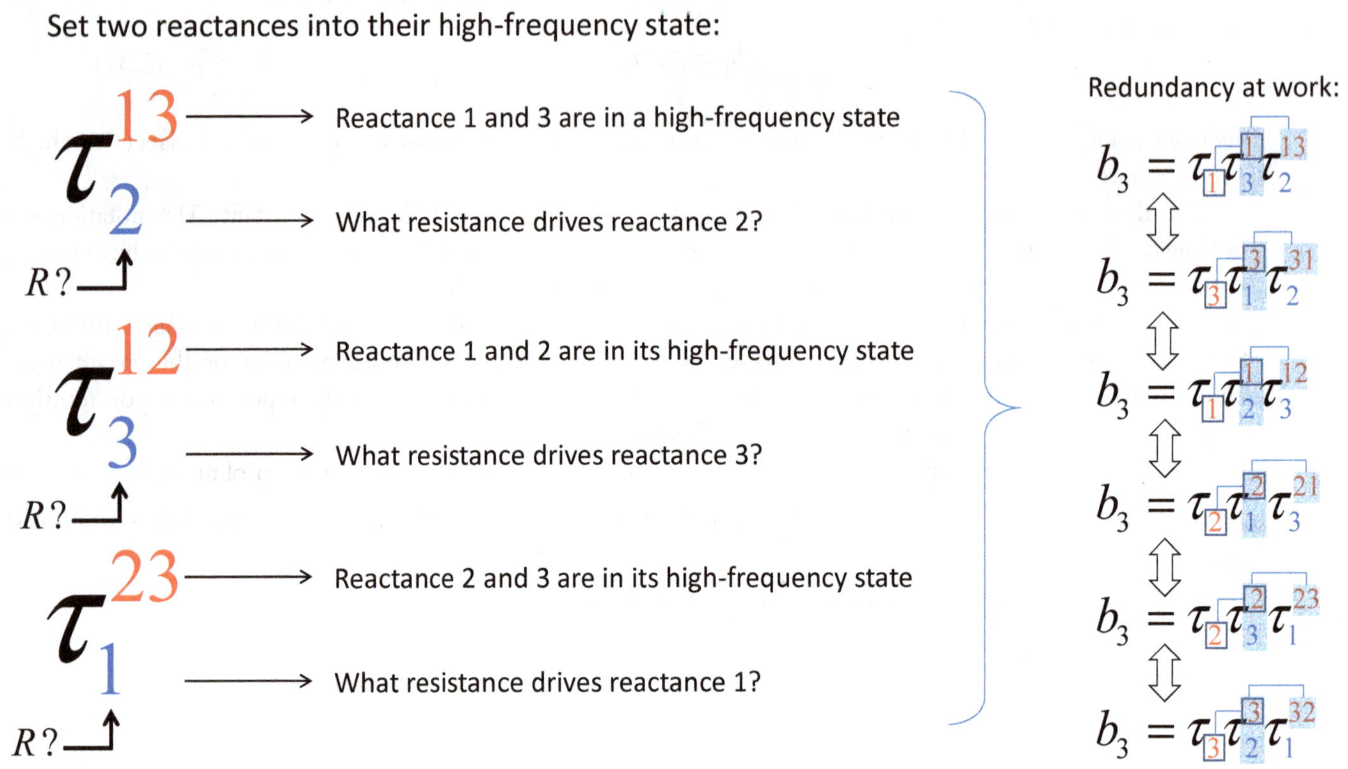
Set two reactances into their high-frequency state:
Reactance 1 and 3 are in a high-frequency state
What resistance drives reactance 2?
R?
Reactance 1 and 2 are in its high-frequency state
What resistance drives reactance 3?
R?
Reactance 2 and 3 are in its high-frequency state
What resistance drives reactance 1?
R?
Redundancy at work:

The Coefficient in b_3 when determining a 3rd-Order Transfer Function Follows the Same Principle with Two Labels in the Exponent-Like Position. As with b_2, Redundancy Helps Rewrite the Original Expression to Explore Different Intermediate Sketches and Diagrams

Figure 2.18

Another tip is when you study circuits featuring an op-amp.

In some situations, having an infinite gain is an advantage for inspecting the circuit, e.g. because both inputs share the same potential and inspection is obvious. Sometimes, it can create difficulties during the analysis and I like to purposely reduce the gain to a finite value labeled A_{OL} and consider the small voltage difference ε between the two inputs.

This helps having meaningful numbers with SPICE dc analyses which can efficiently confirm the intermediate results obtained with the solver. The open-loop gain of the op-amp will then later on be pushed to an infinite value when shaping the final expression.

In the examples, we will apply these tricks.

Let's have a quick exercise with the 3rd-order circuit drawn in Figure 2.19. We start by setting s to zero and place all components in their dc state. This is sketch (b) and the gain is immediate considering a resistive divider involving R_1 and R_2:

$$H_0 = \frac{R_2}{R_1 + R_2} \tag{2.33}$$

The three time constants are found by looking through the connecting terminals of the considered element while the two others are left in their dc state.

This is an easy exercise and if you do it right, you should find:

$$b_1 = C_1\left(R_1 \parallel R_2\right) + \frac{L_2}{R_1 + R_2} + C_3\left(R_1 \parallel R_2\right) \tag{2.34}$$

(a)

(b) $s = 0$

(c) $\tau_1 = \left(R_1 \parallel R_2\right)C_1$
$\tau_3 = \left(R_1 \parallel R_2\right)C_3$

(d) $\tau_2 = \dfrac{L_2}{R_1 + R_2}$

Set C_1 in its HF state (e)

$\tau_2^1 = \dfrac{L_2}{R_2}$

Determine R for L_2

Set C_1 in its HF state (f)

$\tau_3^1 = 0 \cdot C_3$

Determine R for C_3

Set L_2 in its HF state (g)

$\tau_3^2 = R_2 \cdot C_3$

Determine R for C_3

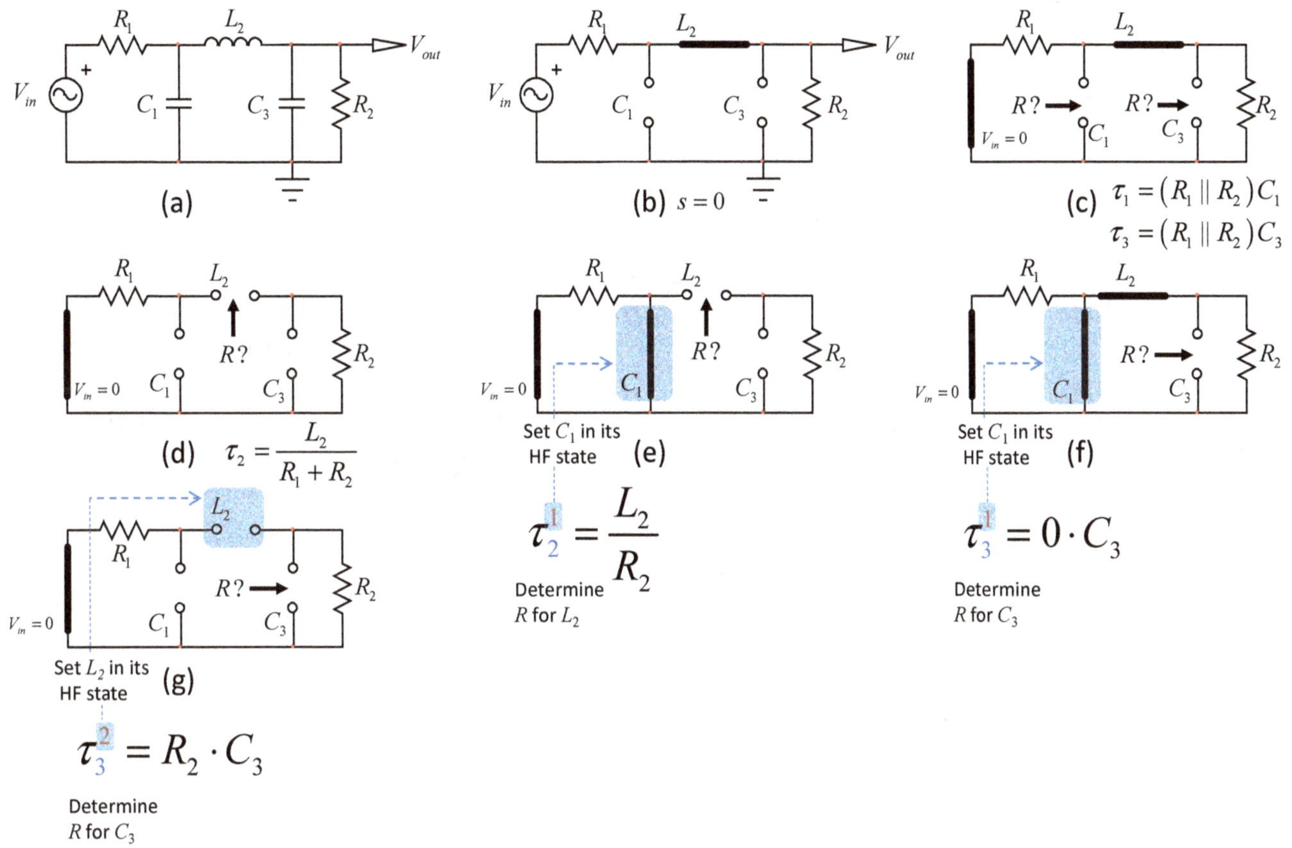

This Circuit Features Three Energy-Storage Elements and is thus a 3rd-Order Network

Figure 2.19

For the higher-order terms, simply apply what Figure 2.18 details and draw three intermediate sketches. If you inspect the circuits correctly, you should obtain the definition for b_2 following (2.30) in few minutes:

$$b_2 = C_1\left(R_1 \parallel R_2\right)\frac{L_2}{R_2} + C_1\left(R_1 \parallel R_2\right)\cdot 0\cdot C_3 + \frac{L_2}{R_1 + R_2}\cdot R_2 C_3 = C_1\left(R_1 \parallel R_2\right)\frac{L_2}{R_2} + \frac{L_2}{R_1 + R_2}R_2 C_3 \tag{2.35}$$

For the final term b_3, we have the product of three time constants defined by (2.31) and illustrated by Figure 2.20.

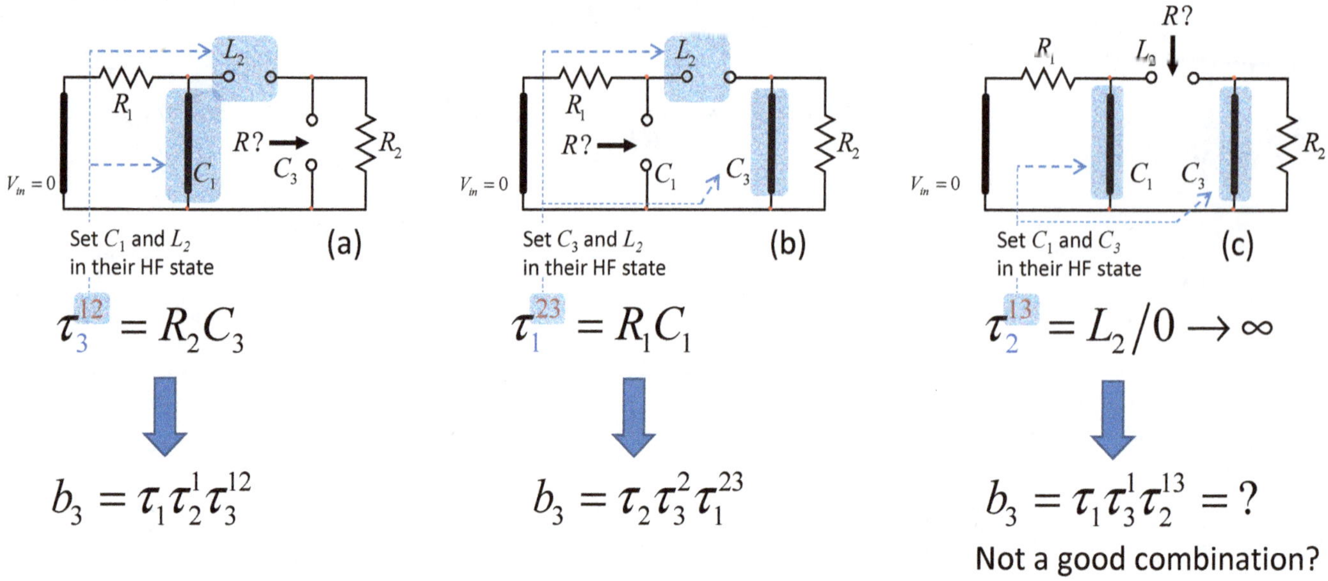

Coefficient b_3 can also be Obtained by Inspection. If the First Option Leads to a Complicated Circuit, Reshuffle the Numbers as Shown in Sketch (b). Sketch (c) is not a Good Option in this Particular Case

Figure 2.20

There are two energy-storing elements set in their high-frequency state while you determine the resistance driving the third one.

Sketch (a) leads to an easy configuration and the final coefficient is:

$$b_3 = C_1\left(R_1 \,||\, R_2\right)\frac{L_2}{R_2}R_2C_3 = C_1\left(R_1 \,||\, R_2\right)L_2C_3 \tag{2.36}$$

In case this first combination leads to an indeterminacy or a complicated circuit to analyze, you still have the choice to reshuffle the combinations and find an easier one to solve.

For instance, as shown in Figure 2.18, all these combinations are equivalent:

$$b_3 = \tau_1\tau_2^1\tau_3^{12} = \tau_2\tau_1^2\tau_3^{21} = \tau_3\tau_1^3\tau_2^{31} = \tau_1\tau_3^1\tau_2^{13} = \tau_2\tau_3^2\tau_1^{23} = \tau_3\tau_2^3\tau_1^{32} \tag{2.37}$$

However, as shown in Figure 2.20 with three possible combinations, the first two ones are giving a valid result while the third one seems to be a dead end.

Is it really?

Let's look how the results combine:

$$b_3 = \tau_1\tau_3^1\tau_2^{13} = C_1\left(R_1 \,||\, R_2\right)\cdot 0\cdot C_3\frac{L_2}{0} \tag{2.38}$$

Looks like we have an indeterminacy as described by the first case in (2.32), right?

Now, assume our solver does not accept 0-Ω resistance but an infinitesimally-small finite value—let's say 1 pΩ—that we call R_s.

In that case, the above equation could be rewritten as:

$$b_3 = \tau_1 \tau_3^1 \tau_2^{13} = C_1 \left(R_1 \parallel R_2 \right) \cdot R_s \cdot C_3 \frac{L_2}{R_s} = C_1 \left(R_1 \parallel R_2 \right) C_3 L_2 \tag{2.39}$$

...and you can now simplify by R_s: thus (2.39) and (2.36) are similar.

We can now assemble the denominator and check the results with a Mathcad® sheet. If we gather the whole set of coefficients following (2.28), we should find:

$$D(s) = 1 + \left[\left(R_1 \parallel R_2 \right)\left(C_1 + C_3 \right) + \frac{L_2}{R_1 + R_2} \right] s + \left[C_1 \left(R_1 \parallel R_2 \right) \frac{L_2}{R_2} + \frac{L_2}{R_1 + R_2} R_2 C_3 \right] s^2 + C_1 \left(R_1 \parallel R_2 \right) L_2 C_3 s^3 \tag{2.40}$$

In this particular example, there are no zeroes—bear with me, I will show why in the next paragraph—and the transfer function is completed:

$$H(s) = \frac{R_2}{R_1 + R_2} \frac{1}{1 + \left[\left(R_1 \parallel R_2 \right)\left(C_1 + C_3 \right) + \frac{L_2}{R_1 + R_2} \right] s + \left[C_1 \left(R_1 \parallel R_2 \right) \frac{L_2}{R_2} + \frac{L_2}{R_1 + R_2} R_2 C_3 \right] s^2 + C_1 \left(R_1 \parallel R_2 \right) L_2 C_3 s^3} \tag{2.41}$$

To verify if this is correct, we can determine the same transfer function using the brute-force approach by involving a Thévenin generator at the junction between R_1 and L_2 of the following characteristics:

$$V_{th}(s) = V_{in}(s) \frac{\frac{1}{sC_1}}{\frac{1}{sC_1} + R_1} \tag{2.42}$$

$$R_{th}(s) = R_1 \parallel \left(\frac{1}{sC_1} \right) \tag{2.43}$$

...then an impedance divider involving L_2 driving the parallel combination of C_3 with R_2:

$$Z_1(s) = R_2 \parallel \left(\frac{1}{sC_3} \right) \tag{2.44}$$

The transfer function is immediate and good luck to develop and rearrange the terms without mistakes!

$$H_{ref}(s) = \frac{\dfrac{1}{sC_1}}{\dfrac{1}{sC_1} + R_1} \cdot \frac{Z_1(s)}{Z_1(s) + R_{th}(s) + sL_2} \tag{2.45}$$

Fortunately, this expression does not imply difficulties for a solver and we have gathered all these formulas in a Mathcad® sheet with results given in Figure 2.21.

$\|(x,y) := \dfrac{x \cdot y}{x + y} \qquad R_{inf} := 10^{23}\,\Omega \qquad R_s := 10^{-12}\,\Omega$

$C_1 := 100\text{nF} \qquad R_1 := 100\Omega \qquad L_2 := 100\mu\text{H} \qquad C_3 := 100\text{nF} \qquad R_2 := 1000\Omega$

$H_0 := \dfrac{R_2}{R_1 + R_2} = 0.90909 \qquad 20 \cdot \log(H_0) = -0.82785 \quad \text{dB}$

$\tau_1 := (R_1 \| R_2) \cdot C_1 = 9.09091\,\mu\text{s}$

$\tau_2 := \dfrac{L_2}{R_1 + R_2} = 90.90909\,\text{ns}$

$\tau_3 := (R_1 \| R_2) \cdot C_3 = 9.09091\,\mu\text{s}$

$b_1 := \tau_1 + \tau_2 + \tau_3 = 18.27273\,\mu\text{s}$

$\tau_{12} := \dfrac{L_2}{R_2} = 100 \cdot \text{ns} \qquad \tau_{13} := C_3 \cdot 0\Omega = 0 \cdot \mu\text{s} \qquad \tau_{23} := C_3 \cdot R_2 = 100\,\mu\text{s}$

$b_2 := \tau_1 \cdot \tau_{12} + \tau_1 \cdot \tau_{13} + \tau_2 \cdot \tau_{23} = 1 \times 10^{-11}\,\text{s}^2$

$\tau_{123} := R_2 \cdot C_3 = 100 \cdot \mu\text{s} \qquad\qquad \tau_{231} := C_1 \cdot R_1$

$b_3 := \tau_1 \cdot \tau_{12} \cdot \tau_{123} = 9.09091 \times 10^{10} \cdot \text{ns}^3 \qquad b_{3a} := \tau_2 \cdot \tau_{23} \cdot \tau_{231} = 9.09091 \times 10^{10} \cdot \text{ns}^3$

$\tau_1 \cdot \tau_{13} \cdot \dfrac{L_2}{0} = \blacksquare \qquad\qquad b_{3b} := \tau_1 \cdot C_3 \cdot R_s \cdot \dfrac{L_2}{R_s} = 9.09091 \times 10^{10} \cdot \text{ns}^3$

indeterminacy

Redundancy at work

$D_1(s) := 1 + b_1 \cdot s + b_2 \cdot s^2 + b_3 \cdot s^3$

$H_1(s) := H_0 \cdot \dfrac{1}{D_1(s)}$

$R_{th}(s) := R_1 \left\| \left(\dfrac{1}{s \cdot C_1}\right) \right. \qquad Z_1(s) := \left(\dfrac{1}{s \cdot C_3}\right) \| R_2$

$H_{ref}(s) := \dfrac{\dfrac{1}{s \cdot C_1}}{R_1 + \dfrac{1}{s \cdot C_1}} \cdot \dfrac{Z_1(s)}{R_{th}(s) + s \cdot L_2 + Z_1(s)}$

Brute-force expression

(a)

Magnitude $|H(f)|$

(dB)

Dominant pole

Double poles

$20 \cdot \log\left(\left|H_{ref}(i \cdot 2\pi \cdot f_k)\right|\right)$

$20 \cdot \log\left(\left|H_1(i \cdot 2\pi \cdot f_k)\right|\right)$

Phase $\angle H(f)$

(°)

$\arg\left(H_{ref}(i \cdot 2\pi \cdot f_k)\right) \cdot \dfrac{180}{\pi}$

$\arg\left(H_1(i \cdot 2\pi \cdot f_k)\right) \cdot \dfrac{180}{\pi}$

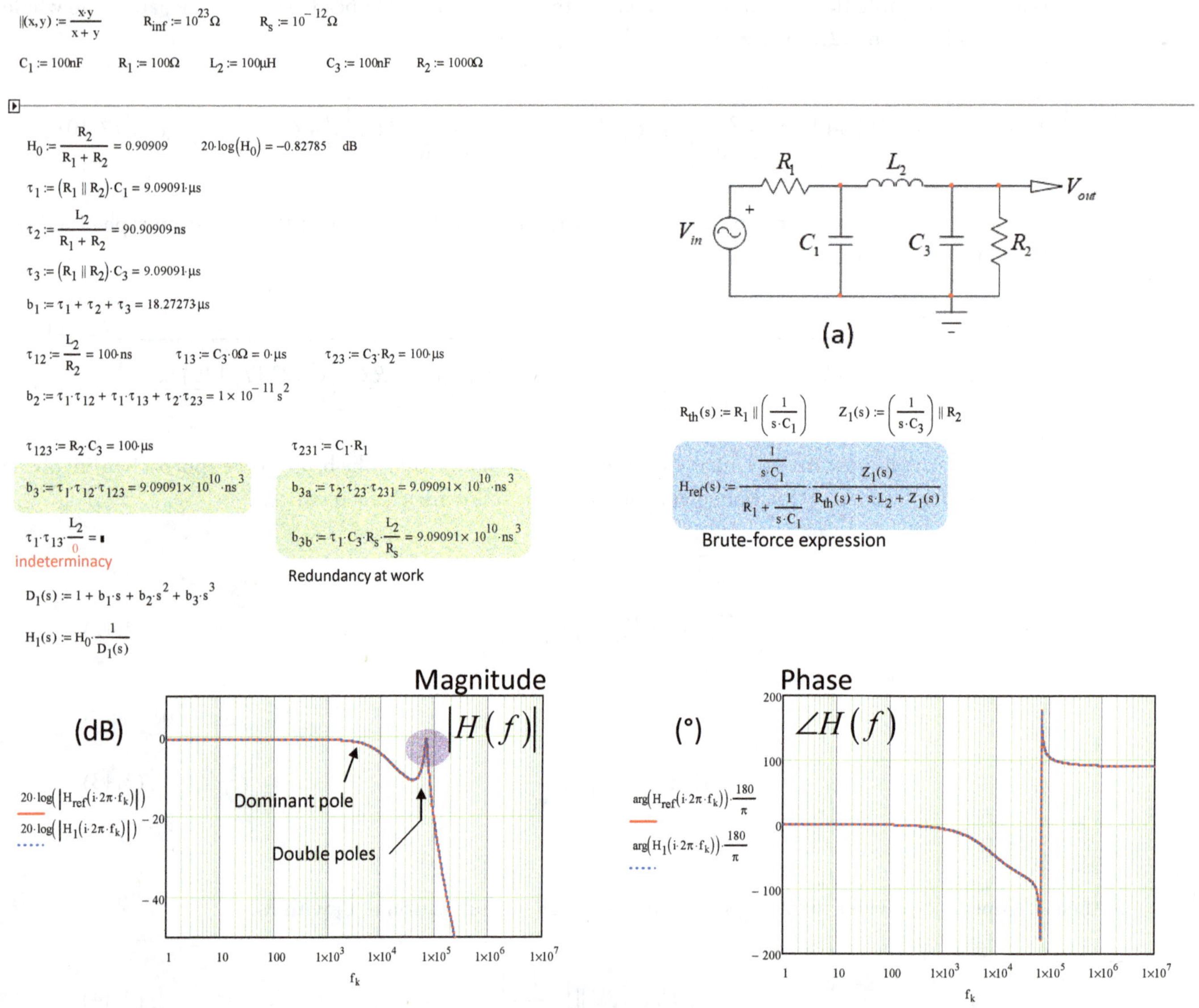

Mathcad® Calculates all Coefficient and Checks Homogeneity in the end

Figure 2.21

By subtracting the magnitude/phase of the brute-force expression from the derived expression given in (2.41), the solver error in magnitude and phase remains in the pico-range, meaning both expressions are rigorously equal. Now, observing the magnitude response, we can see a low-frequency pole—around 2–3 kHz—followed by a peaking located circa 70 kHz: the pole dominates the low-frequency part while the two contiguous poles show up in higher frequencies. To confirm this fact, we can look at the individual time constants [3] and see how they compare with each other. The goal is to rearrange the 3rd-oder polynomial expression in a less complicated formula which would offer more insight in what the frequency response can be.

In this particular case, we could try a low-frequency pole followed by two poles arranged in a 2nd-order expression:

$$H_0\frac{1}{1+sb_1+s^2b_2+s^3b_3}\approx H_0\frac{1}{1+\frac{s}{\omega_p}}\frac{1}{1+\frac{s}{Q\omega_0}+\left(\frac{s}{\omega_0}\right)^2} \tag{2.46}$$

The pole in the first expression is the inverse of b_1 while the newly defined Q and w are described in Figure 2.22.

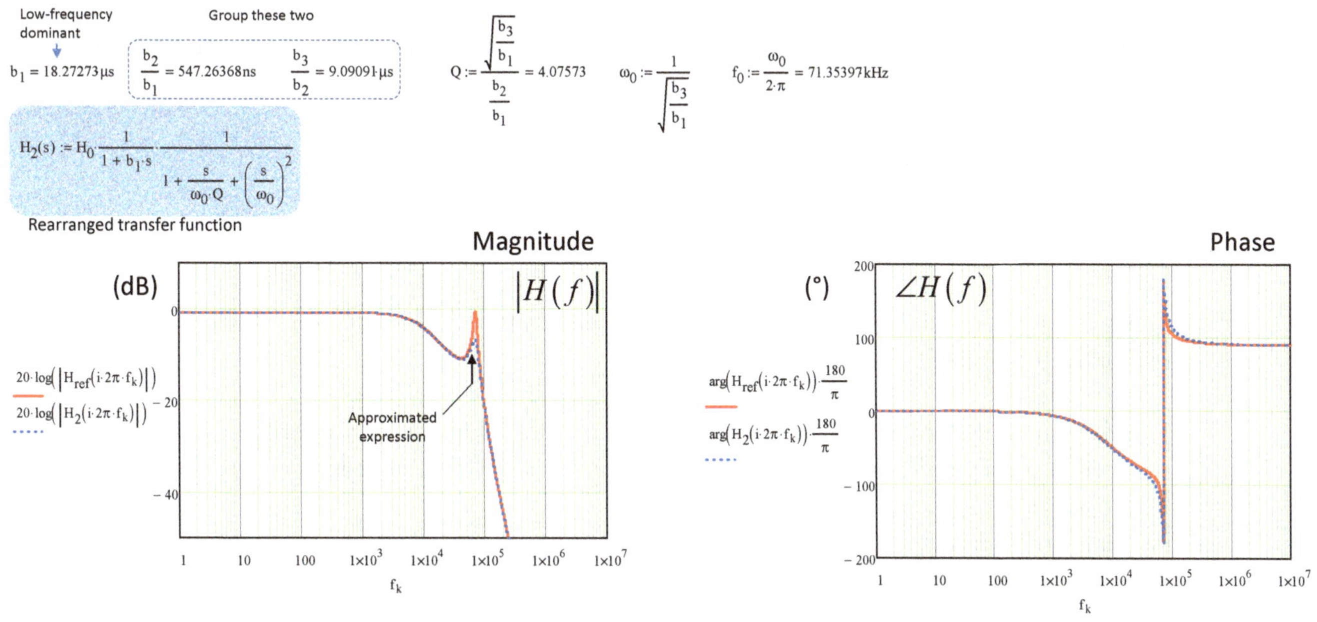

Mathcad® allows the Testing of a New Expression versus the Raw One and Confirms a Good Match Despite Lower Peaking at Resonance

Figure 2.22

We have shown in this section how to determine the coefficients in an nth-order denominator. In Figure 2.23, I gathered the possible combinations up to the 4th-order. You can apply the technique to higher coefficients with the formulas given in [1] but also looking at the 5th- and 6th-order filter examples I posted on my webpage [4]. As the number of energy-storing elements increase, the key lies in organizing all the intermediate steps in clean and well-documented sketches to which you can conveniently go back for correcting errors.

Denominators up to order 4

1st order $D(s) = 1 + \tau_1 s$

2nd order $D(s) = 1 + (\tau_1 + \tau_2)s + (\tau_1 \tau_2^1)s^2 = 1 + (\tau_1 + \tau_2)s + (\tau_2 \tau_1^2)s^2$

3rd order $D(s) = 1 + (\tau_1 + \tau_2 + \tau_3)s + (\tau_1 \tau_2^1 + \tau_1 \tau_3^1 + \tau_2 \tau_3^2)s^2 + (\tau_1 \tau_2^1 \tau_3^{12})s^3$

4th order $$D(s) = 1 + (\tau_1 + \tau_2 + \tau_3 + \tau_4)s + (\tau_1 \tau_2^1 + \tau_1 \tau_3^1 + \tau_1 \tau_4^1 + \tau_2 \tau_3^2 + \tau_2 \tau_4^2 + \tau_3 \tau_4^3)s^2 + (\tau_1 \tau_2^1 \tau_3^{12} + \tau_1 \tau_2^1 \tau_4^{12} + \tau_1 \tau_3^1 \tau_4^{13} + \tau_2 \tau_3^2 \tau_4^{23})s^3 + (\tau_1 \tau_2^1 \tau_3^{12} \tau_4^{123})s^4$$

This Figure Gives the Possible Combinations to Form the Denominator up to the 4th-Order

Figure 2.23

2.2 What Should I Retain from this Chapter?

In this second chapter, we have learned information about the FACTs as summarized below:

1. The poles of a network solely depend on its electrical structure. The excitation—the stimulus—plays no role in the location of these poles.

2. You determine the position of these poles by turning the excitation off: a voltage source set to 0 V is replaced by a wire in the drawing while a zeroed current source becomes an open circuit.

3. In a first-order circuit, the so-called *reference state* can be determined in dc conditions ($s = 0$) or when s approaches infinity.

4. In a second- or higher-order network, the reference state can be obtained by considering various state combinations of each energy-storing element. For a convenient analysis, consider the reference state as SPICE does when it calculates a bias point: all capacitors are open and inductors are shorted.

5. It is important to shape or reorganize the expression you have obtained so that it naturally reveals salient features such as gains, poles and zeroes, useful for design purposes.

6. We have seen that SPICE represents an invaluable tool when checking your calculations. By splitting complicated circuits into many smaller sketches, you can individually verify the steps by running a simple dc operating point and track any mistake. If you spot an error at the end of the process, fix the guilty sketch without restarting from scratch.

2.3 References

1. C. Basso, *Linear Circuit Transfer Functions—An Introduction to Fast Analytical Techniques*, Wiley, 2016.
2. V. Vorpérian, *Fast Analytical Techniques for Electrical and Electronic Circuits*, Cambridge University Press, 2002.
3. R. Erickson, D. Maksimović, *Fundamentals of Power Electronics*, Kluwer Academic Publishers, 2001, pp. 289-293 (https://www.ieee.li/pdf/introduction_to_power_electronics/chapter_08.pdf)
4. See presentations and papers to download section, http://powersimtof.com/Spice.htm

Chapter 3: Zeroes of a Transfer Function

WE HAVE SEEN in the previous chapter how identifying the natural time constant of a network implied a zeroed excitation.

The process is very close here, with the zeroes, except that the stimulus is back in place and the response is nulled.

3.1 Determining the Zeroes

The numerator $N(s)$ hosts the *zeroes* of the transfer function. Mathematically, zeroes are the roots for which the function magnitude is zero. Therefore, in principle, if we could excite a network with a frequency tuned at the zero value, then the transfer function magnitude being zero, there would be no observable ac or small-signal response. In this case, we would say the output is *nulled* at the zero frequency. Because this cannot be observed in the laboratory at first sight, it is a difficult concept for students to grasp.

When the numerator roots lie in the left half-plane, nulling the numerator would imply a negative frequency—which has no physical sense. For instance, assume in Figure 3.1 that you stimulate an *RC* network with a sinusoidal generator whose frequency is exactly tuned at the zero frequency. If you observe the output, a signal is seen where a null response is expected.

This is because the sinusoidal signal delivered by our function generator produces a positive frequency while the numerator root is a negative value. We perform a harmonic analysis in which σ is zero and only explore the imaginary axis in the *s*-plane: it's impossible to visualize a zero in these conditions.

Are there any cases where we could see the manifestation of a zero then? For a 1st-order circuit, yes: if you think of a simple differentiator as the one shown in the bottom of Figure 3.1, then having a zero at the origin ($s = 0$) means that a dc stimulus won't go through the network and you will see 0 V across the output terminals.

The 0-point is common to both real and imaginary axes and we can obviously obtain this condition on the bench.

Can you observe a **zero** in the lab?

response is non-zero

Transfer function:

$$\frac{V_{out}(s)}{V_{in}(s)} = \frac{1 + sr_cC_1}{1 + s(r_C + R_1)C_1} = \frac{1 + \dfrac{s}{\omega_z}}{1 + \dfrac{s}{\omega_p}}$$

zero

pole

No, because this is a harmonic analysis $s = j\omega$
✓ It works for a zero at the origin: dc block

$$V_{out}(0\,\text{Hz}) = 0\,\text{V}$$

Transfer function:

$$\frac{V_{out}(s)}{V_{in}(s)} = \frac{sR_2C_1}{1 + s(R_1 + R_2)C_1} = \frac{\dfrac{s}{\omega_{z0}}}{1 + \dfrac{s}{\omega_p}}$$

zero at origin

pole

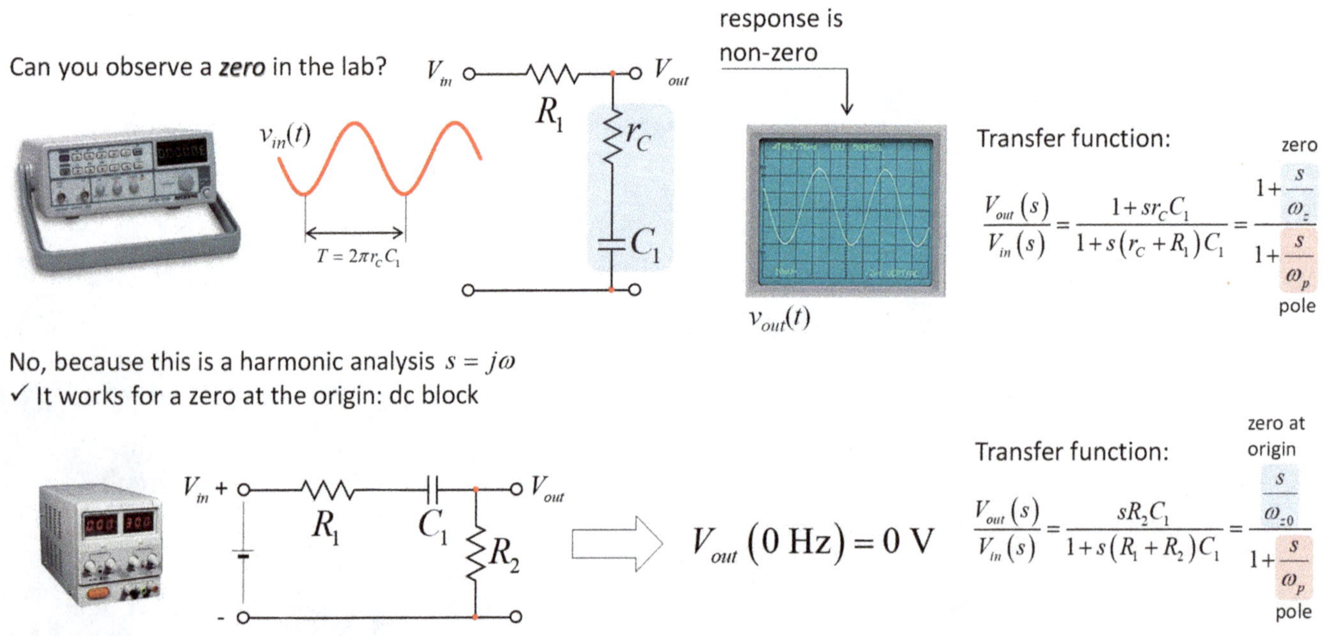

Observing a 1st-Order Zero is not easy in a Practical Experiment Except When the Zero is Located at the Origin

Figure 3.1

Another way of looking at this, is to consider that a negative real pole is associated with an exponentially decaying waveform in the time-domain.

If we use this signal to excite a network whose zero is tuned exactly at the inverse of the exponential time constant, can we illustrate the *null* theory? This is actually a very good way to figure out the action of a zero and it was kindly pointed out by Mr. Gazzoni Filho in correspondence with me.

In Figure 3.2, you see an *RL* circuit driven by a current source whose expression is $e^{-\frac{t}{\tau_1}}$.

This is the decaying time-domain response of a pole located at $s_p = -\frac{1}{\tau_1}$. The time *t* used for computing the exponential term is obtained by integrating a 1-mA current into a 1-µF capacitor and buffered by E_1.

The *RL* network offers the following impedance:

$$Z_1(s) = r_L + sL_1 \tag{3.1}$$

This impedance equals zero if:

$$s_z = -\frac{r_L}{L_1} \tag{3.2}$$

...and leads to a zero located at:

$$\omega_z = \frac{r_L}{L_1} \tag{3.3}$$

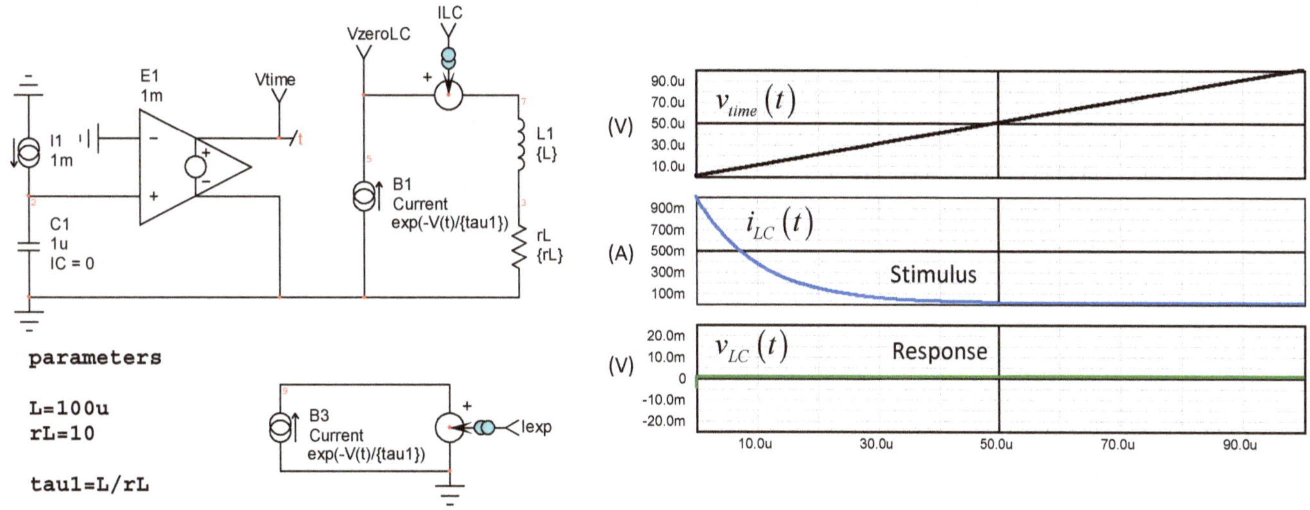

Feed this *RL* Network with a Decaying Exponential Current to Produce a Nulled Response across its Connecting Terminals

Figure 3.2

It is the only zero present in this impedance that can be expressed as:

$$Z_1(s) = 1 + \frac{s}{\omega_z} \tag{3.4}$$

If the time constant of the decaying waveform now exactly matches the inverse of the above zero, then the time-domain response is zero volt as shown in the right side of the figure: the response is nulled as expected.

Bingo!

Now, build a notch filter intended to reject a frequency at a given frequency point.

If the quality factor of this second-order filter approaches infinity, meaning an almost infinite rejection, then the two zeroes are located on the imaginary axis and covered by our harmonic analysis: if you tune your generator exactly to these zeroes frequency ω_0, then, despite the applied stimulus, the output is almost 0 V as shown in Figure 3.3.

With the FACTs, we use a mathematical abstraction to unveil these zeroes.

Rather than solely considering the vertical axis in the s-plane as we normally do in harmonic analysis ($s = j\omega$), we will cover the entire plane allowing for complex roots featuring a negative or positive real component ($s = \sigma + j\omega$). As such, if present in the circuit, a zero will manifest itself by ac-*nulling* the output response when the input signal is tuned to the zero angular frequency s_z.

The output *null* happens because some impedance in the *transformed* circuit blocks the signal propagation despite the presence of an excitation source: a series impedance in the signal path becomes infinite or a branch shunts the stimulus to ground when the transformed circuit is excited at $s = s_z$. Note that this convenient mathematical abstraction offers tremendous help in finding the zeroes by *inspection*, often without writing a line of algebra in passive and active networks.

Build a high-Q notch and you can observe a *nulled* response

$$H(s) = \frac{1 + \dfrac{s}{\omega_0 Q} + \left(\dfrac{s}{\omega_0}\right)^2}{D(s)}$$

$$H(s) = \frac{\left(1 + j\dfrac{\omega}{\omega_0}\right)\left(1 - j\dfrac{\omega}{\omega_0}\right)}{D(s)}$$

$\Im m$

Roots are imaginary

$j\omega_0$

0

$-j\omega_0$

$Q \to \infty$

$$H(s) = \frac{1 + \left(\dfrac{s}{\omega_0}\right)^2}{D(s)}$$

$$\left(1 + j\frac{\omega}{\omega_0}\right) = 0 \to \omega_{z_1} = j\omega_0$$

$$\left(1 - j\frac{\omega}{\omega_0}\right) = 0 \to \omega_{z_2} = -j\omega_0$$

When Q approaches infinity, zeros become imaginary
➤ Roots are along the y axis: harmonic analysis

1 V pp

$T = \dfrac{1}{f_z}$

f_z

-90 dB

Observable null

$$V_{out}(f_z) = 31\,\mu V$$

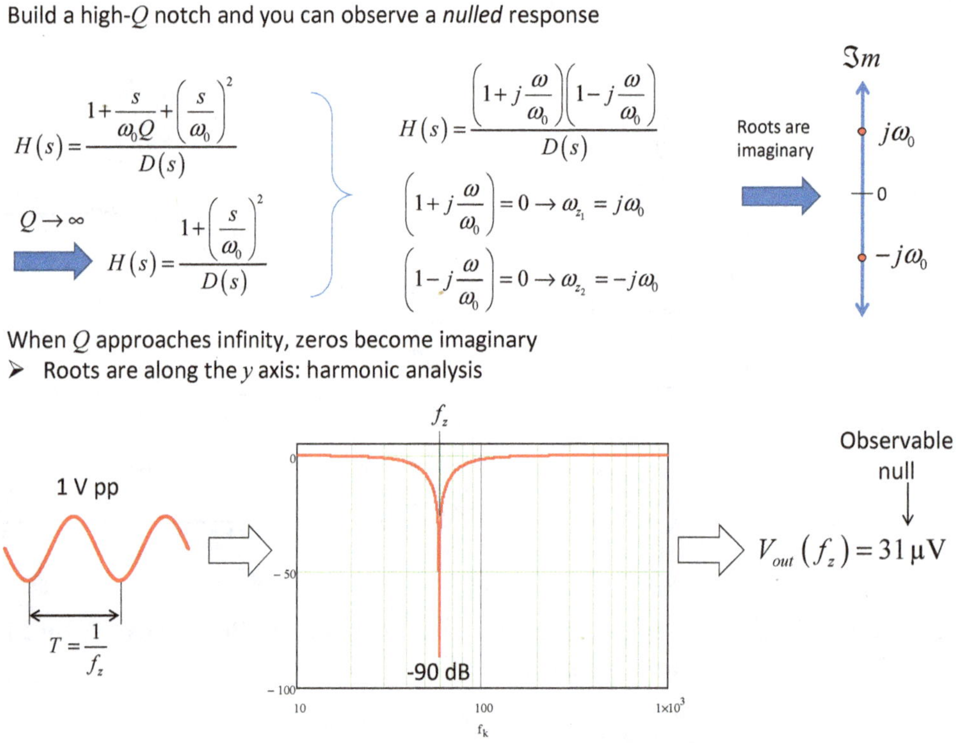

When Tuned to the Notch Frequency, this Filter Delivers almost 0 V: the Response is Nulled

Figure 3.3

A transformed network implies a circuit redrawn with inductor and capacitor featuring their respective impedance expression, sL and $\frac{1}{sC}$ as illustrated in Figure3.4.

The figure offers a simple flow chart which details the procedure to unveil zeroes.

Keep the excitation signal - the stimulus - in place

↓

Consider a null in the output: $V_{out}(s) = 0$ V or $\hat{v}_{out} = 0$ V

↓

Identify in the *transformed* network, one or several impedances combinations that could block the stimulus propagation and create the null:

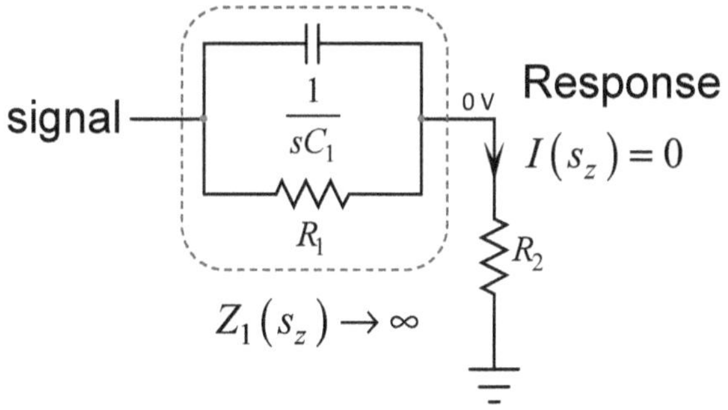

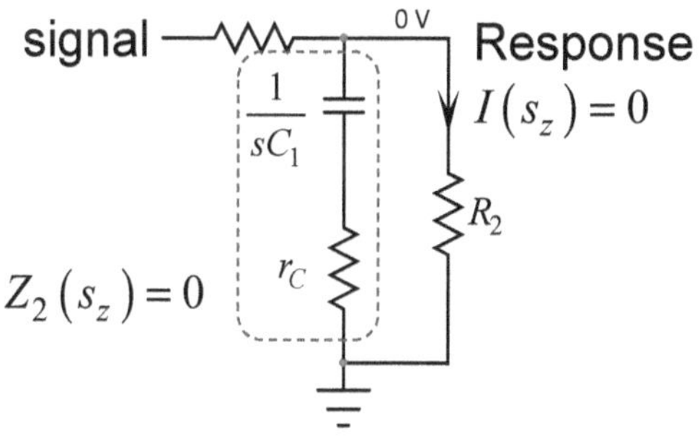

A *transformed* open circuit

$$Z_1(s) = \frac{N(s)}{D(s)} \to \infty \text{ when } D(s) = 0$$

A *transformed* short circuit.

$$Z_2(s) = \frac{N(s)}{D(s)} = 0 \text{ when } N(s) = 0$$

This Simple Flow Chart is a Guides to Determining Zeroes in the Quickest Way—when Inspection does not Work, use *Null Double Injection* or NDI

Figure 3.4

In the upper left section, you see an impedance Z_1 inserted in the signal path. Is there a specific combination for which the magnitude of this impedance could become infinite and bring a null in the response? Express the impedance as:

$$Z_1(s) = R_1 \,||\, \frac{1}{sC_1} = \frac{R_1 \frac{1}{sC_1}}{R_1 + \frac{1}{sC_1}} = R_1 \frac{1}{1 + sR_1C_1} \tag{3.5}$$

The magnitude of this impedance becomes infinite if the denominator *D(s)* equals zero. In other words, the root of this expression or its pole:

$$D(s) = 0 \to s_p = -\frac{1}{R_1C_1} \tag{3.6}$$

…is the *zero* of the entire network we are looking for:

$$\omega_z = \frac{1}{R_1 C_1} \tag{3.7}$$

In the right side of the picture, what are the conditions for which the series combination of r_C and C_1 could become a transformed short circuit and shunt the stimulus to ground during an ac modulation tuned at $s = -s_z$?

$$Z_2(s) = r_C + \frac{1}{sC_1} = \frac{1 + sr_C C_1}{sC_1} \tag{3.8}$$

The numerator equals zero when the stimulus is tuned to:

$$1 + sr_C C_1 = 0 \rightarrow s_z = -\frac{1}{r_C C_1} \tag{3.9}$$

…and it implies a zero located at:

$$\omega_z = \frac{1}{r_C C_1} \tag{3.10}$$

Inspection is a very convenient way to determine zeroes in a network.

A simple trick [1] lets you immediately check if you have one (or several) zeroes in a network, even if inspection did not reveal it at first glance: place the energy-storing element in its high-frequency state (replace the capacitor by a short circuit and open-circuit the inductor) then check if, in this mode, the excitation signal gives a response. What you do is actually verify the presence of a high-frequency gain H^n when s approaches infinity and energy-storing element n is set in its high-frequency state. If that gain exists when C_1 is replaced by a short circuit, e.g., $H^1 \neq 0$, then this capacitor element contributes a zero to the transfer function. On the contrary, $H^1 = 0$ implies there is no zero associated with it and it is why I predict the absence of zero in the circuit of Figure 2.19.

Let's exercise our skill with the four networks shown in Figure 3.5.

In sketch (a), if I replace C by a short circuit in my head, then the stimulus V_{in} can propagate and produces a response V_{out}: I have a zero when r_C and C form a transformed short circuit—see (3.8).

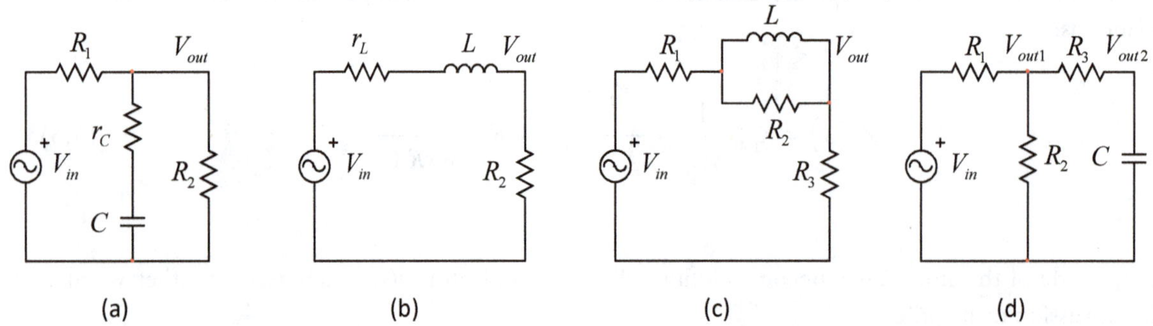

Testing the Existence of the Response when One or Several Energy-Storing Elements are Placed in their High-Frequency State is a Convenient Way to Identify Zeroes

Figure 3.5

In (b), if L is placed in its high-frequency state or open-circuited, then the current is interrupted and there is no response from the stimulus V_{in}. Now place a resistor in parallel with L, as R_2 in sketch (c), and you now have a path to V_{out} when L is open-circuited: there is a zero involving L and R_2. Finally, in (d), when C is a shorted and you consider V_{out2} as the output, there is no zero in the transfer function as the response is 0 V in this mode. Now probe V_{out1} as the new output and, despite the short with C, there is a response, naturally revealing a zero involving R_3 and C.

Once we know that we have zeroes in the circuit, how do we determine their positions? Let's examine a classic example where inspection cannot be beaten in terms of execution speed.

The example is shown in Figure 3.6 where all energy-storing elements are replaced by their equivalent impedance. This is for the sake of illustrating this intermediate step and you won't do it anymore once you master the approach. Here, you need to identify the potential combinations which would stop the stimulus propagation to form a response at V_{out} probe. In other terms, are there infinite-impedance networks which could stop the stimulus propagation and bring a response null or are there branches shunting the signal path to ground when tuned at a zero frequency?

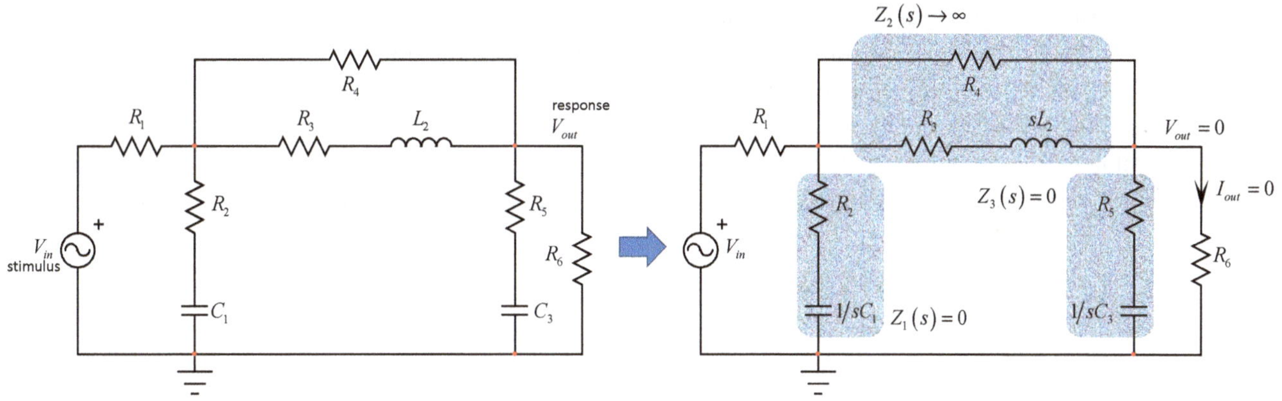

The Exercise Consists of Finding the Network which could Stop Stimulus Propagation

Figure 3.6

A zero-volt ac output or a nullified output is different than a short circuit to ground and it is often a difficult point to grasp for the students. If you look at the right-side of the picture, then you see that no ac or small-signal current is flowing in the load resistance R_6 and it naturally implies the output null.

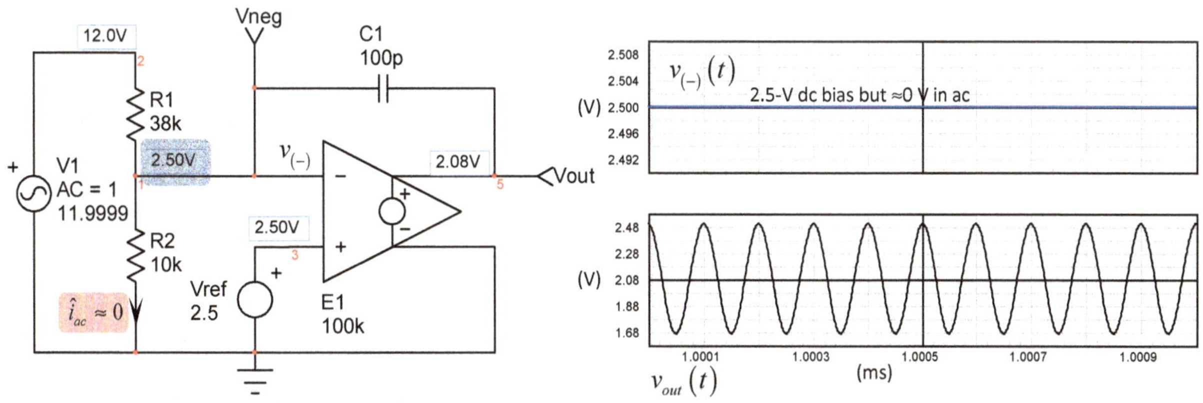

The Low-Side Resistance sees no Ac Current and is Excluded from the Analysis for $s > 0$

Figure 3.7

As already underlined, think of the analogy of the virtual ground in an op-amp-based circuit as the one presented in Figure 3.7.

If you apply a dc bias of 11.9999 V to this circuit, the op-amp produces the expected 2.08 V dc output:

$$V_{out,dc} = \left(2.5 - V_{in,dc}\frac{R_2}{R_1 + R_2}\right)A_{OL} = \left(2.5 - 11.9999 \times \frac{10k}{10k + 38k}\right) \times 100k = 2.083\ \text{V} \quad (3.11)$$

If you superimpose an ac excitation made of a 100-mV peak sinusoidal voltage tuned at 10 kHz, the output delivers an ac level whose amplitude is set by the integrator time constant $\tau = R_1C_1$. It is the 2.5-V peak you can observe in the right-side of the picture.

If you now probe the voltage across R_2, you measure almost 0 V ac. This is because the op-amp strives to maintain an equal voltage between the negative and the positive pins. Considering the amplitude of the 2.5-V reference voltage is 0 V in ac (it does not undergo a modulation), it also imposes a 0-V level—it nulls the (-) pin voltage—or a *virtual ac-ground* at the inverting pin: no ac current flows in R_2.

As a consequence, when calculating the time constant affecting this circuit, only R_1 matters, not R_2. R_1 and R_2 set the dc operating point but R_2 disappears in ac analysis because of the virtual ground. I applied a similar principle in Figure 1.17 but we will come back with more details in a few lines.

Returning to our circuit of Figure 3.6 and starting from the left, you immediately see the series combination of a resistance and a capacitor, R_2 and C_1. When this network impedance becomes zero ohm, otherwise stated when:

$$1 + sR_2C_1 = 0 \rightarrow s_{z_1} = -\frac{1}{R_2C_1} \quad (3.12)$$

...then you have a zero located at:

$$\omega_{z_1} = \frac{1}{R_2C_1} \quad (3.13)$$

If you continue towards the right, you see an inductor associated with two resistors forming a parallel impedance. First, in your mind, what if L_2 is set in its high-frequency state, implying open-circuiting it? Would the stimulus still go through and form a response? Yes, because of R_3 still providing a path: we have a zero involving L_2.

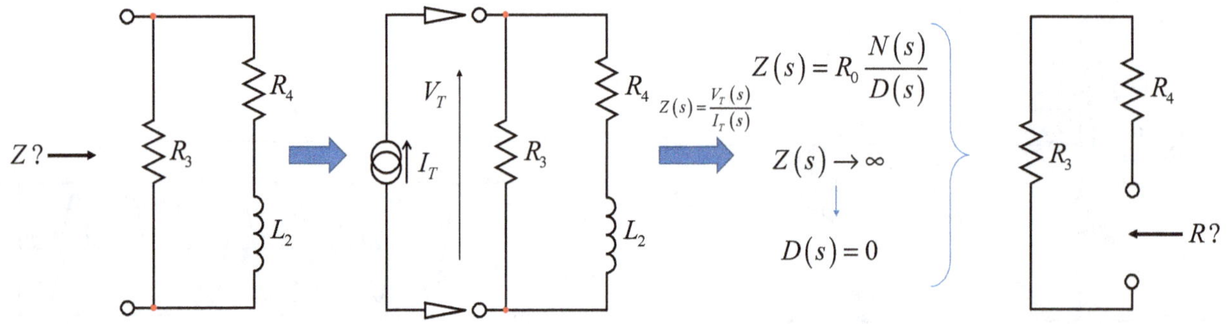

Isolate the Series Impedance and Check the Conditions for where its Magnitude approaches Infinity

Figure 3.8

We need to determine the impedance of this network now isolated in Figure 3.8 and check the condition for which its magnitude approaches infinity. Practically speaking, connect a current source driving the network and express the transfer function $Z(s)$. In reality, simply determine the pole of the denominator $D(s)$ since it naturally makes the impedance infinite when it cancels. The time constant is obvious and determined by turning the current source off, then looking through the inductor's connecting terminals:

$$\tau_2 = \frac{L_2}{R_3 + R_4} \tag{3.14}$$

The pole of this network is the inverse of the time constant and equal to:

$$\omega_p = \frac{R_3 + R_4}{L_2} \tag{3.15}$$

When the frequency of the stimulus is tuned to the above value, the series impedance Z_2 in Figure 3.6 becomes infinite and creates an output null. The pole of this impedance is thus the zero of the entire network:

$$\omega_{z_2} = \frac{R_3 + R_4}{L_2} \tag{3.16}$$

Finally, the last zero is obtained by analyzing impedance Z_3 and immediately writing:

$$1 + sR_5C_3 = 0 \rightarrow s_{z_3} = -\frac{1}{R_5C_3} \tag{3.17}$$

Leading to the third zero located at:

$$\omega_{z_3} = \frac{1}{R_5C_3} \tag{3.18}$$

And this is it! We determined our three zeroes without writing a single line of algebra. The transfer function linking V_{out} to V_{in} is partially determined but its numerator is already perfectly factored:

$$H(s) = H_0 \frac{N(s)}{D(s)} = H_0 \frac{\left(1 + \frac{s}{\omega_{z_1}}\right)\left(1 + \frac{s}{\omega_{z_2}}\right)\left(1 + \frac{s}{\omega_{z_3}}\right)}{D(s)} \tag{3.19}$$

The dc gain would be determined by shorting L_2 and opening the capacitors. If you do the maths, you should find:

$$H_0 = \frac{R_6}{R_6 + R_3 \,||\, R_4 + R_1} \tag{3.20}$$

To determine the denominator, we will see that in the series of solved examples. Now, how do we know our zeroes are correctly located? As exemplified several times in the previous lines, we can simulate a null double injection or NDI as already illustrated in Figure 1.12 of Chapter 1.

The exercise is repeated three times and illustrated in Figure 3.9.

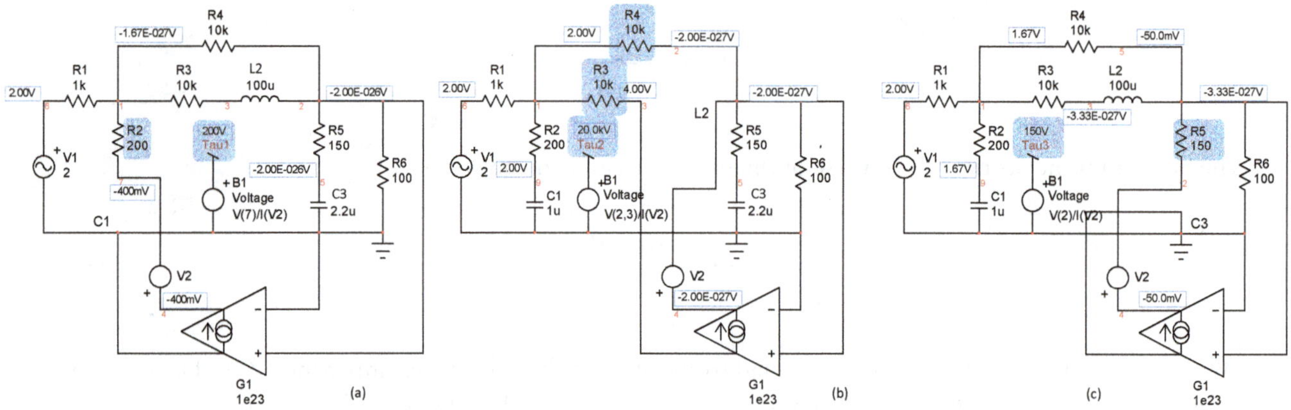

SPICE can help Verify our Calculations are Correct. Here, Force a Second Stimulus to Nullify the Output via an Infinite Transconductance Voltage-Controlled Current-Source

Figure 3.9

If you respect the injection signs while calculating the resistance value, then the dc operating point delivers a correct result. In sketch (a), source B_1 confirms the resistive part of the time constant involving C_1 is R_2 while it is R_5 with C_3 in sketch (c). In sketch (b), the floating injection confirms the series combination of the R_3 and R_4 leading to a 20-kΩ resistance as calculated by B_1.

Please note that the first stimulus source V_1 can take on any value and I arbitrarily selected 2 V in this example.

You could choose another value and it would not change the final result computed by B_1: the stimulus amplitude does not affect the zero position and never factors in the numerator.

3.2 The Null Double Injection

Inspection offers the fastest possible way to determine the zeroes. It usually works well with passive networks but sometimes, complicated arrangements make inspection difficult or simply impossible. This is often the case with active circuits and controlled sources. In these cases, you have no other option than resorting to Kirchhoff's current and voltage laws (KCL and KVL) for analyzing the network.

Fortunately, SPICE dc analyses help confirm your results.

Let's start with a few illustrated examples.

In Figure 3.10, you see that shorting C_5 keeps the stimulus flowing to generate a response at V_{out} but can't figure out how resistors around this capacitor combine to block the excitation. We resort to an NDI as illustrated in the right side of the figure.

Let's figure out the current circulating in R_3 considering its right terminal at a 0-V bias:

$$i_1 = \frac{V_T}{R_3} \tag{3.21}$$

This current flows in R_1 because of the zeroed output current i_{out}. The voltage across R_1 is therefore defined as:

$$V_{R_1} = i_1 R_1 = \frac{V_T}{R_3} R_1 \tag{3.22}$$

The voltage across R_2 is the sum of the voltage across R_1 plus the voltage across R_3, preceded by a - sign:

$$V_{R_2} = -\left(\frac{V_T}{R_3} R_1 + V_T\right) = -V_T\left(\frac{R_1}{R_3} + 1\right) \tag{3.23}$$

From KCL, current i_1 is made of the sum of i_2 and I_T then:

$$i_1 = i_2 + I_T = \frac{V_{R_2}}{R_2} + I_T \tag{3.24}$$

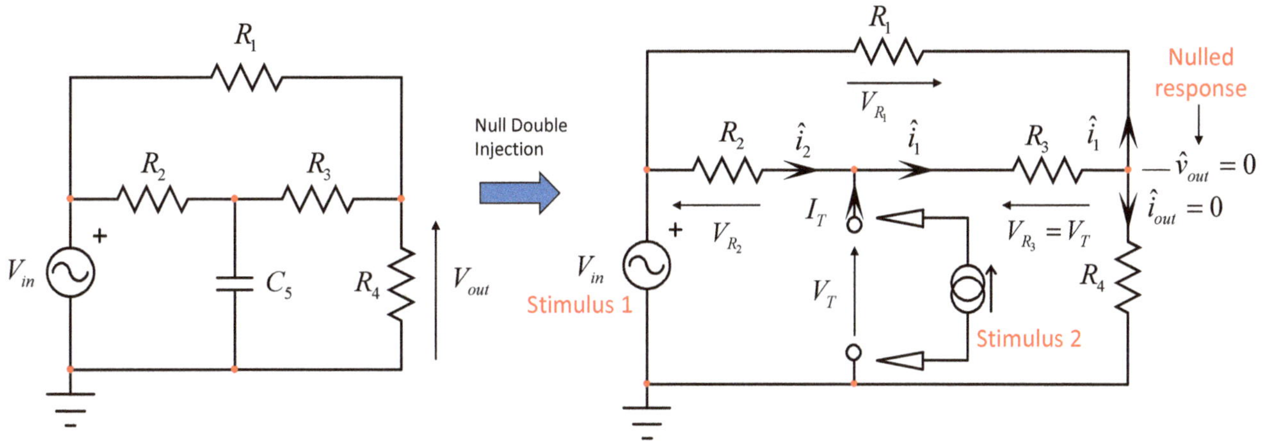

In this First-Order Circuit, Shorting the Capacitor keeps the Response when a Stimulus is Applied: there is a Zero associated with this Capacitor

Figure 3.10

Now substitute (3.23) and (3.21) in (3.24) then rearrange to obtain:

$$\frac{V_T}{R_3} = -\frac{V_T\left(\frac{R_1}{R_3} + 1\right)}{R_2} + I_T \tag{3.25}$$

Solve for V_T and reveal the resistance R_n driving capacitor C_5:

$$\frac{V_T}{I_T} = R = \frac{R_2 R_3}{R_1 + R_2 + R_3} \tag{3.26}$$

The zero we want is then determined by the following expression:

$$\omega_z = \frac{1}{RC_5} = \frac{R_1 + R_2 + R_3}{R_2 R_3 C_5} \tag{3.27}$$

We can now use SPICE and check the dc operating point of this circuit by involving the voltage-controlled current-source represented in Figure 3.11. When run, the resistance found in simulation exactly matches the result obtained via (3.26) and assessed via a few Mathcad® lines: 178.571 Ω is the resistance value driving capacitor C_5.

In case this capacitor is 0.15 µF, then according to (3.27), the zero is located at:

$$f_z = \frac{1}{2\pi \times 178.571 \times 0.15u} \approx 5.9 \text{ kHz} \tag{3.28}$$

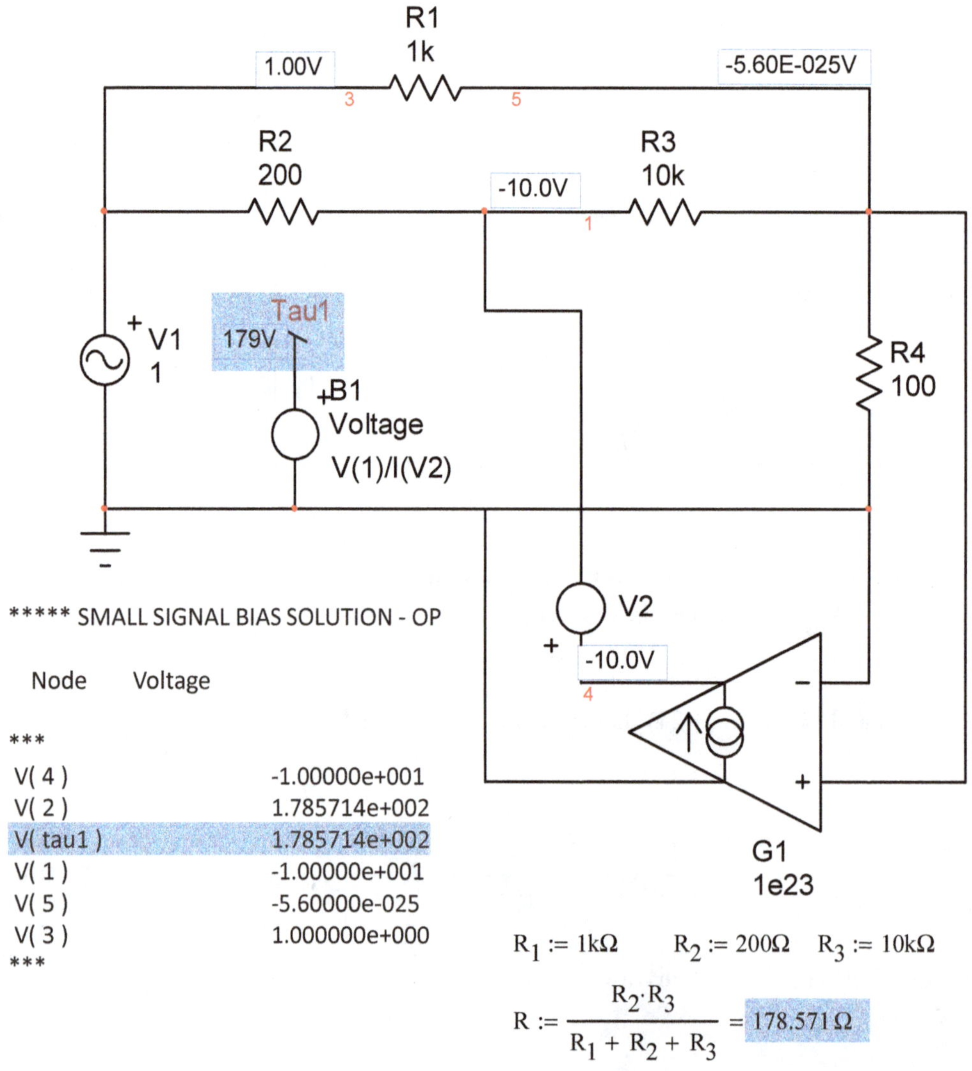

It is Easy and Fast to Check the Result with a Dc Bias Point SPICE Simulation

Figure 3.11

The rest of the transfer function comes easily.

For $s = 0$, open the capacitor and determine the gain linking the input to the output. We have resistive divider and inspecting the circuit in Figure 3.10 leads to:

$$H_0 = \frac{R_4}{R_1 \| (R_2 + R_3) + R_4} \tag{3.29}$$

$$V_{out} = V_1 \times \frac{R_4}{R_1 \| (R_2 + R_3) + R_4} = 0.0989 \text{ V}$$

$R_1 := 1k\Omega \quad R_2 := 200\Omega \quad R_3 := 10k\Omega \quad R_4 := 100\Omega \quad C_1 := 0.47\mu F \quad \|(x,y) := \frac{x \cdot y}{x + y}$

$H_0 := \frac{R_4}{R_1 \| ((R_2 + R_3)) + R_4} = 0.09894 \qquad 20 \cdot \log(H_0) = -20.09257 \text{ dBohms}$

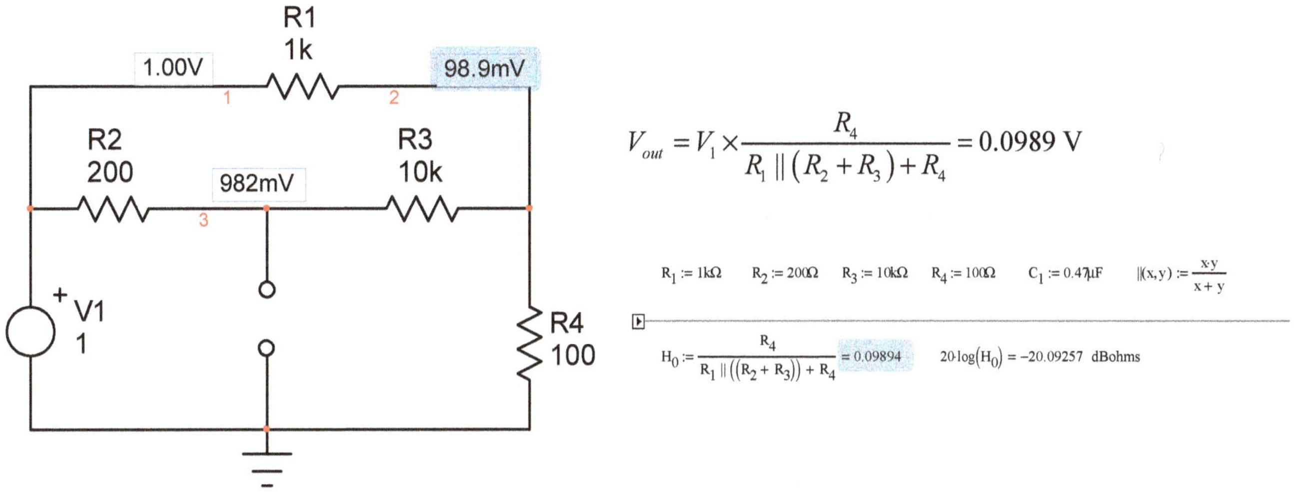

A Dc Bias Point in SPICE Confirms the Dc Gain is Correct

Figure 3.12

A quick SPICE simulation featuring a 1-V input voltage gives us the expected results as shown in Figure 3.12. The dc insertion loss amounts to -20.09 dB. Now that we obtained the zero with the NDI, we can finish the exercise and check the position of the pole. We do that by shorting the input source as recommended by Figure 3.13.

Simulate a bias point

$R := R_2 \| (R_3 + R_1 \| R_4) = 196.11307\Omega$

Parameter	Value
v(3)	196.113

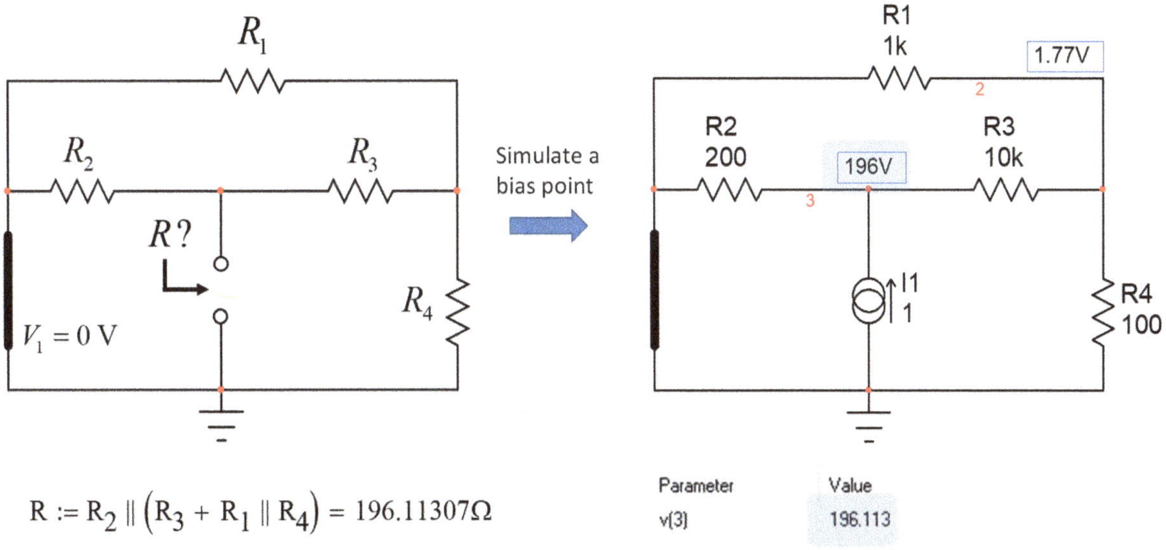

Look through the Connecting Terminals of C_1 to observe the Resistance

Figure 3.13

Inspection at first glance may not be obvious and you might want to use an intermediate sketch. You see R_1 is now in parallel with R_4 while R_2 appears across the capacitor connections. If you get it right, then you should find:

$$R = R_2 \,||\, (R_3 + R_1 \,||\, R_4) \tag{3.30}$$

Following Figure 3.13 steps, we can install a 1-A dc source across the nodes for which we want the resistance and check that the dc value obtained after a dc-bias analysis confirms the analytical results.

We can now express the pole position as:

$$\omega_p = \frac{1}{\left[R_2 \,||\, (R_3 + R_1 \,||\, R_4)\right] C_1} \tag{3.31}$$

...and then write the complete transfer function $H(s)$:

$$H(s) = H_0 \frac{1 + \frac{s}{\omega_z}}{1 + \frac{s}{\omega_p}} \tag{3.32}$$

...with the dc gain, the zero and the pole respectfully obtained in (3.29), (3.27) and (3.31).

The second NDI exercise appears in Figure 3.14 where you see an inductor routing the input stimulus to the output via a resistive divider. If you place L_1 in its high-frequency state—an open circuit—you realize that there is a response in this mode implying the presence of a zero brought by the inductor.

To determine its value, let's implement an NDI by keeping the input stimulus in place while nulling the output with the test generator I_T as in the right side of the figure.

Because the output voltage is a null, the upper connection of R_2 is zeroed, meaning that the current flowing across it is the voltage across R_3:

$$i_1 = \frac{(I_T - i_1) R_3}{R_2} \tag{3.33}$$

Which leads to:

$$i_1 = \frac{I_T R_3}{R_2 + R_3} \tag{3.34}$$

The test voltage V_T is equal to:

$$V_T = i_1 (R_1 + R_2) \tag{3.35}$$

From which we obtain a new definition for i_1:

$$i_1 = \frac{V_T}{R_1 + R_2} \tag{3.36}$$

Equating equations (3.34) and (3.36), we have:

$$\frac{V_T}{I_T} = R = \frac{R_3(R_1 + R_2)}{R_2 + R_3} \tag{3.37}$$

This is the resistance R we want and the zero is immediate:

$$\omega_z = \frac{R}{L_1} = \frac{R_3(R_1 + R_2)}{(R_2 + R_3)L_1} \tag{3.38}$$

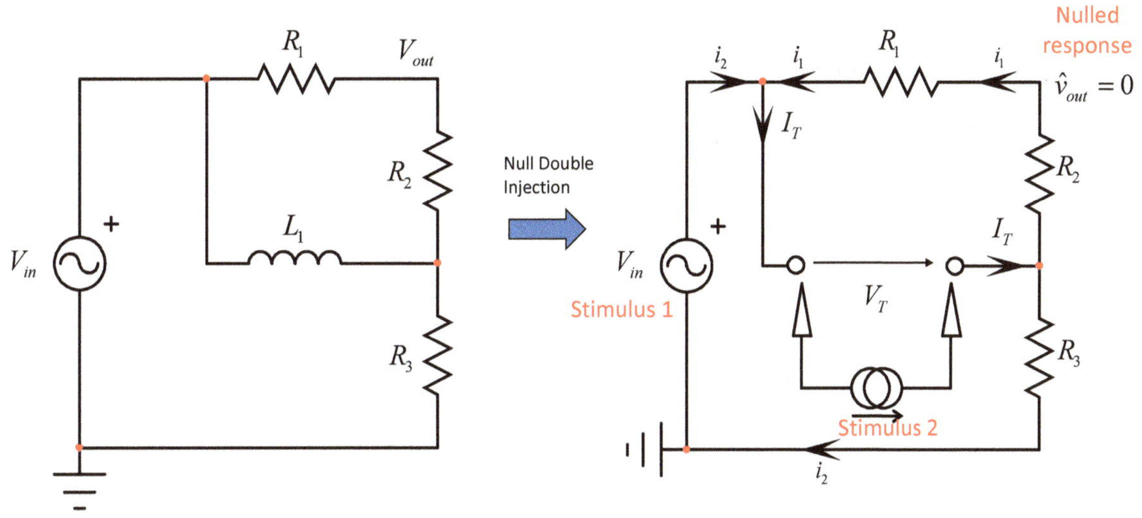

Installing the Test Generator across the Connecting Terminals of L_1 leads to the Resistance we want via an NDI

Figure 3.14

To test if our calculations are correct, the quick SPICE simulation of Figure 3.15 will tell us if we are good to go. The voltage-controlled current source maintains 0 V on the output node effectively nulling the output. An analog behavioral source computes the voltage V_T collected across the inductor's connecting terminals and divides it by the injected current. You obviously need to respect the correct sensing polarities in this operation (the current is positive in V_2 when leaving the source via its negative terminal).

The displayed bias points indicate a nulled response and a resistance of 12.973 kΩ confirming the numerical results of (3.37). By the way, the resistance is positive meaning we have a LHPZ, also designated as a stable zero. Should the simulation return a negative value, then it indicates that you may have a RHP zero in the transfer function.

Now that we have the zero, we can carry on with the pole determination. Reduce the excitation to zero and replace the input stimulus by a wire as drawn in Figure 3.16.

What resistance R do you see between the connecting terminals of the inductor? Because the source is zeroed, the left-side terminal of R_1 is grounded and, together with R_2 in series, they come in parallel with R_3:

$$R = R_3 \| (R_1 + R_2) \tag{3.39}$$

The pole is the inverse of the natural time constant and therefore:

$$\omega_p = \frac{R}{L_1} = \frac{R_3 \,||\, (R_1 + R_2)}{L_1} \tag{3.40}$$

The dc gain H_0 is immediate as the shorted inductance brings the stimulus across R_3 and imposes a 0-A current in R_2 and R_1. Therefore:

$$H_0 = 1 \tag{3.41}$$

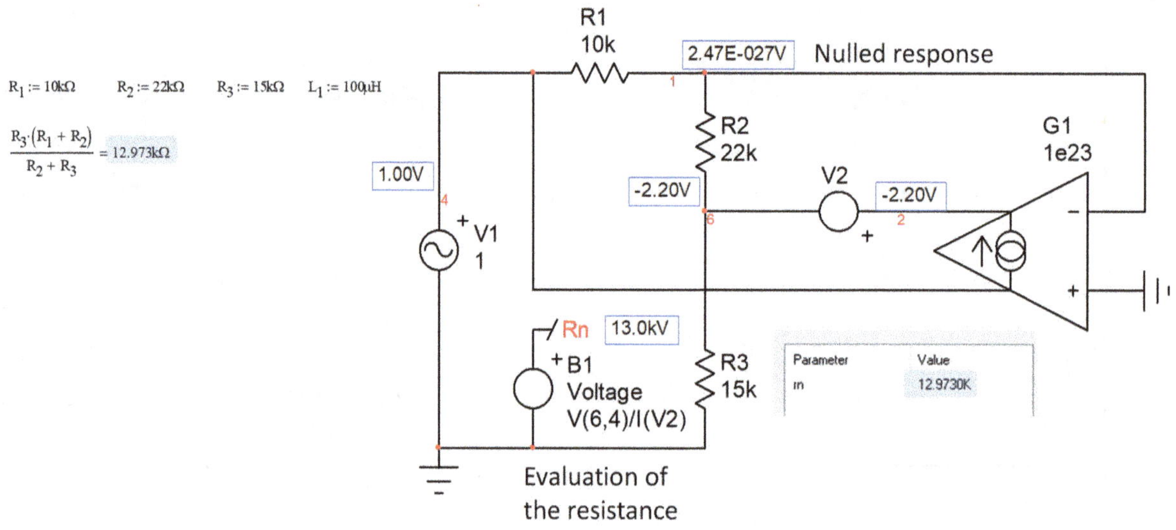

A Simple SPICE Simulation immediately shows that our Results are Correct

Figure 3.15

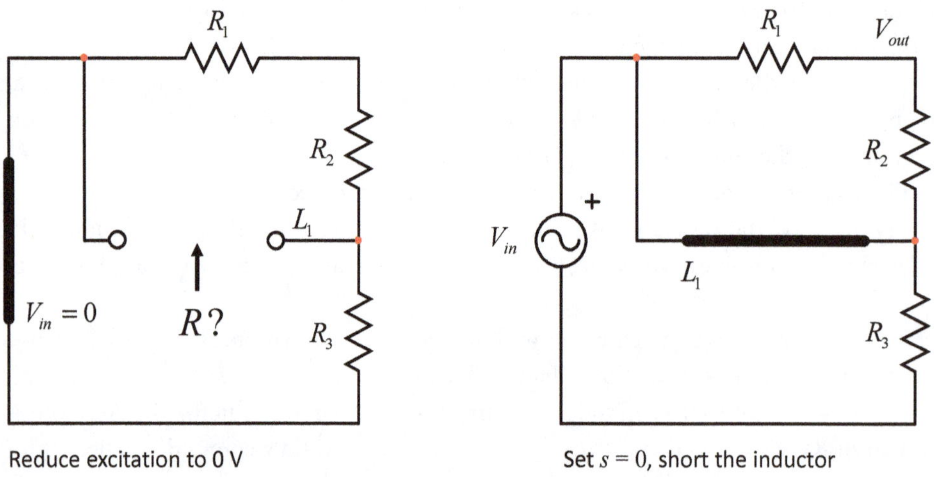

The Pole and Dc Gain are Easily Located

Figure 3.16

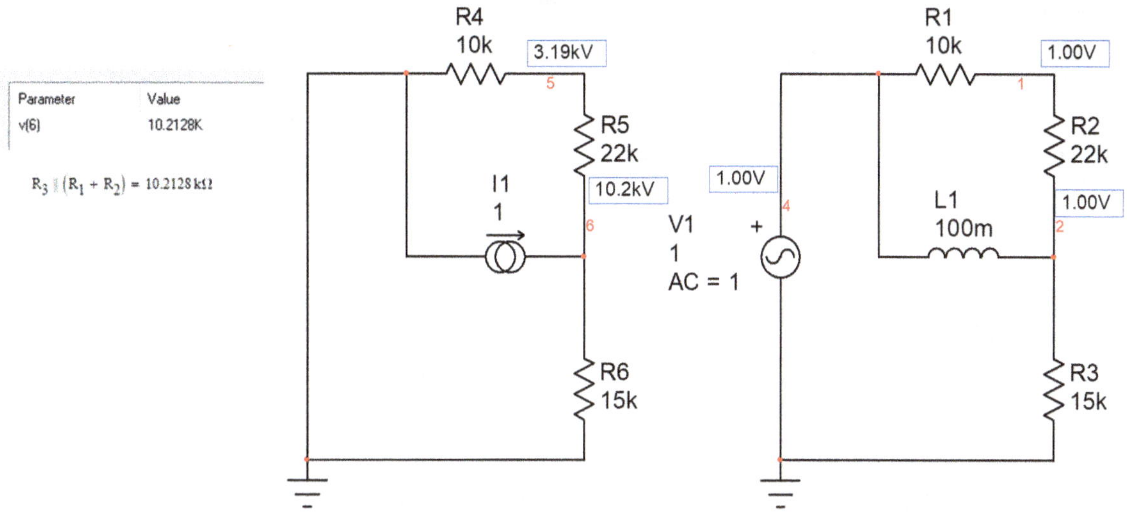

A Quick Dc Operating Point Simulation Confirms our Results

Figure 3.17

The dc operating point from Figure 3.17 shows similar results between the solver and SPICE.

The complete transfer function H of this network is thus:

$$H(s)=\frac{1+\frac{s}{\omega_z}}{1+\frac{s}{\omega_p}} \tag{3.42}$$

...with the zero and the pole respectively expressed by (3.38) and (3.40).

For this third NDI example, we will examine an active circuit featuring a bipolar transistor represented in Figure 3.18.

In this example, to determine the transfer function of this circuit, we must replace the transistor by its equivalent small-signal model—the hybrid-π model—and redraw the circuit.

r_π represents the dynamic base-emitter resistance and could also be labeled h_{11} should you resort to h-parameter notation instead (with h_{21} for the gain).

As is often the case, we consider the dc-block capacitor C_L a short circuit in our ac analysis (we assume it is a large capacitance) and, in this mode, only the two resistors R_1 and R_2 play a role for the dc bias. The V_{cc} line is also 0 V ac and can be shorted, leading to the circuit in the right side of Figure 3.18.

To run an NDI, we install two stimuli and determine the value of stimulus 2 to null the response observed across the collector resistance.

This is what the left side of Figure 3.19 suggests.

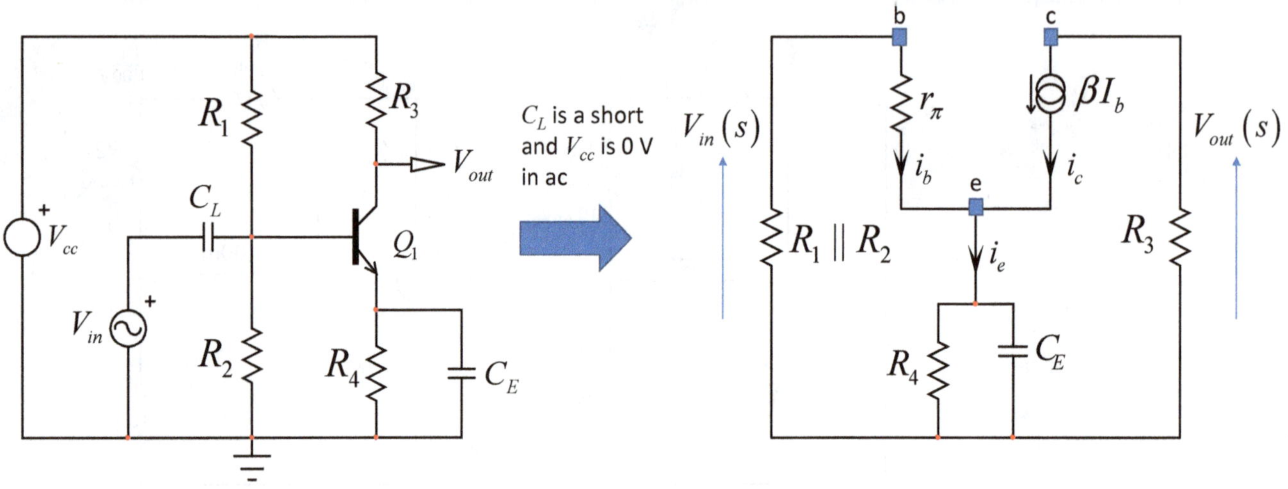

This Transistor-Based Circuit Features a Zero and a Pole. It's Equivalent Small-Signal Circuit is shown on the Right

Figure 3.18

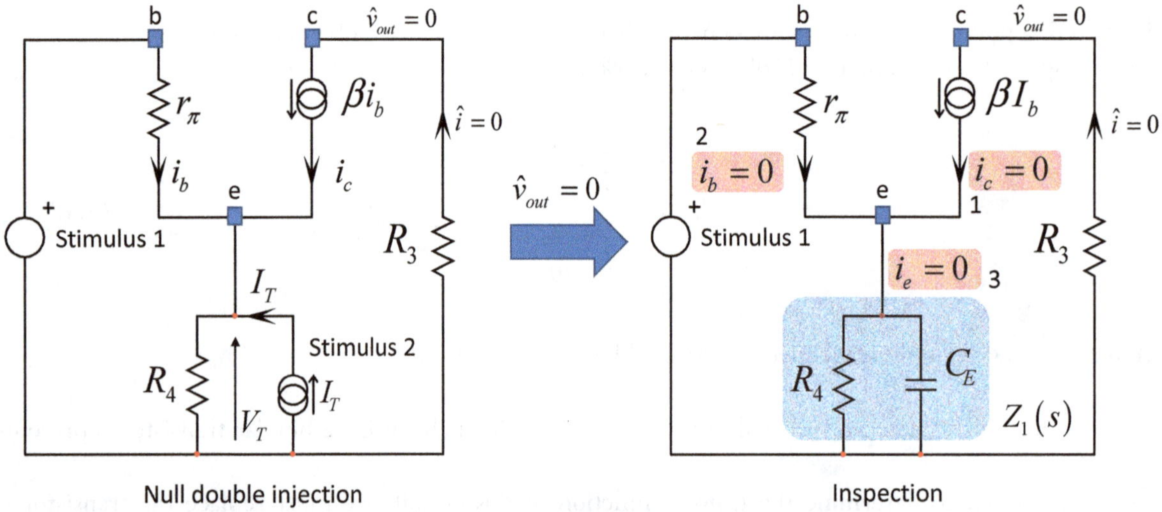

The Response Disappears if the Collector Current is Null

Figure 3.19

You would need to determine the value of the injected current I_T which nulls or zeroes the output current while stimulus 1 delivers a given bias. Then, the value of V_T over I_T would give you the resistance driving capacitor C_E in this configuration, leading to the time constant of the numerator and the zero position.

However, if you take a closer look at the picture then you realize that a nulled output implies a zeroed collector current, correct? If the collector current is zero, then the base current is also zero. What could imply a nulled base current in this configuration? Well, if the emitter is open-ended in ac, then the base current is naturally zeroed. If the impedance of the network grounding the emitter approaches infinity for $s = s_z$, then there is a zero:

$$Z_1(s) = R_4 \| \frac{1}{sC_E} \to \infty \qquad (3.43)$$

If you solve this simple equation, the root of this expression—the s value which cancels the denominator—is actually the zero of our transfer function:

$$\frac{R_4}{1+sR_4C_E} \rightarrow \infty \tag{3.44}$$

The zero is located at:

$$\omega_z = \frac{1}{R_4C_E} \tag{3.45}$$

This is what is called the *null propagation* and the analysis could be formalized this way:

1. The current in the collector is equal to zero
2. Then the base current must be zero
3. The base current can be zeroed if:
   - There is no bias on the base
   - The emitter is open
4. What can ac-open the emitter? $R_4 \,||\, C_E$ becomes an infinite impedance

This approach offers often a very convenient and fast way to determine what causes a zero in the transfer function. To verify this is possible, we ran the SPICE dc operating point of this bipolar circuit shown in Figure 3.20. In this picture, the transistor is replaced by a simple model made of a current-controlled current source featuring a gain of 100.

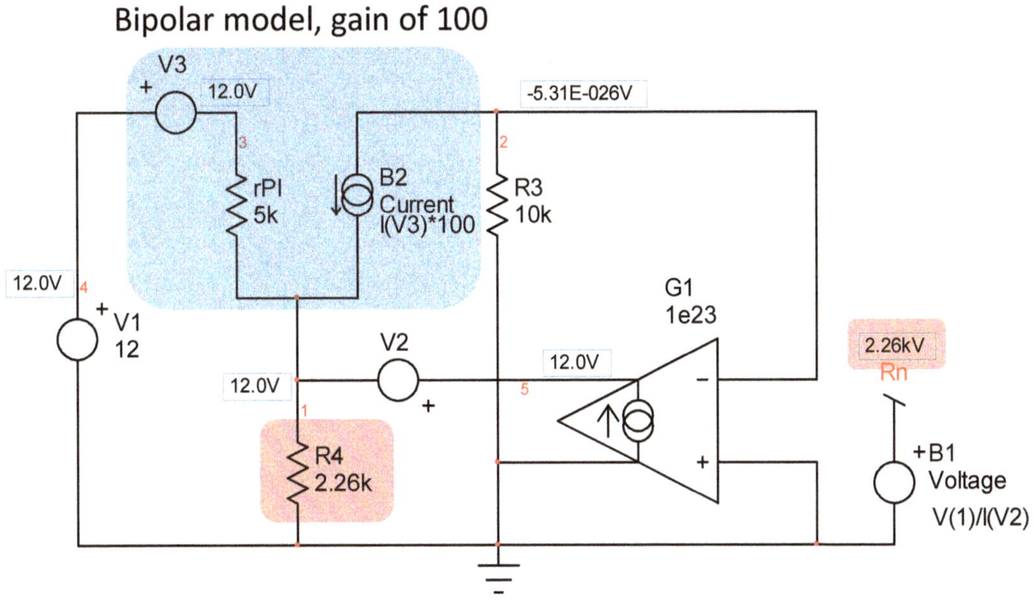

The Voltage-Controlled Current Source Nulls the Output Current and Shows a Resistance Computed by Source B_1 Equal to R_4.

Figure 3.20

The output at node 2 is perfectly zeroed and the resistance seen from the capacitor terminals in this mode is R_4.

Now, to finish the analysis, let's look for the pole. Zero the excitation—meaning r_π is now grounded—and determine the resistance seen from the emitter. The circuit is that of Figure 3.21 in which you see the test current source I_T biasing terminals across R_4. To simplify the analysis, an easy step is to temporarily remove R_4 and determine an intermediate result without it.

The final result will be the intermediate value paralleled with R_4.

We start with the base current i_b:

$$i_b = -\frac{V_T}{r_\pi} \tag{3.46}$$

The collector current is the base current multiplied by the gain:

$$i_c = -\frac{V_T}{r_\pi}\beta \tag{3.47}$$

Finally, I_T is the sum of the base and collector current but flowing in the opposite direction:

$$I_T = -(i_b + i_c) = -\left[-\frac{V_T}{r_\pi}(1+\beta)\right] = \frac{V_T}{r_\pi}(1+\beta) \tag{3.48}$$

The resistance seen from the emitter or the emitter output resistance is:

$$\frac{V_T}{I_T} = \frac{r_\pi}{\beta+1} \tag{3.49}$$

The final resistance R used to determine the pole is therefore:

$$R = \left(\frac{r_\pi}{\beta+1}\right) || R_4 \tag{3.50}$$

Leading to a pole located at:

$$\omega_p = \frac{1}{RC_E} = \frac{1}{\left[\left(\frac{r_\pi}{\beta+1}\right) || R_4\right] C_E} \tag{3.51}$$

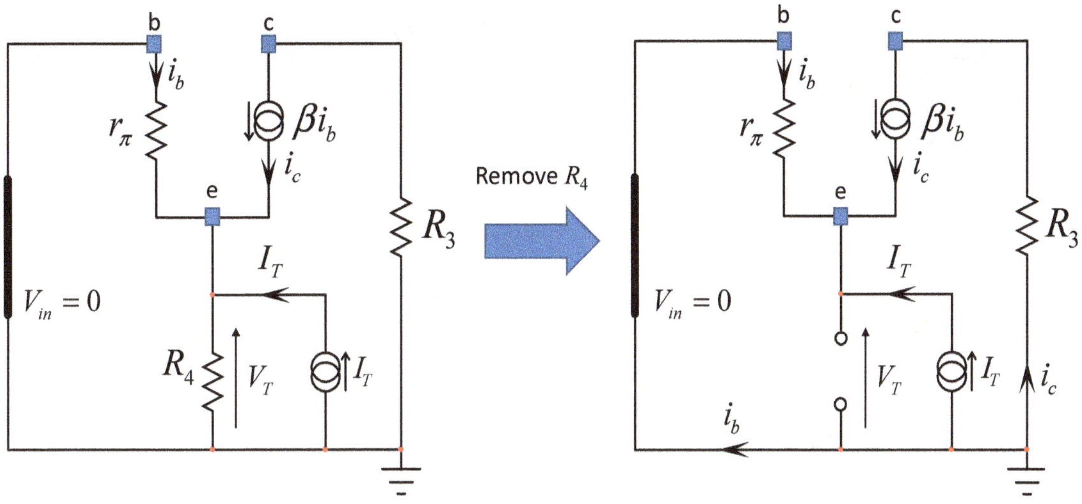

The Excitation is now 0 V—Determine the Resistance Offered by the Emitter

Figure 3.21

Again, the simple SPICE simulation of Figure 3.22 will tell us this is the correct expression.

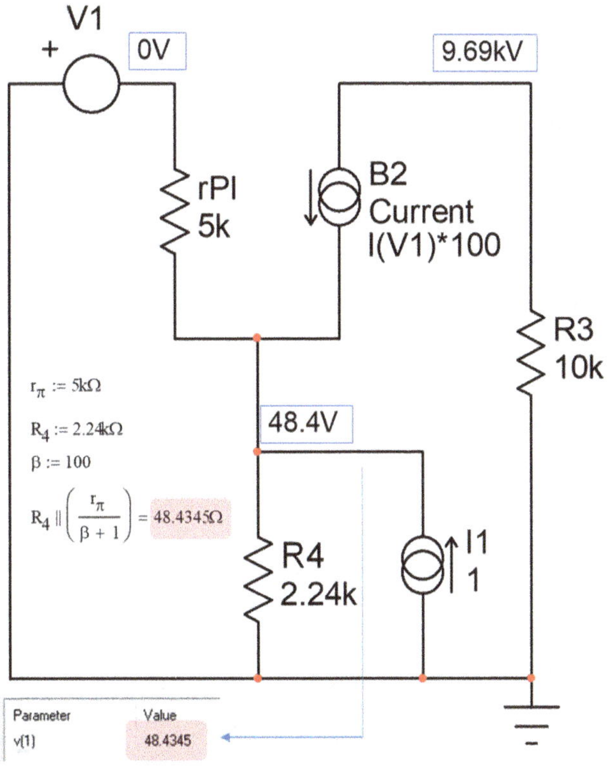

The Resistance Driving the Emitter Capacitor is the Output Impedance of the Emitter in Parallel with R_4

Figure 3.22

To end this exercise, we now compute the dc gain by opening the capacitor and determine the output voltage in this configuration. The circuit appears in Figure 3.23 and can be solved in a few lines of algebra.

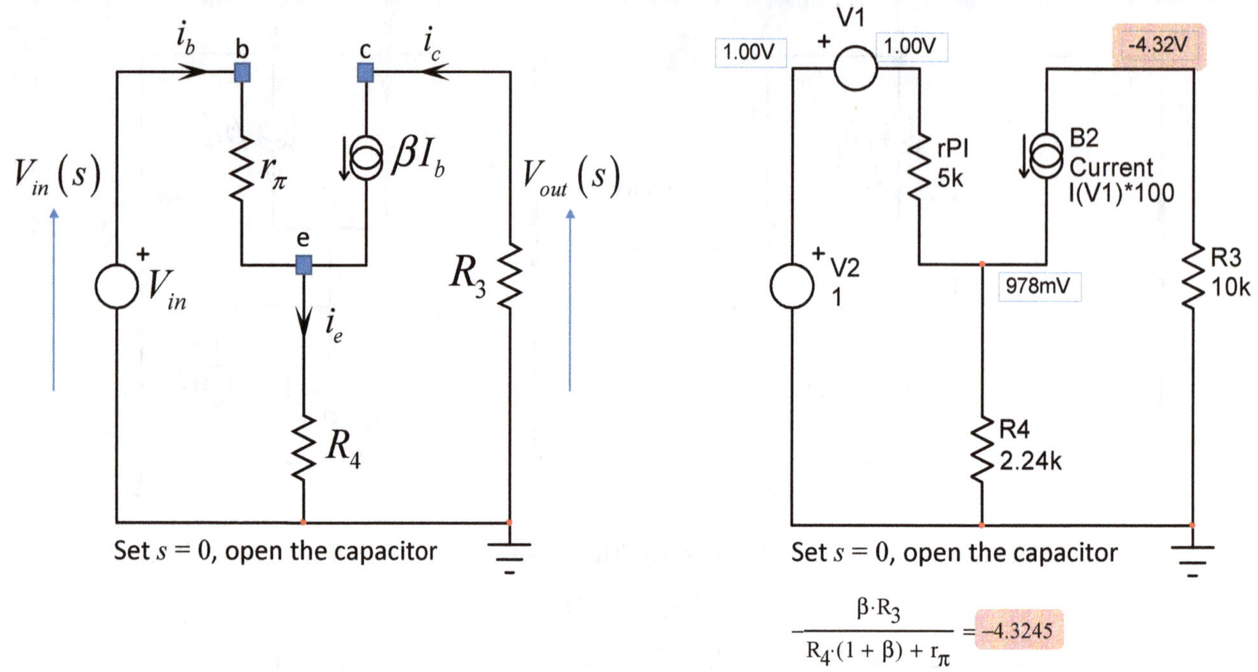

The Dc Gain is Obtained by Opening the Capacitor and Computing the Ratio V_{out}/V_{in}

Figure 3.23

The base current is determined by:

$$i_b = \frac{V_{in} - V_e}{r_\pi} \tag{3.52}$$

The voltage at the emitter is made of the added base and collector currents flowing in R_4:

$$V_e = R_4 i_e = R_4 (i_b + i_c) = R_4 (i_b + \beta i_b) = i_b R_4 (\beta + 1) \tag{3.53}$$

From the two above equations, we can determine the base current i_b equal to:

$$i_b = \frac{V_{in}}{R_4(\beta + 1) + r_\pi} \tag{3.54}$$

The output voltage is obtained easily considering the collector current flowing in R_3:

$$V_{out} = -i_c R_3 = -\beta i_b R_3 \tag{3.55}$$

Substituting (3.54) in the definition of V_{out} and rearranging gives us the gain we want:

$$H_0 = \frac{V_{out}}{V_{in}} = -\frac{\beta R_3}{R_4(\beta + 1) + r_\pi} \tag{3.56}$$

For a high gain value and a small r_π resistance, this expression classically simplifies to:

$$H_0 \approx -\frac{R_3}{R_4} \tag{3.57}$$

There we go, we have determined the complete transfer function of this bipolar stage quite quickly and it is defined as:

$$H(s) = H_0 \frac{1+\frac{s}{\omega_z}}{1+\frac{s}{\omega_p}} \tag{3.58}$$

...with the gain, pole and zero respectively defined by (3.57), (3.51) and (3.45).

For higher-order numerators, the NDI principle remains similar and resembles what we have seen for the denominator. The only difference between the two being that for determining the numerator time constants, the stimulus is back in the circuit and R is obtained considering an output null. For a second-order polynomial—with the N subscript to distinguish the numerator—the expression is given below, including redundancy for the last term:

$$N(s) = 1 + a_1 s + a_2 s^2 = 1 + (\tau_{1N} + \tau_{2N})s + (\tau_{1N}\tau_{2N}^{1})s^2 = 1 + (\tau_{1N} + \tau_{2N})s + (\tau_{2N}\tau_{1N}^{2})s^2 \tag{3.59}$$

The circuit of Figure 3.24 will be used as an example to determine the second-order numerator. However, before diving into the analysis to locate zeroes, are we sure this circuit has some? As usual, place the capacitors in their high-frequency state and check if the stimulus can make its way to the output port and produce an observable response.

If you set C_1 and C_2 alternatively then together in their high-frequency state, does the stimulus go through?

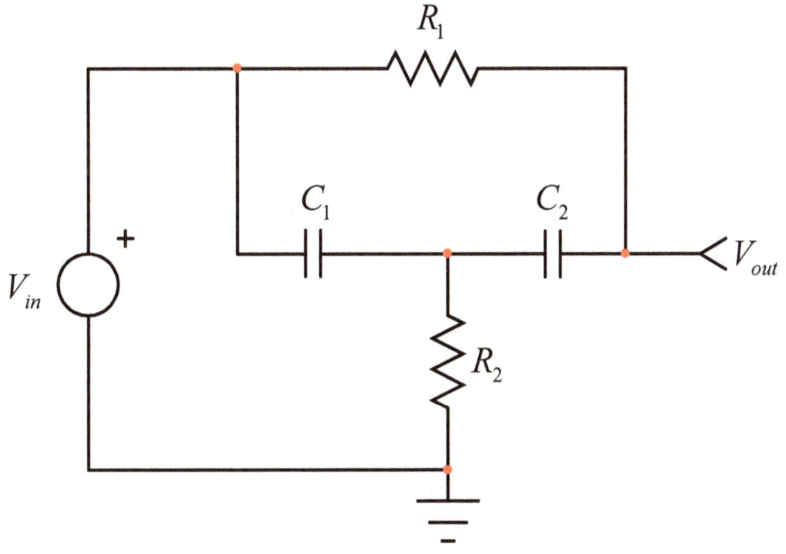

This Two-Capacitor Circuit Represents a Second-Order Network with Two Zeroes

Figure 3.24

You can do the exercise in your head or draw a series of small sketches to take a closer look. This is what I did in Figure 3.25 and you can see that we have two zeroes because a high-frequency gains exist in all three combinations.

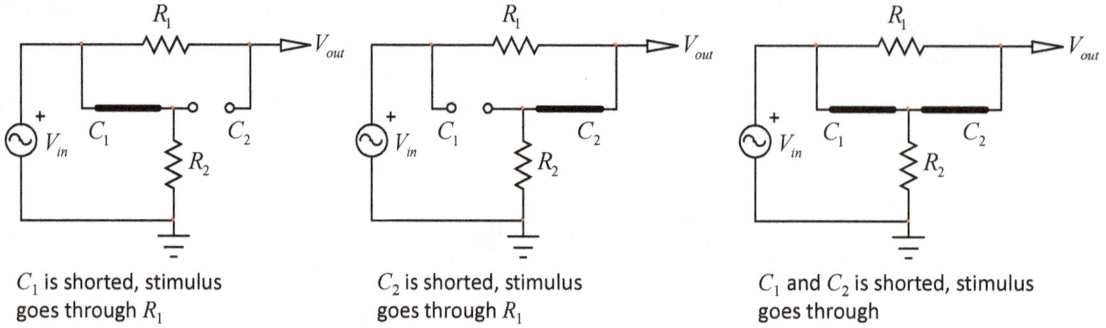
R_1
V_{out}
V_{in}
C_1
C_2
R_2
C_1 is shorted, stimulus goes through R_1
C_2 is shorted, stimulus goes through R_1
C_1 and C_2 is shorted, stimulus goes through

Setting the Energy-Storage Elements in their High-Frequency States Confirms or Denies the Presence of an Associated Zero

Figure 3.25

The zeroes will be found applying NDI three times with the same principle as in Figure 2.14: twice to assemble a_1 with τ_{1N} and τ_{2N}, and a third time for a_2. The steps appear in Figure 3.26.

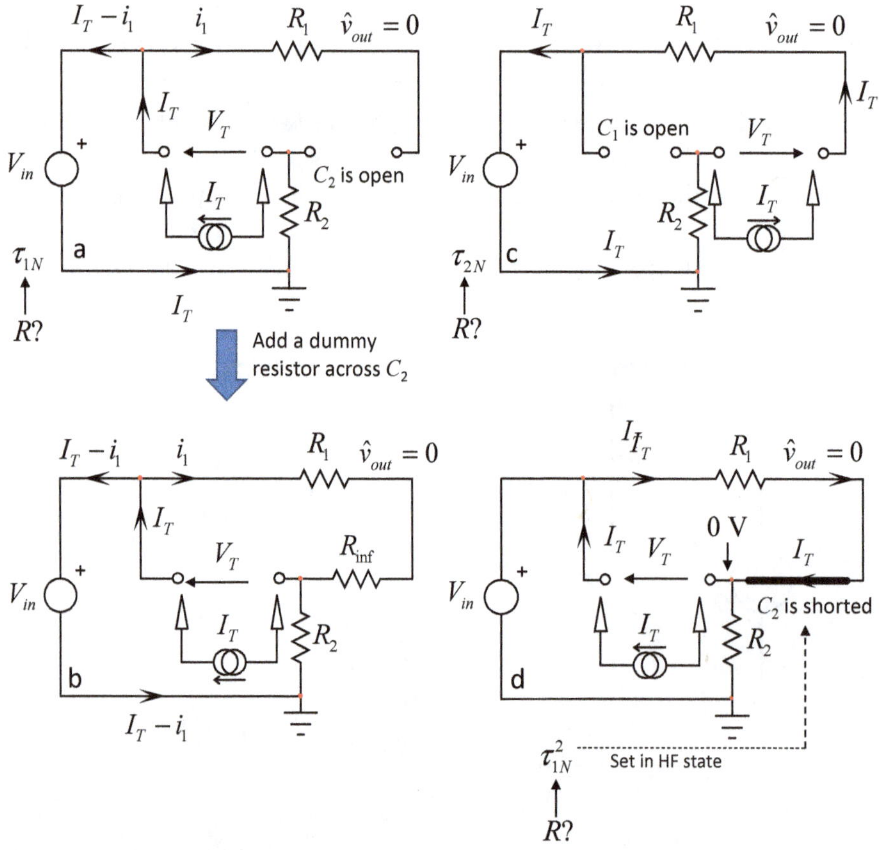

NDI is Applied in Three Different Configurations where the Output is Nulled

Figure 3.26

In Figure 3.26a, the output voltage v_{out} is nulled and the right side of R_1 is at 0 V. However, there is a current i_1 flowing in R_1 imposed by V_{in}. Because C_2 is open, this current has nowhere to go and we have an indeterminacy. The solution simply lies in installing a dummy resistor R_{inf} across C_2 to provide a path for that current.

This is what we did in sketch (b).

From this drawing, we can now express the voltage V_T across the current source:

$$V_T = V_{in} + R_2\left(I_T - i_1\right) = R_1 i_1 + R_2\left(I_T - i_1\right) \tag{3.60}$$

The current i_1 is obtained by equating two expressions:

$$i_1 R_{\text{inf}} = R_2\left(I_T - i_1\right) \tag{3.61}$$

...leading to a current defined as:

$$i_1 = \frac{R_2 I_T}{R_{\text{inf}} - R_2} \tag{3.62}$$

We can now update (3.60) with (3.62):

$$V_T = i_1\left(R_1 - R_2\right) + R_2 I_T = \frac{R_2 I_T}{R_{\text{inf}} - R_2}\left(R_1 - R_2\right) + R_2 I_T \tag{3.63}$$

In this expression, when R_{inf} approaches infinity, then we have:

$$\frac{V_T}{I_T} = R_2 \tag{3.64}$$

...leading to the time constant we want:

$$\tau_{1N} = R_2 C_1 \tag{3.65}$$

This is a particular NDI case where some engineering judgement is necessary.

The issue comes from the input source V_{in} directly feeding the circuit and the current it forces in R_1.

As a side exercise, add a small resistance R_s in series with the source and the left connection of R_1 will now be at 0 V—with no current flowing in the resistor—immediately leading to (3.65).

In Figure 3.26c, the current I_T circulates in R_1 and R_2 however, the right-side of the current generator is grounded as $\hat{v}_{out} = 0$.

So, V_T is the voltage across R_2.

The resistance offered by C_2's terminals is also R_2.

Therefore:

$$\tau_{2N} = R_2 C_2 \tag{3.66}$$

In Figure 3.26d, C_2 is set in its high-frequency state (replaced by a short circuit) while we want to determine C_1's terminals resistance.

Given the output null, the voltage across R_2 is also 0 V and the only remaining resistance in the circuit is R_1.

The time constant is thus:

$$\tau_{1N}^{2} = R_1 C_1 \tag{3.67}$$

We can verify these expressions by capturing a SPICE circuit of the network we studied. This is what we have in Figure 3.27 with the help of transconductance amplifiers. Values computed at nodes R1N, R2N and R21N after a dc bias point analysis confirms our calculations. This is always a good practice to verify calculations with SPICE.

Bias point operation is fast and results are immediate. Please note the presence of a 1-GΩ resistance to ensure all nodes have a dc current to ground—otherwise convergence issues could occur.

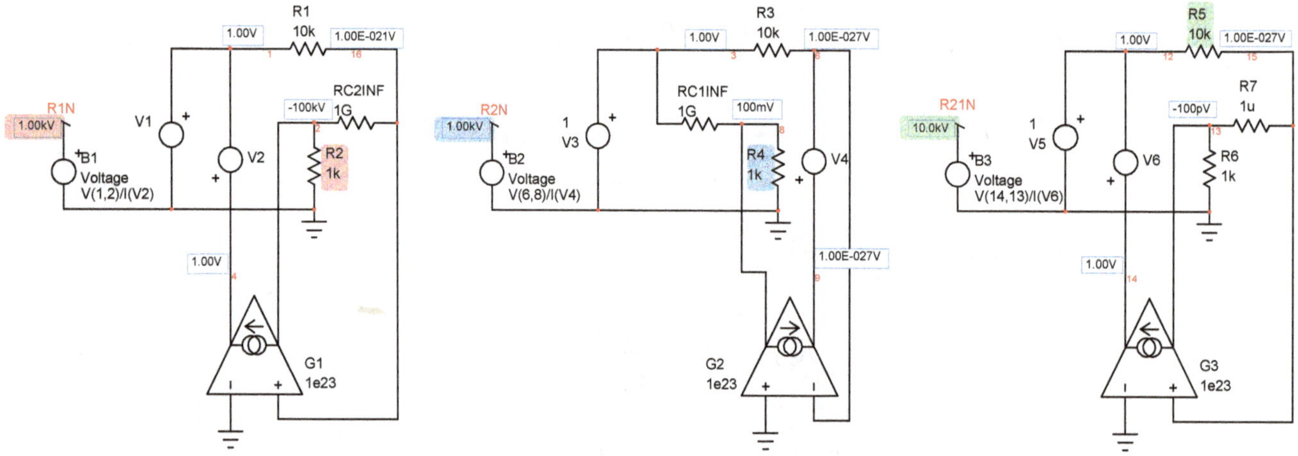

The SPICE Simulation of Figure 3.26 Confirms what we Found Analytically

Figure 3.27

As indicated by the numerical results, the numbers are good and the numerator is defined as:

$$N(s) = 1 + (\tau_{1N} + \tau_{2N})s + (\tau_{1N}\tau_{2N}^{1})s^2 = 1 + R_2(C_1 + C_2)s + (R_1 R_2 C_1 C_2)s^2 \tag{3.68}$$

This second-order expression can be factored in a normalized polynomial form such as:

$$N(s) = 1 + \frac{s}{\omega_{0N} Q_N} + \left(\frac{s}{\omega_{0N}}\right)^2 \tag{3.69}$$

...in which:

$$Q_N = \frac{\sqrt{a_2}}{a_1} = \frac{\sqrt{R_1 R_2 C_1 C_2}}{R_2(C_1 + C_2)} \tag{3.70}$$

...and:

$$\omega_{0N} = \frac{1}{\sqrt{a_2}} = \frac{1}{\sqrt{R_1 R_2 C_1 C_2}} \tag{3.71}$$

Now that we have the zeroes we wanted, we can proceed with the poles by setting the excitation to zero volt as shown in Figure 3.28.

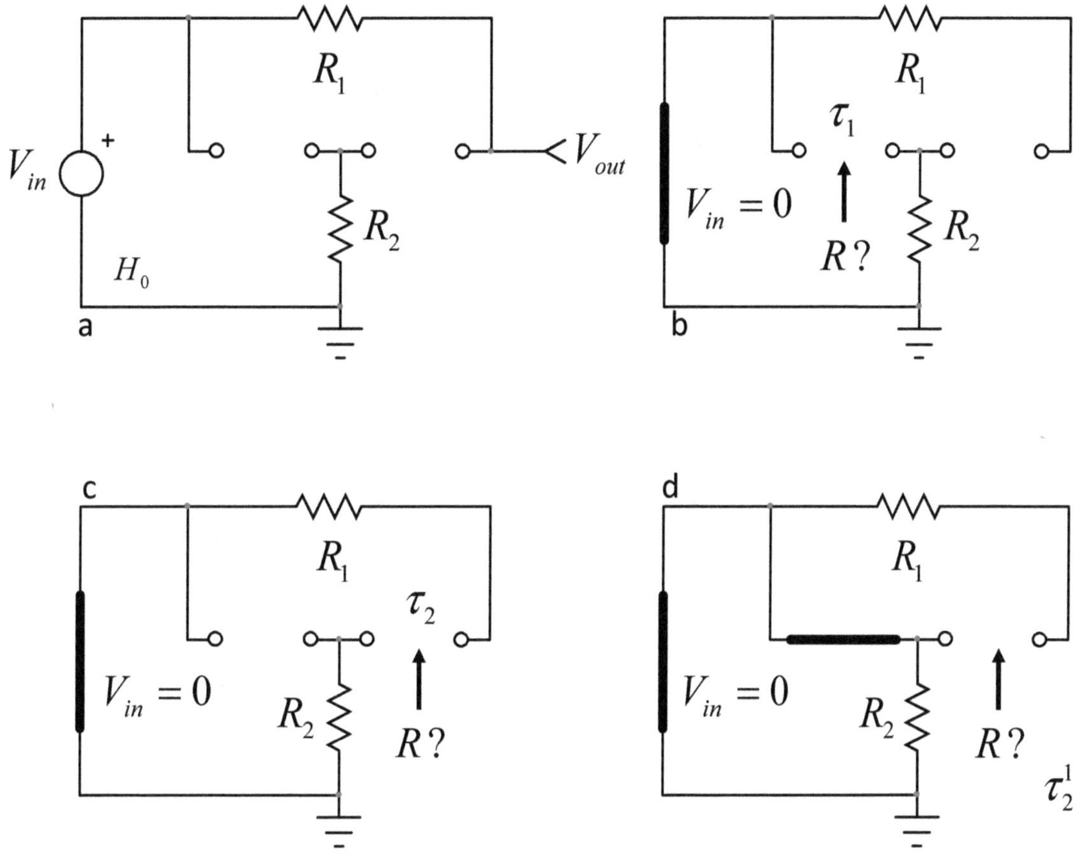

Four Steps are Necessary to Determine the Denominator Coefficients

Figure 3.28

The dc transfer function is found by open-circuiting all capacitors as illustrated in Figure 3.28a.

The dc gain H_0 equals 1 in this mode.

$$H_0 = 1 \tag{3.72}$$

The first time constant τ_1 is obtained by looking into C_1's terminals while the excitation is set to 0 V as in Figure 3.28b. The resistance in this configuration is R_2 as R_1 has one terminal open.

The time constant is:

$$\tau_1 = R_2 C_1 \tag{3.73}$$

In Figure 3.28c, we can determine the second time constant τ_2. It combines R_1 and R_2 in series to give:

$$\tau_2 = (R_1 + R_2) C_2 \tag{3.74}$$

Figure 3.28d shows how to calculate the final part of the b_2 coefficient. R_2 is shorted and R_1 remains alone.

The time constant is therefore:

$$\tau_2^1 = R_1 C_2 \tag{3.75}$$

The denominator gathers all these terms in the following expression:

$$D(s) = 1 + s(\tau_1 + \tau_2) + s^2 \tau_1 \tau_2^1 = 1 + s\left[R_2 C_1 + (R_1 + R_2) C_2\right] + s^2 R_1 R_2 C_1 C_2 \tag{3.76}$$

We can rearrange it according to a normalized second-order polynomial form revealing a quality factor Q and a resonant angular frequency ω_0:

$$D(s) = 1 + \frac{s}{\omega_0 Q} + \left(\frac{s}{\omega_0}\right)^2 \tag{3.77}$$

...in which:

$$Q = \frac{\sqrt{b_2}}{b_1} = \frac{\sqrt{R_1 R_2 C_1 C_2}}{R_2 C_1 + (R_1 + R_2) C_2} \tag{3.78}$$

...and:

$$\omega_0 = \frac{1}{\sqrt{b_2}} = \frac{1}{\sqrt{R_1 R_2 C_1 C_2}} \tag{3.79}$$

We now assemble the numerator and the denominator to obtain the complete transfer function found without writing a single line of algebra beside for determining (3.64):

$$H(s) = \frac{1 + R_2 (C_1 + C_2) s + (R_1 R_2 C_1 C_2) s^2}{1 + s\left[R_2 C_1 + (R_1 + R_2) C_2\right] + s^2 R_1 R_2 C_1 C_2} \tag{3.80}$$

Is there a way we could rearrange this expression in a better, more compact—read *low-entropy*—format? Yes, if you factor $R_2(C_1 + C_2)s$ in the numerator and $s\left[R_2 C_1 + (R_1 + R_2) C_2\right]$ in the denominator.

Then, after some *massage*—as inspector Clouseau would call it—you should find:

$$H(s) = H_{notch} \frac{1 + Q_N \left(\frac{s}{\omega_{0N}} + \frac{\omega_{0N}}{s}\right)}{1 + Q\left(\frac{s}{\omega_0} + \frac{\omega_0}{s}\right)} \tag{3.81}$$

...in which:

$$H_{notch} = \frac{R_2(C_1 + C_2)}{R_2C_1 + (R_1 + R_2)C_2} \quad (3.82)$$

...where Q_N, ω_{0N}, Q and ω_0 are respectively determined by (3.70), (3.71), (3.78) and (3.79).

This is the most compact way of writing this expression in which the leading term represents the attenuation at the resonant point. Of course, this is the ultimate goal here and it would be impossible to derive any value from (3.80) without the D-OA (*Design-Oriented Analysis*) dear to the late Dr. Middlebrook.

To test all these transfer functions and confirm we have done a good job at deriving them, we need a reference transfer function. For that purpose, I applied superposition by redrawing the network but this time driven by two V_{in} sources that I will alternatively set to 0 V.

Figure 3.29 shows the corresponding illustration with the input voltage split in sketch (a):

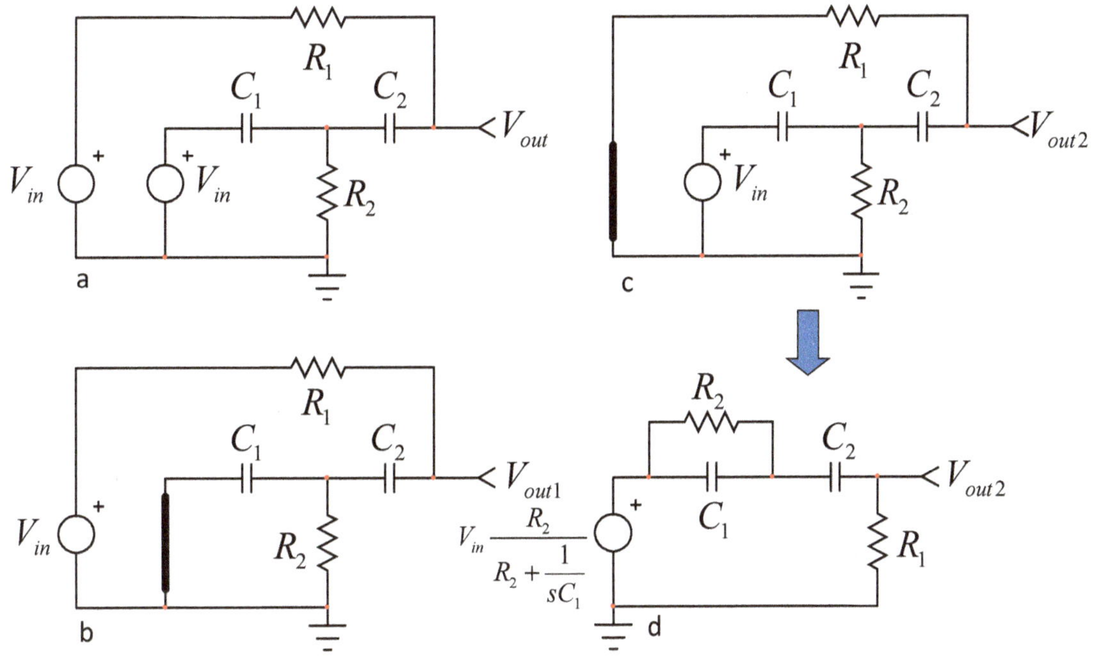

Superposition Helps to Determine a Raw Transfer Function

Figure 3.29

In Figure 3.29b, C_1's left terminal is grounded and the output voltage is obtained by observing a resistive divider:

$$V_{out1} = V_{in} \frac{\frac{1}{sC_2} + \frac{1}{sC_1} \| R_2}{\frac{1}{sC_2} + \frac{1}{sC_1} \| R_2 + R_1} \quad (3.83)$$

Setting the second generator to 0 V by grounding R_1's left terminal, we can rearrange Figure 3.29c in Figure 3.29d, using a Thévenin equivalent circuit featuring R_2 and C_1.

The second output voltage is thus defined as:

$$V_{out2} = V_{in}\frac{R_2}{R_2+\frac{1}{sC_1}}\cdot\frac{R_1}{\frac{1}{sC_1}\|R_2+\frac{1}{sC_2}+R_1} \tag{3.84}$$

Assembling these two equations and factoring V_{in} leads to the transfer function we want:

$$H(s)=\frac{\frac{1}{sC_2}+\frac{1}{sC_1}\|R_2}{\frac{1}{sC_2}+\frac{1}{sC_1}\|R_2+R_1}+\frac{R_2}{R_2+\frac{1}{sC_1}}\cdot\frac{R_1}{\frac{1}{sC_1}\|R_2+\frac{1}{sC_2}+R_1} \tag{3.85}$$

Needless to say, there is absolutely no insight from this expression and trying to expand it would probably end up in algebraic paralysis…and swearwords! We are all set and can confront all our calculations to see how the curves compare. The Mathcad® sheet is shown in Figure 3.30 and expresses the various expressions.

In this illustration, the components values R_1 and R_2 then C_1 and C_2 are respectfully linked via n and k. Equation (3.85) is captured in $H_{ref}(s)$.

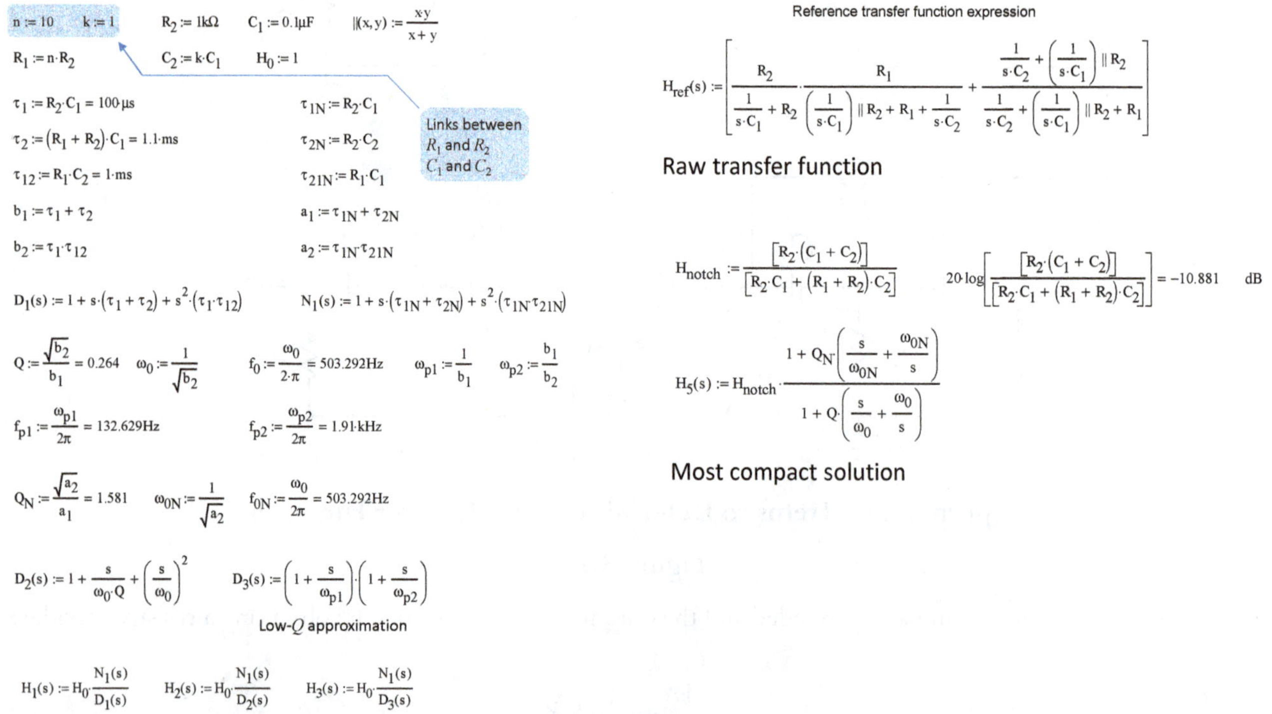

A Mathcad® Sheet allows the Test of all the Derived Expressions

Figure 3.30

The ac plots of these transfer functions are gathered in Figure 3.31 and show how nicely all curves superimpose. The sanity check consists of subtracting magnitudes and phases obtained from the raw expression with those obtained in the intermediate steps. It is important to confirm our results are sound and any significant deviation points to an error. The result oscillating in the pico region acknowledges perfectly similar equations.

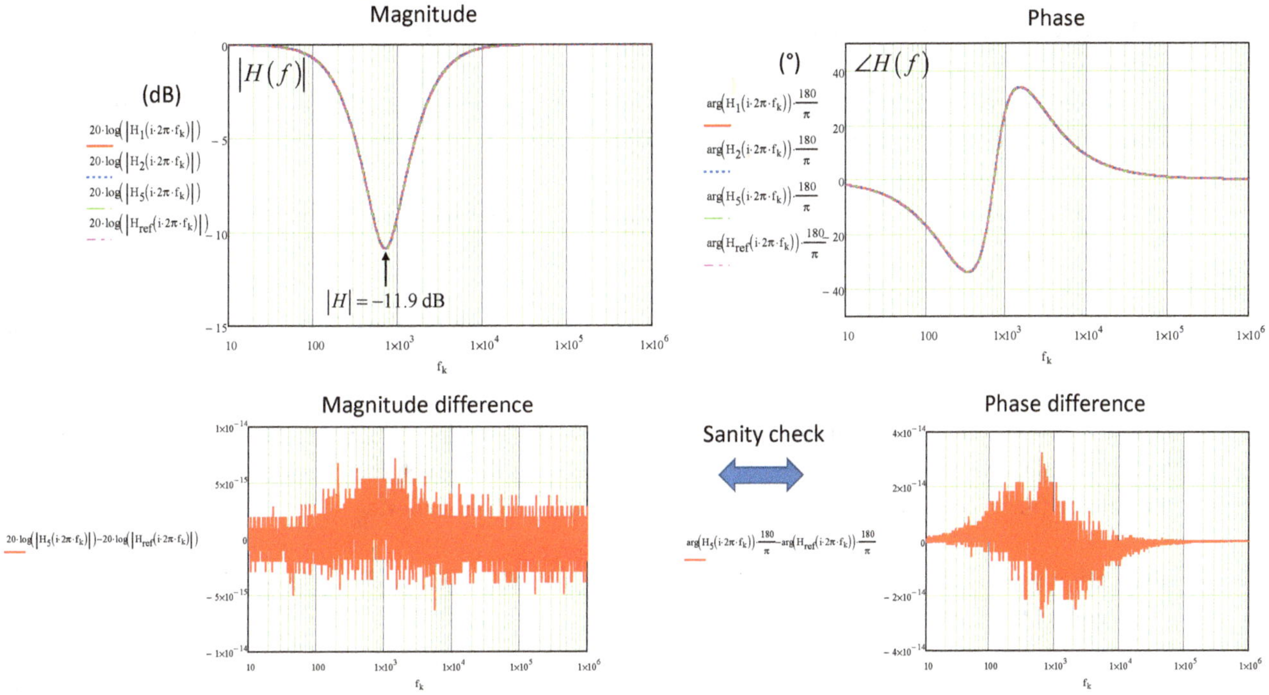

The Magnitude and Phase Responses are Rigorously Identical and Confirm our Analysis

Figure 3.31

Finally, I gathered some possible combinations—reshuffling also works for the numerator coefficients—to let you determine numerators using NDI and up to the 4th-order in Figure 3.32.

Numerators up to order 4

1st order $N(s)=1+\tau_{1N}s$

2nd order $N(s)=1+(\tau_{1N}+\tau_{2N})s+(\tau_{1N}\tau_{2N}^{1})s^2=1+(\tau_{1N}+\tau_{2N})s+(\tau_{2N}\tau_{1N}^{2})s^2$

3rd order $N(s)=1+(\tau_{1N}+\tau_{2N}+\tau_{3N})s+(\tau_{1N}\tau_{2N}^{1}+\tau_{1N}\tau_{3N}^{1}+\tau_{2N}\tau_{3N}^{2})s^2+(\tau_{1N}\tau_{2N}^{1}\tau_{3N}^{12})s^3$

4th order
$$N(s)=1+(\tau_{1N}+\tau_{2N}+\tau_{3N}+\tau_{4N})s+(\tau_{1N}\tau_{2N}^{1}+\tau_{1N}\tau_{3N}^{1}+\tau_{1N}\tau_{4N}^{1}+\tau_{2N}\tau_{3N}^{2}+\tau_{2N}\tau_{4N}^{2}+\tau_{3N}\tau_{4N}^{3})s^2$$
$$+(\tau_{1N}\tau_{2N}^{1}\tau_{3N}^{12}+\tau_{1N}\tau_{2N}^{1}\tau_{4N}^{12}+\tau_{1N}\tau_{3N}^{1}\tau_{4N}^{13}+\tau_{2N}\tau_{3N}^{2}\tau_{4N}^{23})s^3$$
$$+(\tau_{1N}\tau_{2N}^{1}\tau_{3N}^{12}\tau_{4N}^{123})s^4$$

The time constants with an N subscripts are determined when the output is a *null* with a stimulus back in place.

The Numerator can be Determined up to the 4th Order with these Expressions

Figure 3.32

3.3 What Should I Retain from this Chapter?

In this third chapter, we have learned information on the zeroes of a transfer function that is summarized below:

1. In the Laplace domain, when the stimulus is tuned at a zero frequency, the magnitude of the transfer function reduces to zero and the response disappears.

2. When the response is nulled, the current flowing in the load is 0 A and the voltage across its terminals is 0 V.

3. To determine the position of a zero, we assume that some elements inside the network, when tuned at the zero frequency, block the stimulus propagation and null the response.

4. These elements can offer an infinite series impedance at the zero frequency or become a transformed short circuit, shunting one branch to ground. By finding the roots for which the considered impedance becomes infinite or turns into a shunt, you determine the position of the zeroes.

5. Inspection works well for finding the zeroes but you will often need to resort to a null double injection or NDI in which the stimulus is back in place.

3.4 References

1. Ajimiri, Generalized Time- and Transfer-Constant Circuit Analysis, IEEE Transactions on Circuits and Systems, Vol. 57, NO. 6, June 2010 (https://chic.caltech.edu/wp-content/uploads/2014/02/Final-Paper.pdf)

Chapter 4: Generalized Transfer Functions

THIS FOURTH CHAPTER introduces the generalized transfer function which avoids the use of an NDI by reusing the time constant already found the denominator.

4.1 The Generalized Transfer Function

The null double injection exercise can often appear intimidating to the novice designer considering the necessary abstraction to make it work.

Nevertheless, it is the fastest and most efficient way to determine the zeroes in a complicated arrangement when inspection does not work.

Another alternative is to resort to the generalized arrangement proposed in [1] and [2].

This approach reuses the time constants already found for the denominator and shields you from going through an NDI. The obtained transfer function is rigorously equivalent to the one determined by inspection or NDI. However, as more coefficients are assembled, then extra effort can sometimes be necessary to simplify the obtained terms.

In any case, decide which path you prefer: apply the NDI straight away or use the generalized form as described in the following lines.

Without redemonstrating the formulas, the principle is to determine the gain of a 1st-order circuit under study when s approaches infinity: set C or L in their high-frequency state (a short circuit and an open circuit respectfully) and determine the gain or the attenuation in this mode. We started doing it in Figure 3.4 of Chapter 3 but without proceeding to a gain value.

Figure 4.1 gives a graphical illustration of the adopted notation for which the stimulus is back in place because you determine a gain:

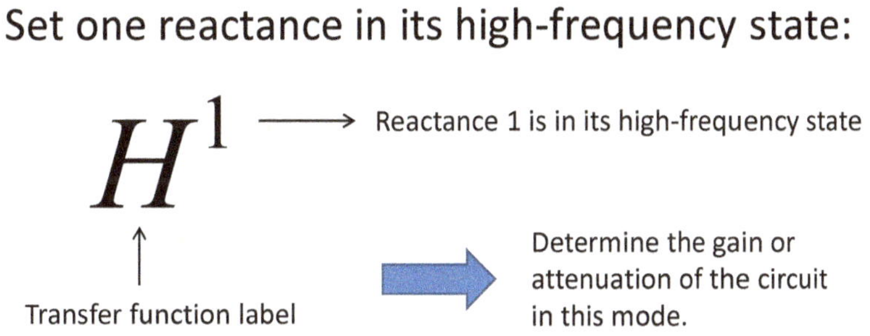

Set one reactance in its high-frequency state:
H
1
Reactance 1 is in its high-frequency state
Transfer function label
Determine the gain or
attenuation of the circuit
in this mode.

The High-Frequency Gain is Indicated by the Energy-Storing Element set as an Exponent in the Gain Label

Figure 4.1

In this figure, we show the notation for a transfer function H whose energy-storing element labeled 1—C_1 or L_1—is respectively replaced by a short circuit or an open circuit in the circuit under study.

As the stimulus is back in this exercise, simply determine the transfer function in this mode.

This gain can be null, equal to one or take on a more complicated value depending on the arrangement.

It is easy to verify the result with a SPICE simulation.

The generalized formula to determine the transfer function while reusing the time constant determined for the denominator is:

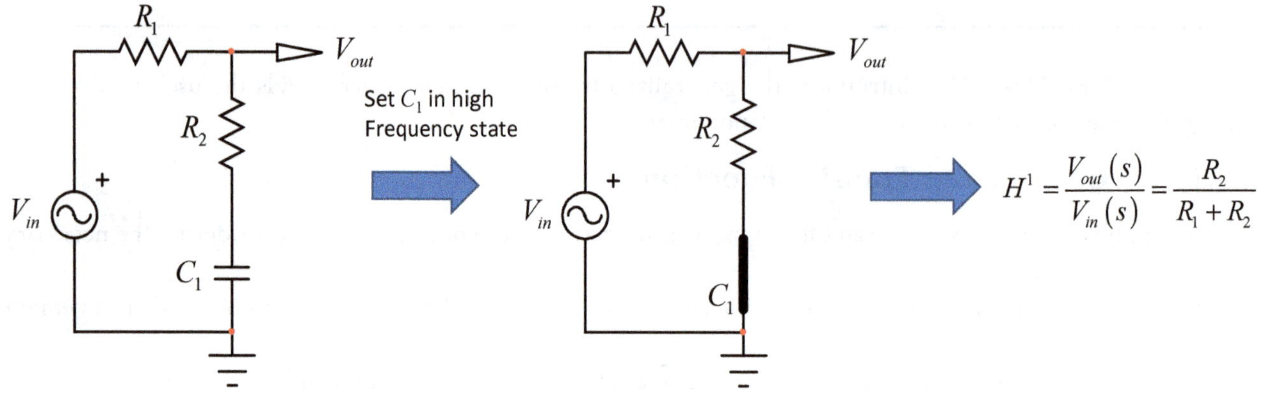

In this Circuit, Determine the High-Frequency Gain by Shorting the Capacitor

Figure 4.2

$$H(s) = \frac{H_0 + H^1\tau_1 s}{1+s\tau_1} \tag{4.1}$$

If the dc gain H_0 is different than zero, then you can advantageously rewrite this expression in the *low-entropy* form we favor:

$$H(s) = H_0 \frac{1+\frac{H^1}{H_0}\tau_1 s}{1+s\tau_1} = H_0 \frac{1+\frac{s}{\omega_z}}{1+\frac{s}{\omega_p}} \tag{4.2}$$

With a zero defined as:

$$\omega_z = \frac{H_0}{H^1\tau_1} \tag{4.3}$$

Because H_0 and H^1 are of similar dimension, the expression is homogenous.

The pole is obtained by computing the inverse of the natural time constant:

$$\omega_p = \frac{1}{\tau_1} \tag{4.4}$$

Let's apply this expression immediately to the circuit of Figure 4.2.

In your head, without drawing an intermediate sketch, you can see that the dc gain H_0 is 1 (C_1 is open circuited) and after setting the input source to 0 V (a short circuit), the pole involving C_1 is simply:

$$\omega_p = \frac{1}{\tau_1} = \frac{1}{C_1(R_1+R_2)} \tag{4.5}$$

Now, for the zero, bringing the stimulus back and shorting the capacitor to determine H^1 leads to:

$$H^1 = \frac{R_2}{R_1 + R_2} \tag{4.6}$$

The complete transfer function is then:

$$H(s) = H_0 \frac{1 + \frac{H^1}{H_0}\tau_1 s}{1 + s\tau_1} = 1 \cdot \frac{1 + \frac{\frac{R_2}{R_1+R_2}}{1} C_1 (R_1 + R_2) s}{1 + C_1 (R_1 + R_2) s} = \frac{1 + sR_2C_1}{1 + s(R_2 + R_1)C_1} \tag{4.7}$$

Of course, when skilled in the art, you could also immediately inspect the circuit and see the transformed short brought by R_2 and C_1 leading to the zero position straight away. This is a simple example to illustrate the application of (4.1).

Let's see another quick example in which the dc gain is zero. This is the classical *CR* circuit shown in Figure 4.3. If you set *s* to zero, then you can see $H_0 = 0$.

Now setting the source to zero and determining the time constant leads to the same pole position as with (4.5):

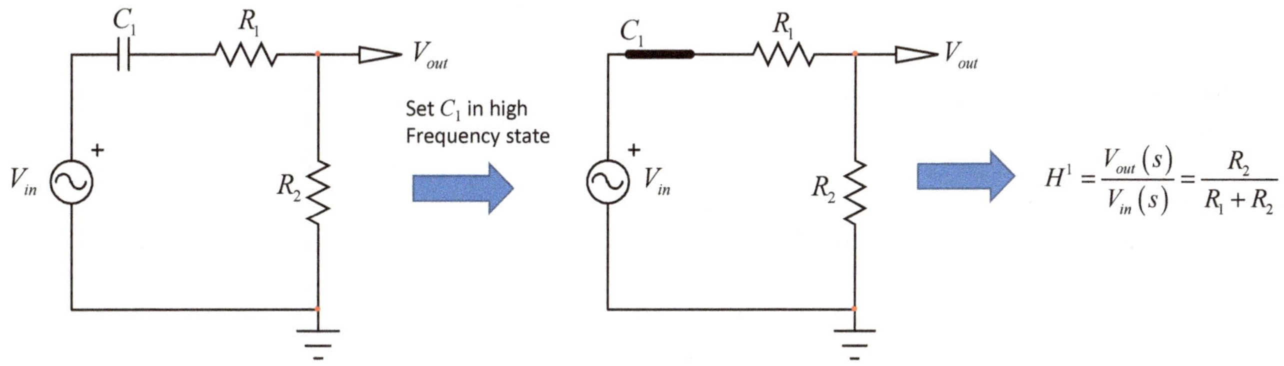

There is a Zero at the Origin in this Circuit, Blocking the Dc Component from the Source

Figure 4.3

$$\omega_p = \frac{1}{\tau_1} = \frac{1}{C_1(R_1 + R_2)} \tag{4.8}$$

When the capacitor is set to its high-frequency state, the transfer function H^1 is also identical to (4.6) and leads to:

$$H^1 = \frac{R_2}{R_1 + R_2} \tag{4.9}$$

The final transfer function of this circuit is then:

$$H(s)=\frac{H_0+H^1\tau_1 s}{1+s\tau_1}=\frac{0+\dfrac{R_2}{R_1+R_2}C_1(R_1+R_2)s}{1+C_1(R_1+R_2)s}=\frac{sR_2C_1}{1+s(R_2+R_1)C_1} \tag{4.10}$$

...and shows a zero at the origin as expected.

This transfer function can be rewritten using an *inverted pole* by factoring the numerator and the denominator:

$$H(s)=\frac{sR_2C_1}{s(R_2+R_1)C_1}\frac{1}{1+\dfrac{1}{s(R_1+R_2)C_1}}=H_\infty\frac{1}{1+\dfrac{\omega_p}{s}} \tag{4.11}$$

In this formula, H^1 is replaced by H_∞ to match our normalized expression.

After this simple 1st-order circuit example, let's see how we could apply our generalized approach to a second-order system.

Before we look at the electrical diagram, we first reproduce the generalized expression given in [1]:

$$H(s)=\frac{H_0+s\left(H^1\tau_1+H^2\tau_2\right)+s^2H^{12}\tau_1\tau_2^1}{1+s\left(\tau_1+\tau_2\right)+s^2\tau_1\tau_2^1} \tag{4.12}$$

In this expression, you recognize the notation we previously used like H^1 or H^2 which designates gains respectfully obtained with energy-storing element 1 or 2 placed in their high-frequency state.

Now, because we have two energy-storing elements, when one of them is placed in its high-frequency state when determining H^1, for instance, the second is always left in its dc state. Then, for the higher-order term like H^{12}, both energy-storing elements are placed in their high-frequency state. Figure 4.4 illustrates this simple principle.

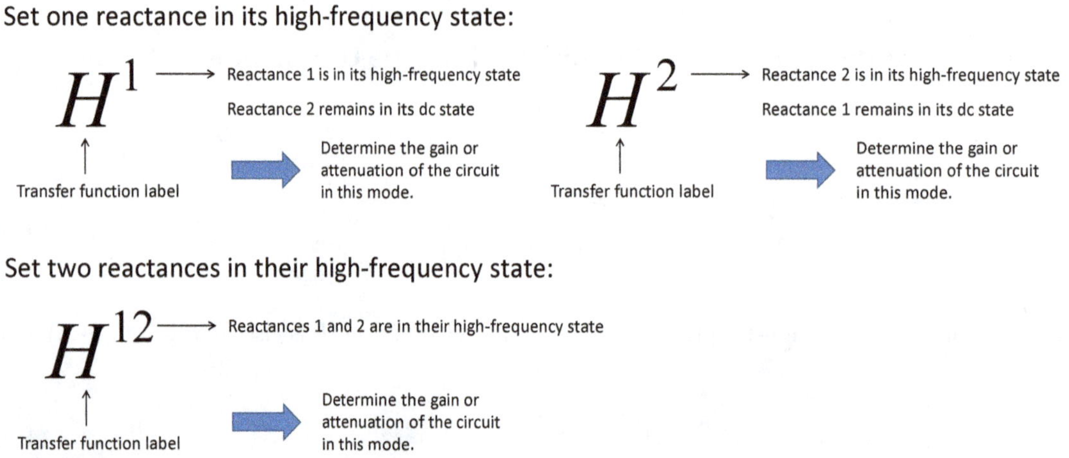
Set one reactance in its high-frequency state:
H1
Reactance 1 is in its high-frequency state
Reactance 2 remains in its dc state
Transfer function label
Determine the gain or attenuation of the circuit in this mode.
H2
Reactance 2 is in its high-frequency state
Reactance 1 remains in its dc state
Transfer function label
Determine the gain or attenuation of the circuit in this mode.
Set two reactances in their high-frequency state:
H12
Reactances 1 and 2 are in their high-frequency state
Transfer function label
Determine the gain or attenuation of the circuit in this mode.

Each Energy-Storing Element is Alternatively Placed in its High-Frequency State

Figure 4.4

The circuit proposed for exercising the formula appears in Figure 4.5 and shows a perfect op-amp with two capacitors. The first thing is to set s to zero and open all capacitors.

In this mode, the gain H_0 is simply equal to zero since no signal can reach the inverting pin.

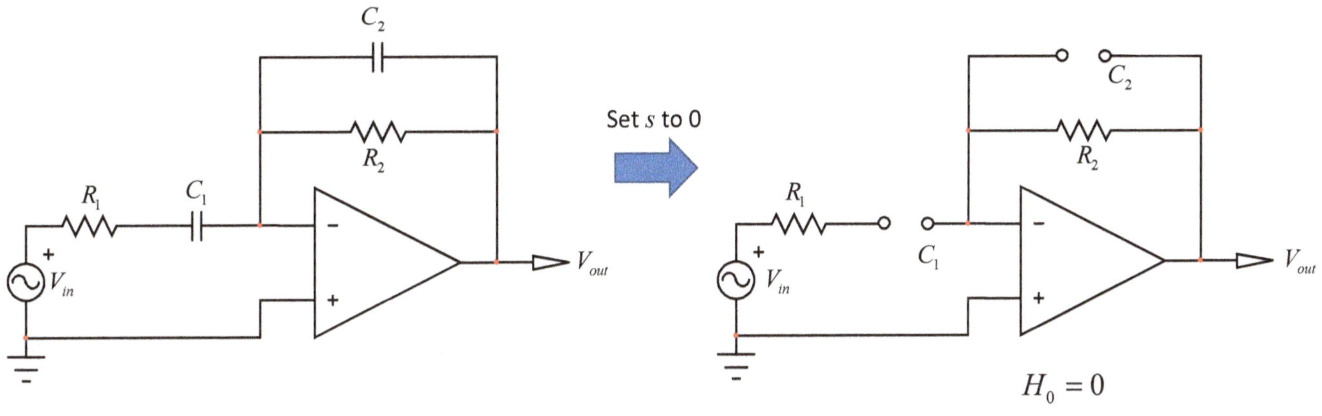

There are Two Energy-Storing Elements in this Op-Amp Circuit. When Both are Open-Circuited in Dc, the Gain is Zero

Figure 4.5

For the next step, we are going to determine the two time constants involving C_1 and C_2 after turning the input source off to 0 V (a short circuit). The sketch appears in the left side of Figure 4.6. For C_1, with C_2 open-circuited, because of the virtual ground and the large gain of the op-amp, the (-) is almost at 0 V.

Therefore, the only resistance seen by C_1 in this mode is R_1, leading to the first time constant:

$$\tau_1 = R_1 C_1 \quad (4.13)$$

In the second case in which C_1 is open circuited, there is no current entering the (-) pin.

Therefore, if you install a test current source I_T across R_2, all the current circulates in R_2 making it the resistance we want:

$$\tau_2 = R_2 C_2 \quad (4.14)$$

In the right side of Figure 4.6, we are now determining τ_2^1 by shorting C_1 and looking through C_2's terminals.

Again, because of the virtual ground no current flows in R_1, the only resistance seen by the capacitor is R_2:

$$\tau_2^1 = R_2 C_2 \quad (4.15)$$

This is it; we have determined the denominator by inspection, without writing a line of algebra.

Assembling the time constants leads to:

$$D(s) = 1 + s(\tau_1 + \tau_2) + s\tau_1\tau_2^1 = 1 + s(R_1C_1 + R_2C_2) + s^2 R_1C_1R_2C_2 \quad (4.16)$$

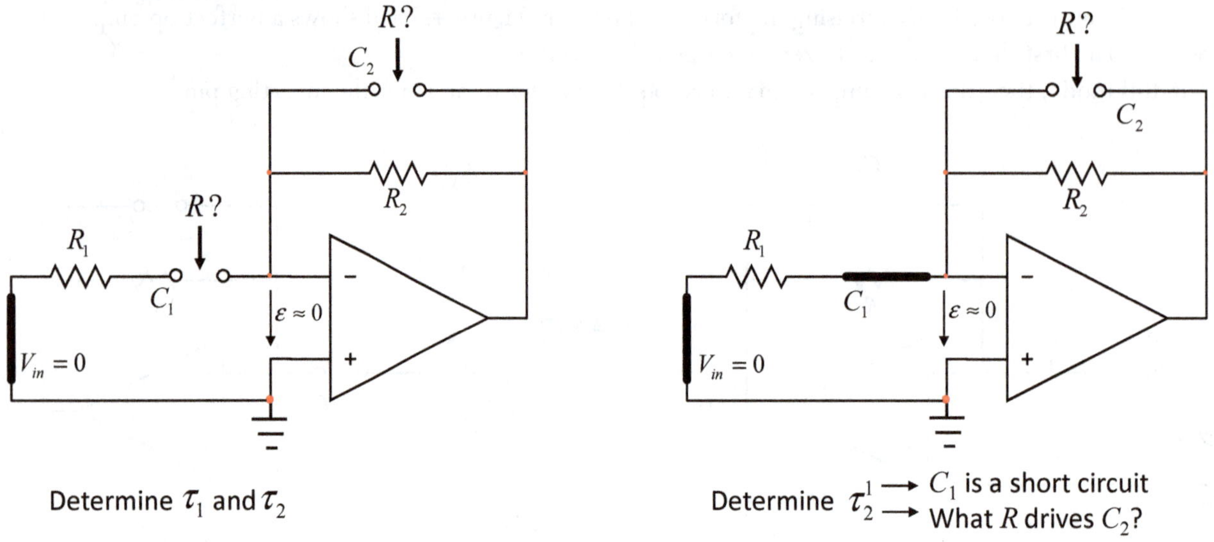
R?
C_2
R_2
R_1
C_1
$\varepsilon \approx 0$
$V_{in} = 0$
Determine τ_1 and τ_2
Determine τ_2^1
C_1 is a short circuit
What R drives C_2?

The Perfect Op-Amp brings a Virtual Ground to make Both Inputs share a Common 0 V

Figure 4.6

To verify if our analysis is correct, we can quickly simulate a perfect op-amp with a voltage-controlled voltage source (E primitive) as illustrated in Figure 4.7.

The schematic is simple: reproduce the electrical network with a zeroed input and install a 1 A test generator across the capacitor's connecting terminals. Then measure the voltage across the current source and it will give a voltage image of the resistance you want. Make sure the polarity is correct when measuring a floating voltage as V_T and I_T point towards the same direction.

In sketches (a), (b) and (c), you can see that the measured value is the resistance we did find by inspection.

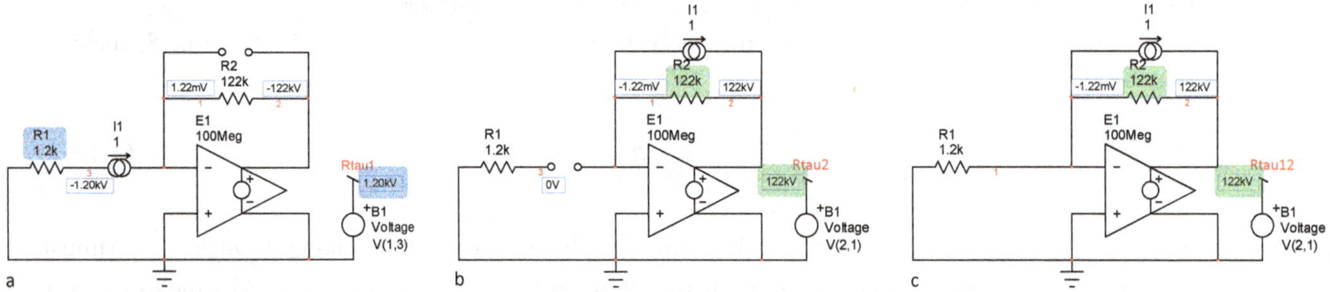

SPICE Quickly Checks our Calculations via a Dc Bias Point Simulation

Figure 4.7

Some readers may not be at ease with the concept of virtual ground and I have purposely run a dc analysis considering an op-amp with a finite gain A_{OL}.

Look at Figure 4.8 to see how a portion of the test generator current circulates in R_2 while the rest flows through the op-amp.

We can write:

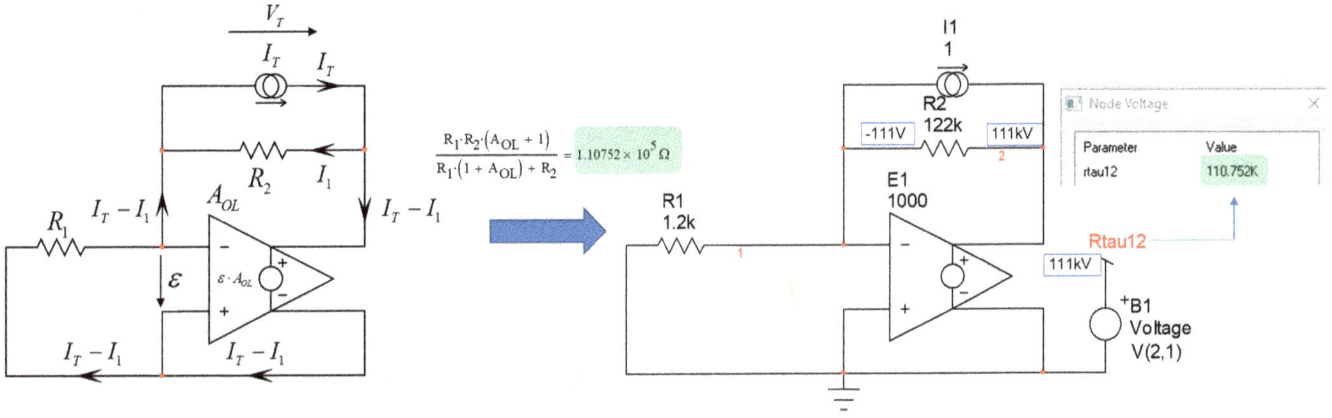

With an Imperfect Op-Amp, what is the Resistance seen by C_1's Terminals?

Figure 4.8

$$\varepsilon = R_1 (I_T - I_1) \tag{4.17}$$

The voltage V_T across the current source is the op-amp output voltage minus the voltage at the (-) pin:

$$V_T = \varepsilon A_{OL} - (-\varepsilon) = \varepsilon (1 + A_{OL}) \tag{4.18}$$

The current I_1 flowing in R_2 is simply:

$$I_1 = \frac{V_T}{R_2} \tag{4.19}$$

Now combining these equations leads to:

$$V_T = R_1 \left(I_T - \frac{V_T}{R_2} \right) (1 + A_{OL}) \tag{4.20}$$

Solving for V_T and rearranging gives:

$$R = \frac{V_T}{I_T} = \frac{R_1 R_2 (1 + A_{OL})}{R_1 (1 + A_{OL}) + R_2} \tag{4.21}$$

If the open-loop gain A_{OL} is very large, then the result simplifies to $R = R_2$. To verify this result, we set the open-loop gain to 60 dB or 1000, leading to a resistance of 110.752 kΩ and confirmed by the bias point simulation.

Our time constants have been determined and the denominator expressed in (4.16). Following (4.12), we need to determine three high-frequency gains: H^1, H^2 and H^{12}. The sketches are shown in Figure 4.9 and do not present particular difficulties to find the gain linking V_{out} to V_{in}.

From these drawings, we have:

$$H^1 = -\frac{R_2}{R_1} \tag{4.22}$$

$$H^2 = 0 \tag{4.23}$$

$$H^{12} = 0 \tag{4.24}$$

The numerator can now be determined according to (4.12) by combining the simple high-frequency gains we found with the time constants obtained for the denominator:

$$H(s) = H_0 + s\left(H^1\tau_1 + H^2\tau_2\right) + s^2 H^{12}\tau_1\tau_2^1 = 0 + s\left(-\frac{R_2}{R_1}C_1R_1 + 0\cdot\tau_2\right) + s^2 0\cdot\tau_1\tau_2^1 = -sR_2C_1 \tag{4.25}$$

The complete transfer is then immediately assembled as:

$$H(s) = -\frac{sR_2C_1}{1 + s\left(R_1C_1 + R_2C_2\right) + s^2 R_1C_1R_2C_2} \tag{4.26}$$

This equation is accurate but brings no insight to what it does. We can rework it by factoring sR_2C_1 in the numerator and $s(R_1C_1+R_2C_2)$ in the denominator:

$$H(s) = -\frac{sR_2C_1}{s\left(R_1C_1 + R_2C_2\right)}\frac{1}{1 + \frac{1}{s\left(R_1C_1 + R_2C_2\right)} + s^2\frac{R_1C_1R_2C_2}{s\left(R_1C_1 + R_2C_2\right)}} = -\frac{R_2C_1}{R_1C_1 + R_2C_2}\frac{1}{1 + \frac{1}{s\left(R_1C_1 + R_2C_2\right)} + s\frac{R_1C_1R_2C_2}{R_1C_1 + R_2C_2}} \tag{4.27}$$

This expression is that of a bandpass filter [1] whose mid-band gain is determined by the leading term.

The normalized polynomial follows the below expression and its terms can be obtained by simple identification:

$$H(s) = -H_0\frac{1}{1 + Q\left(\frac{\omega_0}{s} + \frac{s}{\omega_0}\right)} \tag{4.28}$$

...in which:

$$H_0 = \frac{R_2C_1}{R_1C_1 + R_2C_2} \tag{4.29}$$

$$Q = \frac{\sqrt{R_1C_1C_2R_2}}{R_1C_1 + R_2C_2} \tag{4.30}$$

$$\omega_0 = \frac{1}{\sqrt{R_1 C_1 R_2 C_2}} \tag{4.31}$$

We now assign arbitrary values to components and check all expressions in the Mathcad® sheet presented in Figure 4.10.

A brute-force expression is also derived and is simple in this case. The goal is to compare the response of both transfer functions—the *low-entropy* version in (4.28) versus the brute-force expression—and detect any discrepancy. You can see how I successively simplified and rearranged the formulas to end up with the final version. Each step is carefully tested versus the raw expression to highlight any typo during the process.

As confirmed in Figure 4.11, magnitude and phase responses between the final version and the reference transfer function are rigorously similar.

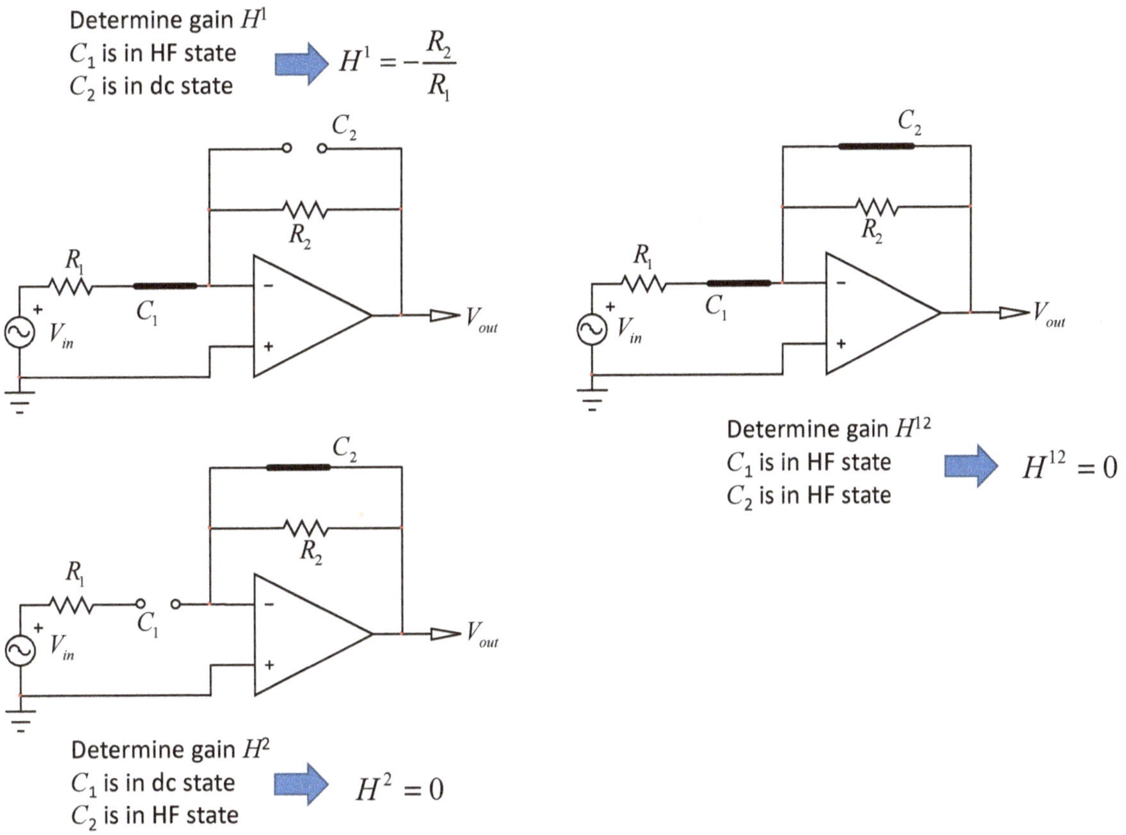

The Three High-Frequency Gains can be Inferred from Circuit Inspection

Figure 4.9

$R_1 := 10k\Omega$ $\quad C_1 := 2.2nF$ $\quad C_3 := 100nF$ $\quad \|(x,y) := \frac{x \cdot y}{x+y}$

$R_2 := 10k\Omega$ $\quad C_2 := 22nF$ $\quad R_3 := 10k\Omega$

$H_0 := 0$

$\tau_1 := R_1 \cdot C_1 = 22\cdot\mu s$ $\quad \tau_3 := R_3 \cdot C_3$

$\tau_2 := R_2 \cdot C_2 = 220\cdot\mu s$

$b_1 := \tau_1 + \tau_2 = 2.42\times 10^{-4}\cdot s$

$\tau_{12} := C_2 \cdot R_2 = 220\cdot\mu s$

$b_2 := \tau_1 \cdot \tau_{12} = 4.84\times 10^{3}\cdot\mu s^2$

$D_1(s) := 1 + s\cdot b_1 + s^2\cdot b_2$

$H_1 := -\frac{R_2}{R_1}$ $\quad H_2 := 0$ $\quad H_{12} := 0$ High-frequency gains

$N_1(s) := H_0 + s\cdot(H_1\cdot\tau_1 + H_2\cdot\tau_2) + s^2\cdot(H_{12}\cdot\tau_1\cdot\tau_{12})$

$H_{10}(s) := \frac{N_1(s)}{D_1(s)}$ $\quad H_{11}(s) := -\frac{s\cdot(R_2\cdot C_1)}{1 + s\cdot(R_1\cdot C_1 + R_2\cdot C_2) + s^2\cdot[R_1\cdot C_1\cdot(C_2\cdot R_2)]}$

$$H_{110}(s) := -\frac{(R_2\cdot C_1)}{(R_1\cdot C_1 + R_2\cdot C_2)}\cdot\frac{1}{1 + \frac{1}{[s\cdot(R_1\cdot C_1 + R_2\cdot C_2)]} + s^2\cdot\left[\frac{R_1\cdot C_1\cdot(C_2\cdot R_2)}{s\cdot(R_1\cdot C_1 + R_2\cdot C_2)}\right]}$$

$Q_1 := \frac{\sqrt{b_2}}{b_1} = 0.28748$ $\quad \omega_{00} := \frac{1}{\sqrt{b_2}}$ $\quad f_0 := \frac{\omega_{00}}{2\cdot\pi} = 2.28769kHz$

$Q := \frac{\sqrt{R_1\cdot C_1\cdot(C_2\cdot R_2)}}{R_1\cdot C_1 + R_2\cdot C_2} = 0.28748$ $\quad \omega_0 := \frac{1}{\sqrt{R_1\cdot C_1\cdot(C_2\cdot R_2)}}$

$H_0 := \frac{(R_2\cdot C_1)}{(R_1\cdot C_1 + R_2\cdot C_2)}$ $\quad 20\cdot\log(H_0) = -20.82785$ Mid-band gain

$H_{30}(s) := -H_0\cdot\frac{1}{1 + Q\cdot\left(\frac{\omega_0}{s} + \frac{s}{\omega_0}\right)}$ Final low-entropy transfer function

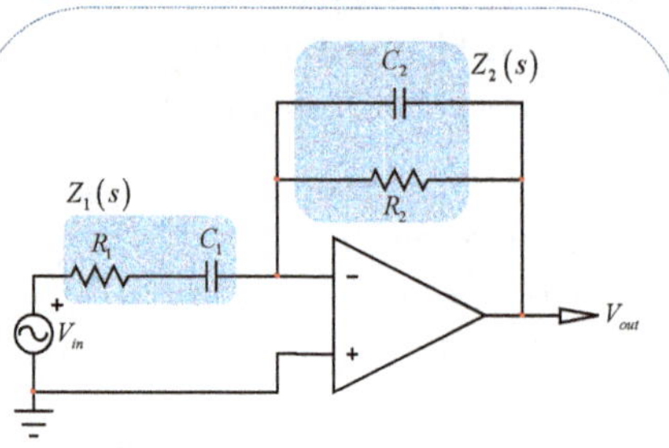

$Z_1(s) := R_1 + \frac{1}{s\cdot C_1}$ $\quad Z_2(s) := R_2 \,\|\left(\frac{1}{s\cdot C_2}\right)$

$H_{ref}(s) := -\frac{Z_2(s)}{Z_1(s)}$

Brute-force expression

A Simple Mathcad® Sheet Confirms our Calculations

Figure 4.10

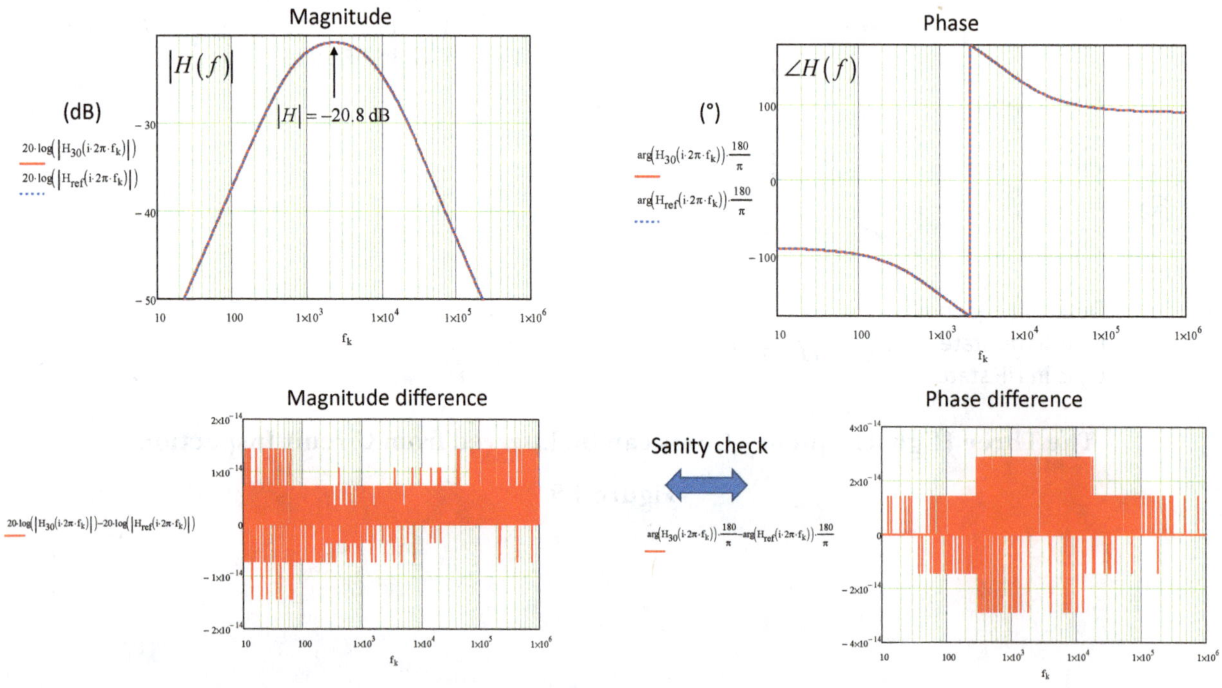

The Magnitude Plot Confirms this is a Bandpass Filter

Figure 4.11

We will close this section on the generalized transfer functions with a 3rd-order example.

The expression describing this type of structure [1] is given by:

$$H(s) = \frac{H_0 + s\left(\tau_1 H^1 + \tau_2 H^2 + \tau_3 H^3\right) + s^2\left(\tau_1\tau_2^1 H^{12} + \tau_1\tau_3^1 H^{13} + \tau_2\tau_3^2 H^{23}\right) + s^3\tau_1\tau_2^1\tau_3^{12} H^{123}}{1 + s\left(\tau_1 + \tau_2 + \tau_3\right) + s^2\left(\tau_1\tau_2^1 + \tau_1\tau_3^1 + \tau_2\tau_3^2\right) + s^3\tau_1\tau_2^1\tau_3^{12}} \quad (4.32)$$

To exercise this formula, let's have a look at the circuit shown in Figure 4.12 which associates three energy-storing elements.

The denominator will be of degree three, but how many zeroes are there?

If we set, in our head, each energy-storing element in its high-frequency state, we immediately see that only when C_1 and L_2 are respectively short-circuited and open-circuited does the stimulus reaches the output: we have a pair of zeroes involving these elements.

And because these two components act as dc blocks, the zeroes will be placed at the origin. This observation will become clearer when determining the high-frequency gains. Let's start for now with the dc gain H_0 obtained for $s = 0$.

As confirmed by Figure 4.13 it is zero considering the open-circuited capacitor C_1 and inductor L_2 shorting the node to ground.

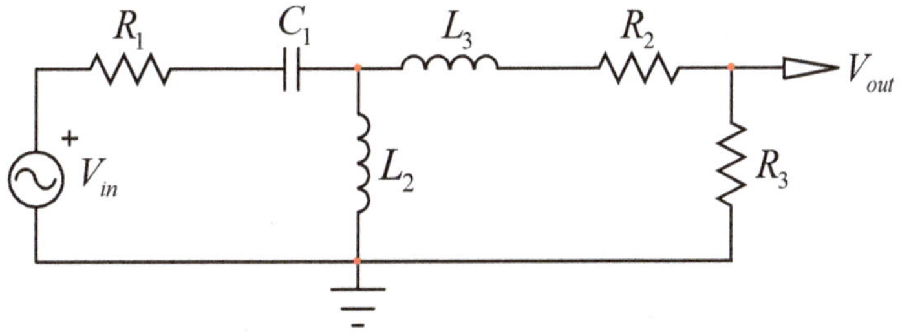

This 3rd-Order Filter Associates One Capacitor and Two Inductors

Figure 4.12

$$H_0 = 0 \quad (4.33)$$

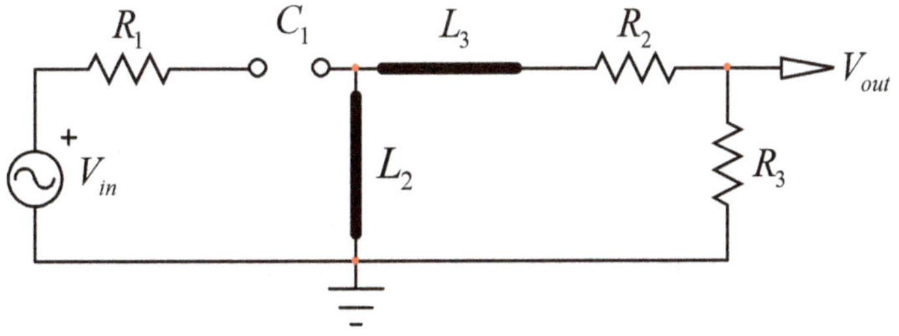

The Dc Gain is clearly Zero considering L_2 Shorting Path to Ground and C_1 which is Open Circuited

Figure 4.13

We will now determine the time constants of this circuit by turning the stimulus off: replace V_{in} by a wire and look through all energy-storing elements connecting terminals to determine the resistance R which drives them.

Nothing unsurmountable here and inspection works perfectly as long as you carefully isolate all individual cases. This is what I did in Figure 4.14, leading to the following denominator:

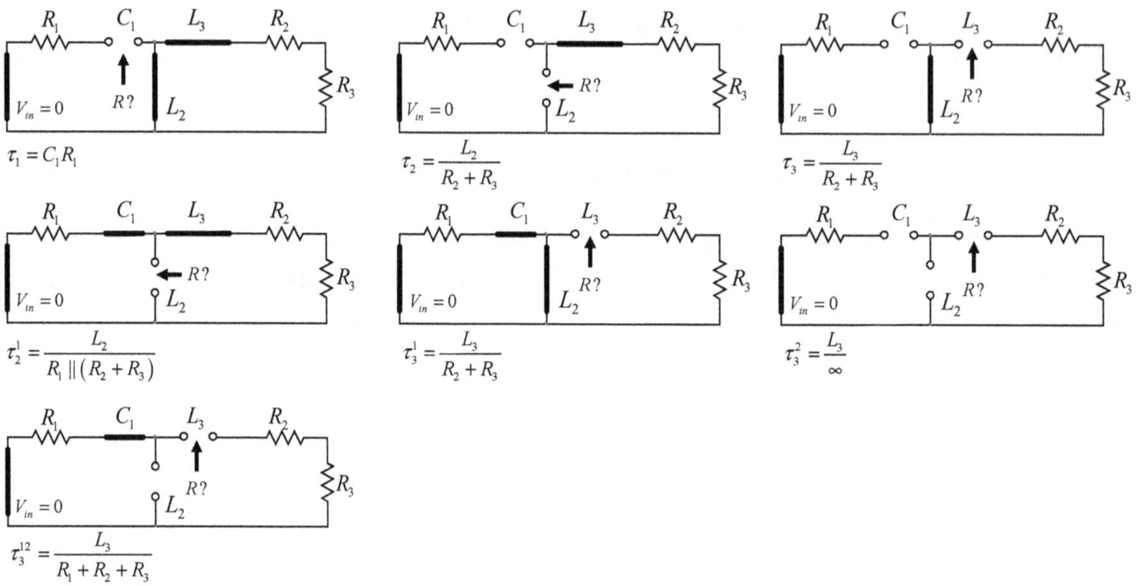

The Time Constants Can Swiftly be Determined by Inspection

Figure 4.14

$$\begin{aligned}
D(s) &= 1+s(\tau_1+\tau_2+\tau_3)+s^2\left(\tau_1\tau_2^1+\tau_1\tau_3^1+\tau_2\tau_3^2\right)+s^3\tau_1\tau_2^1\tau_3^{12} \\
&= 1+s\left(R_1C_1+\frac{L_2}{R_2+R_3}+\frac{L_3}{R_2+R_3}\right)+s^2\left(R_1C_1\frac{L_2}{R_1\|(R_2+R_3)}+R_1C_1\frac{L_3}{R_2+R_3}+\frac{L_2}{R_2+R_3}\cdot 0\right) \\
&\quad +s^3R_1C_1\frac{L_2}{R_1\|(R_2+R_3)}\frac{L_3}{R_1+R_2+R_3} \\
&= 1+s\left(R_1C_1+\frac{L_2+L_3}{R_2+R_3}\right)+s^2\left(R_1C_1\frac{L_2}{R_1\|(R_2+R_3)}+R_1C_1\frac{L_3}{R_2+R_3}\right) \\
&\quad +s^3R_1C_1\frac{L_2}{R_1\|(R_2+R_3)}\frac{L_3}{R_1+R_2+R_3}
\end{aligned} \tag{4.34}$$

Now that we have the denominator, we can determine the high-frequency gains H in which some of the elements are set in their high-frequency state while the rest of them remain in dc state.

The complete process is shown in Figure 4.15 and follows the guidance given in Figure 4.4. Considering the network arrangement, it appears that only one combination provides a gain. This is when C_1 is a short and L_2 is set in its high-frequency state.

This is the gain designated as H^{12}:

$$H^{12} = \frac{R_3}{R_1+R_2+R_3} \tag{4.35}$$

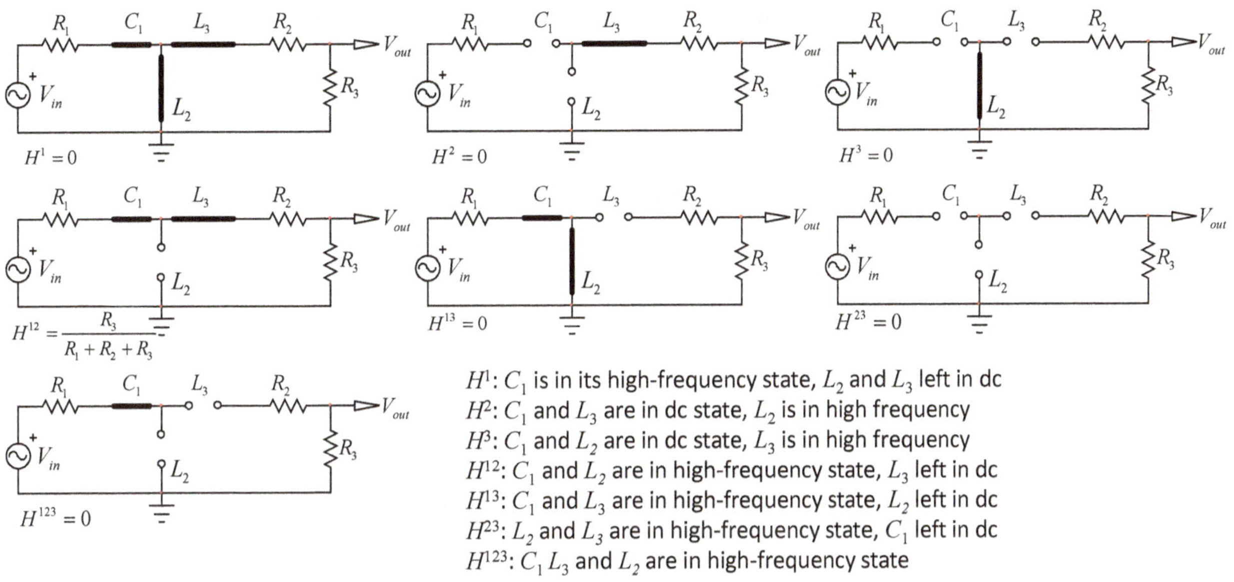
$H^1 = 0$
$H^2 = 0$
$H^3 = 0$
$H^{12} = \frac{R_3}{R_1 + R_2 + R_3}$
$H^{13} = 0$
$H^{23} = 0$
$H^{123} = 0$
H^1: C_1 is in its high-frequency state, L_2 and L_3 left in dc
H^2: C_1 and L_3 are in dc state, L_2 is in high frequency
H^3: C_1 and L_2 are in dc state, L_3 is in high frequency
H^{12}: C_1 and L_2 are in high-frequency state, L_3 left in dc
H^{13}: C_1 and L_3 are in high-frequency state, L_2 left in dc
H^{23}: L_2 and L_3 are in high-frequency state, C_1 left in dc
H^{123}: C_1 L_3 and L_2 are in high-frequency state

The High-Frequency Gains are all Zero except when C_1 and L_2 are in their High-Frequency State

Figure 4.15

The numerator is quickly assembled considering all the other gains equal to zero:

$$\begin{aligned} N(s) &= H_0 + s\left(\tau_1 H^1 + \tau_2 H^2 + \tau_3 H^3\right) + s^2\left(\tau_1\tau_2^1 H^{12} + \tau_1\tau_3^1 H^{13} + \tau_2\tau_3^2 H^{23}\right) + s^3\tau_1\tau_2^1\tau_3^{12} H^{123} \\ &= s^2\tau_1\tau_2^1 H^{12} = s^2 R_1 C_1 \frac{L_2}{R_1 \| (R_2 + R_3)} \frac{R_3}{R_1 + R_2 + R_3} \end{aligned} \tag{4.36}$$

And if you now develop and simplify this expression, you should obtain:

$$N(s) = s^2 \frac{C_1 L_2 R_3}{R_2 + R_3} \tag{4.37}$$

It confirms our initial guess that there are two zeroes at the origin which involve C_1 and L_2 in the numerator expression.

The complete transfer function of this filter is obtained by combining (4.34) and (4.37):

$$H(s) = \frac{s^2 \frac{C_1 L_2 R_3}{R_2 + R_3}}{1 + s\left(R_1 C_1 + \frac{L_2 + L_3}{R_2 + R_3}\right) + s^2\left(R_1 C_1 \frac{L_2}{R_1 \| (R_2 + R_3)} + R_1 C_1 \frac{L_3}{R_2 + R_3}\right) + s^3 R_1 C_1 \frac{L_2}{R_1 \| (R_2 + R_3)} \frac{L_3}{R_1 + R_2 + R_3}} \tag{4.38}$$

From this point, we could look at the various time constants in the denominator and see if $D(s)$ could be factored as a dominant pole followed by a second-order network, but this is out of the scope for now. What can be done, though, is factor the numerator and the s^2 terms in $D(s)$ to form a leading term. This value will be the gain at the bandpass frequency. And if you now reduce R_2 to 0 Ω, you find the expression given in the 6th example of [2].

All these expressions have been gathered in a Mathcad® sheet to check the validity of the approach. As usual, the transfer function magnitude and phase are tested against a brute-force expression involving Thévenin's theorem.

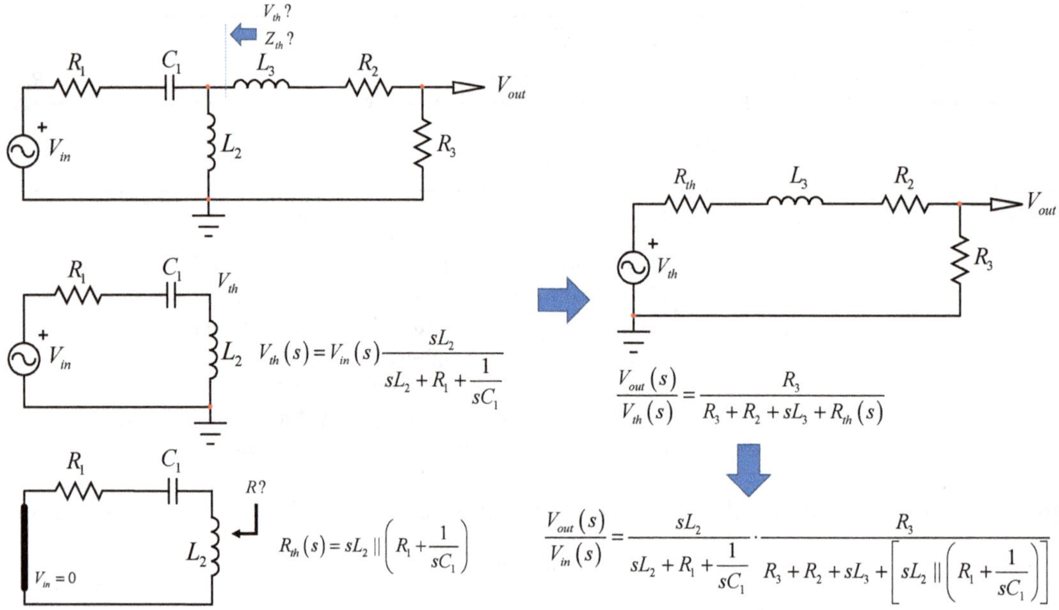

The Brute-Force Expression is Determined using Thévenin's Theorem

Figure 4.16

To obtain this latter, simply determine the unloaded voltage V_{th} and the output resistance R_{th} across L_2. Then consider an impedance divider involving the other elements as shown in Figure 4.16.

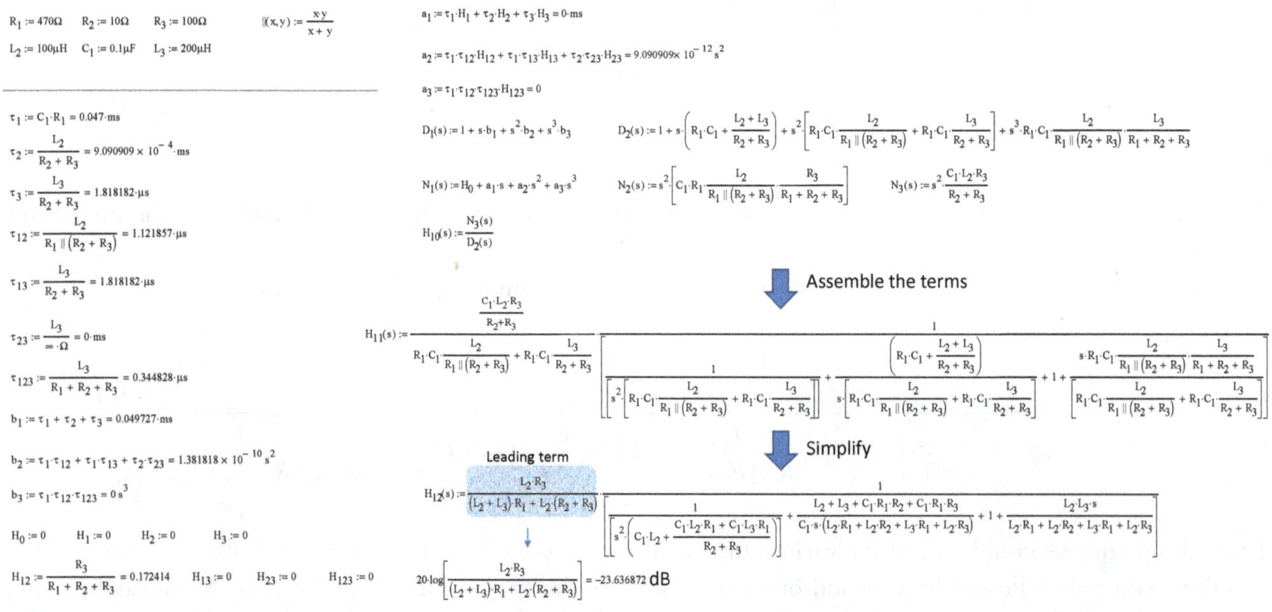

This Sheet Confirms our Approach is Correct and Leads to the Right Expression

Figure 4.17

Needless to say, trying to expanding this expression would require some work and then even more to rearrange it properly.

On the contrary, as you've seen in the lines before, the transfer function of this 3rd-order network was obtained painlessly with the FACTs.

No lines of algebra were written and the result from (4.38) is already factored.

Figure 4.17 gathers all the expressions and gives all necessary details, including the leading term factorization.

Finally, the ac response appears in Figure 4.18 and confirms the results are valid.

The generalized transfer function offers a powerful approach compared to NDI.

I have applied it in many occasions where NDI was too complicated or less obvious in the first place.

Figure 4.19 shows how to build numerators up to the order 4.

When H_0 is different than zero, of course, it has to be factored and placed as a leading term.

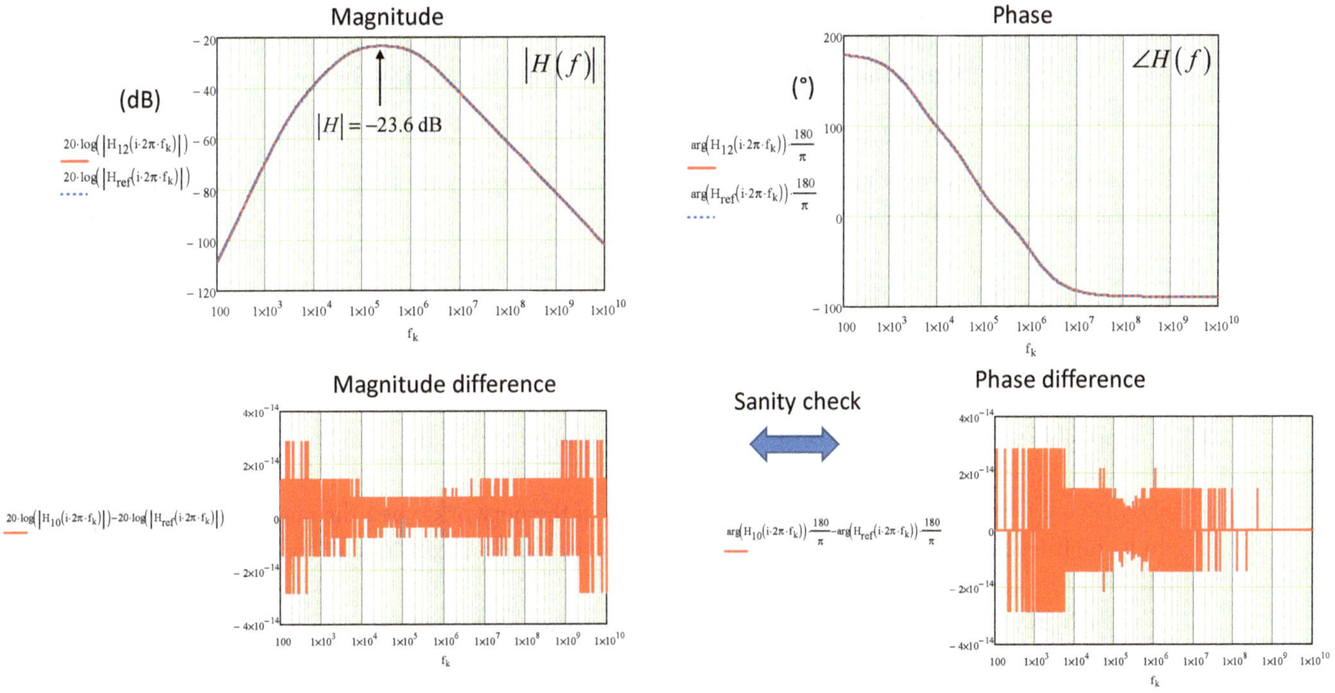

All Curves Superimpose, so Responses from the Factored Equation and the Brute-Force Expression are Similar

Figure 4.18

1st order $N(s) = H_0 + H^1\tau_1 s$ with $D(s) = 1 + s\tau_1$

2nd order $N(s) = H_0 + \left(\tau_1 H^1 + \tau_2 H^2\right)s + \left(\tau_1 \tau_2^1 H^{12}\right)s^2 = H_0 + \left(\tau_1 H^1 + \tau_2 H^2\right)s + \left(\tau_2 \tau_1^2 H^{21}\right)s^2$

3rd order $N(s) = H_0 + s\left(\tau_1 H^1 + \tau_2 H^2 + \tau_3 H^3\right) + s^2\left(\tau_1 \tau_2^1 H^{12} + \tau_1 \tau_3^1 H^{13} + \tau_2 \tau_3^2 H^{23}\right) + s^3\left(\tau_1 \tau_2^1 \tau_3^{12} H^{123}\right)$

4th order
$$\begin{aligned} N(s) = {} & H_0 + s\left(\tau_1 H^1 + \tau_2 H^2 + \tau_3 H^3 + \tau_4 H^4\right) \\ & + s^2\left(\tau_1 \tau_2^1 H^{12} + \tau_1 \tau_3^1 H^{13} + \tau_1 \tau_4^1 H^{14} + \tau_2 \tau_3^2 H^{23} + \tau_2 \tau_4^2 H^{24} + \tau_3 \tau_4^3 H^{34}\right) \\ & + s^3\left(\tau_1 \tau_2^1 \tau_3^{12} H^{123} + \tau_1 \tau_2^1 \tau_4^{12} H^{124} + \tau_1 \tau_3^1 \tau_4^{13} H^{134} + \tau_2 \tau_3^2 \tau_4^{23} H^{234}\right) \\ & + s^4\left(\tau_1 \tau_2^1 \tau_3^{12} \tau_4^{123} H^{1234}\right) \end{aligned}$$

Generalized Transfer Functions up to the 4th-Order

Figure 4.19

This last example ends our introduction on the fast analytical circuits techniques and the next chapters will go through numerous solved examples.

4.2 What Should I Retain from this Chapter?

In this last chapter, we learned key information:

1. It is possible to reuse the time constant already determined for the denominator to express the numerator.

2. To proceed with this approach, determine the gain of the network when the energy-storing element is set in its high-frequency state: a short circuit for a capacitor and an open circuit for an inductor.

3. This high-frequency gain is then combined with the natural time constant to form the generalized transfer function expression.

4. If this method indeed helps determining the zero position without resorting to an NDI, the final result may need further energy to simplify the obtained result. This penalty is associated with the method.

5. The generalized expression can be extended to n^{th}-order networks without difficulty.

4.3 References

1. C. Basso, *Linear Circuit Transfer Functions—An Introduction to Fast Analytical Techniques*, Wiley, 2016.
2. A. Ajimiri, *Generalized Time- and Transfer-Constant Circuit Analysis*, IEEE Transactions on Circuits and Systems, Vol. 57, NO. 6, June 2010 (https://chic.caltech.edu/wp-content/uploads/2014/02/Final-Paper.pdf)

Chapter 5: First-Order Transfer Functions

IN THE PREVIOUS chapter, I laid the foundations for identifying time constants and determining transfer functions in active and passive linear circuits featuring energy-storing elements. In this new chapter, I will document many examples without going through the details of each step in the text for the sake of keeping a compact format. Rather, the presentation will be a simple picture including all the iterations and sketches for a self-guided exploration of the derivation process. Additional comments, if necessary, will be provided in the text below the figure. The goal here is not to provide an exhaustive study of 1st-order circuits—there is an infinity of them—but rather to acquaint you with the technique so you can solve your specific problem.

Please note that, sometimes, the FACTs are not *always* the fastest and most convenient way for determining the transfer function of some very simple circuits. For this reason—and only for the simplest arrangements—I added some extra figures where rearranging the brute-force expression is faster than the FACTs. I however purposely applied the method to these basic networks anyway, so you acquire the skill *petit à petit* (step-by-step) and later excel with more complicated examples. Remember that obtaining an expression is one thing, but rearranging it the right way, in a *low-entropy* format, for design-oriented purposes (D-OA) represents the ultimate goal of FACTs.

As always, engineering judgement is important to assess which method leads to the response in the most straightforward way.

5.1 A Set of Three Expressions

When solving our 1st-order examples, we will have the choice between the three *low-entropy* expressions illustrated in Figure 5.1:

$$H(s)=H_0\frac{1+s\tau_N}{1+s\tau_D}=H_0\frac{1+\frac{s}{\omega_z}}{1+\frac{s}{\omega_p}}$$

H_0 represents the dc gain of the transfer function H obtained for $s = 0$. For an impedance Z, you would use R_0, for an admittance Y, Y_0 and so on.

τ_N represents the time constant for the zero
τ_D represents the time constant for the pole

Time constants are linked with zero and poles positions:

$$\omega_z=\frac{1}{\tau_N}\qquad \omega_p=\frac{1}{\tau_D}$$

$$H(s)=H_\infty\frac{1+\frac{1}{s\tau_N}}{1+\frac{1}{s\tau_D}}=H_\infty\frac{1+\frac{\omega_z}{s}}{1+\frac{\omega_p}{s}}$$

In cases where there is no dc value, because there is a zero at the origin for example, you may resort to this expression which uses H_∞ obtained for $s\to\infty$. In this approach, the time constants remain the same, it is just the expression which now uses *inverted* poles and zeroes. Some circuits can have a gain in dc but also in high frequency. In this case, both expressions are identical depending on the feature you want to highlight.

$$H(s)=\frac{H_0+sH^1\tau_D}{1+s\tau_D}$$

There are examples where it is difficult to determine the time constant τ_N using an NDI. It is then advantageous to use the generalized version which factors the time constant τ_D already determined for the denominator. The high-frequency gain H^1 is determined by setting the energy-storing element in its high-frequency mode. When H_0 is non-null, it can be factored as the leading term:

$$H(s)=H_0\frac{1+s\frac{H^1}{H_0}\tau_D}{1+s\tau_D}=H_0\frac{1+s\tau_N}{1+s\tau_D}=H_0\frac{1+\frac{s}{\omega_z}}{1+\frac{s}{\omega_p}}$$

$$\omega_z=\frac{1}{\tau_N}=\frac{H_0}{H^1\tau_D}\qquad \omega_p=\frac{1}{\tau_D}$$

The Three Expressions Describe a 1st-Order Transfer Function

Figure 5.1

The left-side expression depicts a circuit where the dc gain H_0 exists as in a low-pass filter for instance. In this case, you can determine the gain for $s = 0$ and the time constants are assembled to build the transfer function. In examples where the dc gain is zero, for instance in circuits where a capacitor appears in series with the stimulus, H_0 does not exist. In this case, it is simpler to determine the reference gain when s approaches infinity and use the middle expression. The latter implements so-called *inverted* pole and zero, factoring the time constants differently. If you remember, we have seen this expression in Chapter 1—equation (1.61)—when determining an impedance and when required in our list of examples.

Finally, when the determination of the zero is complicated or not obvious with a null double injection (NDI), the right-side generalized transfer function may be the quickest way to go. In that case, if the dc gain exists, you will advantageously factor it as a leading term as with the other expressions. In the end, you may need to rework the zero expression as this approach naturally complicates it by combining terms absent from in an NDI. However, all zeroes—determined by NDI, inspection or the generalized transfer function—are rigorously similar.

Once you acquire the skill, you will opportunistically select the formula which suits you the best in terms of simplicity and speed for the circuit you study. This is what I did in the documented examples. You can exercise the different formulas as part of the learning process.

5.2   Circuits with one Energy-Storing Element

This is the simplest *RC* circuit shown in Figure 5.2. I used the classical impedance divider expression in the first transfer function then paralleled and summed impedances respectively for the output and input expressions. You can see how I rearranged the expressions with a leading term when it was possible.

Please note the presence of the inverted zero in the input impedance expression which helps obtaining the most compact *low-entropy* expression.

Voltage gain, V_{out}/V_{in}

$$\frac{V_{out}(s)}{V_{in}(s)} = \frac{\frac{1}{sC_1}}{\frac{1}{sC_1} + R_1} = \frac{1}{1 + sR_1C_1} = \frac{1}{1 + \frac{s}{\omega_p}}$$

$$\omega_p = \frac{1}{R_1C_1}$$

(a)

Output impedance: short V_{in}, look at the resistance offered by the output port

$$Z_{out}(s) = \left(\frac{1}{sC_1}\right) \| R_1 = \frac{\frac{1}{sC_1} \cdot R_1}{\frac{1}{sC_1} + R_1} = \frac{R_1}{1 + sR_1C_1} = R_0 \frac{1}{1 + \frac{s}{\omega_p}}$$

$$\omega_p = \frac{1}{R_1C_1}$$

$$R_0 = R_1$$

(b)

Input impedance: look at the impedance offered by the input port

$$Z_{in}(s) = \left(\frac{1}{sC_1}\right) + R_1 = \frac{1 + sR_1C_1}{sR_1C_1} = \frac{sR_1C_1}{sR_1C_1} \cdot \frac{1 + \frac{1}{sR_1C_1}}{1} = R_\infty \left(1 + \frac{\omega_z}{s}\right)$$

$$\omega_z = \frac{1}{R_1C_1}$$

$$R_\infty = R_1$$

(c)

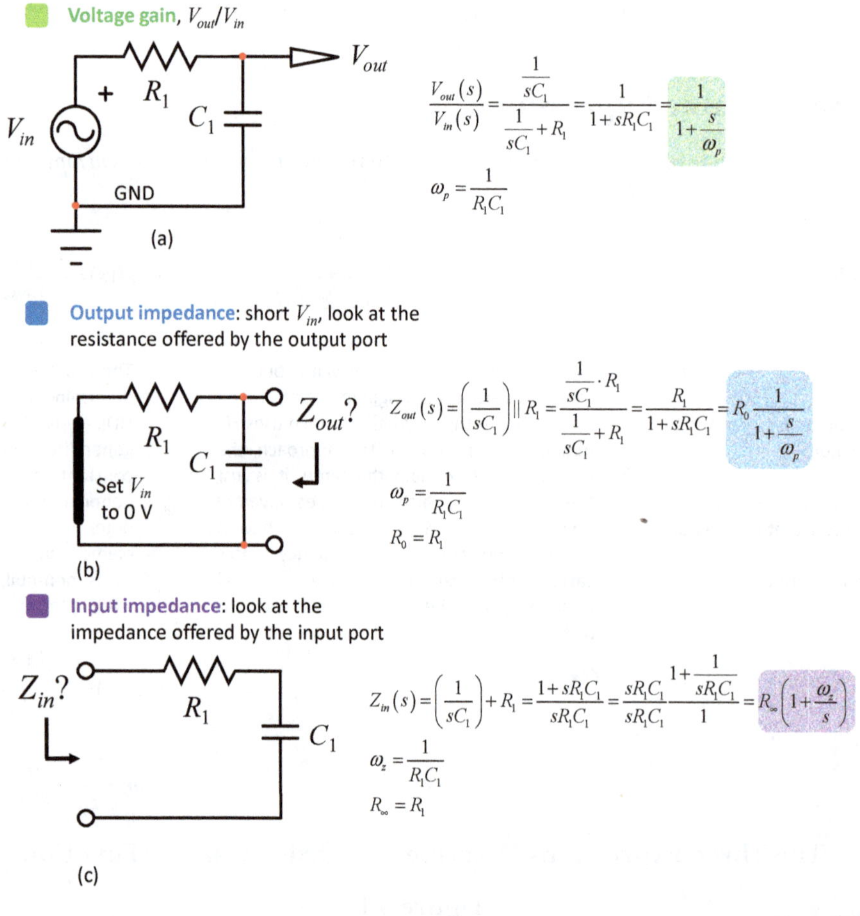

This Simple *RC* Filter Starts Analyses with Three Associated Transfer Functions

Figure 5.2

Let's now apply the FACTs instead in Figure 5.3. We start with sketch (a) by finding the dc gain and proceed with the time constant in (c). Then we check for the presence of a zero by setting C_1 in its high-frequency state. The response disappears in (d) with an input stimulus back in place: there is no zero. We find the pole by setting the source to 0 V and write our first transfer function in the correct format. You recognize a low-pass filter where the gain goes to zero as s approaches infinity.

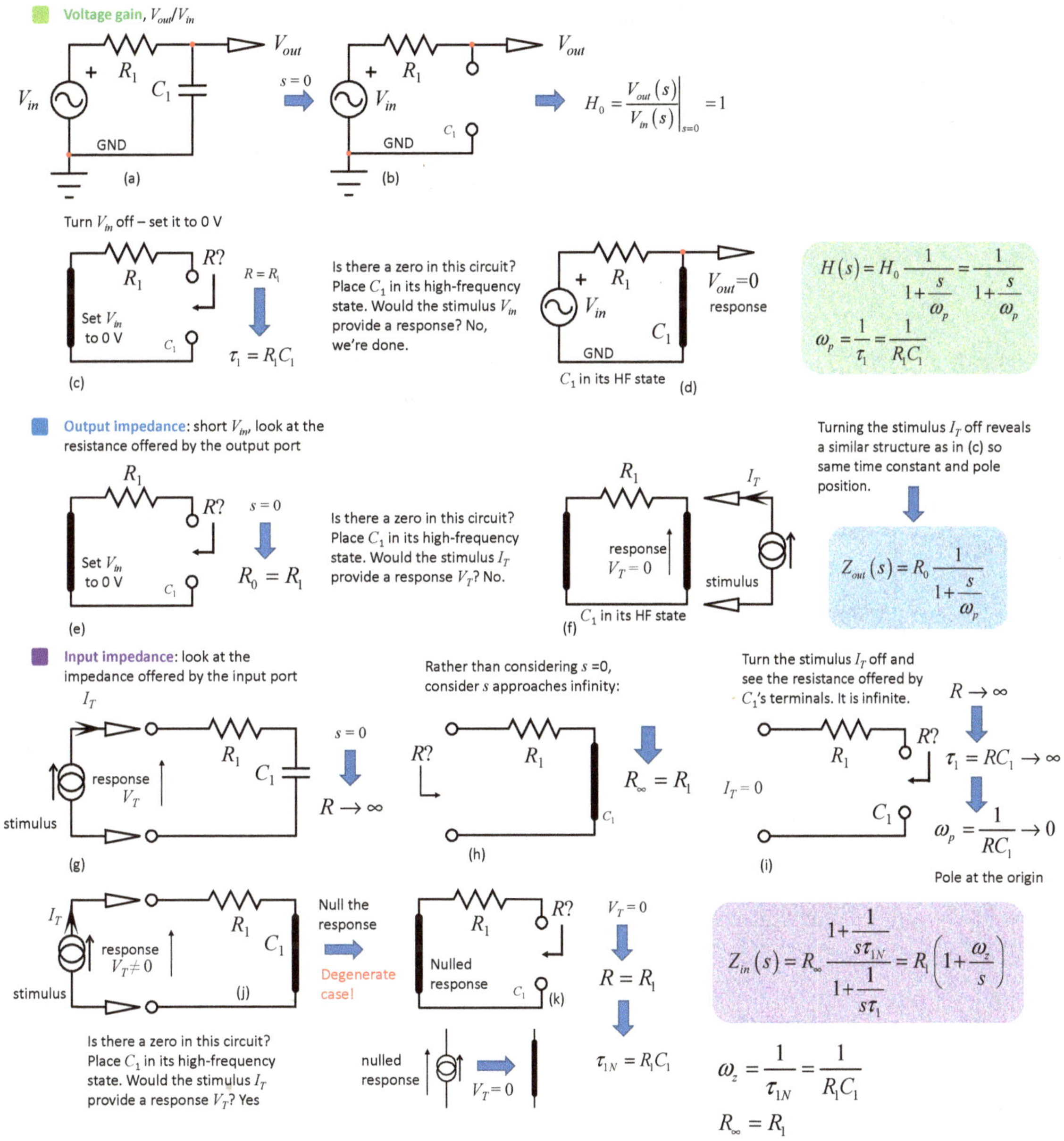

Voltage gain, V_{out}/V_{in}
$H_0 = \left.\frac{V_{out}(s)}{V_{in}(s)}\right|_{s=0} = 1$
Turn V_{in} off – set it to 0 V
Set V_{in} to 0 V
$\tau_1 = R_1C_1$
Is there a zero in this circuit? Place C_1 in its high-frequency state. Would the stimulus V_{in} provide a response? No, we're done.
C_1 in its HF state
$H(s) = H_0\frac{1}{1+\frac{s}{\omega_p}} = \frac{1}{1+\frac{s}{\omega_p}}$
$\omega_p = \frac{1}{\tau_1} = \frac{1}{R_1C_1}$
Output impedance: short V_{in}, look at the resistance offered by the output port
$R_0 = R_1$
Is there a zero in this circuit? Place C_1 in its high-frequency state. Would the stimulus I_T provide a response V_T? No.
Turning the stimulus I_T off reveals a similar structure as in (c) so same time constant and pole position.
$Z_{out}(s) = R_0\frac{1}{1+\frac{s}{\omega_p}}$
Input impedance: look at the impedance offered by the input port
Rather than considering s =0, consider s approaches infinity:
Turn the stimulus I_T off and see the resistance offered by C_1's terminals. It is infinite.
$R \to \infty$
$R_\infty = R_1$
$\tau_1 = RC_1 \to \infty$
$\omega_p = \frac{1}{RC_1} \to 0$
Pole at the origin
Null the response
Degenerate case!
Nulled response
$\tau_{1N} = R_1C_1$
$Z_{in}(s) = R_\infty\frac{1+\frac{1}{s\tau_{1N}}}{1+\frac{1}{s\tau_1}} = R_1\left(1+\frac{\omega_z}{s}\right)$
Is there a zero in this circuit? Place C_1 in its high-frequency state. Would the stimulus I_T provide a response V_T? Yes
nulled response
$\omega_z = \frac{1}{\tau_{1N}} = \frac{1}{R_1C_1}$
$R_\infty = R_1$

Applying the FACTs to this First Simple *RC* Filter Leads to the Result by Inspection

Figure 5.3

The output impedance is then observed by shorting the input port. In this mode, the dc output resistance offered when $s = 0$ is simply R_1. We check for the presence of a zero but, this time, the stimulus is the test generator I_T applied at the output port and the response is the voltage V_T developed across its connecting terminals. There is no response and voltage V_T is 0 V when C_1 becomes a short circuit: no zero. By turning off I_T to determine the pole, we return the structure to that of sketch (c) and we can thus reuse the previous denominator. We have our output impedance in which the leading term R_0 carries the unit in ohm as it should be.

For the input impedance, the stimulus I_T is now applied at the input port. In dc, because the output capacitor is open, the resistance is infinite. We then resort to the formula given in the middle of Figure 5.1 and find that the input resistance determined when C_1 is shorted (s approaches infinity) is simply R_1. The zero is obtained when the response V_T is nulled which implies a 0 V across the current generator. This is a degenerate case and the generator can be replaced by a wire in sketch (k). The time constant is immediate in this mode. The pole is different than with the two other expressions because turning off the generator leaves the left terminal of R_1 unconnected. You see an infinite resistance when the stimulus is turned off (a 0 A current source is an open circuit) leading to a pole set a 0 Hz: this is a pole at the origin. The transfer function is immediate and given in a compact form.

$R_1 := 2k\Omega \qquad C_1 := 10nF \qquad R_{inf} := 10^{12}\Omega \qquad \|(x,y) := \dfrac{xy}{x+y}$

$H_0 := 1 \qquad R_i := R_1 \qquad H_1(s) := H_0 \dfrac{1}{1+\frac{s}{\omega_p}} \qquad Z_{out}(s) := R_1 \dfrac{1}{1+\frac{s}{\omega_p}} \qquad Z_{in}(s) := R_i\left(1+\frac{\omega_z}{s}\right)$

$\tau_1 := R_1\cdot C_1 = 20\,\mu s \qquad \omega_p := \dfrac{1}{\tau_1} \qquad f_p := \dfrac{\omega_p}{2\cdot\pi} = 7.95775 kHz$

$\tau_{1N} := C_1\cdot R_1 \qquad \omega_z := \dfrac{1}{\tau_{1N}} \qquad f_z := \dfrac{\omega_z}{2\cdot\pi} = 7.95775 kHz \qquad H_{ref}(s) := \dfrac{\frac{1}{s\cdot C_1}}{R_1+\frac{1}{s\cdot C_1}} \qquad Z_{outR}(s) := R_1 \| \left(\frac{1}{s\cdot C_1}\right) \qquad Z_{inR}(s) := R_1 + \frac{1}{s\cdot C_1}$

Brute-force expressions

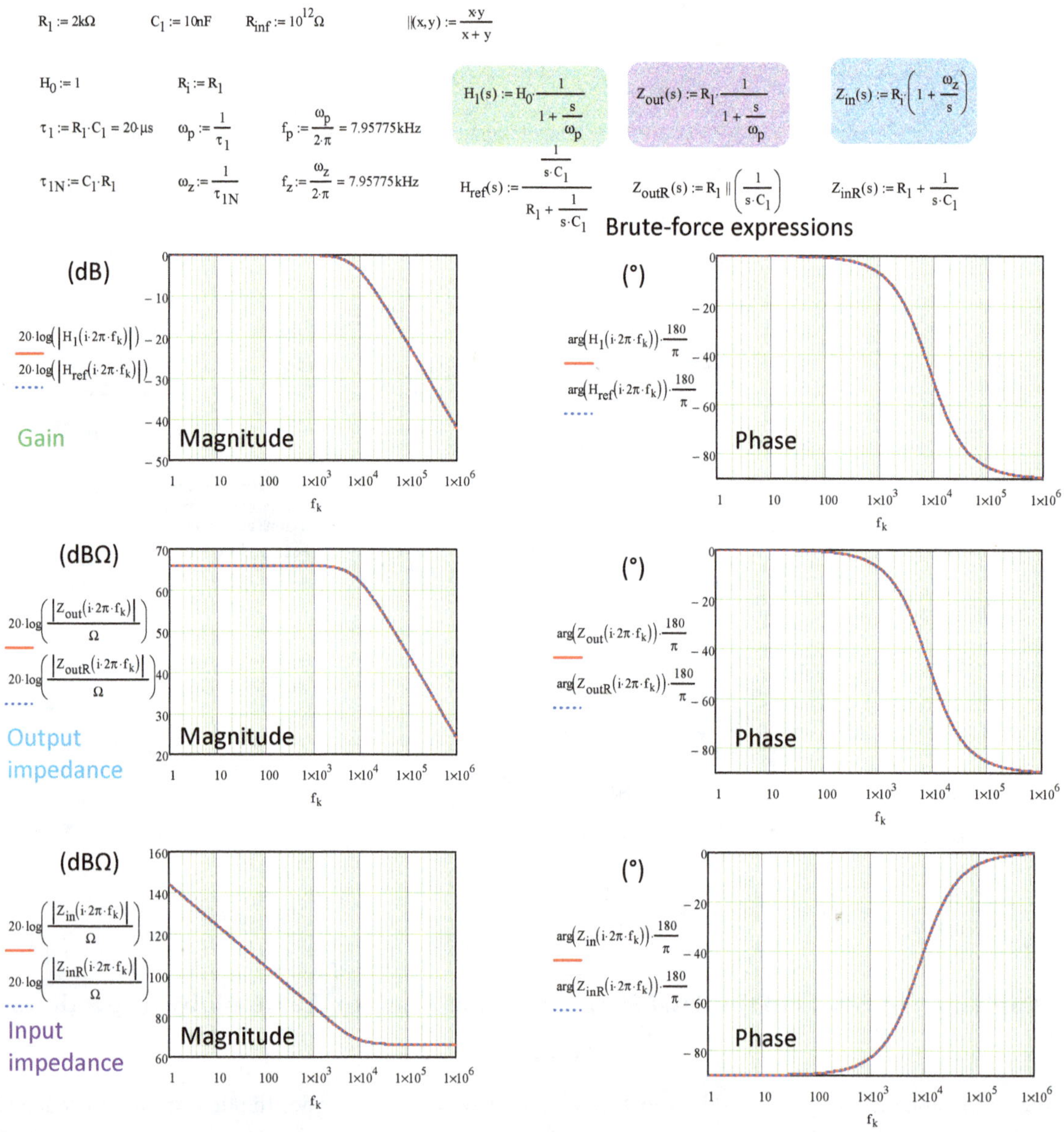

The Three Transfer Functions are Easily Plotted with a Mathematical Solver

Figure 5.4

Figure 5.4 plots the frequency responses of these three first transfer functions where all results are gathered.

The graphs plot FACTs and brute-force expressions which should be identical in magnitude and phase.

The second example swaps the resistance and the capacitor. It is a *CR* differentiator and appears in Figure 5.6. The classical method is applied first but you can already see in the first transfer function how an inverted pole helps factoring the expression in a real compact way.

In this circuit, output and input impedance are similar to the *RC* circuit we have tackled in our first example.

Voltage gain, V_{out}/V_{in}

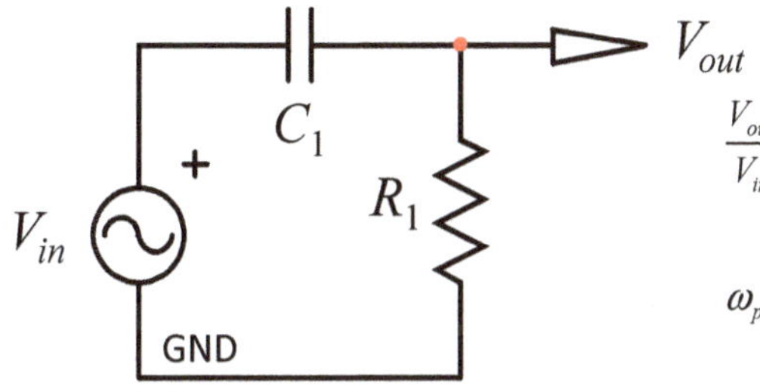

$$\frac{V_{out}(s)}{V_{in}(s)}=\frac{R_1}{\frac{1}{sC_1}+R_1}=\frac{sC_1R_1}{1+sR_1C_1}=\frac{sR_1C_1}{sR_1C_1}\frac{1}{1+\frac{1}{sR_1C_1}}=\frac{1}{1+\frac{\omega_p}{s}}$$

$$\omega_p=\frac{1}{R_1C_1}$$

(a)

Output impedance: short V_{in}, look at the resistance offered by the output port

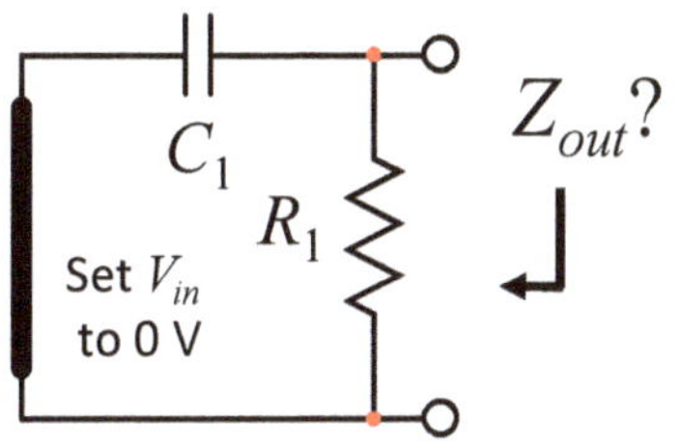

$$Z_{out}(s)=\left(\frac{1}{sC_1}\right)\|R_1=\frac{\frac{1}{sC_1}\cdot R_1}{\frac{1}{sC_1}+R_1}=\frac{R_1}{1+sR_1C_1}=R_0\frac{1}{1+\frac{s}{\omega_p}}$$

$$\omega_p=\frac{1}{R_1C_1}$$

$$R_0=R_1$$

(b)

Input impedance: look at the impedance offered by the input port

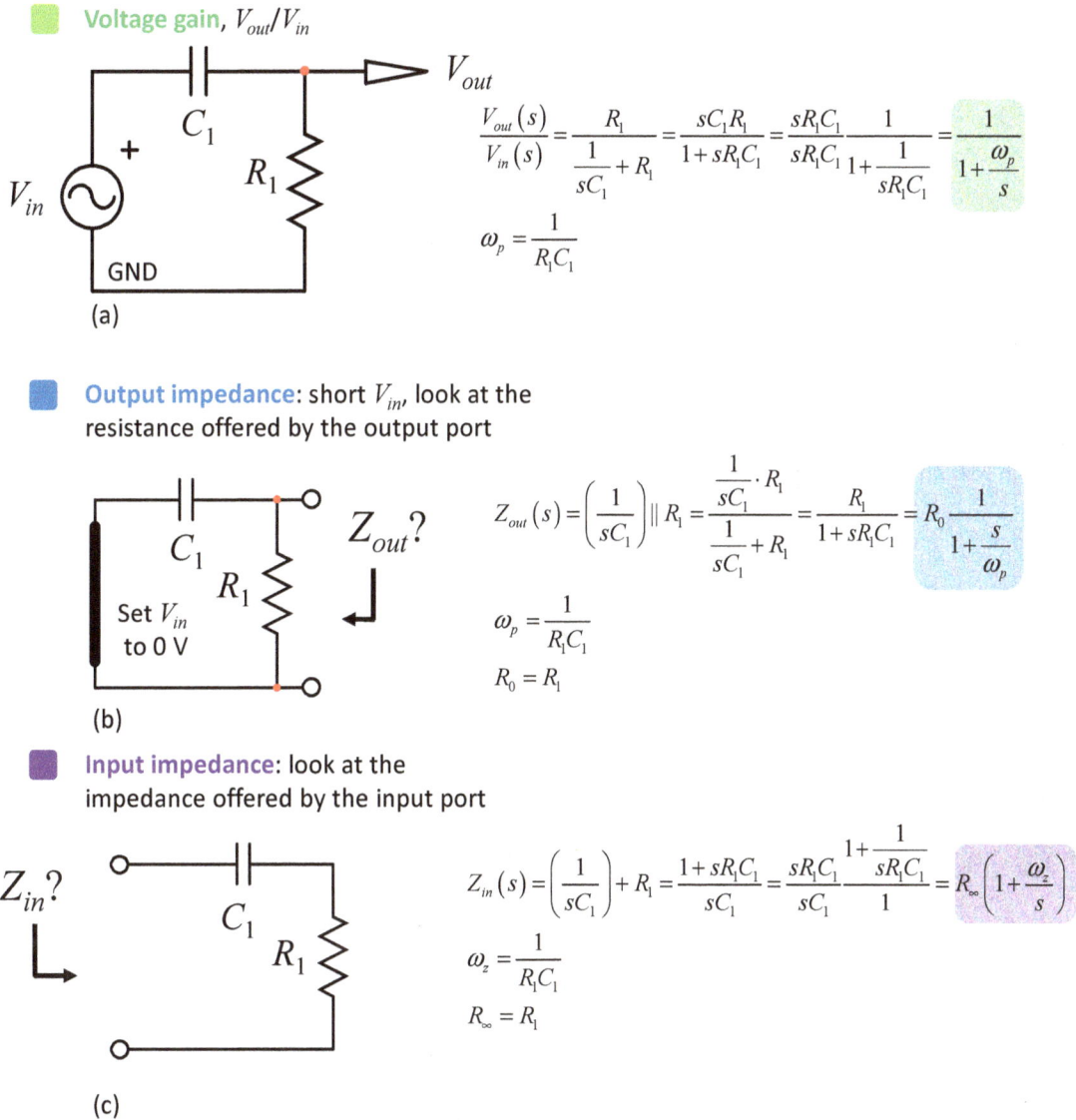

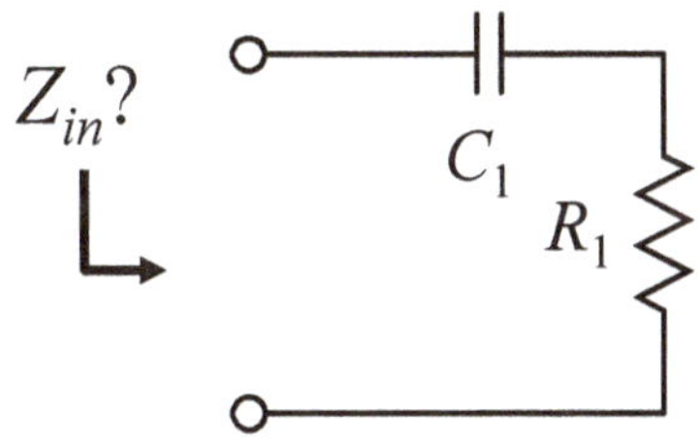

$$Z_{in}(s)=\left(\frac{1}{sC_1}\right)+R_1=\frac{1+sR_1C_1}{sC_1}=\frac{sR_1C_1}{sC_1}\frac{1+\frac{1}{sR_1C_1}}{1}=R_\infty\left(1+\frac{\omega_z}{s}\right)$$

$$\omega_z=\frac{1}{R_1C_1}$$

$$R_\infty=R_1$$

(c)

The Capacitor is Placed First and Blocks the Dc Component, Placing a Zero at the Origin

Figure 5.5

The FACTs approach follows a similar path as before, starting from dc conditions in which the series capacitor is open. The series capacitor sets the gain to zero in dc and we use the generalized transfer function expression from Figure 5.1. Here the expression is obtained quickly considering a high-frequency gain of 1. Once factored, we have an inverted pole illustrating a magnitude increasing from dc and zeroing when the pole kicks in. This is the zero at the origin blocking the dc component. For the output impedance, a current generator is placed across the output terminals. A quick examination does not show a zero but a pole similar to what we originally found in sketch (a). For the input impedance, considering the infinite value at dc, I observe the circuit for s approaching infinity. The impedance in this mode is R_1 and the rest comes down easily with an inverted pole.

All frequency responses are shown in Figure 5.7.

Voltage gain, V_{out}/V_{in}

$$H_0 = \frac{V_{out}(s)}{V_{in}(s)}\Big|_{s=0} = 0$$

0-dc gain, we can use the generalized transfer function.

Turn V_{in} off – set it to 0 V

$R = R_1$

$\tau_1 = R_1 C_1$

Is there a zero in this circuit? Place C_1 in its high-frequency state. Would the stimulus V_{in} provide a response? Yes, there is a zero. What is the gain in this mode? $H^1=1$

(d) C_1 in its HF state

$$H(s) = \frac{H_0 + sH^1\tau_1}{1+s\tau_1} = \frac{0+s\tau_1}{1+s\tau_1}$$
$$= \frac{s\tau_1}{s\tau_1}\frac{1}{1+\frac{1}{s\tau_1}} = \frac{1}{1+\frac{\omega_p}{s}}$$
$$\omega_p = \frac{1}{\tau_1} = \frac{1}{R_1 C_1}$$

Output impedance: short V_{in}, look at the resistance offered by the output port

Set V_{in} to 0 V

$s=0$

$R_0 = R_1$

Is there a zero in this circuit? Place C_1 in its high-frequency state. Would the stimulus I_T provide a response V_T? No.

(f) C_1 in its HF state

Turning the stimulus I_T off reveals a similar structure as in (c) so same time constant and pole position.

$$Z_{out}(s) = R_0 \frac{1}{1+\frac{s}{\omega_p}}$$

Input impedance: look at the impedance offered by the input port

(g)

$s=0$

$\frac{V_T}{I_T} \to \infty$

Check R when s approaches infinity

$R_\infty = R_1$

(h)

Is there a zero in this circuit? Place C_1 in its high-frequency state. Would the stimulus I_T provide a response V_T? Yes

Null the response

Degenerate case!

Nulled response

nulled response $V_T=0$

$V_T=0$

$R = R_1$

$\tau_{1N} = R_1 C_1$

(i)

Turn the stimulus I_T off and see the resistance offered by C_1's terminals.

(j)

$R \to \infty$

$\tau_1 = RC_1 \to \infty$

$\omega_p = \frac{1}{RC_1} \to 0$

Pole at the origin

$$Z_{in}(s) = R_\infty \frac{1+\frac{1}{s\tau_{1N}}}{1+\frac{1}{s\tau_1}} = R_\infty\left(1+\frac{\omega_z}{s}\right)$$

$\omega_z = \frac{1}{\tau_{1N}} = \frac{1}{R_1 C_1}$ $R_\infty = R_1$

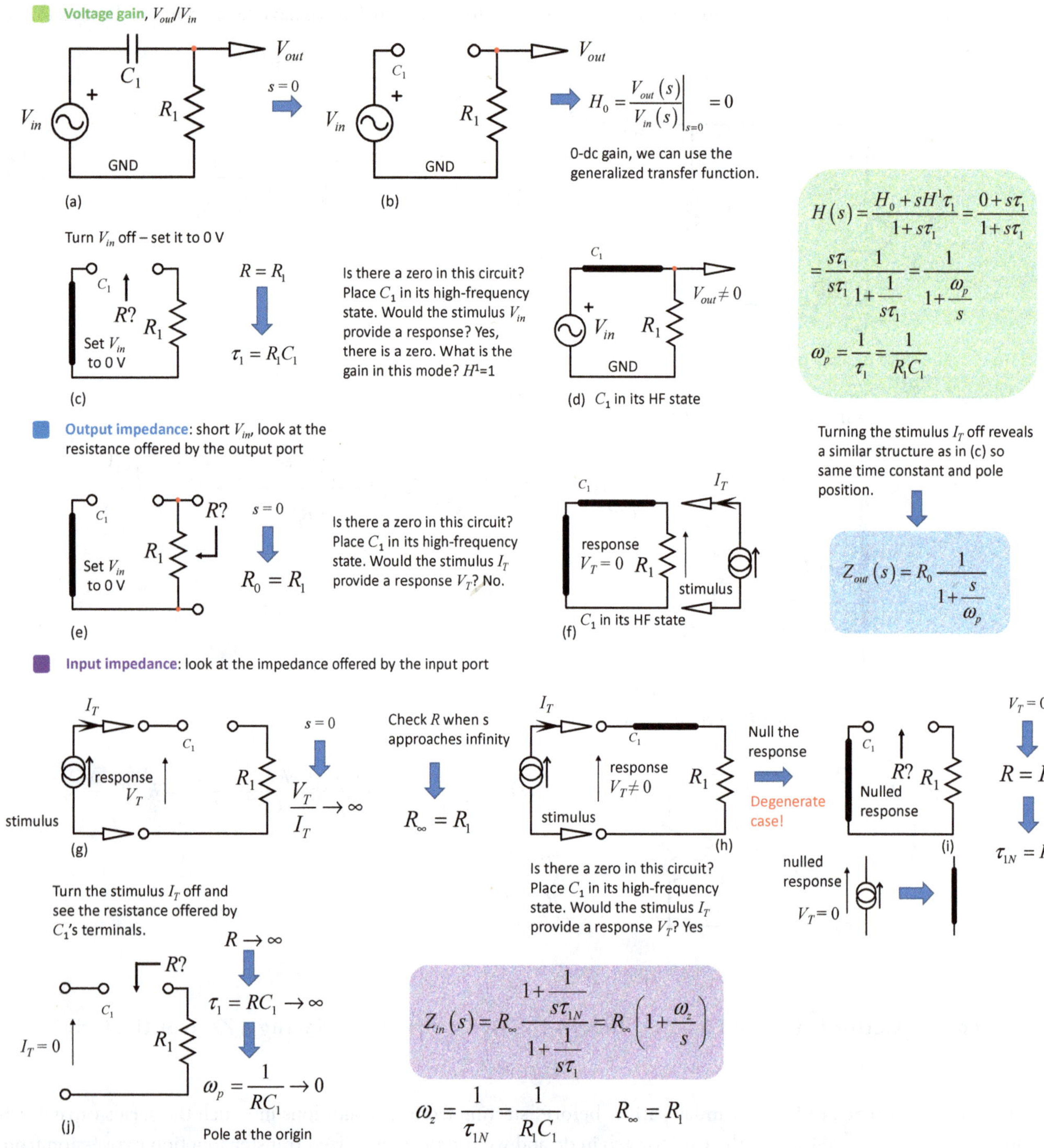

The Capacitor is Now Placed in Series with the Input Source

Figure 5.6

$R_1 := 2k\Omega$ $C_1 := 10nF$ $R_{inf} := 10^{12}\Omega$ $\|(x,y) := \dfrac{x \cdot y}{x + y}$

$H_1(s) := H_0 \cdot \dfrac{1}{1 + \dfrac{\omega_p}{s}}$ $Z_{out}(s) := R_0 \cdot \dfrac{1}{1 + \dfrac{s}{\omega_p}}$ $Z_{in}(s) := R_i \cdot \left(1 + \dfrac{\omega_p}{s}\right)$

$H_0 := 1$ $\tau_1 := R_1 \cdot C_1 = 20\,\mu s$ $\omega_p := \dfrac{1}{\tau_1}$ $f_p := \dfrac{\omega_p}{2 \cdot \pi} = 7.95775\,kHz$

$H_{ref}(s) := \dfrac{R_1}{R_1 + \dfrac{1}{s \cdot C_1}}$ $Z_{outR}(s) := R_1 \parallel \left(\dfrac{1}{s \cdot C_1}\right)$ $Z_{inR}(s) := R_1 + \dfrac{1}{s \cdot C_1}$

$R_0 := R_1$ $\tau_{1N} := R_1 \cdot C_1$ $\omega_z := \dfrac{1}{\tau_{1N}}$ $f_z := \dfrac{\omega_z}{2 \cdot \pi} = 7.95775\,kHz$

Brute-force expressions

$R_i := R_1$

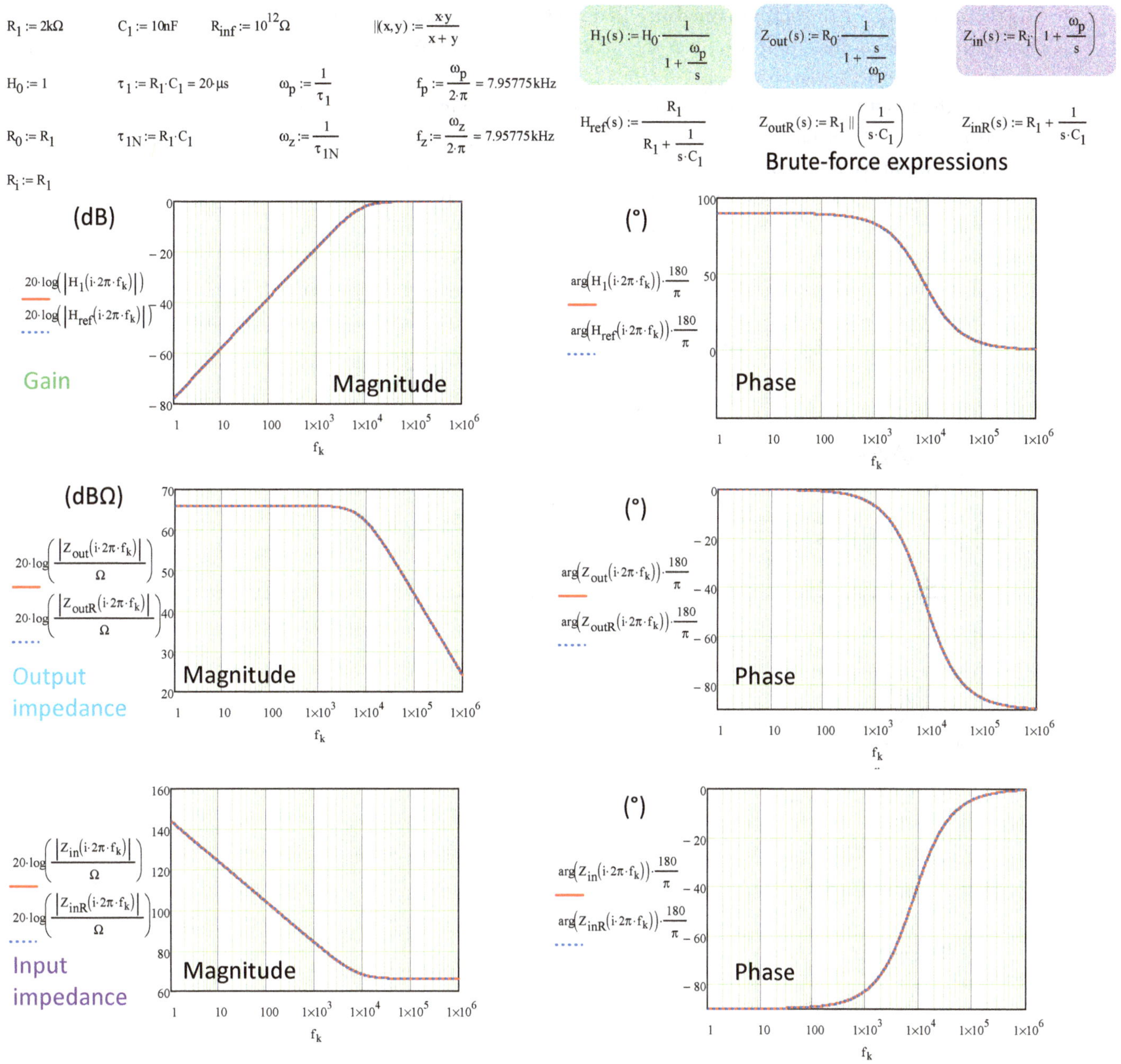

This New Transfer Function Features a Zero at the Origin Set by the Series Capacitor

Figure 5.7

Now that you have seen the principles in action, we start complicating things by adding a load resistance as shown in Figure 5.8. In this first approach, the parallel combination of the capacitor and resistor R_2 makes things less obvious, but Thévenin helps determine the expression in a swift way. Please note that expansion is now needed with its possible mistakes. The FACTs detailed in Figure 5.9 start making sense in that respect.

We begin with the dc gain H_0. By removing capacitor C_1, we have a resistive divider. The time constant is found immediately by reducing the input source to 0 V and looking through C_1's terminals. There is no zero since the response disappears if C_1 is set in its high-frequency state. The transfer function is assembled after sketch (d).

It's the same path for the output impedance first studied in dc when C_1 is open-circuited. There is no zero in Z_{out} and setting the stimulus I_T to 0 A returns the structure to that of sketch (c): the time constant and thus the pole is already determined, we are done.

For the input impedance, a current generator excites the input terminals and reveals a series connection of R_1 and R_2 for $s = 0$. A quick drawing reveals the presence of a zero—we will forget these small intermediate sketches as we move forward—and its location is found by nulling the response.

It is similar to replacing the current source by a wire: the resistance for the zero is easy and corresponds to the paralleling of R_1 and R_2. Finally, reduce the stimulus I_T to 0 A and determine the time constant for the pole position.

This is it; we have our three transfer functions obtained via a series of small sketches. We can now plot the ac response of our expressions and compare it to those of the brute-force approach determined in Figure 5.8.

Results shown in Figure 5.10 confirm all is good.

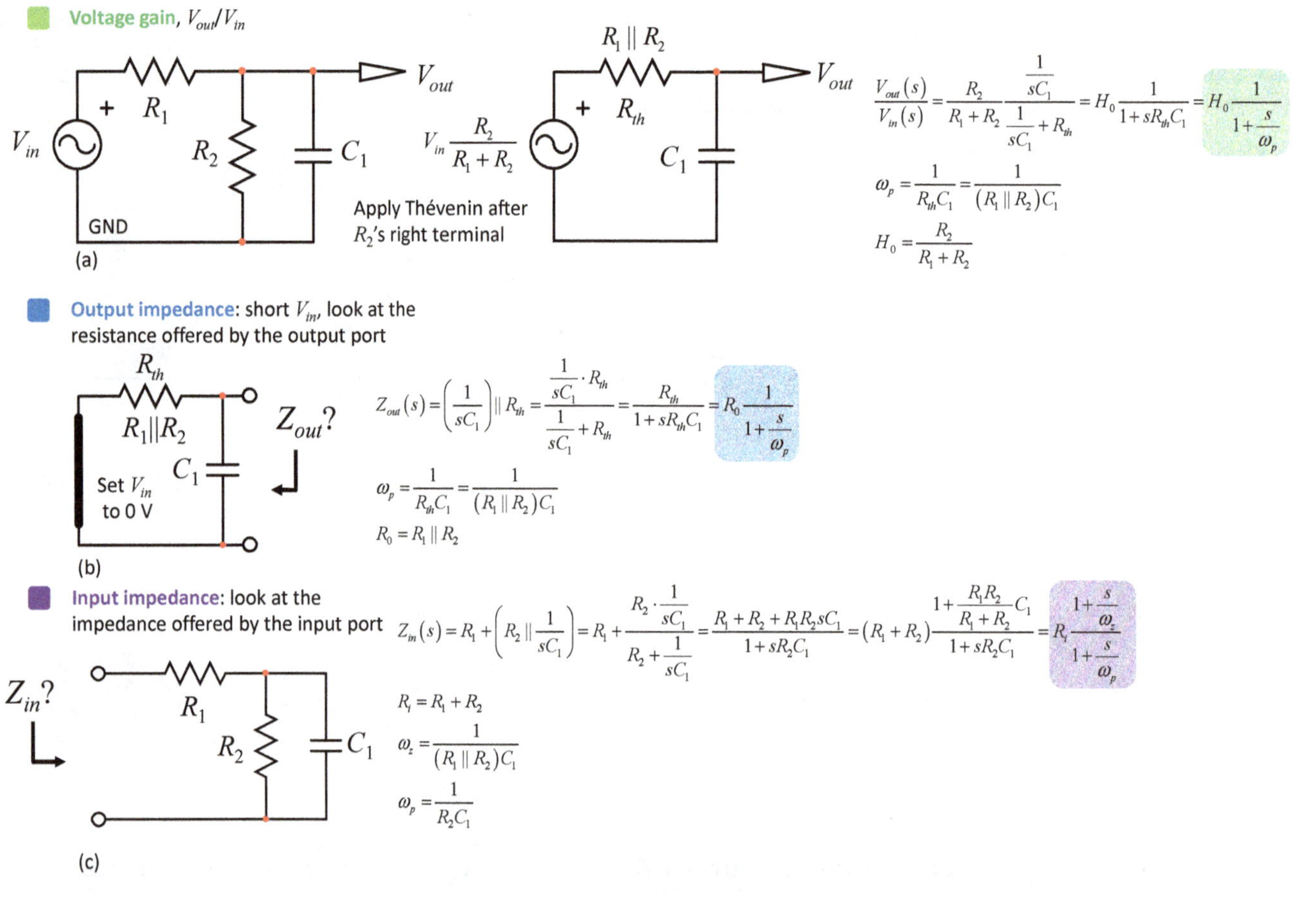

The Added Resistor Complicates the Network, but Thévenin Remains a Convenient Tool to Determine these Transfer Functions

Figure 5.8

Voltage gain, V_{out}/V_{in}

(a)

(b)

$$H_0 = \frac{V_{out}(s)}{V_{in}(s)}\Bigg|_{s=0} = \frac{R_2}{R_1 + R_2}$$

Turn V_{in} off – set it to 0 V

Set V_{in} to 0 V

$R = R_1 \| R_2$

$\tau_1 = (R_1 \| R_2)C_1$

(c)

Is there a zero in this circuit? Place C_1 in its high-frequency state. Would the stimulus V_{in} provide a response? No, we're done.

(d) C_1 in its HF state

$$H(s) = H_0 \frac{1}{1+s\tau_1} = H_0 \frac{1}{1+\dfrac{s}{\omega_p}}$$

$$\omega_p = \frac{1}{\tau_1} = \frac{1}{(R_1 \| R_2)C_1}$$

Output impedance: short V_{in}, look at the resistance offered by the output port

Set V_{in} to 0 V

$R?$

$R_0 = R_1 \| R_2$

(e)

Is there a zero in this circuit? Place C_1 in its high-frequency state. Would the stimulus I_T provide a response V_T? No.

(f) C_1 in its HF state

Turning the stimulus I_T off reveals a similar structure as in (c) so same time constant and pole position.

$$Z_{out}(s) = R_0 \frac{1}{1+\dfrac{s}{\omega_p}}$$

Input impedance: look at the impedance offered by the input port

stimulus / response V_T

$R_i = R_1 + R_2$

(g)

Is there a zero in this circuit? Place C_1 in its high-frequency state. Would the stimulus I_T provide a response V_T? Yes.

stimulus / response $V_T \neq 0$

(h)

Null the response

Degenerate case!

Nulled response

$R?$

$R = R_1 \| R_2$

$\tau_{1N} = (R_1 \| R_2)C_1$

nulled response $V_T = 0$

(i)

Turn the stimulus I_T off and see the resistance offered by C_1's terminals.

$I_T = 0$

$R?$

$R = R_2$

$\tau_1 = R_2 C_1$

$$Z_{in}(s) = R_i \frac{1+s\tau_{1N}}{1+s\tau_1}$$

(j)

$$Z_{in}(s) = R_i \frac{1+\dfrac{s}{\omega_z}}{1+\dfrac{s}{\omega_p}}$$

$$\omega_z = \frac{1}{\tau_{1N}} = \frac{1}{(R_1 \| R_2)C_1}$$

$$\omega_p = \frac{1}{\tau_1} = \frac{1}{R_2 C_1}$$

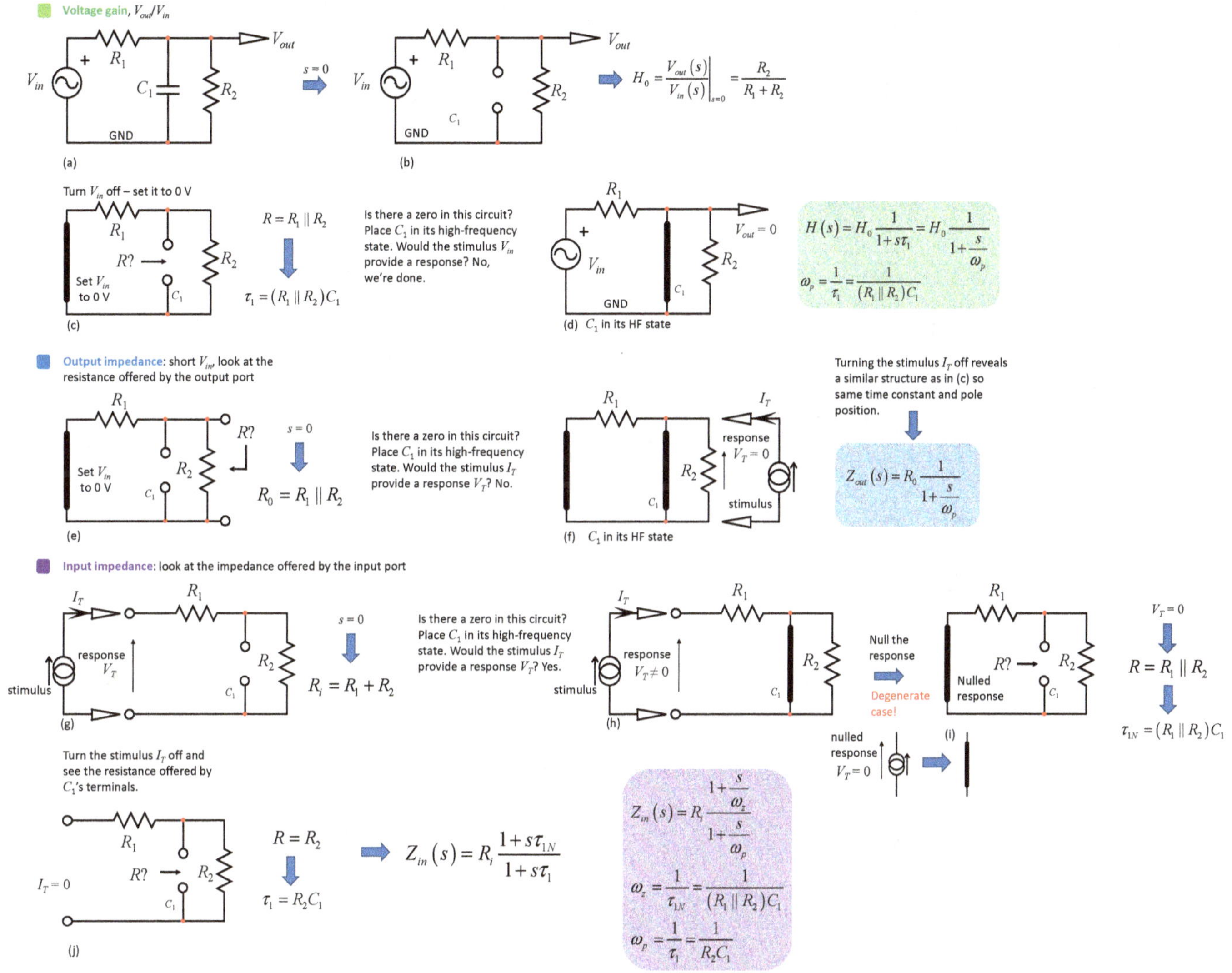

The FACTs are More Efficient than the Brute-Force with This Network—No Expression Expansion is Needed

Figure 5.9

$R_1 := 2k\Omega \quad R_2 := 5k\Omega \quad C_1 := 10nF \quad R_{inf} := 10^{12}\Omega \qquad \|(x,y) := \dfrac{x \cdot y}{x+y}$

$H_0 := \dfrac{R_2}{R_2 + R_1} \qquad \tau_{1a} := (R_1 \| R_2) \cdot C_1 = 14.28571\mu s \qquad \omega_{pa} := \dfrac{1}{\tau_{1a}} \qquad f_p := \dfrac{\omega_{pa}}{2 \cdot \pi} = 11.14085kHz$

$R_0 := R_1 \| R_2 \qquad \tau_{1N} := (R_1 \| R_2) \cdot C_1 \qquad \omega_z := \dfrac{1}{\tau_{1N}} \qquad f_z := \dfrac{\omega_z}{2 \cdot \pi} = 11.14085kHz$

$R_i := R_1 + R_2 \qquad \tau_{1b} := R_2 \cdot C_1 \qquad \omega_{pb} := \dfrac{1}{\tau_{1b}} \qquad f_{pb} := \dfrac{\omega_{pb}}{2 \cdot \pi} = 3.1831kHz$

$H_1(s) := H_0 \cdot \dfrac{1}{1 + \dfrac{s}{\omega_{pa}}}$

$Z_{out}(s) := R_0 \cdot \dfrac{1}{1 + \dfrac{s}{\omega_{pa}}}$

$Z_{in}(s) := R_i \cdot \dfrac{1 + \dfrac{s}{\omega_z}}{1 + \dfrac{s}{\omega_{pb}}}$

$H_{ref}(s) := \dfrac{R_2 \| \left(\dfrac{1}{s \cdot C_1}\right)}{R_1 + R_2 \| \left(\dfrac{1}{s \cdot C_1}\right)} \qquad Z_{outR}(s) := R_1 \| \left(\dfrac{1}{s \cdot C_1}\right) \| R_2 \qquad Z_{inR}(s) := R_1 + R_2 \| \left(\dfrac{1}{s \cdot C_1}\right)$

Brute-force expressions

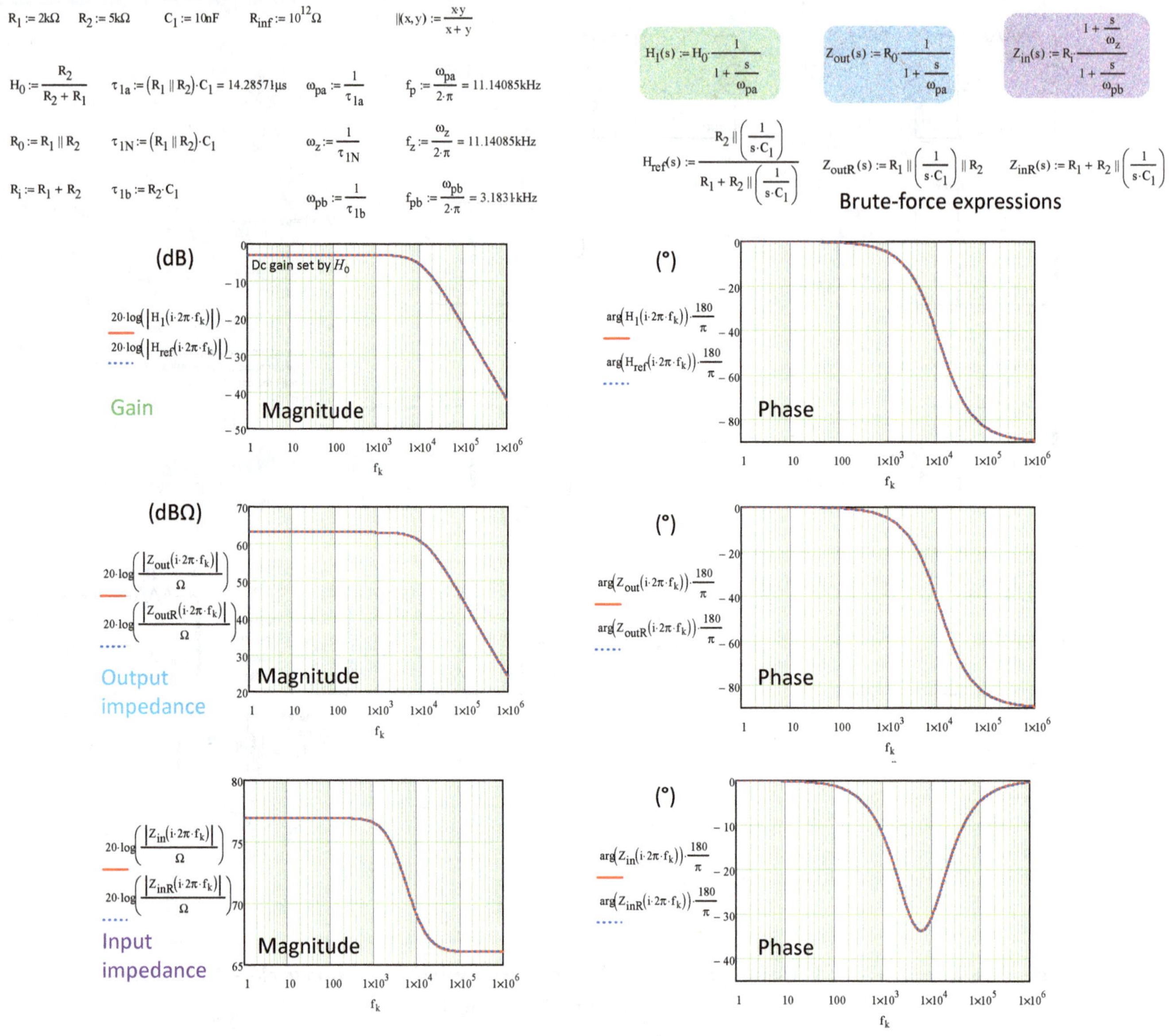

These Graphs Confirm our Results are Correct

Figure 5.10

As a next example, we now add resistor R_1 in parallel with capacitor C_1 from Figure 5.6 and carry on with the three transfer functions of the circuit in Figure 5.11.

We go straight to the FACTs this time.

Adding a resistor in parallel to C_1 brings a zero to the transfer function. Its location can be quickly determined by inspection, considering the arrangement of C_1 paralleled with R_1 which offers an infinite impedance at the zero frequency. Resistor R_1 also provides a path in dc which, together with R_2 form a resistive divider setting the attenuation for $s = 0$.

The output impedance does not feature a zero and is easily derived.

Voltage gain, V_{out}/V_{in}

R_1

C_1

V_{out}

V_{in}

GND

(a)

$s = 0$

R_1

C_1

V_{out}

V_{in}

R_2

GND

(b)

$$H_0 = \frac{V_{out}(s)}{V_{in}(s)}\bigg|_{s=0} = \frac{R_2}{R_1 + R_2}$$

Turn V_{in} off – set it to 0 V

R_1

C_1

$R?$

R_2

Set V_{in} to 0 V

(c)

$R = R_1 \parallel R_2$

$\tau_1 = (R_1 \parallel R_2)C_1$

Is there a zero in this circuit? Place C_1 in its high-frequency state. Would the stimulus V_{in} provide a response? Yes. What could possibly stop the stimulus propagation?

$$Z_1(s) = R_1 \parallel \frac{1}{sC_1} = R_1 \frac{1}{1 + sR_1C_1}$$

$$1 + sR_1C_1 = 0 \rightarrow s_z = -\frac{1}{R_1C_1}$$

$Z_1(s) \rightarrow \infty$

R_1

C_1

$V_{out} = 0$

V_{in}

R_2

GND

(d) What condition nulls V_{out}?

$$H(s) = H_0 \frac{1 + \dfrac{s}{\omega_z}}{1 + \dfrac{s}{\omega_p}}$$

$$\omega_p = \frac{1}{\tau_1} = \frac{1}{(R_1 \parallel R_2)C_1}$$

$$\omega_z = \frac{1}{R_1C_1}$$

Output impedance: short V_{in}, look at the resistance offered by the output port

R_1

C_1

$R?$

R_2

Set V_{in} to 0 V

(e)

$s = 0$

$R_0 = R_1 \parallel R_2$

Is there a zero in this circuit? Place C_1 in its high-frequency state. Would the stimulus I_T provide a response V_T? No.

R_1

C_1

R_2

response $V_T = 0$

I_T

stimulus

(f) C_1 in its HF state

Turning the stimulus I_T off reveals a similar structure as in (c) so same time constant and pole position.

$$Z_{out}(s) = R_0 \frac{1}{1 + \dfrac{s}{\omega_p}}$$

Input impedance: look at the impedance offered by the input port

I_T

R_1

C_1

response V_T

R_2

stimulus

(g)

$s = 0$

$R_i = R_1 + R_2$

Turn the stimulus I_T off and see the resistance offered by C_1's terminals.

R_1

C_1

$R?$

R_2

$I_T = 0$

(j)

$R = R_1$

$\tau_1 = R_1C_1$

Is there a zero in this circuit? Place C_1 in its high-frequency state. Would the stimulus I_T provide a response V_T? Yes.

I_T

R_1

C_1

response $V_T \neq 0$

R_2

stimulus

(h)

Null the response

Degenerate case!

nulled response $V_T = 0$

R_1

C_1

$R?$

R_2

Nulled response

(i)

$V_T = 0$

$R = R_1 \parallel R_2$

$\tau_{1N} = (R_1 \parallel R_2)C_1$

$$Z_{in}(s) = R_i \frac{1 + s\tau_{1N}}{1 + s\tau_1}$$

$$Z_{in}(s) = R_i \frac{1 + \dfrac{s}{\omega_z}}{1 + \dfrac{s}{\omega_p}}$$

$$\omega_z = \frac{1}{\tau_{1N}} = \frac{1}{(R_1 \parallel R_2)C_1}$$

$$\omega_p = \frac{1}{\tau_1} = \frac{1}{R_1C_1}$$

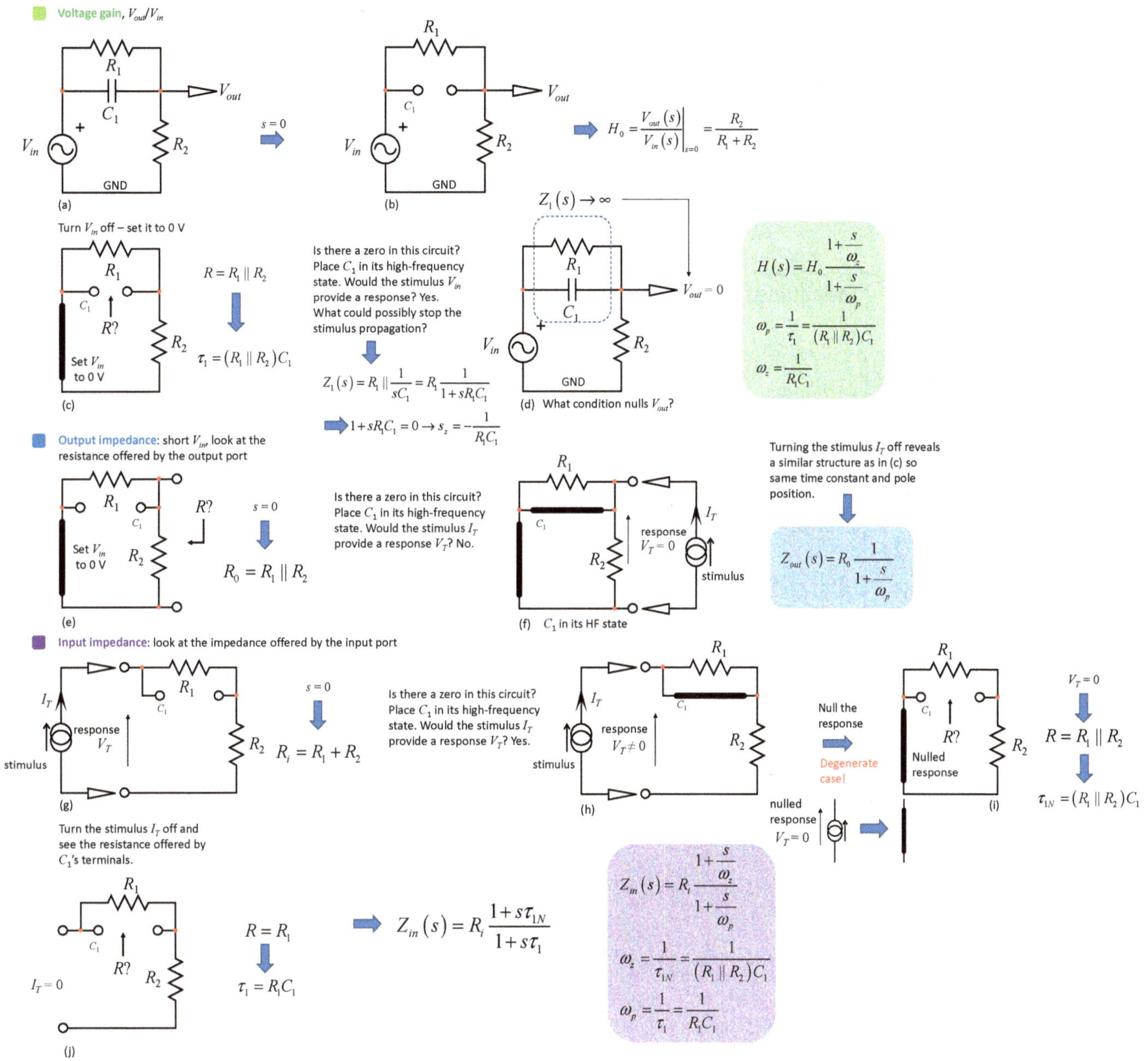

Resistor R_1 now Offers a Parallel Path to C_1: Do You see the Zero?

Figure 5.11

The input impedance hosts a zero and a pole also quickly obtained by inspection only. Please note that we applied a convenient NDI in sketch (i) but we could also determine the conditions for which $R_2 + R_1 \parallel \dfrac{1}{sC_1} = 0$.

Figure 5.12 compares responses obtained with the FACTs and the brute-force expressions. They are identical.

$R_1 := 2k\Omega \quad R_2 := 5k\Omega \quad\quad C_1 := 10nF \quad\quad R_{inf} := 10^{12}\Omega \quad\quad \|(x,y) := \dfrac{x \cdot y}{x+y}$

$H_0 := \dfrac{R_2}{R_2 + R_1} \quad \tau_{1a} := (R_1 \| R_2) \cdot C_1 = 14.28571\mu s \quad \omega_{pa} := \dfrac{1}{\tau_{1a}} \quad f_p := \dfrac{\omega_{pa}}{2\cdot\pi} = 11.14085 kHz$

$R_0 := R_1 \| R_2 \quad \tau_{1Na} := R_1 \cdot C_1 \quad\quad\quad \omega_{za} := \dfrac{1}{\tau_{1Na}} \quad f_z := \dfrac{\omega_{za}}{2\cdot\pi} = 7.95775 kHz$

$R_i := R_1 + R_2 \quad \tau_{1b} := R_1 \cdot C_1 \quad\quad\quad\quad \omega_{pb} := \dfrac{1}{\tau_{1b}} \quad f_{pb} := \dfrac{\omega_{pb}}{2\cdot\pi} = 7.95775 kHz$

$\tau_{1Nb} := (R_1 \| R_2) \cdot C_1 \quad\quad \omega_{zb} := \dfrac{1}{\tau_{1Nb}} \quad f_{zb} := \dfrac{\omega_{zb}}{2\cdot\pi} = 11.14085 kHz$

$H_1(s) := H_0 \cdot \dfrac{1 + \dfrac{s}{\omega_{za}}}{1 + \dfrac{s}{\omega_{pa}}}$

$Z_{out}(s) := R_0 \cdot \dfrac{1}{1 + \dfrac{s}{\omega_{pa}}}$

$Z_{in}(s) := R_i \cdot \dfrac{1 + \dfrac{s}{\omega_{zb}}}{1 + \dfrac{s}{\omega_{pb}}}$

$H_{ref}(s) := \dfrac{R_2}{R_2 + R_1 \left\| \left(\dfrac{1}{s \cdot C_1}\right)\right.}$

$Z_{outR}(s) := R_1 \left\| \left(\dfrac{1}{s \cdot C_1}\right)\right. \| R_2$

$Z_{inR}(s) := R_2 + R_1 \left\| \left(\dfrac{1}{s \cdot C_1}\right)\right.$

Brute-force expressions

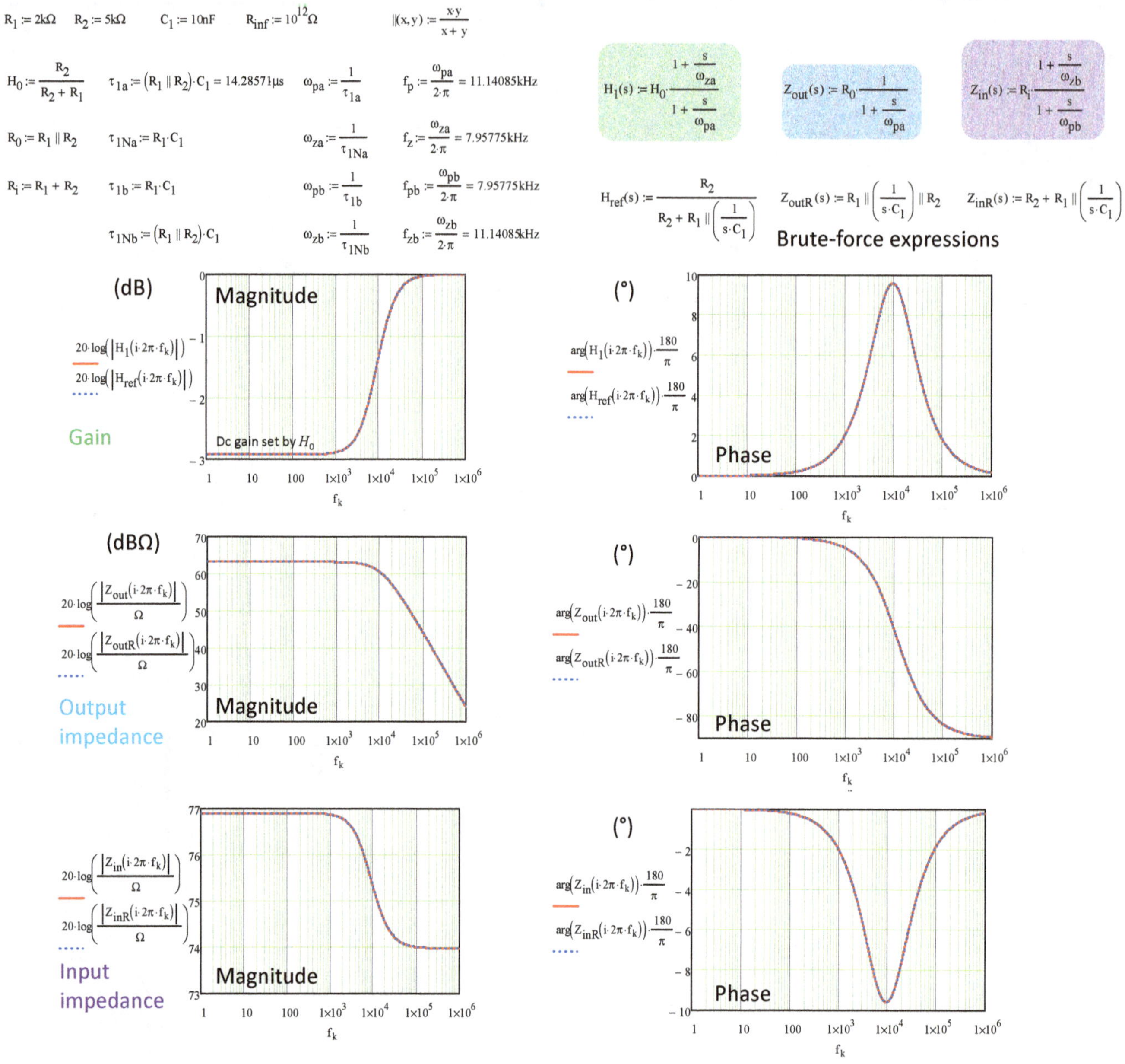

The Addition of Resistance Across the Capacitor Moves the Zero from the Origin to a Higher Value

Figure 5.12

We carry on by inserting a resistance in series with capacitor C_1 now grounded as shown in Figure 5.13. Resistance R_3 could be seen as the equivalent series resistance or ESR of the capacitor. The analysis remains similar and, by adding resistor R_3, we included a zero in the voltage gain transfer function $H(s)$ as seen without resorting to an equation. This zero is found in sketch (d) by identifying a possible impedance combination which would null the response when tuned at a zero frequency. This is the series combination of R_3 and C_1 which becomes a transformed short. By finding the root of this impedance, you have the zero. The output impedance features a similar pole and zero than the voltage gain transfer function. For the input impedance, R_1 is out of the picture when the stimulus current I_T is reduced to 0 A and it eases the pole determination. For the zero in these input impedance exercises, as you have perhaps noticed, because of the shorted stimulus in the nulled response exercise in sketch (h), we always come back to the resistance determination in sketch (c), leading to a similar structure for determining R. It can be a real timesaver in complicated structure where several transfer functions need to be obtained.

We check our results against brute-force expressions in Figure 5.14 and all is correct.

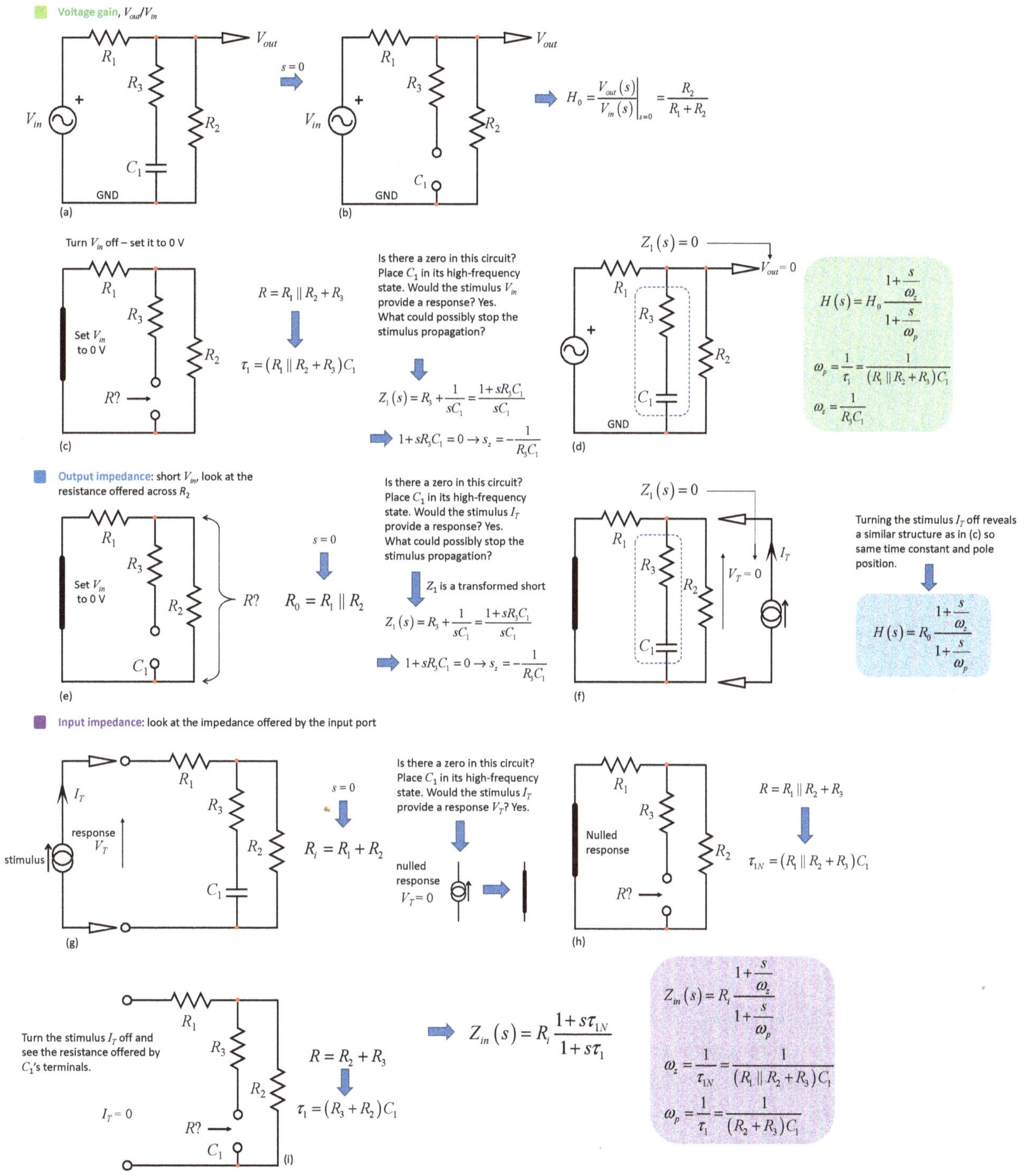
Voltage gain, V_{out}/V_{in}
V_{out}
R_1
R_3
R_2
C_1
V_{in}
GND
$s = 0$
(a)
(b)
$H_0 = \left.\frac{V_{out}(s)}{V_{in}(s)}\right|_{s=0} = \frac{R_2}{R_1 + R_2}$
Turn V_{in} off – set it to 0 V
Set V_{in} to 0 V
$R?$
$R = R_1 \| R_2 + R_3$
$\tau_1 = (R_1 \| R_2 + R_3)C_1$
(c)
Is there a zero in this circuit? Place C_1 in its high-frequency state. Would the stimulus V_{in} provide a response? Yes. What could possibly stop the stimulus propagation?
$Z_1(s) = R_3 + \frac{1}{sC_1} = \frac{1 + sR_3C_1}{sC_1}$
$1 + sR_3C_1 = 0 \rightarrow s_z = -\frac{1}{R_3C_1}$
$Z_1(s) = 0$
$V_{out} = 0$
(d)
$H(s) = H_0 \frac{1 + \frac{s}{\omega_z}}{1 + \frac{s}{\omega_p}}$
$\omega_p = \frac{1}{\tau_1} = \frac{1}{(R_1 \| R_2 + R_3)C_1}$
$\omega_z = \frac{1}{R_3C_1}$
Output impedance: short V_{in}, look at the resistance offered across R_2
$R_0 = R_1 \| R_2$
(e)
Is there a zero in this circuit? Place C_1 in its high-frequency state. Would the stimulus I_T provide a response? Yes. What could possibly stop the stimulus propagation?
Z_1 is a transformed short
I_T
$V_T = 0$
(f)
Turning the stimulus I_T off reveals a similar structure as in (c) so same time constant and pole position.
$H(s) = R_0 \frac{1 + \frac{s}{\omega_z}}{1 + \frac{s}{\omega_p}}$
Input impedance: look at the impedance offered by the input port
response
V_T
stimulus
$R_i = R_1 + R_2$
(g)
Is there a zero in this circuit? Place C_1 in its high-frequency state. Would the stimulus I_T provide a response V_T? Yes.
nulled response
$V_T = 0$
Nulled response
(h)
$\tau_{1N} = (R_1 \| R_2 + R_3)C_1$
Turn the stimulus I_T off and see the resistance offered by C_1's terminals.
$I_T = 0$
$R = R_2 + R_3$
$\tau_1 = (R_3 + R_2)C_1$
(i)
$Z_{in}(s) = R_i \frac{1 + s\tau_{1N}}{1 + s\tau_1}$
$Z_{in}(s) = R_i \frac{1 + \frac{s}{\omega_z}}{1 + \frac{s}{\omega_p}}$
$\omega_z = \frac{1}{\tau_{1N}} = \frac{1}{(R_1 \| R_2 + R_3)C_1}$
$\omega_p = \frac{1}{\tau_1} = \frac{1}{(R_2 + R_3)C_1}$

A Resistor is Inserted in Series with the Capacitor and Reveals Another Configuration

Figure 5.13

$R_1 := 2k\Omega \quad R_2 := 5k\Omega \quad R_3 := 1.5k\Omega \quad C_1 := 10nF \quad R_{inf} := 10^{12}\Omega \quad \|(x,y) := \frac{x \cdot y}{x+y}$

$H_0 := \frac{R_2}{R_2 + R_1} \quad \tau_{1a} := [(R_1 \| R_2) + R_3] \cdot C_1 = 29.28571\mu s \quad \omega_{pa} := \frac{1}{\tau_{1a}} \quad f_p := \frac{\omega_{pa}}{2 \cdot \pi} = 5.43456kHz$

$R_0 := R_1 \| R_2 \quad \tau_{1Na} := R_3 \cdot C_1 \quad \omega_{za} := \frac{1}{\tau_{1Na}} \quad f_z := \frac{\omega_{za}}{2 \cdot \pi} = 10.61033kHz$

$R_i := R_1 + R_2 \quad \tau_{1b} := (R_2 + R_3) \cdot C_1 \quad \omega_{pb} := \frac{1}{\tau_{1b}} \quad f_{pb} := \frac{\omega_{pb}}{2 \cdot \pi} = 2.44854kHz$

$\tau_{1Nb} := [(R_1 \| R_2) + R_3] \cdot C_1 \quad \omega_{zb} := \frac{1}{\tau_{1Nb}} \quad f_{zb} := \frac{\omega_{zb}}{2 \cdot \pi} = 5.43456kHz$

$$H_1(s) := H_0 \cdot \frac{1 + \frac{s}{\omega_{za}}}{1 + \frac{s}{\omega_{pa}}} \qquad Z_{out}(s) := R_0 \cdot \frac{1 + \frac{s}{\omega_{za}}}{1 + \frac{s}{\omega_{pa}}} \qquad Z_{in}(s) := R_i \cdot \frac{1 + \frac{s}{\omega_{zb}}}{1 + \frac{s}{\omega_{pb}}}$$

$$H_{ref}(s) := \frac{R_2 \| \left(\frac{1}{s \cdot C_1} + R_3\right)}{R_1 + R_2 \| \left(\frac{1}{s \cdot C_1} + R_3\right)} \qquad Z_{outR}(s) := R_1 \| \left(\frac{1}{s \cdot C_1} + R_3\right) \| R_2 \qquad Z_{inR}(s) := R_1 + R_2 \| \left(\frac{1}{s \cdot C_1} + R_3\right)$$

Brute-force expressions

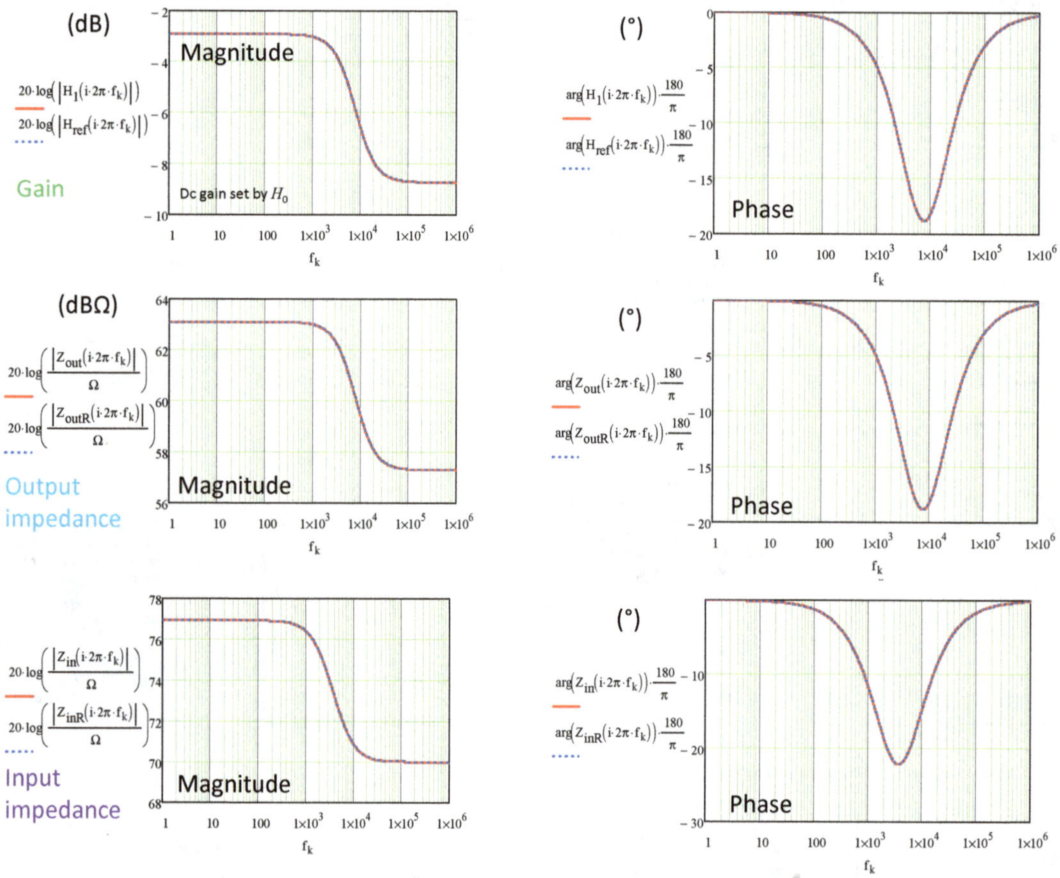

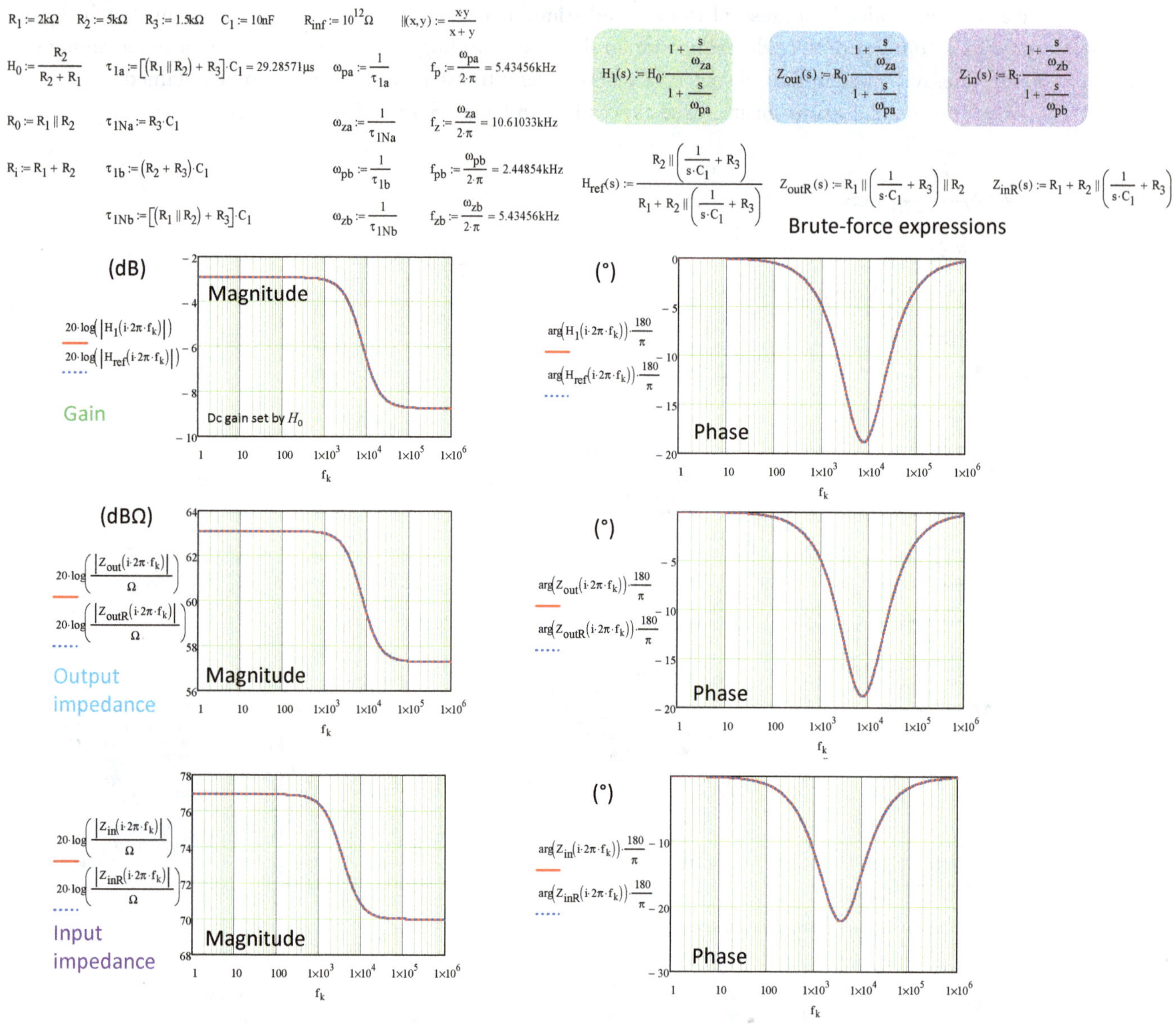

Mathcad Plots Confirm the Derivations in Figure 5.13 are Correct

Figure 5.14

Let's now add another resistor, R_4, across capacitor C_1 to slightly complicate the circuit now shown in Figure 5.15.

The process starts by opening the capacitor to determine the dc attenuation H_0. You can see how keeping paralleled resistors naturally brings a compact format. Do not expand this expression as insight would be lost. For instance, you can see that having R_2 of very high value as in an unloaded configuration, then its effect of the paralleling with R_3+R_4 could be neglected and H_0 immediately simplifies. With a fully expanded expression, then this type of observation would be more difficult to conduct. The time constant can be obtained by inspection and a quick intermediate drawing in (d) helps finding the resistance driving the capacitor. The output impedance does not present any complication and the transfer function is quickly obtained. The capacitor in the input impedance exercise sees the same driving resistance as in sketch (c) when the response is nulled. This is because shorting the current source—a degenerate case with the voltage across the generator is nulled—brings the circuit back to the sketch used when V_{in} was zeroed.

Then, conduct an NDI in sketch (i) and go to the final transfer function which hosts a zero and a pole. The *low-entropy* expressions are tested against the brute-force expressions in Figure 5.16 and confirm they are correct.

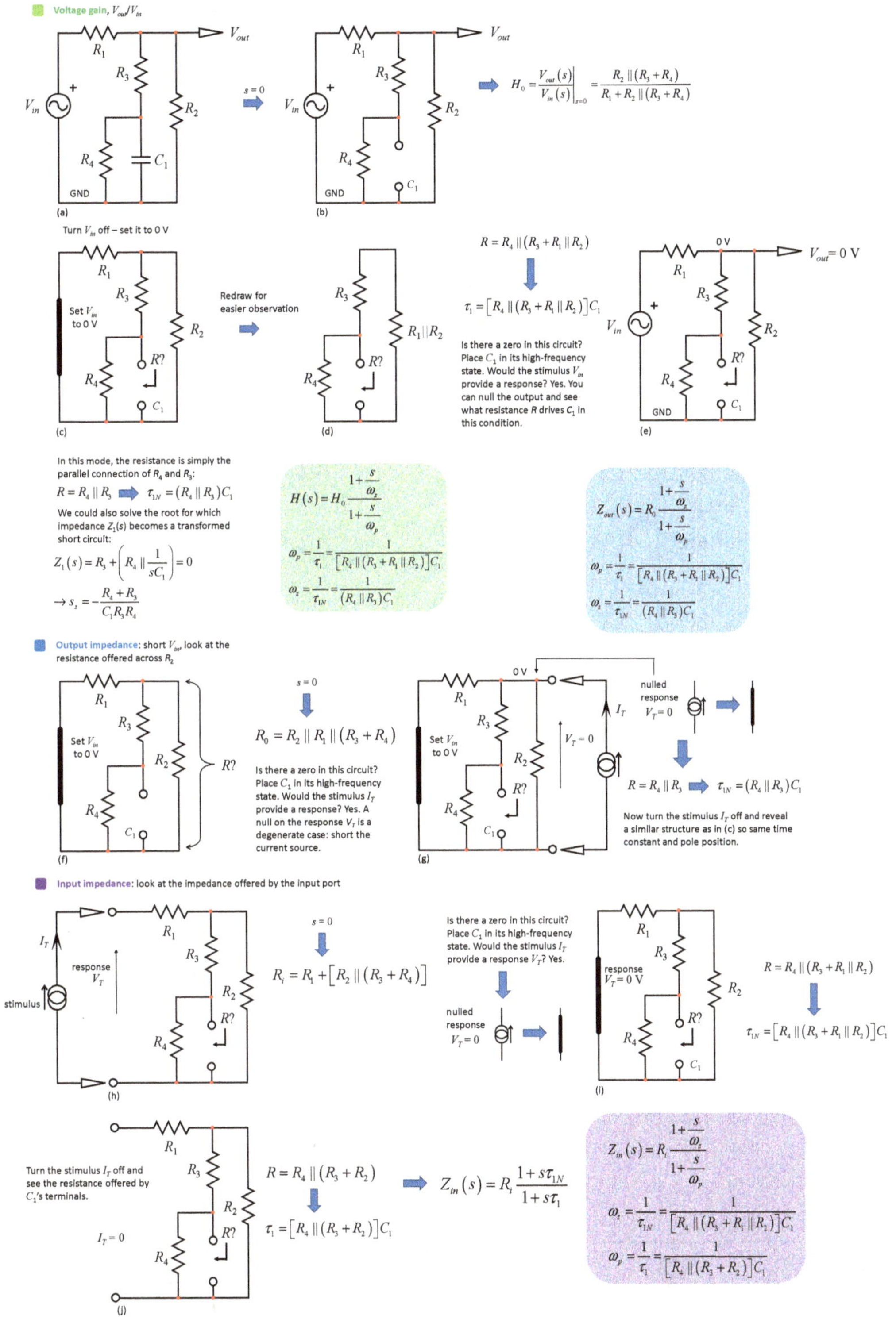

Voltage gain, V_{out}/V_{in}
$H_0 = \left.\frac{V_{out}(s)}{V_{in}(s)}\right|_{s=0} = \frac{R_2 \| (R_3 + R_4)}{R_1 + R_2 \| (R_3 + R_4)}$
Turn V_{in} off – set it to 0 V
Redraw for easier observation
$R = R_4 \| (R_3 + R_1 \| R_2)$
$\tau_1 = [R_4 \| (R_3 + R_1 \| R_2)]C_1$
Is there a zero in this circuit? Place C_1 in its high-frequency state. Would the stimulus V_{in} provide a response? Yes. You can null the output and see what resistance R drives C_1 in this condition.
$V_{out} = 0$ V
In this mode, the resistance is simply the parallel connection of R_4 and R_3:
$R = R_4 \| R_3 \Rightarrow \tau_{1N} = (R_4 \| R_3)C_1$
We could also solve the root for which impedance $Z_1(s)$ becomes a transformed short circuit:
$Z_1(s) = R_3 + \left(R_4 \| \frac{1}{sC_1}\right) = 0$
$\rightarrow s_z = -\frac{R_4 + R_3}{C_1 R_3 R_4}$
$H(s) = H_0 \frac{1 + \frac{s}{\omega_z}}{1 + \frac{s}{\omega_p}}$
$\omega_p = \frac{1}{\tau_1} = \frac{1}{[R_4 \| (R_3 + R_1 \| R_2)]C_1}$
$\omega_z = \frac{1}{\tau_{1N}} = \frac{1}{(R_4 \| R_3)C_1}$
$Z_{out}(s) = R_0 \frac{1 + \frac{s}{\omega_z}}{1 + \frac{s}{\omega_p}}$
Output impedance: short V_{in}, look at the resistance offered across R_2
$R_0 = R_2 \| R_1 \| (R_3 + R_4)$
Is there a zero in this circuit? Place C_1 in its high-frequency state. Would the stimulus I_T provide a response? Yes. A null on the response V_T is a degenerate case: short the current source.
nulled response $V_T = 0$
$R = R_4 \| R_3 \Rightarrow \tau_{1N} = (R_4 \| R_3)C_1$
Now turn the stimulus I_T off and reveal a similar structure as in (c) so same time constant and pole position.
Input impedance: look at the impedance offered by the input port
$R_i = R_1 + [R_2 \| (R_3 + R_4)]$
Is there a zero in this circuit? Place C_1 in its high-frequency state. Would the stimulus I_T provide a response V_T? Yes.
response $V_T = 0$ V
$R = R_4 \| (R_3 + R_1 \| R_2)$
$\tau_{1N} = [R_4 \| (R_3 + R_1 \| R_2)]C_1$
Turn the stimulus I_T off and see the resistance offered by C_1's terminals.
$R = R_4 \| (R_3 + R_2)$
$\tau_1 = [R_4 \| (R_3 + R_2)]C_1$
$Z_{in}(s) = R_i \frac{1 + s\tau_{1N}}{1 + s\tau_1}$
$Z_{in}(s) = R_i \frac{1 + \frac{s}{\omega_z}}{1 + \frac{s}{\omega_p}}$
$\omega_z = \frac{1}{\tau_{1N}} = \frac{1}{[R_4 \| (R_3 + R_1 \| R_2)]C_1}$
$\omega_p = \frac{1}{\tau_1} = \frac{1}{[R_4 \| (R_3 + R_2)]C_1}$

An Extra Resistor R_4 Complicates the Circuit but Nothing Unsurmountable for the FACTs

Figure 5.15

$R_1 := 2k\Omega \quad R_2 := 5k\Omega \quad R_3 := 1.5k\Omega \quad R_4 := 10k\Omega \quad C_1 := 10nF \quad R_{inf} := 10^{12}\Omega \quad \|(x,y) := \frac{x \cdot y}{x+y}$

$H_0 := \frac{R_2 \| (R_3 + R_4)}{R_2 \| (R_3+R_4) + R_1} = 0.63536$

$\tau_{1a} := [R_4 \| (R_3 + R_1 \| R_2)] \cdot C_1 = 22.65193\mu s$

$\omega_{pa} := \frac{1}{\tau_{1a}}$

$f_p := \frac{\omega_{pa}}{2 \cdot \pi} = 7.02611 \cdot kHz$

$R_0 := R_1 \| R_2 \| (R_3 + R_4) = 1.27072k\Omega$

$\tau_{1Na} := (R_4 \| R_3) \cdot C_1 = 13.04348\mu s$

$\omega_{za} := \frac{1}{\tau_{1Na}}$

$f_z := \frac{\omega_{za}}{2 \cdot \pi} = 12.20188kHz$

$R_i := R_1 + [R_2 \| (R_3 + R_4)] = 5.48485k\Omega$

$\tau_{1b} := [R_4 \| (R_3 + R_2)] \cdot C_1 = 39.39394\mu s$

$\omega_{pb} := \frac{1}{\tau_{1b}}$

$f_{pb} := \frac{\omega_{pb}}{2 \cdot \pi} = 4.04009kHz$

$\tau_{1Nb} := [R_4 \| (R_3 + R_1 \| R_2)] \cdot C_1 = 22.65193\mu s$

$\omega_{zb} := \frac{1}{\tau_{1Nb}}$

$f_{zb} := \frac{\omega_{zb}}{2 \cdot \pi} = 7.02611 \cdot kHz$

$H_1(s) := H_0 \cdot \frac{1 + \frac{s}{\omega_{za}}}{1 + \frac{s}{\omega_{pa}}} \qquad Z_{out}(s) := R_0 \cdot \frac{1 + \frac{s}{\omega_{za}}}{1 + \frac{s}{\omega_{pa}}} \qquad Z_{in}(s) := R_i \cdot \frac{1 + \frac{s}{\omega_{zb}}}{1 + \frac{s}{\omega_{pb}}}$

$H_{ref}(s) := \frac{R_2 \| \left[\left(\frac{1}{s \cdot C_1}\right) \| R_4 + R_3\right]}{R_1 + R_2 \| \left[\left(\frac{1}{s \cdot C_1}\right) \| R_4 + R_3\right]} \qquad Z_{outR}(s) := R_1 \| \left[\left(\frac{1}{s \cdot C_1}\right) \| R_4 + R_3\right] \| R_2 \qquad Z_{inR}(s) := R_1 + R_2 \| \left[\left(\frac{1}{s \cdot C_1}\right) \| R_4 + R_3\right]$

Brute-force expressions

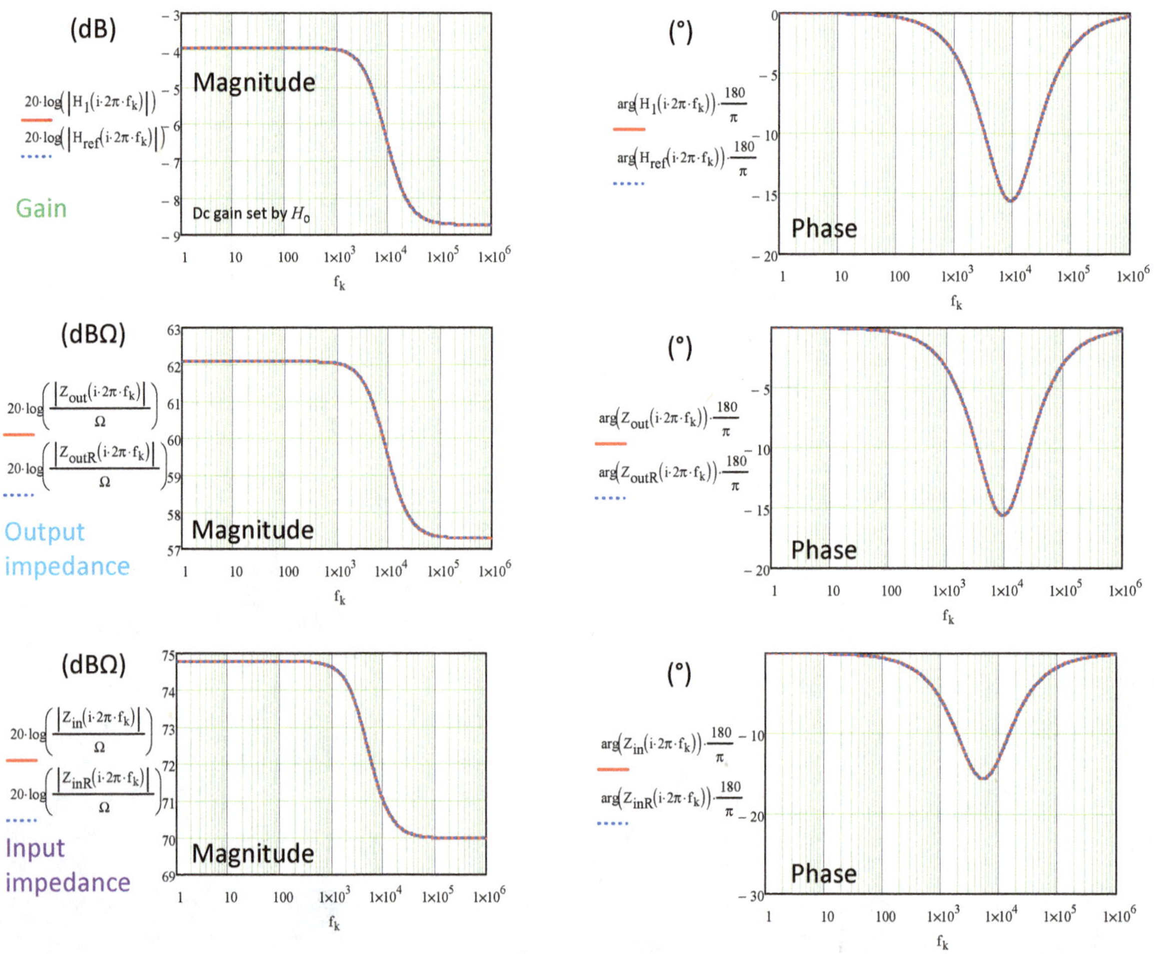

Mathcad Plots Confirm our Derivations are Correct

Figure 5.16

We can mark a pause with capacitor-based circuits for now and replace C_1 by L_1 as in the simplest RL circuit shown in Figure 5.17. We go back to the classical approach, letting the newcomers to the FACTs find their marks. By inspection, you should see that when dc is applied for the stimulus, L_1 is a short circuit and there is a zero at the origin.

This is quickly confirmed in sketch (a) and an inverted pole in the transfer function. The output impedance also features a zero at the origin and can also be expressed with an inverted pole in which the leading term represents the asymptote as s approaches infinity.

The input impedance is simply the sum of the two components and is easily rearranged with a zero.

Voltage gain, V_{out}/V_{in}

$$\frac{V_{out}(s)}{V_{in}(s)} = \frac{sL_1}{R_1 + sL_1} = \frac{sL_1}{sL_1}\frac{1}{1 + \frac{R_1}{sL_1}} = \frac{1}{1 + \frac{\omega_p}{s}}$$

$$\omega_p = \frac{R_1}{L_1}$$

(a)

Output impedance: short V_{in}, look at the resistance offered by the output port

$$Z_{out}(s) = (sL_1) \| R_1 = \frac{sL_1 \cdot R_1}{sL_1 + R_1} = \frac{sL_1}{sL_1}\frac{R_1}{1 + \frac{R_1}{sL_1}} = R_0 \frac{1}{1 + \frac{\omega_p}{s}}$$

$$\omega_p = \frac{R_1}{L_1}$$

$$R_0 = R_1$$

(b)

Input impedance: look at the impedance offered by the input port

$$Z_{in}(s) = sL_1 + R_1 = R_1\left(1 + s\frac{L_1}{R_1}\right) = R_1\left(1 + \frac{s}{\omega_z}\right)$$

$$\omega_z = \frac{R_1}{L_1}$$

$$R_i = R_1$$

(c)

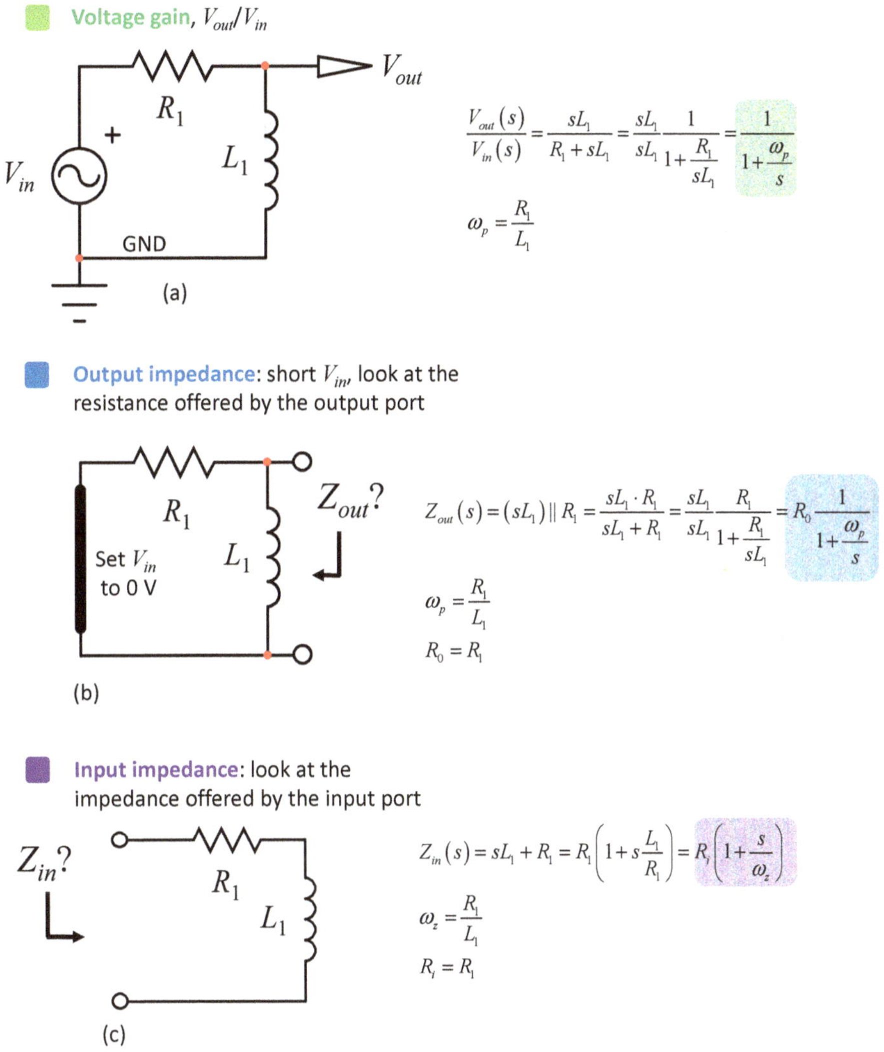

An Inductor Loads the Resistance

Figure 5.17

Whether you deal with one or several capacitors and inductors, the FACTs process in Figure 5.18 remains the same: determine the resistance R driving each energy-storing element to find the associated time constant. When the inductor is set in its dc state, for $s = 0$, you see a short circuit across the output terminals, implying a zero located at the origin in sketch (b). The pole is then easily determined and leads to a transfer function showing an inverted pole. We used the generalized transfer function for the sake of the example but the middle expression in

Figure 5.1 would have led us there straight away.

This is the most possible compact expression you can find for this simple circuit. The output impedance also includes this zero at the origin. Figure 5.19 confirms our approach is correct by comparing magnitude and phase between the rearranged expressions and the brute-force ones from Figure 5.17.

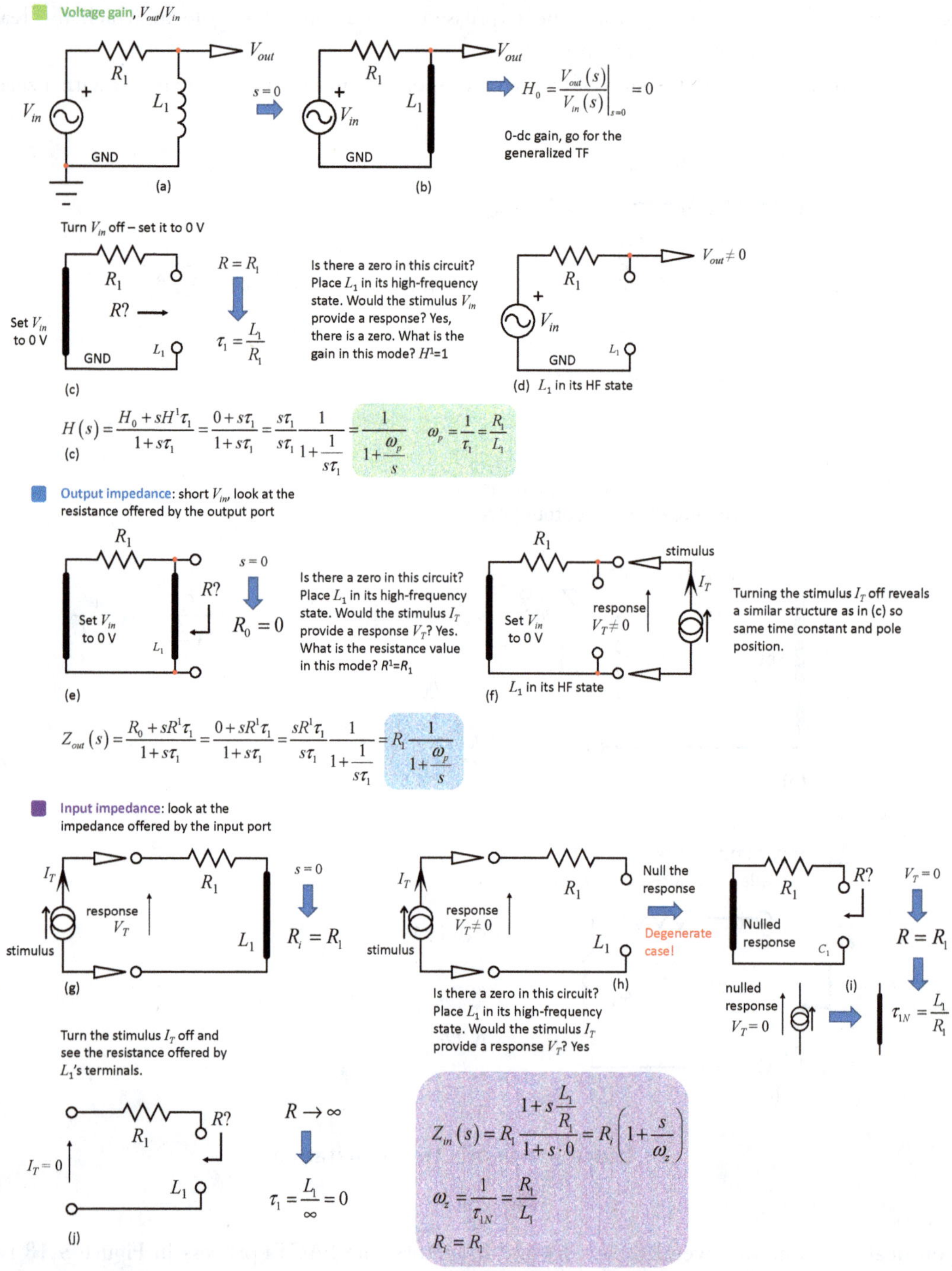

Voltage gain, V_{out}/V_{in}
$s=0$
$H_0 = \left.\frac{V_{out}(s)}{V_{in}(s)}\right|_{s=0} = 0$
0-dc gain, go for the generalized TF
Turn V_{in} off – set it to 0 V
$R = R_1$
$\tau_1 = \frac{L_1}{R_1}$
Is there a zero in this circuit? Place L_1 in its high-frequency state. Would the stimulus V_{in} provide a response? Yes, there is a zero. What is the gain in this mode? $H^1=1$
Set V_{in} to 0 V
$V_{out} \neq 0$
(d) L_1 in its HF state
$H(s) = \frac{H_0 + sH^1\tau_1}{1+s\tau_1} = \frac{0+s\tau_1}{1+s\tau_1} = \frac{s\tau_1}{s\tau_1}\frac{1}{1+\frac{1}{s\tau_1}} = \frac{1}{1+\frac{\omega_p}{s}}$ $\omega_p = \frac{1}{\tau_1} = \frac{R_1}{L_1}$
Output impedance: short V_{in}, look at the resistance offered by the output port
$R_0 = 0$
Is there a zero in this circuit? Place L_1 in its high-frequency state. Would the stimulus I_T provide a response V_T? Yes. What is the resistance value in this mode? $R^1=R_1$
stimulus
response
$V_T \neq 0$
(f) L_1 in its HF state
Turning the stimulus I_T off reveals a similar structure as in (c) so same time constant and pole position.
$Z_{out}(s) = \frac{R_0 + sR^1\tau_1}{1+s\tau_1} = \frac{0+sR^1\tau_1}{1+s\tau_1} = \frac{sR^1\tau_1}{s\tau_1}\frac{1}{1+\frac{1}{s\tau_1}} = R_1\frac{1}{1+\frac{\omega_p}{s}}$
Input impedance: look at the impedance offered by the input port
$R_i = R_1$
Null the response
Degenerate case!
Nulled response
Is there a zero in this circuit? Place L_1 in its high-frequency state. Would the stimulus I_T provide a response V_T? Yes
nulled response $V_T=0$
$R = R_1$
$\tau_{1N} = \frac{L_1}{R_1}$
Turn the stimulus I_T off and see the resistance offered by L_1's terminals.
$R \to \infty$
$\tau_1 = \frac{L_1}{\infty} = 0$
$Z_{in}(s) = R_1\frac{1+s\frac{L_1}{R_1}}{1+s\cdot 0} = R_i\left(1+\frac{s}{\omega_z}\right)$
$\omega_z = \frac{1}{\tau_{1N}} = \frac{R_1}{L_1}$
$R_i = R_1$

FACTs are a Simple Way to Quickly Obtain Transfer Functions without Complicated Manipulation

Figure 5.18

$R_1 := 1k\Omega$ $\quad L_1 := 50mH$ $\quad R_{inf} := 10^{12}\Omega$ $\quad \|(x,y) := \frac{x \cdot y}{x+y}$

$H_0 := 0$ $\quad R_0 := R_1$ $\quad R_i := R_1$

$\tau_1 := \frac{L_1}{R_1} = 50 \cdot \mu s$ $\quad \omega_p := \frac{1}{\tau_1}$ $\quad f_p := \frac{\omega_p}{2 \cdot \pi} = 3.183 \cdot kHz$

$\tau_{1N} := \frac{L_1}{R_1}$ $\quad \omega_z := \frac{1}{\tau_{1N}}$ $\quad f_z := \frac{\omega_z}{2 \cdot \pi} = 3.183 \cdot kHz$

$H_1(s) := \frac{1}{1+\frac{\omega_p}{s}}$ $\quad Z_{out}(s) := R_0 \cdot \frac{1}{1+\frac{\omega_p}{s}}$ $\quad Z_{in}(s) := R_i \cdot \left(1+\frac{s}{\omega_z}\right)$

$H_{ref}(s) := \frac{s \cdot L_1}{R_1 + s \cdot L_1}$ $\quad Z_{outR}(s) := R_1 \| (s \cdot L_1)$ $\quad Z_{inR}(s) := R_1 + s \cdot L_1$

Brute-force expressions

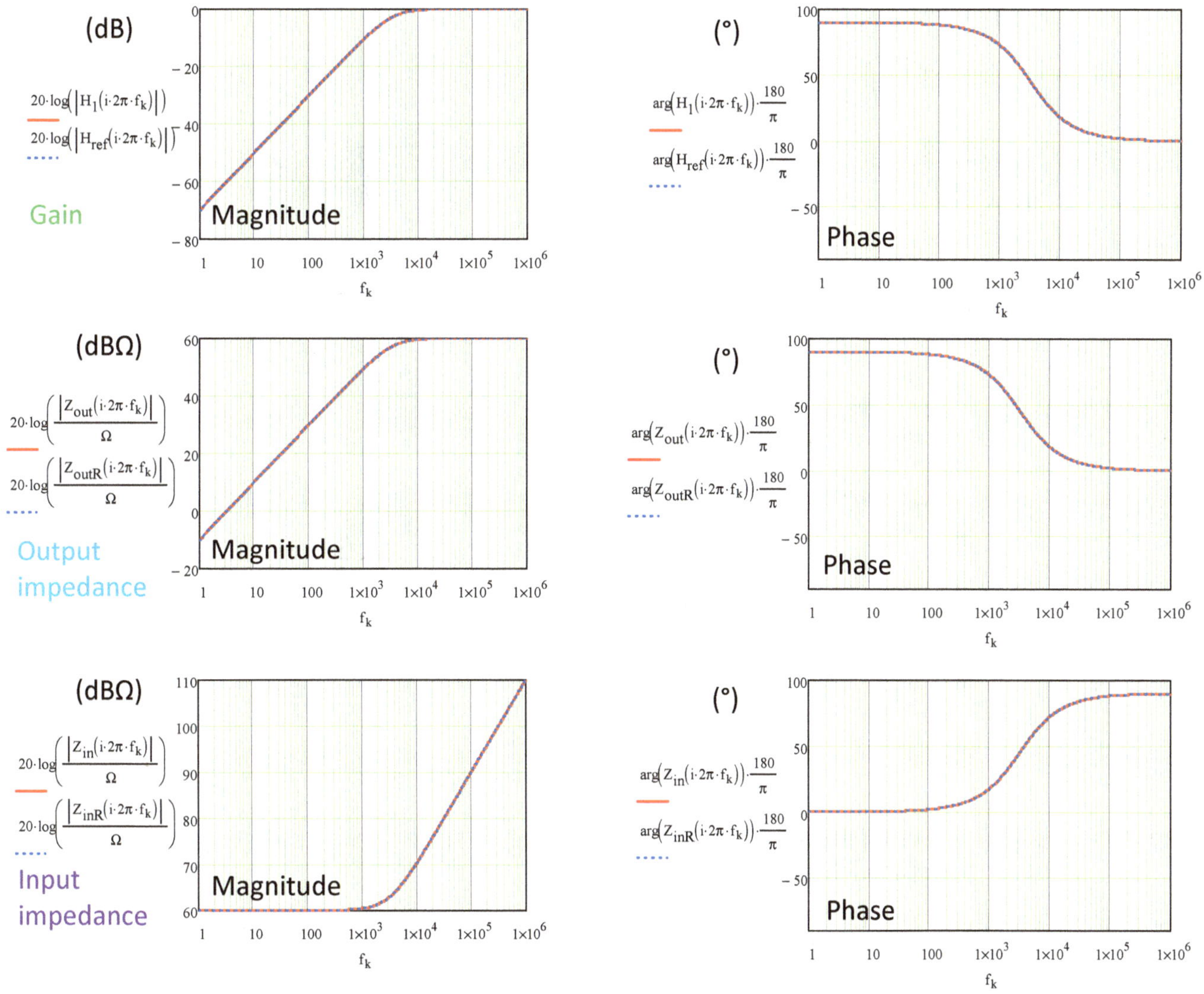
(dB)
$20 \cdot \log(|H_1(i \cdot 2\pi \cdot f_k)|)$
$20 \cdot \log(|H_{ref}(i \cdot 2\pi \cdot f_k)|)$
Gain
Magnitude
(°)
$\arg(H_1(i \cdot 2\pi \cdot f_k)) \cdot \frac{180}{\pi}$
$\arg(H_{ref}(i \cdot 2\pi \cdot f_k)) \cdot \frac{180}{\pi}$
Phase
(dBΩ)
$20 \cdot \log\left(\frac{|Z_{out}(i \cdot 2\pi \cdot f_k)|}{\Omega}\right)$
$20 \cdot \log\left(\frac{|Z_{outR}(i \cdot 2\pi \cdot f_k)|}{\Omega}\right)$
Output impedance
Magnitude
(°)
$\arg(Z_{out}(i \cdot 2\pi \cdot f_k)) \cdot \frac{180}{\pi}$
$\arg(Z_{outR}(i \cdot 2\pi \cdot f_k)) \cdot \frac{180}{\pi}$
Phase
(dBΩ)
$20 \cdot \log\left(\frac{|Z_{in}(i \cdot 2\pi \cdot f_k)|}{\Omega}\right)$
$20 \cdot \log\left(\frac{|Z_{inR}(i \cdot 2\pi \cdot f_k)|}{\Omega}\right)$
Input impedance
Magnitude
(°)
$\arg(Z_{in}(i \cdot 2\pi \cdot f_k)) \cdot \frac{180}{\pi}$
$\arg(Z_{inR}(i \cdot 2\pi \cdot f_k)) \cdot \frac{180}{\pi}$
Phase
f_k

The Gain Transfer Function shows a Zero at the Origin as does the Output Impedance

Figure 5.19

The inductor can now be placed in series with the resistor to form an *LR* filter. Here also, we will use the classical approach using an impedance divider to determine the transfer functions. This is shown in Figure 5.20.

The calculations do not present any particular difficulty and the impedance divider is at work here also. The transfer function now uses a classical pole considering the position of the inductor. The output and input impedances are similar to those already found in Figure 5.17 as *R* and *L* have simply swapped their position, leading to similar parallel or series arrangements.

The FACTs applied to this simple *LR* circuit are detailed in Figure 5.21. In dc, the inductor is a short circuit and the gain H_0 is unity. The pole is easily found by setting the source to 0 V. The output impedance is 0 Ω in dc considering the shorted inductor. Once properly arranged, the equation shows a pole which makes the impedance reach an asymptote set by R_1 at high frequencies. The input impedance is not more difficult to obtain and shows a dc resistance equal to R_1 and then increases towards infinity as frequency increases.

Figure 5.22 confirms our equations are correct.

Voltage gain, V_{out}/V_{in}

$$\frac{V_{out}(s)}{V_{in}(s)} = \frac{R_1}{R_1 + sL_1} = \frac{R_1}{R_1} \frac{1}{1 + \frac{sL_1}{R_1}} = \frac{1}{1 + \frac{s}{\omega_p}}$$

$$\omega_p = \frac{R_1}{L_1}$$

(a)

Output impedance: short V_{in}, look at the resistance offered by the output port

$$Z_{out}(s) = (sL_1) \| R_1 = \frac{sL_1 \cdot R_1}{sL_1 + R_1} = \frac{sL_1}{sL_1} \frac{R_1}{1 + \frac{R_1}{sL_1}} = R_0 \frac{1}{1 + \frac{\omega_p}{s}}$$

$$\omega_p = \frac{R_1}{L_1}$$

$$R_0 = R_1$$

(b)

Input impedance: look at the impedance offered by the input port

$$Z_{in}(s) = sL_1 + R_1 = R_1\left(1 + s\frac{L_1}{R_1}\right) = R_i\left(1 + \frac{s}{\omega_z}\right)$$

$$\omega_z = \frac{R_1}{L_1}$$

$$R_i = R_1$$

(c)

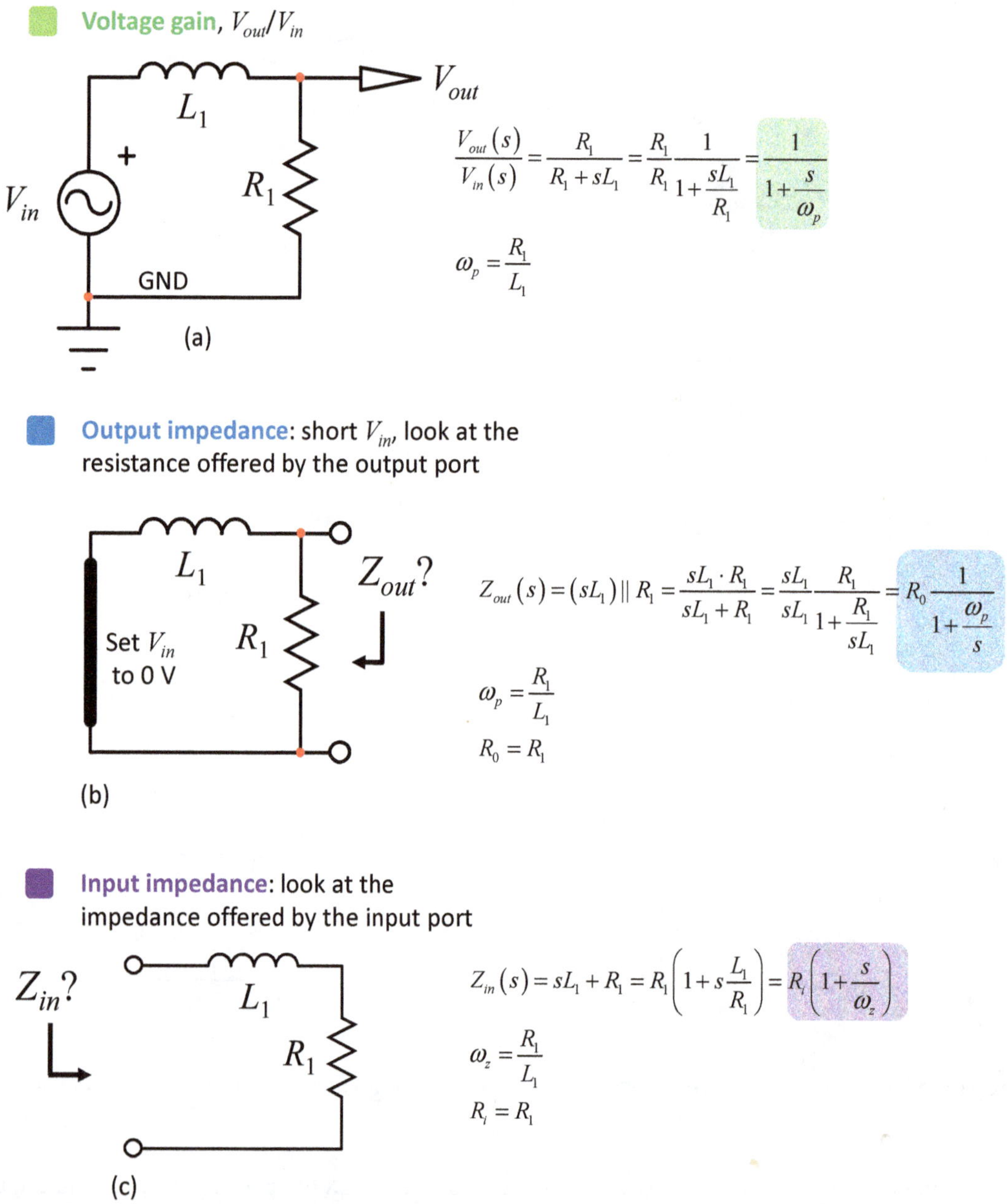

The Determination of the Transfer Functions is Quick with only Two Circuit Elements

Figure 5.20

Voltage gain, V_{out}/V_{in}

$$H_0 = \frac{V_{out}(s)}{V_{in}(s)}\bigg|_{s=0} = 1$$

(a) (b)

Turn V_{in} off – set it to 0 V

Set V_{in} to 0 V

$R = R_1$

$\tau_1 = \frac{L_1}{R_1}$

(c)

Is there a zero in this circuit? Place L_1 in its high-frequency state. Would the stimulus V_{in} provide a response? No, we're done.

$V_{out}=0$ response

L_1 in its HF state (d)

$$H(s) = H_0 \frac{1}{1+\dfrac{s}{\omega_p}} = \frac{1}{1+\dfrac{s}{\omega_p}}$$

$$\omega_p = \frac{1}{\tau_1} = \frac{R_1}{L_1}$$

Output impedance: short V_{in}, look at the resistance offered by the output port

Set V_{in} to 0 V

$R?$

$s = 0$

$R_0 = 0$

Go for generalized TF

(e)

Is there a zero in this circuit? Place L_1 in its high-frequency state. Would the stimulus I_T provide a response V_T? Yes.

response $V_T \neq 0$

stimulus

L_1 in its HF state (f)

What is the high-frequency resistance R^1 in this mode?

$R^1 = R_1$

Turning the stimulus I_T off reveals a similar structure as in (c) so same time constant and pole position.

$$Z_{out}(s) = \frac{R_0 + sR^1\tau_1}{1+s\tau_1} = \frac{0 + sR_1\tau_1}{1+s\tau_1} = \frac{sR_1\tau_1}{s\tau_1}\frac{1}{1+\dfrac{1}{s\tau_1}} = R_1\frac{1}{1+\dfrac{\omega_p}{s}}$$

Input impedance: look at the impedance offered by the input port

I_T stimulus

response V_T

$s = 0$

$R_i = R_1$

(g)

I_T stimulus

response $V_T \neq 0$

Is there a zero in this circuit? Place L_1 in its high-frequency state. Would the stimulus I_T provide a response V_T? Yes

Degenerate case!

(h)

Null the response

$R?$

$V_T = 0$

Nulled response

$R = R_1$

$\tau_{1N} = \frac{L_1}{R_1}$

nulled response $V_T = 0$

(i)

Turn the stimulus I_T off and see the resistance offered by L_1's terminals.

$I_T = 0$

$R?$

$R \rightarrow \infty$

$\tau_1 = \frac{L_1}{\infty} = 0$

(j)

$$Z_{in}(s) = R_1\frac{1+s\dfrac{L_1}{R_1}}{1+s\cdot 0} = R_1\left(1+\dfrac{s}{\omega_z}\right)$$

$$\omega_z = \frac{1}{\tau_{1N}} = \frac{R_1}{L_1}$$

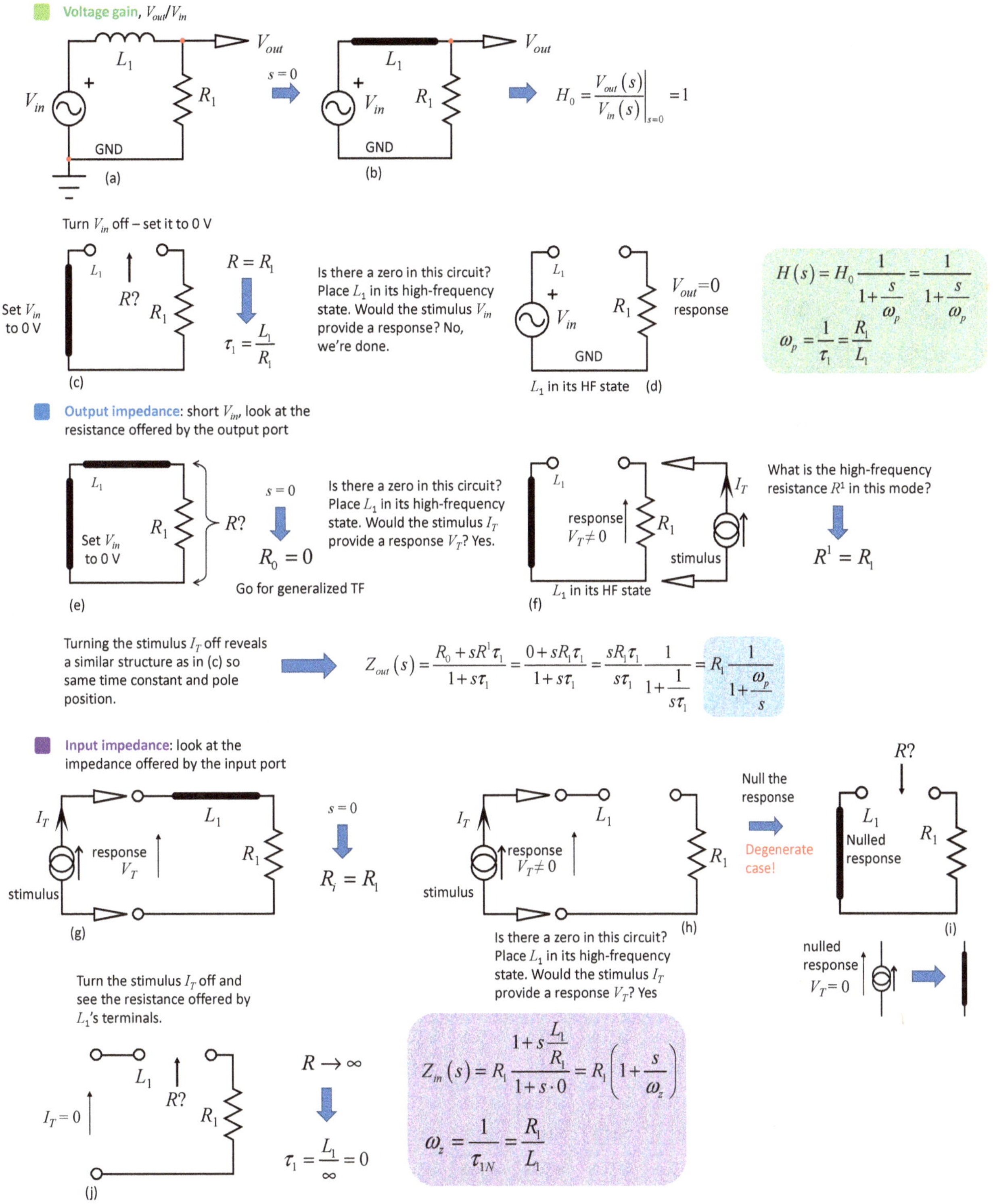

The Inductor is Now in Series with the Source

Figure 5.21

$R_1 := 1k\Omega$ $L_1 := 50mH$ $R_{inf} := 10^{12}\Omega$ $\|(x,y) := \dfrac{x \cdot y}{x+y}$

$H_0 := 1$ $R_0 := 0\Omega$ $R_i := R_1$

$H_1(s) := \dfrac{1}{1 + \dfrac{s}{\omega_p}}$ $Z_{out}(s) := R_1 \cdot \dfrac{1}{1 + \dfrac{\omega_p}{s}}$ $Z_{in}(s) := R_i \cdot \left(1 + \dfrac{s}{\omega_z}\right)$

$\tau_1 := \dfrac{L_1}{R_1} = 50 \cdot \mu s$ $\omega_p := \dfrac{1}{\tau_1}$ $f_p := \dfrac{\omega_p}{2 \cdot \pi} = 3.183 \cdot kHz$

$\tau_{1N} := \dfrac{L_1}{R_1}$ $\omega_z := \dfrac{1}{\tau_{1N}}$ $f_z := \dfrac{\omega_z}{2 \cdot \pi} = 3.183 \cdot kHz$ $H_{ref}(s) := \dfrac{R_1}{R_1 + s \cdot L_1}$ $Z_{outR}(s) := R_1 \| (s \cdot L_1)$ $Z_{inR}(s) := s \cdot L_1 + R_1$

Brute-force expressions

Gain

$20 \cdot \log\left(\left|H_1(i \cdot 2\pi \cdot f_k)\right|\right)$ ──
$20 \cdot \log\left(\left|H_{ref}(i \cdot 2\pi \cdot f_k)\right|\right)$ ·····

$\arg\left(H_1(i \cdot 2\pi \cdot f_k)\right) \cdot \dfrac{180}{\pi}$ ──
$\arg\left(H_{ref}(i \cdot 2\pi \cdot f_k)\right) \cdot \dfrac{180}{\pi}$ ·····

$20 \cdot \log\left(\dfrac{\left|Z_{out}(i \cdot 2\pi \cdot f_k)\right|}{\Omega}\right)$ ──
$20 \cdot \log\left(\dfrac{\left|Z_{outR}(i \cdot 2\pi \cdot f_k)\right|}{\Omega}\right)$ ·····

Output impedance

$\arg\left(Z_{out}(i \cdot 2\pi \cdot f_k)\right) \cdot \dfrac{180}{\pi}$ ──
$\arg\left(Z_{outR}(i \cdot 2\pi \cdot f_k)\right) \cdot \dfrac{180}{\pi}$ ·····

$20 \cdot \log\left(\dfrac{\left|Z_{in}(i \cdot 2\pi \cdot f_k)\right|}{\Omega}\right)$ ──
$20 \cdot \log\left(\dfrac{\left|Z_{inR}(i \cdot 2\pi \cdot f_k)\right|}{\Omega}\right)$ ·····

Input impedance

$\arg\left(Z_{in}(i \cdot 2\pi \cdot f_k)\right) \cdot \dfrac{180}{\pi}$ ──
$\arg\left(Z_{inR}(i \cdot 2\pi \cdot f_k)\right) \cdot \dfrac{180}{\pi}$ ·····

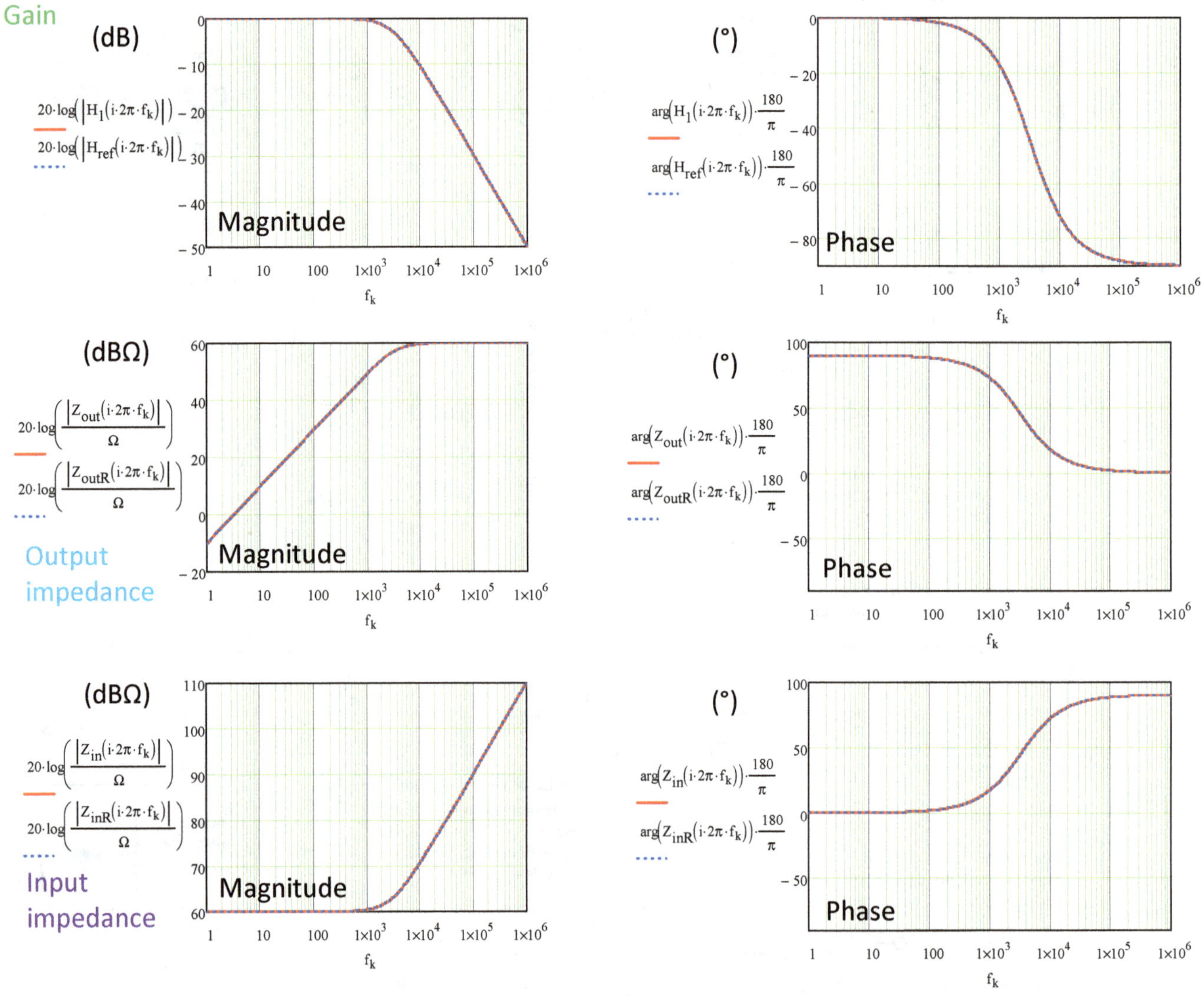

The Pole at the Origin and the Zero have Disappeared in the First Transfer Function

Figure 5.22

A resistor now loads the inductor and forms a voltage divider in Figure 5.23.

The transfer functions are easily found and Figure 5.24 confirms our findings. Again, you could indistinctly use the generalized transfer function expression or the one from the middle in Figure 5.1, whichever is the simplest for you.

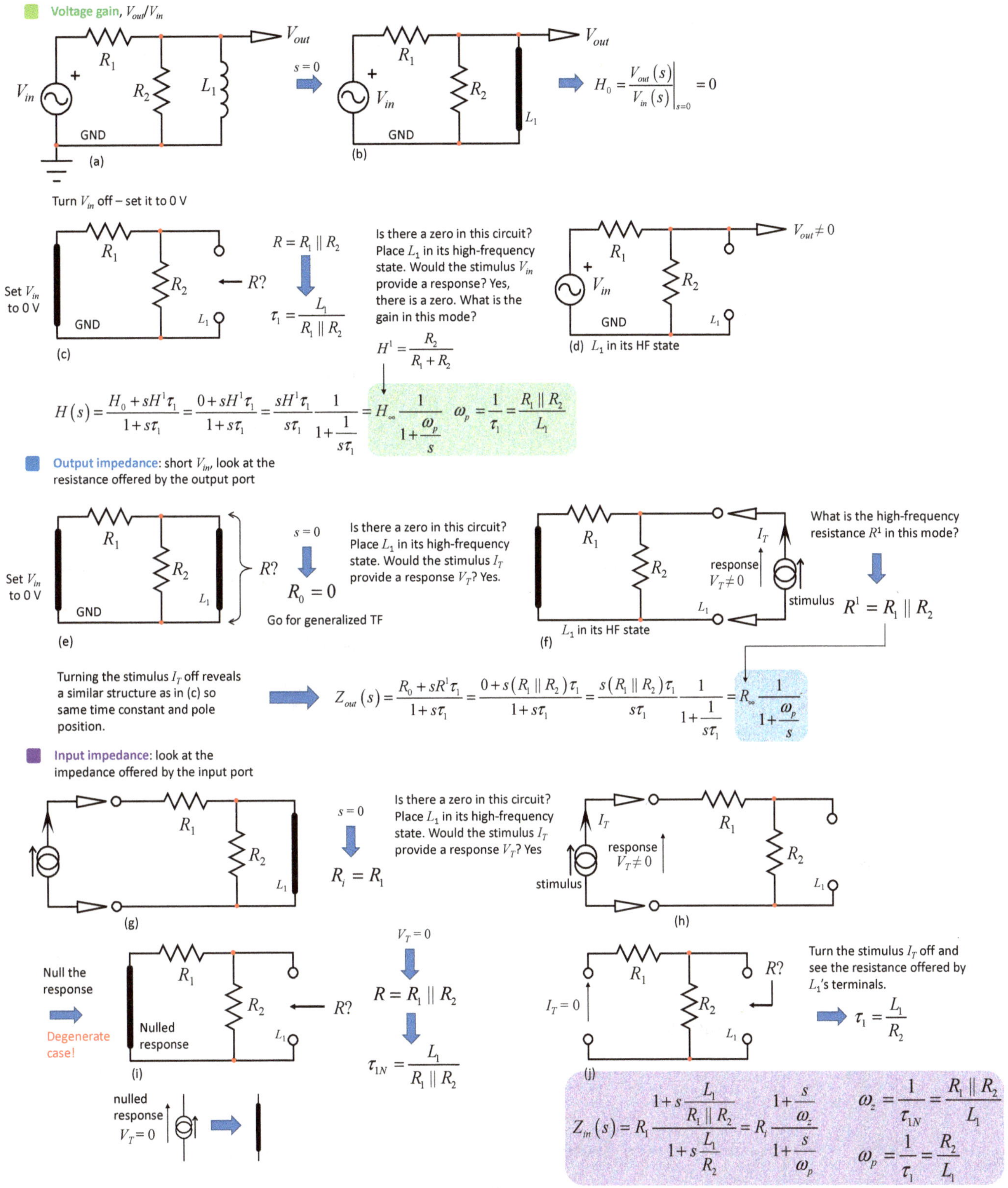
Voltage gain, V_{out}/V_{in}
V_{in}
R_1
R_2
L_1
V_{out}
GND
(a)
$s=0$
(b)
$H_0 = \left.\frac{V_{out}(s)}{V_{in}(s)}\right|_{s=0} = 0$
Turn V_{in} off – set it to 0 V
Set V_{in} to 0 V
(c)
$R?$
$R = R_1 \| R_2$
$\tau_1 = \frac{L_1}{R_1 \| R_2}$
Is there a zero in this circuit? Place L_1 in its high-frequency state. Would the stimulus V_{in} provide a response? Yes, there is a zero. What is the gain in this mode?
$H^1 = \frac{R_2}{R_1 + R_2}$
$V_{out} \neq 0$
(d) L_1 in its HF state
$H(s) = \frac{H_0 + sH^1\tau_1}{1+s\tau_1} = \frac{0+sH^1\tau_1}{1+s\tau_1} = \frac{sH^1\tau_1}{s\tau_1}\frac{1}{1+\frac{1}{s\tau_1}} = H_\infty \frac{1}{1+\frac{\omega_p}{s}}$ $\omega_p = \frac{1}{\tau_1} = \frac{R_1 \| R_2}{L_1}$
Output impedance: short V_{in}, look at the resistance offered by the output port
(e)
$R_0 = 0$
Go for generalized TF
Is there a zero in this circuit? Place L_1 in its high-frequency state. Would the stimulus I_T provide a response V_T? Yes.
I_T
response $V_T \neq 0$
stimulus
L_1 in its HF state
(f)
What is the high-frequency resistance R^1 in this mode?
$R^1 = R_1 \| R_2$
Turning the stimulus I_T off reveals a similar structure as in (c) so same time constant and pole position.
$Z_{out}(s) = \frac{R_0 + sR^1\tau_1}{1+s\tau_1} = \frac{0+s(R_1 \| R_2)\tau_1}{1+s\tau_1} = \frac{s(R_1 \| R_2)\tau_1}{s\tau_1}\frac{1}{1+\frac{1}{s\tau_1}} = R_\infty \frac{1}{1+\frac{\omega_p}{s}}$
Input impedance: look at the impedance offered by the input port
(g)
$R_i = R_1$
Is there a zero in this circuit? Place L_1 in its high-frequency state. Would the stimulus I_T provide a response V_T? Yes
(h)
Null the response
Degenerate case!
Nulled response
(i)
$V_T = 0$
$R = R_1 \| R_2$
$\tau_{1N} = \frac{L_1}{R_1 \| R_2}$
nulled response $V_T = 0$
$I_T = 0$
(j)
Turn the stimulus I_T off and see the resistance offered by L_1's terminals.
$\tau_1 = \frac{L_1}{R_2}$
$Z_{in}(s) = R_i \frac{1+s\frac{L_1}{R_1 \| R_2}}{1+s\frac{L_1}{R_2}} = R_i \frac{1+\frac{s}{\omega_z}}{1+\frac{s}{\omega_p}}$
$\omega_z = \frac{1}{\tau_{1N}} = \frac{R_1 \| R_2}{L_1}$
$\omega_p = \frac{1}{\tau_1} = \frac{R_2}{L_1}$

Despite the Addition of resistor R_2, the Zero at the Origin Remains due to the Short-Circuit of L_1 in Dc

Figure 5.23

$$R_1 := 1k\Omega \quad R_2 := 10k\Omega \quad L_1 := 50mH \quad \|(x,y) := \frac{x \cdot y}{x+y}$$

$$H_{inf} := \frac{R_2}{R_1 + R_2} = 0.90909 \quad R_{inf} := R_1 \| R_2 \quad R_i := R_1$$

$$\tau_{1a} := \frac{L_1}{R_1 \| R_2} = 55 \cdot \mu s \quad \omega_p := \frac{1}{\tau_{1a}} \quad f_p := \frac{\omega_p}{2 \cdot \pi} = 2.89373kHz$$

$$\tau_{1N} := \frac{L_1}{R_1 \| R_2} \quad \omega_z := \frac{1}{\tau_{1N}} \quad f_z := \frac{\omega_z}{2 \cdot \pi} = 2.89373kHz$$

$$\tau_{1b} := \frac{L_1}{R_2} = 5 \cdot \mu s \quad \omega_{pb} := \frac{1}{\tau_{1b}} \quad f_{pb} := \frac{\omega_{pb}}{2 \cdot \pi} = 31.83099kHz$$

$$H_1(s) := H_{inf} \cdot \frac{1}{1 + \frac{\omega_p}{s}} \quad Z_{out}(s) := R_{inf} \cdot \frac{1}{1 + \frac{\omega_p}{s}} \quad Z_{in}(s) := R_i \cdot \frac{1 + \frac{s}{\omega_z}}{1 + \frac{s}{\omega_{pb}}}$$

$$H_{ref}(s) := \frac{R_2 \| (s \cdot L_1)}{R_1 + R_2 \| (s \cdot L_1)} \quad Z_{outR}(s) := R_1 \| R_2 \| (s \cdot L_1) \quad Z_{inR}(s) := R_1 + R_2 \| (s \cdot L_1)$$

Brute-force expressions

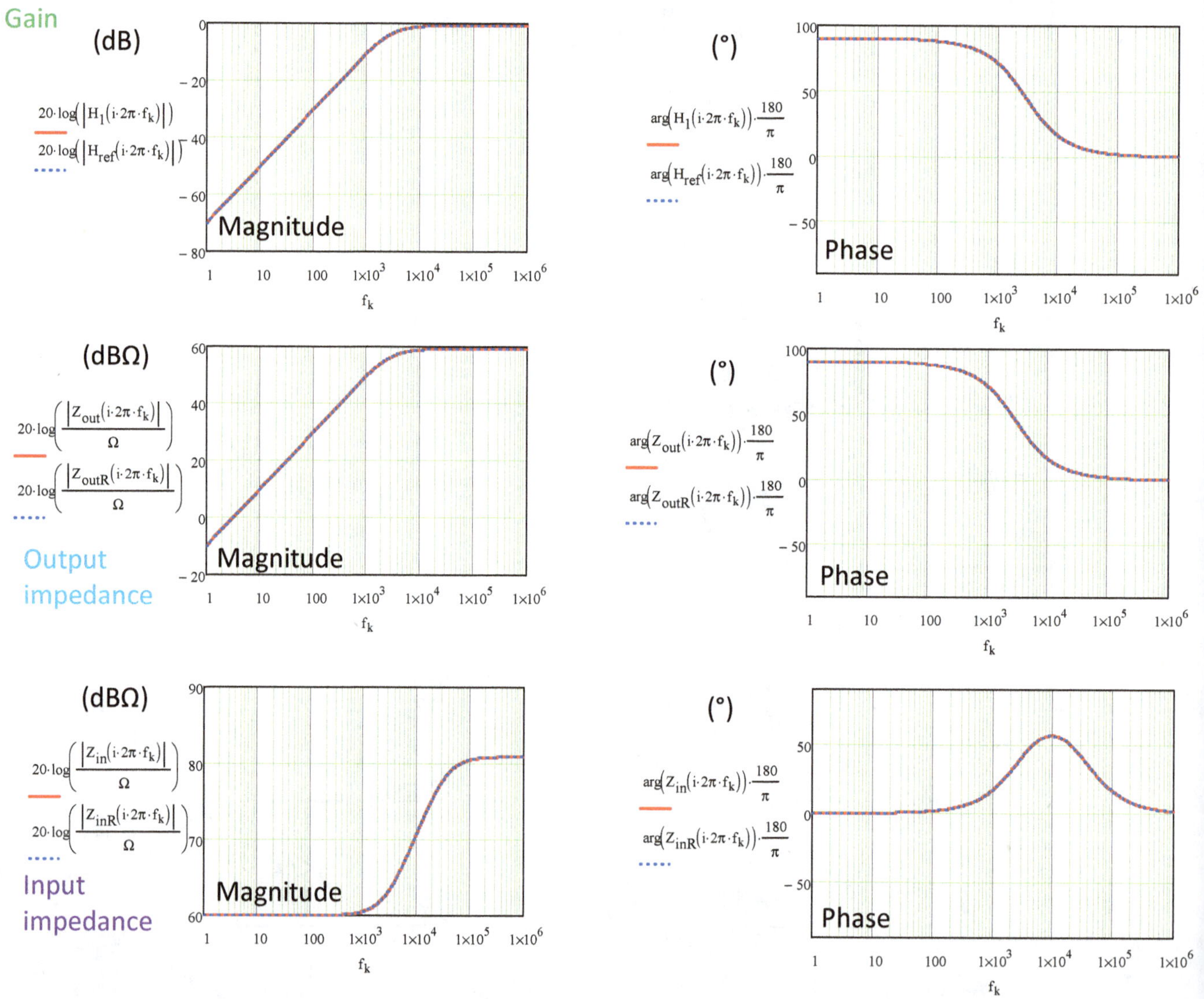

Responses Between Brute-Force and FACTs Expressions are Similar

Figure 5.24

Voltage gain, V_{out}/V_{in}

(a)

(b)

$s = 0$

$$H_0 = \frac{V_{out}(s)}{V_{in}(s)}\bigg|_{s=0} = 1$$

Turn V_{in} off – set it to 0 V

(c)

Set V_{in} to 0 V

$R = R_1 \| R_2$

$$\tau_1 = \frac{L_1}{R_1 \| R_2}$$

Is there a zero in this circuit? Place L_1 in its high-frequency state. Would the stimulus V_{in} provide a response? Yes. What could possibly stop the stimulus propagation?

$$Z_1(s) = sL_1 \| R_1 = \frac{sL_1 R_1}{sL_1 + R_1} \to \infty$$

$$sL_1 + R_1 = 0 \to s_z = -\frac{R_1}{L_1}$$

$Z_1(s) \to \infty$

(d)

$V_{out} = 0$

$$H(s) = 1 \cdot \frac{1 + s\dfrac{L_1}{R_1}}{1 + s\dfrac{L_1}{R_1 \| R_2}} = \frac{1 + \dfrac{s}{\omega_z}}{1 + \dfrac{s}{\omega_p}}$$

$$\omega_p = \frac{1}{\tau_1} = \frac{R_1 \| R_2}{L_1} \qquad \omega_z = \frac{R_1}{L_1}$$

$$H(s) = \frac{s\dfrac{L_1}{R_1}}{s\dfrac{L_1}{R_1 \| R_2}} \cdot \frac{1 + \dfrac{R_1}{sL_1}}{1 + \dfrac{R_1 \| R_2}{sL_1}} = H_\infty \frac{1 + \dfrac{\omega_z}{s}}{1 + \dfrac{\omega_p}{s}}$$

$$H_\infty = \frac{R_2}{R_1 + R_2} \qquad \omega_p = \frac{1}{\tau_1} = \frac{R_1 \| R_2}{L_1} \qquad \omega_z = \frac{R_1}{L_1}$$

Output impedance: short V_{in} look at the resistance offered by the output port

(e)

Set V_{in} to 0 V

$s = 0$

$R_0 = 0$

Go for generalized TF

$R?$

Is there a zero in this circuit? Place L_1 in its high-frequency state. Would the stimulus I_T provide a response V_T? Yes.

Determine high-frequency gain R^1.

(f) L_1 in its HF state

response $V_T \neq 0$

stimulus

I_T

What is the high-frequency resistance R^1 in this mode?

$$R^1 = R_1 \| R_2$$

Turning the stimulus I_T off reveals a similar structure as in (c) so same time constant and pole position.

$$Z_{out}(s) = \frac{R_0 + sR^1\tau_1}{1 + s\tau_1} = \frac{0 + s(R_1 \| R_2)\dfrac{L_1}{R_1 \| R_2}}{1 + s\dfrac{L_1}{R_1 \| R_2}} = \frac{s\dfrac{L_1}{R_1 \| R_2}}{s\dfrac{L_1}{R_1 \| R_2}} \cdot \frac{1}{1 + \dfrac{1}{sL_1}\dfrac{R_1 + R_2}{R_1 R_2}} = R_\infty \frac{1}{1 + \dfrac{\omega_p}{s}}$$

$$R_\infty = R_1 \| R_2$$

Input impedance: look at the impedance offered by the input port

(g)

stimulus

I_T

response V_T

$s = 0$

$R_i = R_2$

(h)

I_T

response $V_T \neq 0$

stimulus

Null the response

Degenerate case!

Nulled response

(i)

$V_T = 0$

$R = R_1 \| R_2$

$$\tau_{1N} = \frac{L_1}{R_1 \| R_2}$$

$R?$

Turn the stimulus I_T off and see the resistance offered by L_1's terminals.

(j)

$I_T = 0$

$R?$

$R = R_1$

$$\tau_1 = \frac{L_1}{R_1}$$

Is there a zero in this circuit? Place L_1 in its high-frequency state. Would the stimulus I_T provide a response V_T? Yes

nulled response $V_T = 0$

$$Z_{in}(s) = R_2 \frac{1 + s\dfrac{L_1}{R_1 \| R_2}}{1 + s\dfrac{L_1}{R_1}} = R_i \frac{1 + \dfrac{s}{\omega_z}}{1 + \dfrac{s}{\omega_p}}$$

$$\omega_p = \frac{L_1}{R_1} \qquad \omega_z = \frac{1}{\tau_{1N}} = \frac{L_1}{R_1 \| R_2}$$

$$Z_{in}(s) = R_2 \frac{s\dfrac{L_1}{R_1 \| R_2}}{s\dfrac{L_1}{R_1}} \cdot \frac{1 + \dfrac{R_1 \| R_2}{sL_1}}{1 + \dfrac{R_1}{sL_1}} = R_\infty \frac{1 + \dfrac{\omega_z}{s}}{1 + \dfrac{\omega_p}{s}}$$

$$\omega_p = \frac{L_1}{R_1} \qquad \omega_z = \frac{L_1}{R_1 \| R_2} \qquad R_\infty = R_1 + R_2$$

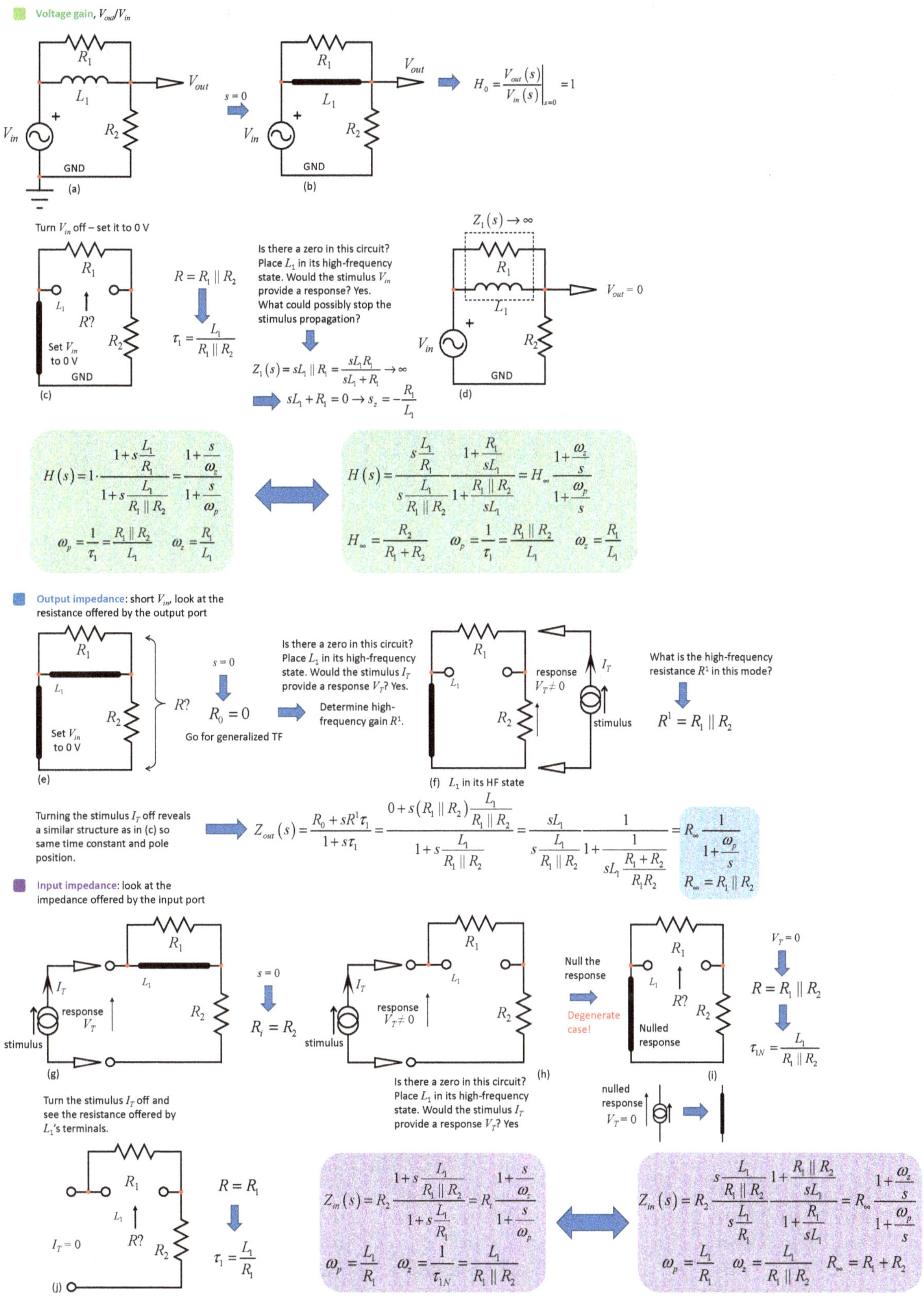

A Resistor in Parallel with the Inductor adds Complexity to the Network

Figure 5.25

We now add a resistor in parallel with the inductor and apply the FACTs as described in Figure 5.25.

The voltage transfer function is easy to find and the two paralleled resistors now set the pole. There is a zero and it is quickly identified by inspection. The transfer function is then assembled with a unity leading term, obtained for $s = 0$. We can also easily transform this expression having a leading term describing the gain when s approaches infinity. Obtain the expression with proper factoring and the use of inverted zero and pole.

Both expressions are similar but serve different design purposes, depending on what is of interest to you, low- or high-frequency gain. The output impedance is quickly obtained with the generalized transfer function and an inverted pole. Again, a very compact notation. Finally, the input impedance does not present particular difficulties and you can also rearrange the expression to highlight different characteristics.

$R_1 := 10\text{k}\Omega \quad R_2 := 5\text{k}\Omega \quad L_1 := 50\text{mH} \quad \|(x,y) := \dfrac{x \cdot y}{x + y}$

$H_0 := 1 \quad H_{inf} := \dfrac{R_2}{R_2 + R_1} = 0.33333 \quad R_{inf} := R_1 \| R_2 \quad R_i := R_2 \quad R_{infi} := R_1 + R_2$

$\tau_{1a} := \dfrac{L_1}{R_1 \| R_2} = 15 \cdot \mu s \quad \omega_p := \dfrac{1}{\tau_{1a}} \quad f_p := \dfrac{\omega_p}{2 \cdot \pi} = 10.61033\text{kHz}$

$\tau_{1N} := \dfrac{L_1}{R_1} = 5 \cdot \mu s \quad \omega_{za} := \dfrac{1}{\tau_{1N}} \quad f_{za} := \dfrac{\omega_{za}}{2 \cdot \pi} = 31.83099\text{kHz}$

$\tau_{1Nb} := \dfrac{L_1}{R_1 \| R_2} = 15 \cdot \mu s \quad \omega_{zb} := \dfrac{1}{\tau_{1Nb}} \quad f_{zb} := \dfrac{\omega_{zb}}{2 \cdot \pi} = 10.61033\text{kHz}$

$\tau_{1b} := \dfrac{L_1}{R_1} = 5 \cdot \mu s \quad \omega_{pb} := \dfrac{1}{\tau_{1b}} \quad f_{pb} := \dfrac{\omega_{pb}}{2 \cdot \pi} = 31.83099\text{kHz}$

$H_1(s) := H_0 \dfrac{1 + \dfrac{s}{\omega_{za}}}{1 + \dfrac{s}{\omega_p}}$

$H_2(s) := H_{inf} \dfrac{1 + \dfrac{\omega_{za}}{s}}{1 + \dfrac{\omega_p}{s}}$

$Z_{out}(s) := R_{inf} \dfrac{1}{1 + \dfrac{\omega_p}{s}}$

$Z_{in}(s) := R_1 \dfrac{1 + \dfrac{s}{\omega_{zb}}}{1 + \dfrac{s}{\omega_{pb}}}$

$Z_{in2}(s) := R_{infi} \dfrac{1 + \dfrac{\omega_{zb}}{s}}{1 + \dfrac{\omega_{pb}}{s}}$

$H_{ref}(s) := \dfrac{R_2}{R_2 + R_1 \| (s \cdot L_1)} \quad Z_{outR}(s) := R_1 \| R_2 \| (s \cdot L_1) \quad Z_{inR}(s) := R_2 + R_1 \| (s \cdot L_1)$

Brute-force expressions

Gain

Output impedance

Input impedance

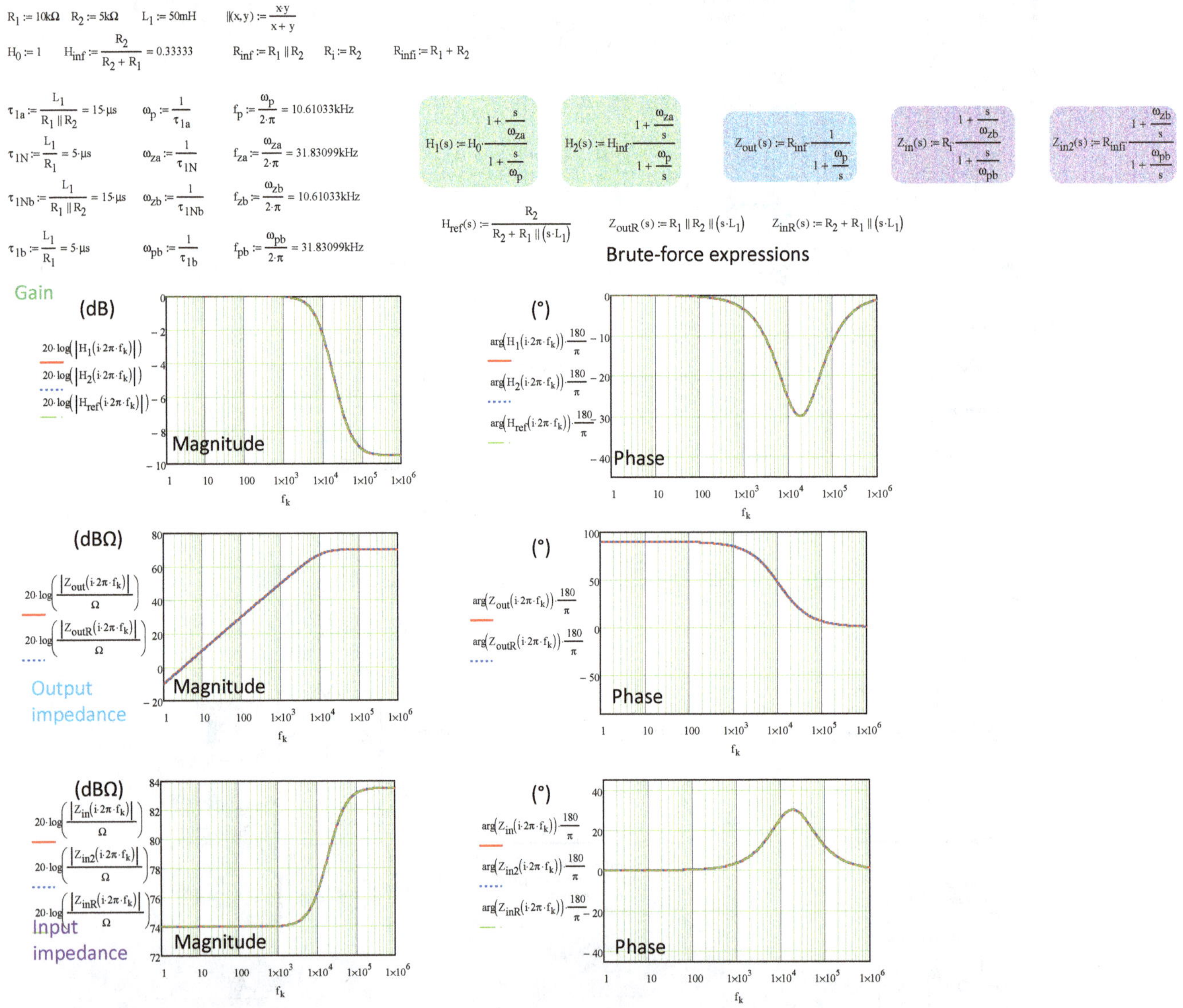

All Transfer Functions—Including the Rearranged Versions—via Brute Force or FACTs are Similar in Magnitude and Phase

Figure 5.26

Figure 5.26 confirms our expressions are correct.

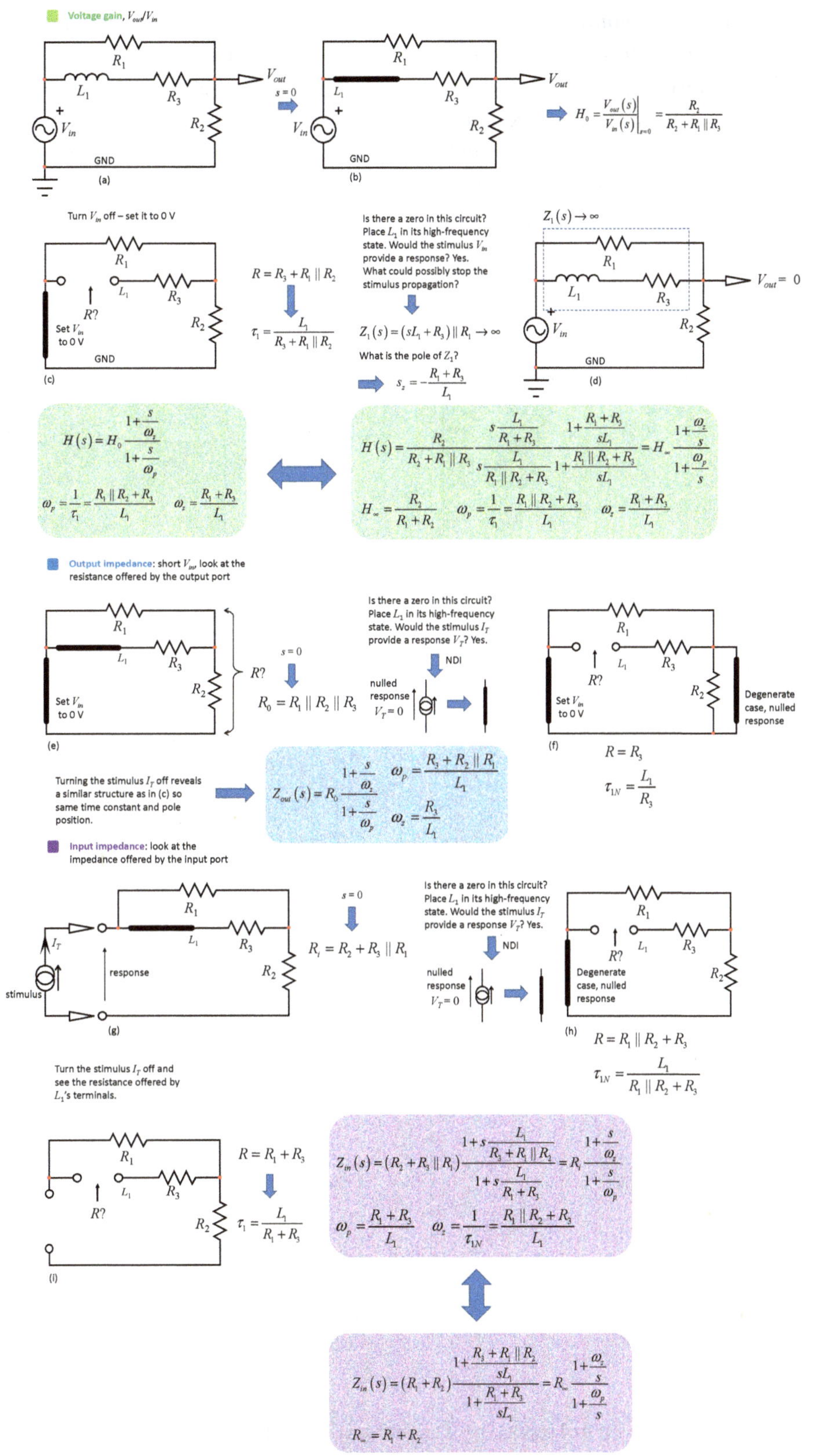
Voltage gain, V_{out}/V_{in}
(a)
(b)
GND
$s = 0$
$H_0 = \frac{V_{out}(s)}{V_{in}(s)}\Big|_{s=0} = \frac{R_2}{R_2 + R_1 \| R_3}$
Turn V_{in} off – set it to 0 V
Set V_{in} to 0 V
$R?$
(c)
$R = R_3 + R_1 \| R_2$
$\tau_1 = \frac{L_1}{R_3 + R_1 \| R_2}$
Is there a zero in this circuit? Place L_1 in its high-frequency state. Would the stimulus V_{in} provide a response? Yes. What could possibly stop the stimulus propagation?
$Z_1(s) = (sL_1 + R_3) \| R_1 \to \infty$
What is the pole of Z_1?
$s_z = -\frac{R_1 + R_3}{L_1}$
$Z_1(s) \to \infty$
$V_{out} = 0$
(d)
$H(s) = H_0 \frac{1 + \frac{s}{\omega_z}}{1 + \frac{s}{\omega_p}}$
$\omega_p = \frac{1}{\tau_1} = \frac{R_3 \| R_2 + R_1}{L_1}$
$\omega_z = \frac{R_1 + R_3}{L_1}$
$H(s) = \frac{R_2}{R_2 + R_1 \| R_3} \frac{1 + s\frac{L_1}{R_1 + R_3}}{1 + s\frac{L_1}{R_1 \| R_2 + R_3}} = H_\infty \frac{1 + \frac{R_1 + R_3}{sL_1}}{1 + \frac{R_1 \| R_2 + R_3}{sL_1}} = H_\infty \frac{1 + \frac{\omega_z}{s}}{1 + \frac{\omega_p}{s}}$
$H_\infty = \frac{R_2}{R_1 + R_2}$
$\omega_p = \frac{1}{\tau_1} = \frac{R_1 \| R_2 + R_3}{L_1}$
$\omega_z = \frac{R_1 + R_3}{L_1}$
Output impedance: short V_{in}, look at the resistance offered by the output port
(e)
$R_0 = R_1 \| R_2 \| R_3$
Is there a zero in this circuit? Place L_1 in its high-frequency state. Would the stimulus I_T provide a response V_T? Yes.
NDI
nulled response
$V_T = 0$
(f)
Degenerate case, nulled response
$R = R_3$
$\tau_{1N} = \frac{L_1}{R_3}$
Turning the stimulus I_T off reveals a similar structure as in (c) so same time constant and pole position.
$Z_{out}(s) = R_0 \frac{1 + \frac{s}{\omega_z}}{1 + \frac{s}{\omega_p}}$
$\omega_p = \frac{R_3 + R_2 \| R_1}{L_1}$
$\omega_z = \frac{R_3}{L_1}$
Input impedance: look at the impedance offered by the input port
stimulus
response
(g)
$R_i = R_2 + R_3 \| R_1$
Is there a zero in this circuit? Place L_1 in its high-frequency state. Would the stimulus I_T provide a response V_T? Yes.
(h)
$R = R_1 \| R_2 + R_3$
$\tau_{1N} = \frac{L_1}{R_1 \| R_2 + R_3}$
Turn the stimulus I_T off and see the resistance offered by L_1's terminals.
(i)
$R = R_1 + R_3$
$\tau_1 = \frac{L_1}{R_1 + R_3}$
$Z_{in}(s) = (R_2 + R_3 \| R_1) \frac{1 + s\frac{L_1}{R_3 + R_1 \| R_2}}{1 + s\frac{L_1}{R_1 + R_3}} = R_i \frac{1 + \frac{s}{\omega_z}}{1 + \frac{s}{\omega_p}}$
$\omega_p = \frac{R_1 + R_3}{L_1}$
$\omega_z = \frac{1}{\tau_{1N}} = \frac{R_1 \| R_2 + R_3}{L_1}$
$Z_{in}(s) = (R_1 + R_2) \frac{1 + \frac{R_3 + R_1 \| R_2}{sL_1}}{1 + \frac{R_1 + R_3}{sL_1}} = R_\infty \frac{1 + \frac{\omega_z}{s}}{1 + \frac{\omega_p}{s}}$
$R_\infty = R_1 + R_2$

R_3 added in Series with the Inductor Models its Equivalent Series Resistance

Figure 5.27

Let's make things a little more complicated and add resistor R_3 in series with the inductor as if you wanted to include its equivalent series resistance (ESR). You are now looking at Figure 5.27. Applying the technique to this circuit reveals the three transfer functions with the help of a few simple sketches. Some of the final expressions can be rearranged to factor a leading term representing the asymptotic dc or high-frequency value of the considered transfer function. You can also directly use the middle expression in Figure 5.1 to get there straight away.

Inverted poles and zeroes are very handy to that respect and lead to the most compact expressions.

We test all of these expressions in Figure 5.28.

$R_1 := 10k\Omega \quad R_2 := 5k\Omega \quad L_1 := 50mH \quad R_3 := 500\Omega \qquad \|(x,y) := \frac{xy}{x+y}$

$H_0 := \frac{R_2}{R_2 + R_1 \| R_3} = 0.913 \qquad H_{inf} := \frac{R_2}{R_2 + R_1} = 0.333 \qquad R_0 := R_1 \| R_2 \| R_3 = 434.783\Omega$

$R_1 := R_2 + R_3 \| R_1 = 5.476k\Omega \qquad R_{inf} := R_1 + R_2 = 15k\Omega$ R_{inf} and H_{inf} are obtained when s approaches infinity

$\tau_{1a} := \frac{L_1}{R_1 \| R_2 + R_3} = 13.043\mu s \qquad \omega_p := \frac{1}{\tau_{1a}} \qquad f_p := \frac{\omega_p}{2 \cdot \pi} = 12.202kHz$

$\tau_{1N} := \frac{L_1}{R_1 + R_3} = 4.762\mu s \qquad \omega_{za} := \frac{1}{\tau_{1N}} \qquad f_{za} := \frac{\omega_{za}}{2 \cdot \pi} = 33.423kHz$

$\tau_{1Nb} := \frac{L_1}{R_3} = 100\mu s \qquad \omega_{zb} := \frac{1}{\tau_{1Nb}} \qquad f_{zb} := \frac{\omega_{zb}}{2 \cdot \pi} = 1.592kHz$

$\tau_{1Nc} := \frac{L_1}{R_1 \| R_2 + R_3} = 13.043\mu s \qquad \omega_{zc} := \frac{1}{\tau_{1Nc}} \qquad f_{zc} := \frac{\omega_{zc}}{2 \cdot \pi} = 12.202kHz$

$\tau_{1c} := \frac{L_1}{R_1 + R_3} = 4.762\mu s \qquad \omega_{pc} := \frac{1}{\tau_{1c}} \qquad f_{pc} := \frac{\omega_{pc}}{2 \cdot \pi} = 33.423kHz$

$H_1(s) := H_0 \dfrac{1 + \frac{s}{\omega_{za}}}{1 + \frac{s}{\omega_p}} \qquad H_2(s) := H_{inf} \dfrac{1 + \frac{\omega_{za}}{s}}{1 + \frac{\omega_p}{s}} \qquad Z_{out}(s) := R_0 \dfrac{1 + \frac{s}{\omega_{zb}}}{1 + \frac{s}{\omega_p}} \qquad Z_{in}(s) := R_1 \dfrac{1 + \frac{s}{\omega_{zc}}}{1 + \frac{s}{\omega_{pc}}} \qquad Z_{in2}(s) := R_{inf} \dfrac{1 + \frac{\omega_{zc}}{s}}{1 + \frac{\omega_{pc}}{s}}$

$H_{ref}(s) := \dfrac{R_2}{R_2 + R_1 \| (s \cdot L_1 + R_3)} \qquad Z_{outR}(s) := R_1 \| R_2 \| (s \cdot L_1 + R_3) \qquad Z_{inR}(s) := R_2 + R_1 \| (s \cdot L_1 + R_3)$

Brute-force expressions

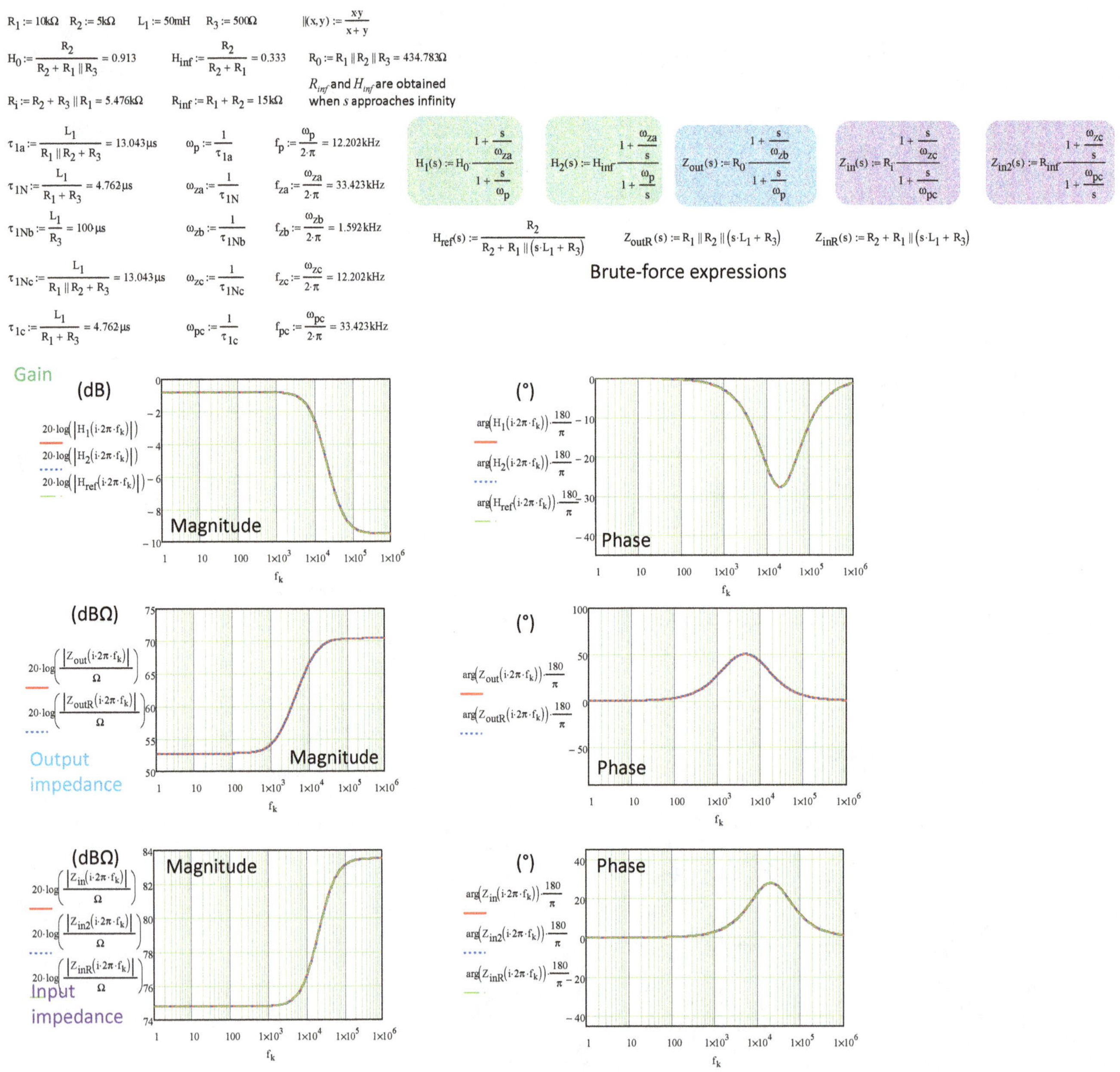

All Transfer Functions are Tested and Produce the Same Plots

Figure 5.28

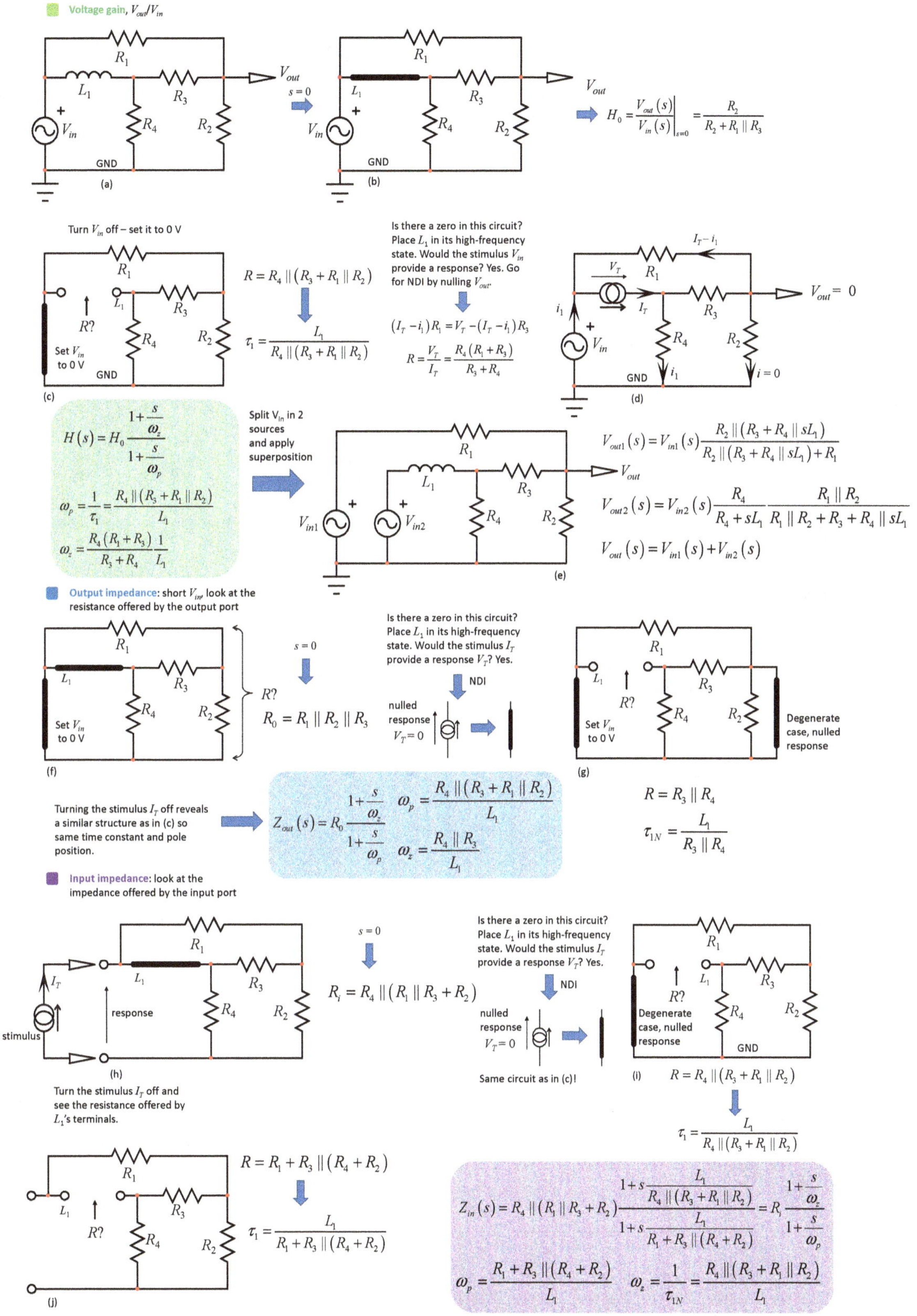

Voltage gain, V_{out}/V_{in}
R_1 L_1 R_3 R_4 R_2 V_{in} GND V_{out}
(a)
$s = 0$
(b)
$H_0 = \left.\frac{V_{out}(s)}{V_{in}(s)}\right|_{s=0} = \frac{R_2}{R_2 + R_1 \| R_3}$
Turn V_{in} off – set it to 0 V
$R?$
Set V_{in} to 0 V
(c)
$R = R_4 \| (R_3 + R_1 \| R_2)$
$\tau_1 = \frac{L_1}{R_4 \| (R_3 + R_1 \| R_2)}$
Is there a zero in this circuit? Place L_1 in its high-frequency state. Would the stimulus V_{in} provide a response? Yes. Go for NDI by nulling V_{out}.
$(I_T - i_1)R_1 = V_T - (I_T - i_1)R_3$
$R = \frac{V_T}{I_T} = \frac{R_4(R_1 + R_3)}{R_3 + R_4}$
$I_T - i_1$ V_T I_T i_1 $V_{out} = 0$ $i = 0$
(d)
$H(s) = H_0 \frac{1 + \frac{s}{\omega_z}}{1 + \frac{s}{\omega_p}}$
$\omega_p = \frac{1}{\tau_1} = \frac{R_4 \| (R_3 + R_1 \| R_2)}{L_1}$
$\omega_z = \frac{R_4(R_1 + R_3)}{R_3 + R_4} \frac{1}{L_1}$
Split V_{in} in 2 sources and apply superposition
V_{in1} V_{in2}
(e)
$V_{out1}(s) = V_{in1}(s) \frac{R_2 \| (R_3 + R_4 \| sL_1)}{R_2 \| (R_3 + R_4 \| sL_1) + R_1}$
$V_{out2}(s) = V_{in2}(s) \frac{R_4}{R_4 + sL_1} \frac{R_1 \| R_2}{R_1 \| R_2 + R_3 + R_4 \| sL_1}$
$V_{out}(s) = V_{in1}(s) + V_{in2}(s)$
Output impedance: short V_{in}, look at the resistance offered by the output port
Set V_{in} to 0 V
(f)
$s = 0$
$R_0 = R_1 \| R_2 \| R_3$
Is there a zero in this circuit? Place L_1 in its high-frequency state. Would the stimulus I_T provide a response V_T? Yes.
NDI
nulled response $V_T = 0$
Degenerate case, nulled response
(g)
$R = R_3 \| R_4$
$\tau_{1N} = \frac{L_1}{R_3 \| R_4}$
Turning the stimulus I_T off reveals a similar structure as in (c) so same time constant and pole position.
$Z_{out}(s) = R_0 \frac{1 + \frac{s}{\omega_z}}{1 + \frac{s}{\omega_p}}$
$\omega_p = \frac{R_4 \| (R_3 + R_1 \| R_2)}{L_1}$
$\omega_z = \frac{R_4 \| R_3}{L_1}$
Input impedance: look at the impedance offered by the input port
I_T stimulus response
(h)
Turn the stimulus I_T off and see the resistance offered by L_1's terminals.
$s = 0$
$R_i = R_4 \| (R_1 \| R_3 + R_2)$
Is there a zero in this circuit? Place L_1 in its high-frequency state. Would the stimulus I_T provide a response V_T? Yes.
NDI
nulled response $V_T = 0$
Same circuit as in (c)!
Degenerate case, nulled response
GND
(i)
$R = R_4 \| (R_3 + R_1 \| R_2)$
$\tau_1 = \frac{L_1}{R_4 \| (R_3 + R_1 \| R_2)}$
$R = R_1 + R_3 \| (R_4 + R_2)$
$\tau_1 = \frac{L_1}{R_1 + R_3 \| (R_4 + R_2)}$
(j)
$Z_{in}(s) = R_4 \| (R_1 \| R_3 + R_2) \frac{1 + s \frac{L_1}{R_4 \| (R_3 + R_1 \| R_2)}}{1 + s \frac{L_1}{R_1 + R_3 \| (R_4 + R_2)}} = R_i \frac{1 + \frac{s}{\omega_z}}{1 + \frac{s}{\omega_p}}$
$\omega_p = \frac{R_1 + R_3 \| (R_4 + R_2)}{L_1}$ $\omega_z = \frac{1}{\tau_{1N}} = \frac{R_4 \| (R_3 + R_1 \| R_2)}{L_1}$

Resistor R_4 Seems to Make Things Difficult, but does it?

Figure 5.29

$R_1 := 10k\Omega \quad R_2 := 4.7k\Omega \quad L_1 := 50mH \qquad \|(x,y) := \frac{x\cdot y}{x+y}$

$R_4 := 5k\Omega \quad R_3 := 2.2k\Omega$

$H_0 := \frac{R_2}{R_2 + R_1 \| R_3} = 0.723 \qquad R_0 := R_1 \| R_2 \| R_3 = 1.303\times 10^3\,\Omega$

$R_i := R_4 \| (R_1 \| R_3 + R_2) = 2.827k\Omega$

$\tau_{1a} := \frac{L_1}{R_4 \| (R_3 + R_1 \| R_2)} = 19.264\mu s \qquad \omega_p := \frac{1}{\tau_{1a}} \qquad f_p := \frac{\omega_p}{2\cdot\pi} = 8.262kHz$

$\tau_{1N} := \frac{L_1}{\frac{R_4\cdot(R_1+R_3)}{R_3+R_4}} = 5.902\mu s \qquad \omega_{za} := \frac{1}{\tau_{1N}} \qquad f_{za} := \frac{\omega_{za}}{2\cdot\pi} = 26.968kHz$

$\tau_{1Nb} := \frac{L_1}{R_3 \| R_4} = 32.727\mu s \qquad \omega_{zb} := \frac{1}{\tau_{1Nb}} \qquad f_{zb} := \frac{\omega_{zb}}{2\cdot\pi} = 4.863kHz$

$\tau_{1Nc} := \frac{L_1}{R_4 \| (R_3 + R_1 \| R_2)} = 19.264\mu s \qquad \omega_{zc} := \frac{1}{\tau_{1Nc}} \qquad f_{zc} := \frac{\omega_{zc}}{2\cdot\pi} = 8.262kHz$

$\tau_{1c} := \frac{L_1}{R_1 + R_3 \| (R_4 + R_2)} = 4.24\mu s \qquad \omega_{pc} := \frac{1}{\tau_{1c}} \qquad f_{pc} := \frac{\omega_{pc}}{2\cdot\pi} = 37.539kHz$

$H_1(s) := H_0\cdot\frac{1+\frac{s}{\omega_{za}}}{1+\frac{s}{\omega_p}} \qquad Z_{out}(s) := R_0\cdot\frac{1+\frac{s}{\omega_{zb}}}{1+\frac{s}{\omega_p}} \qquad Z_{in}(s) := R_i\cdot\frac{1+\frac{s}{\omega_{zc}}}{1+\frac{s}{\omega_{pc}}}$

$H_{ref}(s) := \frac{R_2 \| [R_3 + R_4 \| (s\cdot L_1)]}{R_2 \| [R_3 + R_4 \| (s\cdot L_1)] + R_1} + \frac{R_4}{R_4 + s\cdot L_1}\cdot\frac{R_1 \| R_2}{R_1 \| R_2 + R_3 + R_4 \| (s\cdot L_1)}$

$Z_{outR}(s) := R_1 \| R_2 \| [R_4 \| (s\cdot L_1) + R_3] \qquad H_{zin} := READPRN("Zin.txt")$ SIMetrix data import

Brute-force expressions

SIMetrix circuit for $Z_{in}(s)$ →

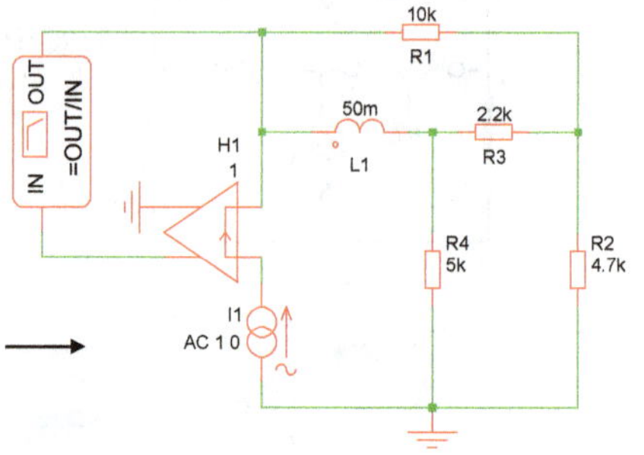

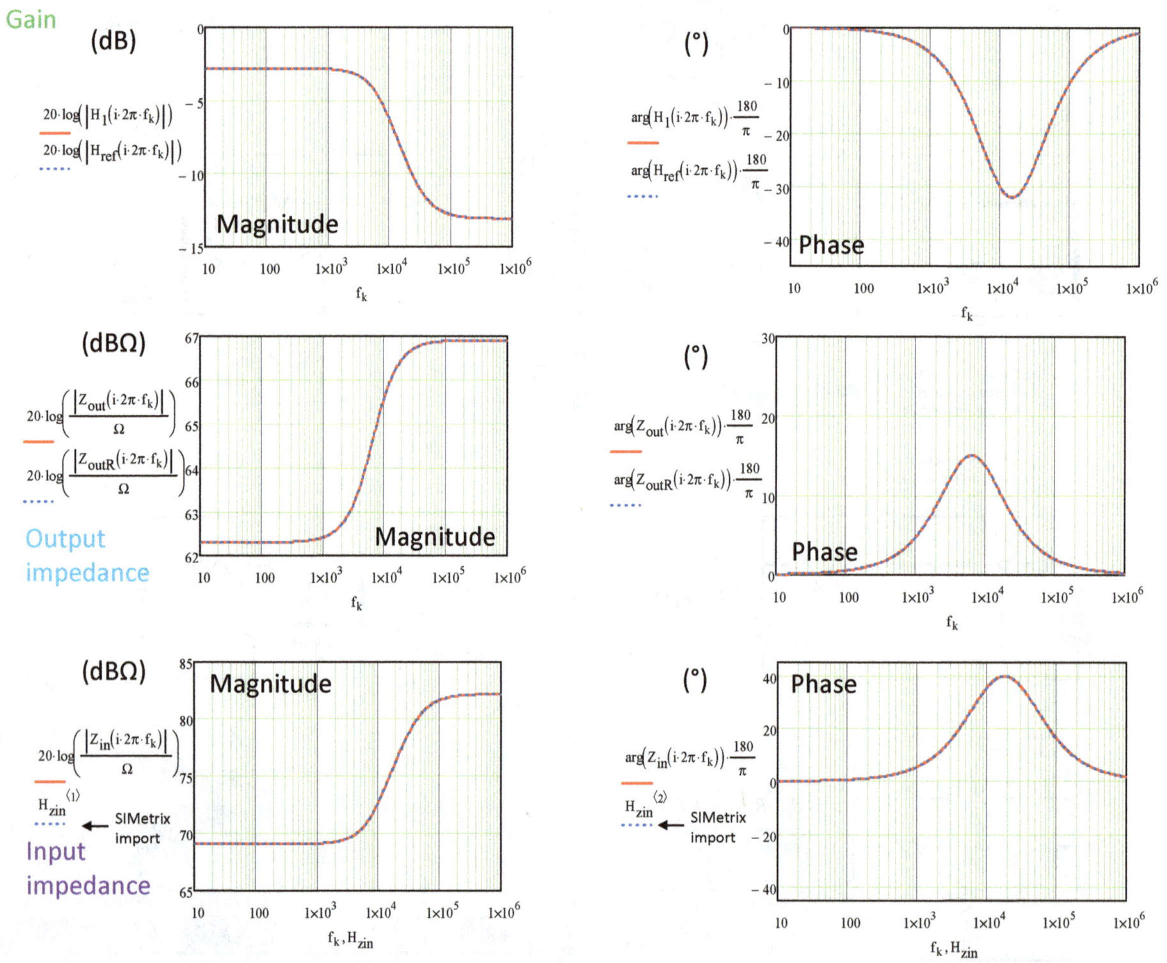

All Transfer Functions are Tested and Create the Same Plots

Figure 5.30

What if we add a resistor loading the junction between L_1 and R_3 to ground?

This is what Figure 5.29 shows.

Resistor R_4 clearly complicates the circuit, but not a problem for the FACTs.

I used superposition to derive the gain transfer function but other options are possible.

For the input impedance, this time, rather than solving the circuit using Kirchhoff voltage and current laws (KCL and KVL), I used SIMetrix® which is a SPICE engine whose data were imported into Mathcad® then pasted in the magnitude and phase graphs. Perfect match as shown in Figure 5.30.

The inductor is now replaced by a capacitor as shown in Figure 5.31.

What is interesting is that the resistance driving the energy-storing element—*L* or *C*—remains the same and we can reuse most of what we previously calculated.

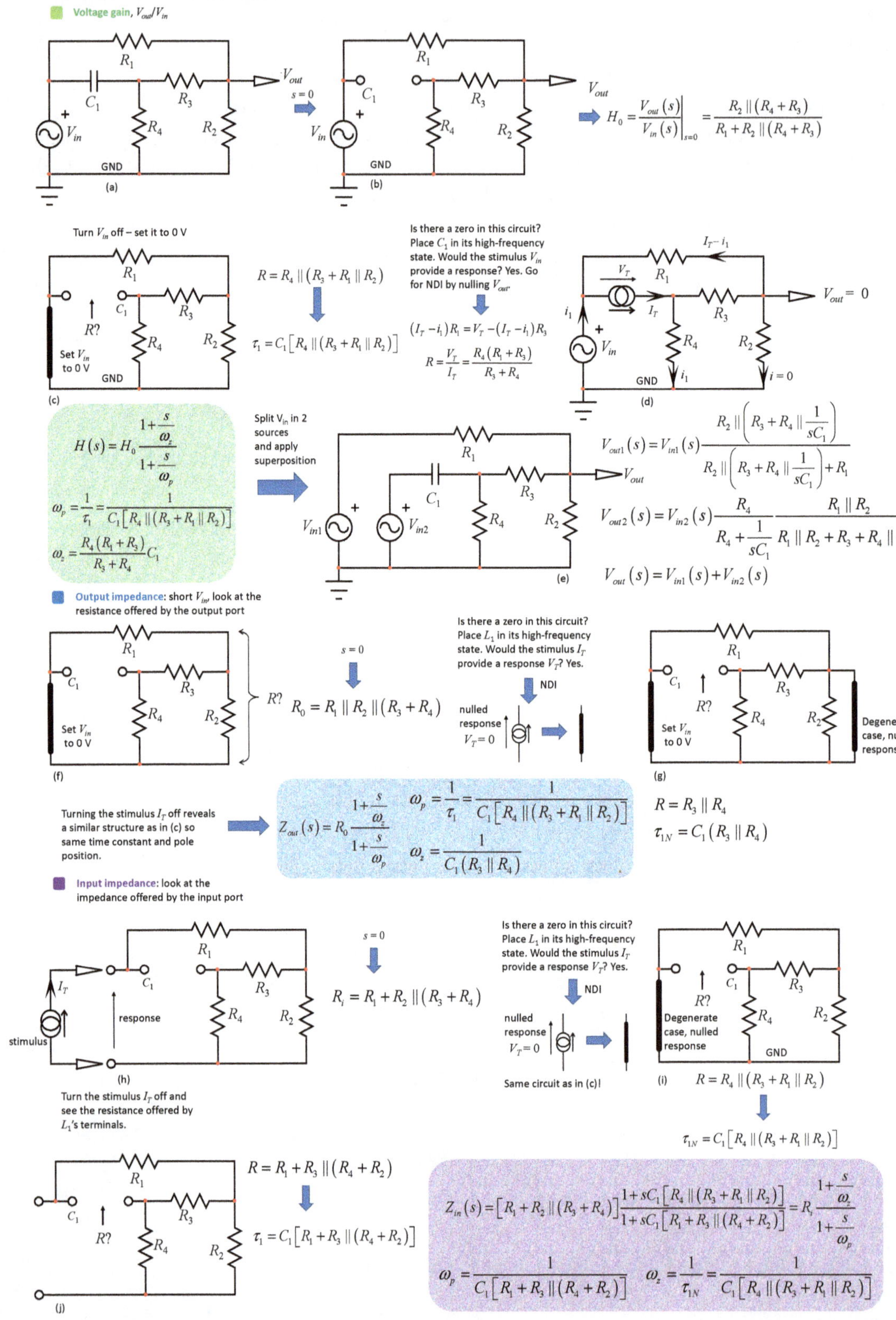

Voltage gain, V_{out}/V_{in}
R_1 C_1 R_3 R_4 R_2 V_{in} V_{out} GND (a)
$s = 0$
(b)
$H_0 = \left.\frac{V_{out}(s)}{V_{in}(s)}\right|_{s=0} = \frac{R_2 \| (R_4 + R_3)}{R_1 + R_2 \| (R_4 + R_3)}$
Turn V_{in} off – set it to 0 V
R? Set V_{in} to 0 V (c)
$R = R_4 \| (R_3 + R_1 \| R_2)$
$\tau_1 = C_1 [R_4 \| (R_3 + R_1 \| R_2)]$
Is there a zero in this circuit? Place C_1 in its high-frequency state. Would the stimulus V_{in} provide a response? Yes. Go for NDI by nulling V_{out}.
$(I_T - i_1)R_1 = V_T - (I_T - i_1)R_3$
$R = \frac{V_T}{I_T} = \frac{R_4(R_1 + R_3)}{R_3 + R_4}$
$I_T - i_1$ V_T I_T i_1 $V_{out} = 0$ $i = 0$ (d)
$H(s) = H_0 \frac{1 + \frac{s}{\omega_z}}{1 + \frac{s}{\omega_p}}$
$\omega_p = \frac{1}{\tau_1} = \frac{1}{C_1 [R_4 \| (R_3 + R_1 \| R_2)]}$
$\omega_z = \frac{R_4(R_1 + R_3)}{R_3 + R_4} C_1$
Split V_{in} in 2 sources and apply superposition
V_{in1} V_{in2} (e)
$V_{out1}(s) = V_{in1}(s) \frac{R_2 \| \left(R_3 + R_4 \| \frac{1}{sC_1}\right)}{R_2 \| \left(R_3 + R_4 \| \frac{1}{sC_1}\right) + R_1}$
$V_{out2}(s) = V_{in2}(s) \frac{R_4}{R_4 + \frac{1}{sC_1}} \frac{R_1 \| R_2}{R_1 \| R_2 + R_3 + R_4 \| \frac{1}{sC_1}}$
$V_{out}(s) = V_{in1}(s) + V_{in2}(s)$
Output impedance: short V_{in}, look at the resistance offered by the output port
Set V_{in} to 0 V (f)
$s = 0$
$R_0 = R_1 \| R_2 \| (R_3 + R_4)$
Is there a zero in this circuit? Place L_1 in its high-frequency state. Would the stimulus I_T provide a response V_T? Yes.
NDI
nulled response $V_T = 0$
Degenerate case, nulled response (g)
$R = R_3 \| R_4$
$\tau_{1N} = C_1 (R_3 \| R_4)$
Turning the stimulus I_T off reveals a similar structure as in (c) so same time constant and pole position.
$Z_{out}(s) = R_0 \frac{1 + \frac{s}{\omega_z}}{1 + \frac{s}{\omega_p}}$
$\omega_p = \frac{1}{\tau_1} = \frac{1}{C_1 [R_4 \| (R_3 + R_1 \| R_2)]}$
$\omega_z = \frac{1}{C_1 (R_3 \| R_4)}$
Input impedance: look at the impedance offered by the input port
stimulus response (h)
$s = 0$
$R_i = R_1 + R_2 \| (R_3 + R_4)$
Turn the stimulus I_T off and see the resistance offered by L_1's terminals.
Is there a zero in this circuit? Place L_1 in its high-frequency state. Would the stimulus I_T provide a response V_T? Yes.
NDI
nulled response $V_T = 0$
Same circuit as in (c)!
Degenerate case, nulled response (i)
$R = R_4 \| (R_3 + R_1 \| R_2)$
$\tau_{1N} = C_1 [R_4 \| (R_3 + R_1 \| R_2)]$
(j)
$R = R_1 + R_3 \| (R_4 + R_2)$
$\tau_1 = C_1 [R_1 + R_3 \| (R_4 + R_2)]$
$Z_{in}(s) = [R_1 + R_2 \| (R_3 + R_4)] \frac{1 + sC_1 [R_4 \| (R_3 + R_1 \| R_2)]}{1 + sC_1 [R_1 + R_3 \| (R_4 + R_2)]} = R_i \frac{1 + \frac{s}{\omega_z}}{1 + \frac{s}{\omega_p}}$
$\omega_p = \frac{1}{C_1 [R_1 + R_3 \| (R_4 + R_2)]}$
$\omega_z = \frac{1}{\tau_{1N}} = \frac{1}{C_1 [R_4 \| (R_3 + R_1 \| R_2)]}$

The Capacitor Replaces the Inductor and the Results from the Previous Example can be Reused

Figure 5.31

$R_1 := 10k\Omega \quad R_2 := 4.7k\Omega \quad C_1 := 22nF \qquad \|(x,y) := \frac{x \cdot y}{x+y}$

$R_4 := 5k\Omega \quad R_3 := 2.2k\Omega$

$H_0 := \frac{R_2 \| (R_3 + R_4)}{R_1 + R_2 \| (R_3 + R_4)} = 0.221 \qquad R_0 := R_1 \| R_2 \| (R_3 + R_4) = 2.214 \times 10^3\,\Omega$

$R_i := R_2 \| (R_3 + R_4) + R_1 = 12.844k\Omega$

$\tau_{1a} := C_1 \cdot [R_4 \| (R_3 + R_1 \| R_2)] = 0.057ms \qquad \omega_p := \frac{1}{\tau_{1a}} \qquad f_p := \frac{\omega_p}{2 \cdot \pi} = 2.787kHz$

$\tau_{1N} := C_1 \cdot \left[\frac{R_4 \cdot (R_1 + R_3)}{R_3 + R_4}\right] = 0.186ms \qquad \omega_{za} := \frac{1}{\tau_{1N}} \qquad f_{za} := \frac{\omega_{za}}{2 \cdot \pi} = 0.854kHz$

$\tau_{1Nb} := C_1 \cdot (R_4 \| R_3) = 0.034ms \qquad \omega_{zb} := \frac{1}{\tau_{1Nb}} \qquad f_{zb} := \frac{\omega_{zb}}{2 \cdot \pi} = 4.735kHz$

$\tau_{1Nc} := C_1 \cdot [R_4 \| (R_3 + R_1 \| R_2)] = 0.057ms \qquad \omega_{zc} := \frac{1}{\tau_{1Nc}} \qquad f_{zc} := \frac{\omega_{zc}}{2 \cdot \pi} = 2.787kHz$

$\tau_{1c} := C_1 \cdot [R_1 + R_3 \| (R_4 + R_2)] = 0.259ms \qquad \omega_{pc} := \frac{1}{\tau_{1c}} \qquad f_{pc} := \frac{\omega_{pc}}{2 \cdot \pi} = 0.613kHz$

$H_1(s) := H_0 \cdot \frac{1 + \frac{s}{\omega_{za}}}{1 + \frac{s}{\omega_p}} \qquad Z_{out}(s) := R_0 \cdot \frac{1 + \frac{s}{\omega_{zb}}}{1 + \frac{s}{\omega_p}} \qquad Z_{in}(s) := R_i \cdot \frac{1 + \frac{s}{\omega_{zc}}}{1 + \frac{s}{\omega_{pc}}}$

$$H_{ref}(s) := \frac{R_2 \| \left[R_3 + R_4 \| \left(\frac{1}{s \cdot C_1}\right)\right]}{R_2 \| \left[R_3 + R_4 \| \left(\frac{1}{s \cdot C_1}\right)\right] + R_1} + \frac{R_4}{R_4 + \frac{1}{s \cdot C_1}} \cdot \frac{R_1 \| R_2}{R_1 \| R_2 + R_3 + R_4 \| \left(\frac{1}{s \cdot C_1}\right)}$$

$Z_{outR}(s) := R_1 \| R_2 \| \left[R_4 \| \left(\frac{1}{s \cdot C_1}\right) + R_3\right] \qquad H_{zin} := READPRN(".\backslash Zin.txt")$ SIMetrix data import

Brute-force expressions

SIMetrix circuit for $Z_{in}(s)$ →

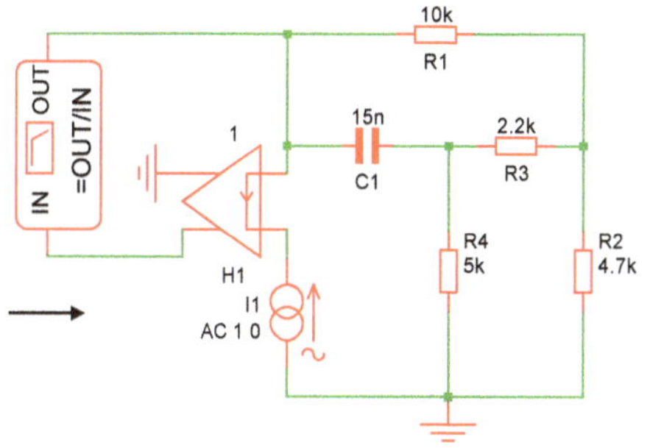

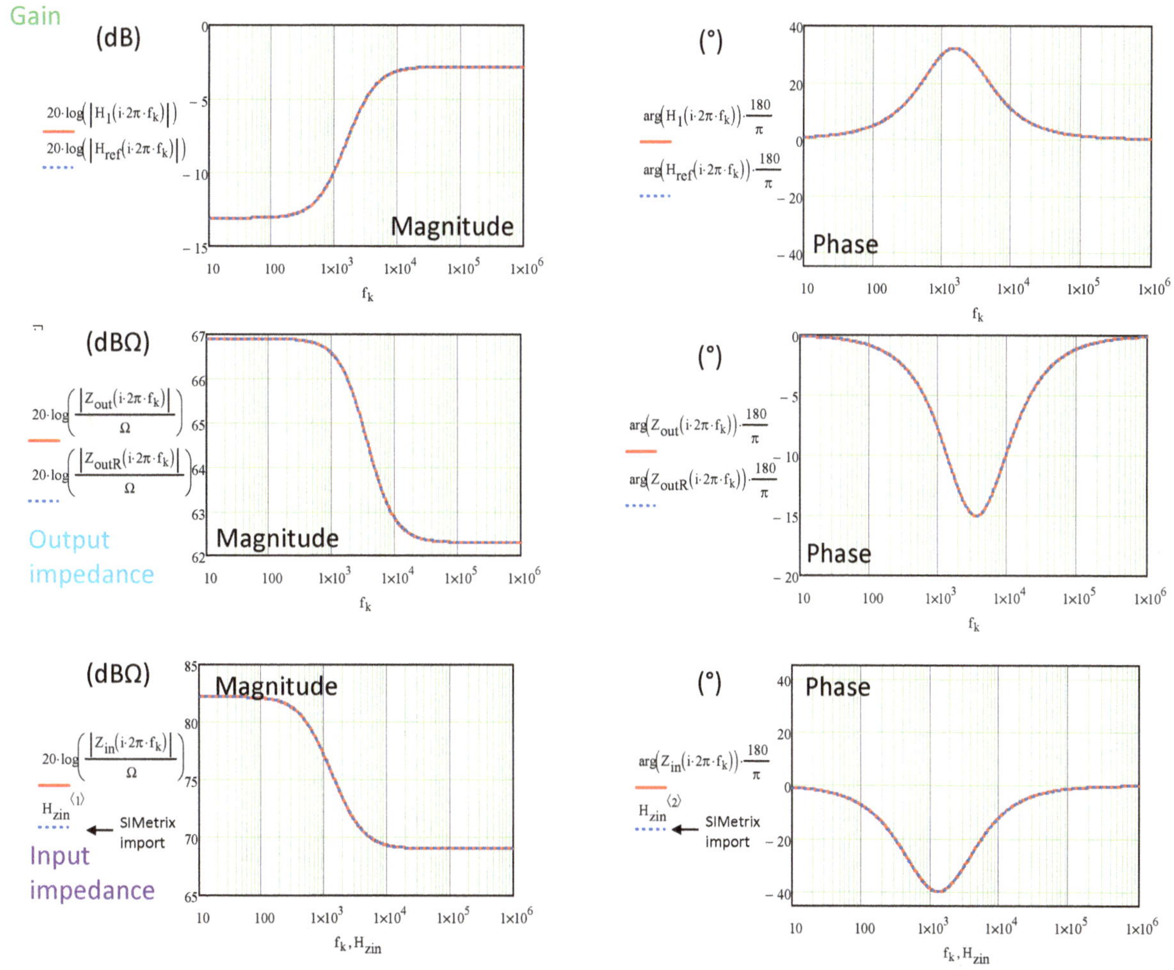

All Transfer Functions are Tested and Create the Same Plots

Figure 5.32

We will now explore a few circuits featuring an active element such as an operation amplifier (op-amp). The first one appears in Figure 5.33. You see a low-pass filter followed by an amplifier. The FACTs can lead you to the fastest answer without writing a line of algebra. The results are given in Figure 5.34. The added 1-V dc bias in V_2 is there to display the dc gain as an operating bias point on the output of E_1. It works in this linear circuit and confirms the value highlighted for H_0.

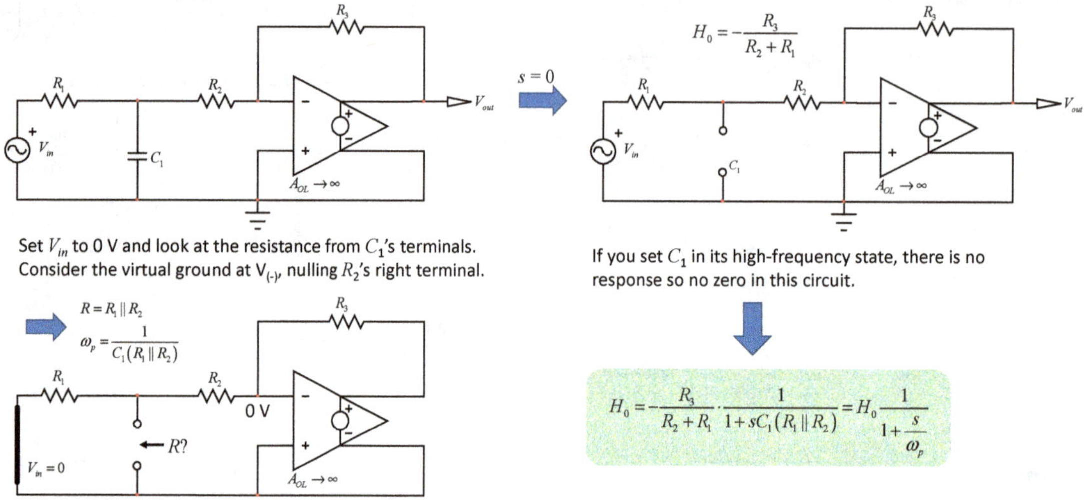

An Operational Amplifier is Inserted in the Circuit

Figure 5.33

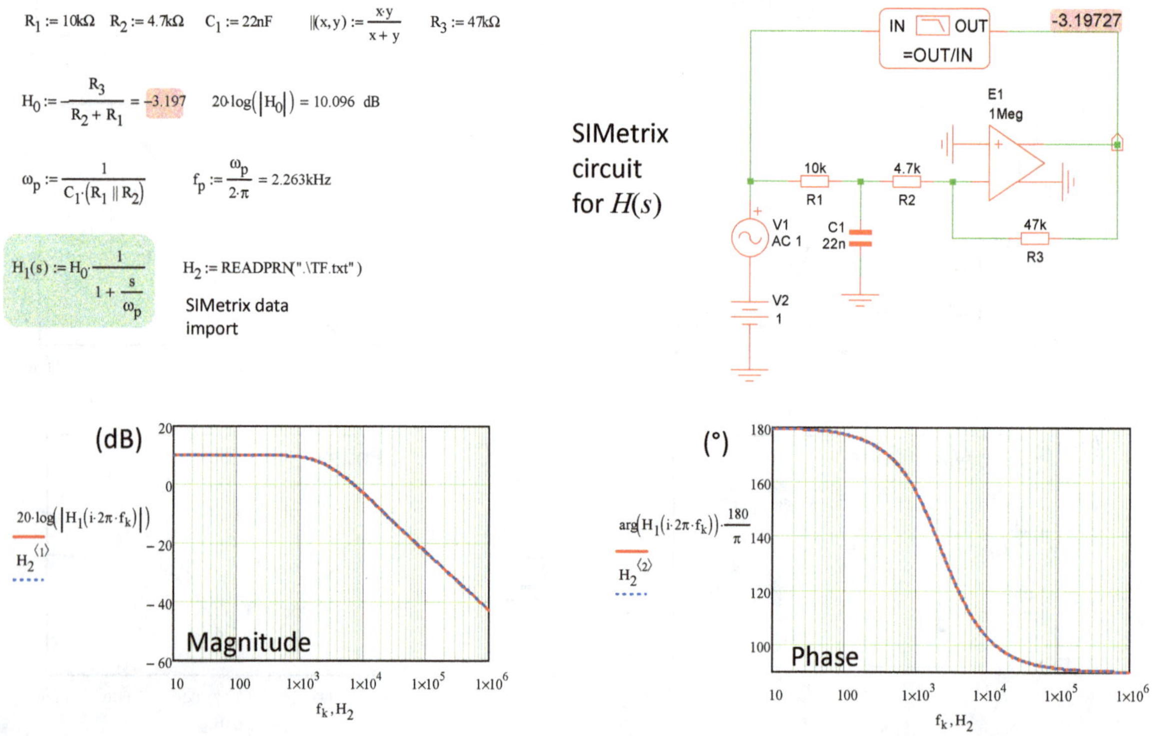

The Plots from SIMetrix® and Mathcad® show Good Agreement

Figure 5.34

Let's add an ESR to the capacitor as in Figure 5.35. You should now immediately see the zero brought by this extra resistor as it creates conditions to null the output. The pole is also affected by the additional element but it does not complicate the exercise whose results are given in Figure 5.36.

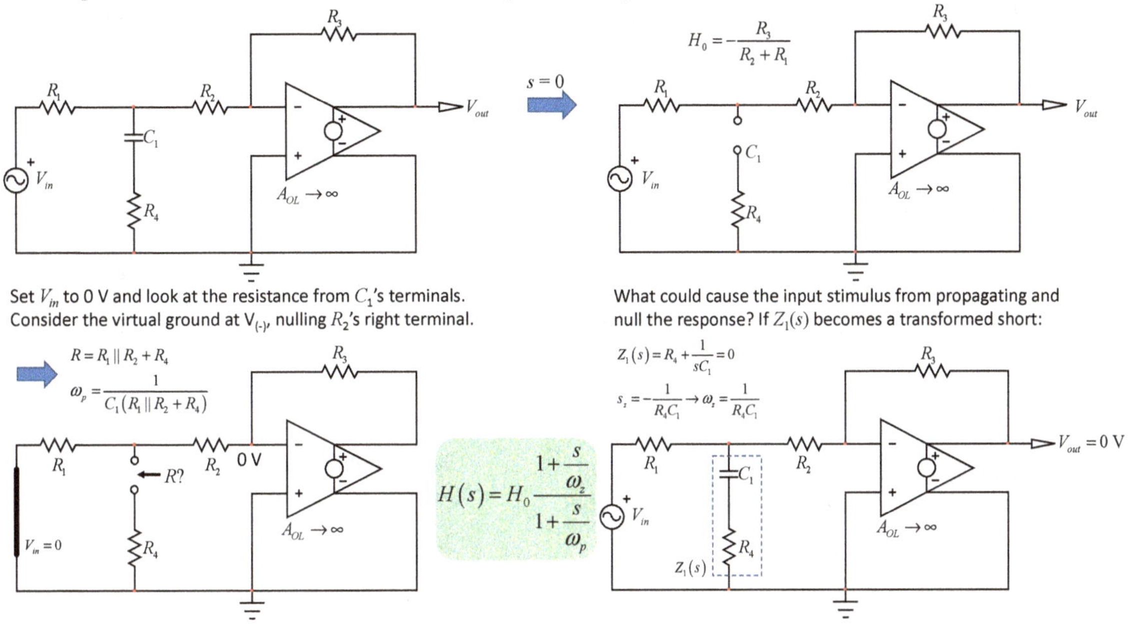

The Addition of the ESR Creates a Zero and Slightly Changes the Pole Position

Figure 5.35

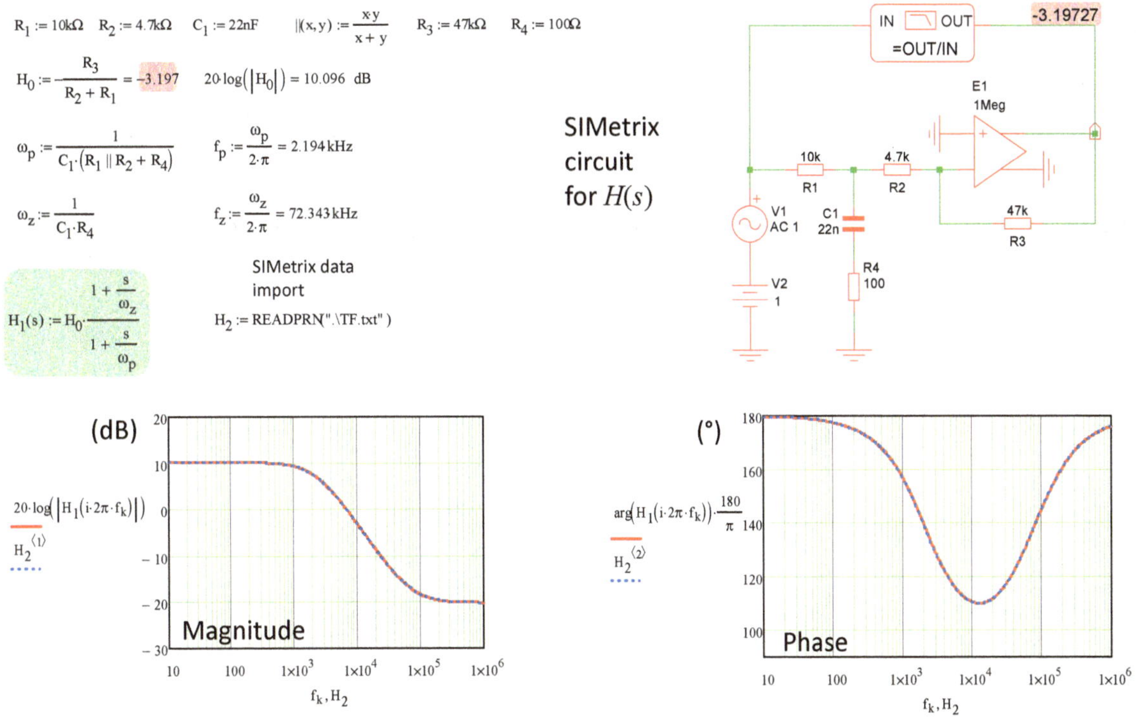

Resistor R_4 adds a Zero to the Transfer Function and Alters the Pole Position

Figure 5.36

The capacitor now moves across the input resistance R_1 and it changes the transfer function (Figure 5.37). The virtual ground excludes R_2 in the ac analysis and a single zero is created with C_1. The frequency response shown in Figure 5.38 indicates a theoretical constant $+1$-slope which, in reality, could not be realized. A real op-amp would naturally roll-off the magnitude at higher frequency considering its natural response including one or several poles.

$$s = 0$$

$$H_0 = -\frac{R_3}{R_1}$$

Set V_{in} to 0 V and look at the resistance from C_1's terminals. Considering the virtual ground at $V_{(-)}$, the resistance is 0 ohm:

$R = 0$ $\omega_p = \dfrac{1}{C_1 \cdot 0} \to \infty$ There is no pole in this circuit

$$H(s) = H_0\left(1+\frac{s}{\omega_z}\right)$$

What could cause the input stimulus from propagating and null the response? If $Z_1(s)$ becomes an transformed open:

$$Z_1(s) = R_1 \| \frac{1}{sC_1} \to \infty$$

$$s_z = -\frac{1}{R_1 C_1} \to \omega_z = \frac{1}{R_1 C_1}$$

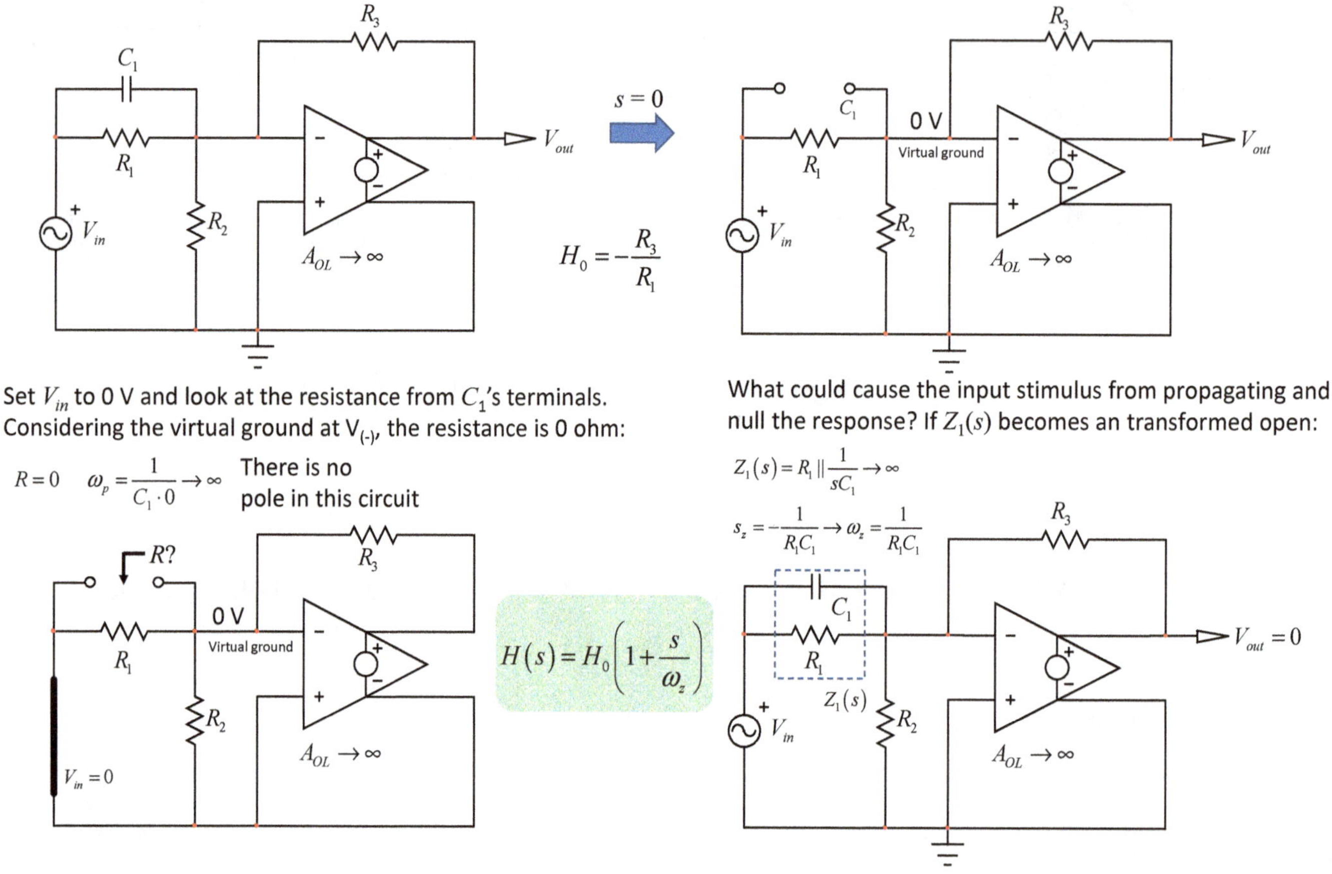

The Capacitor Connected across the Input Resistance Adds a Zero to the Transfer Function

Figure 5.37

$R_1 := 10\text{k}\Omega \quad R_2 := 4.7\text{k}\Omega \quad C_1 := 22\text{nF} \qquad \|(x,y) := \dfrac{x \cdot y}{x+y} \qquad R_3 := 47\text{k}\Omega$

$H_0 := -\dfrac{R_3}{R_1} = -4.7 \qquad\qquad 20 \cdot \log\left(\left|H_0\right|\right) = 13.442 \;\; \text{dB}$

$\omega_p := \dfrac{1}{C_1 \cdot 10^{-6}\Omega} \qquad\qquad f_p := \dfrac{\omega_p}{2 \cdot \pi} = 7.234 \times 10^9 \cdot \text{kHz}$

$\omega_z := \dfrac{1}{C_1 \cdot R_1} \qquad\qquad f_z := \dfrac{\omega_z}{2 \cdot \pi} = 723.432 \,\text{Hz}$

$H_1(s) := H_0 \cdot \left(1 + \dfrac{s}{\omega_z}\right) \qquad H_2 := \text{READPRN}(".\backslash\text{TF.txt}")$

SIMetrix data import

SIMetrix circuit for $H(s)$

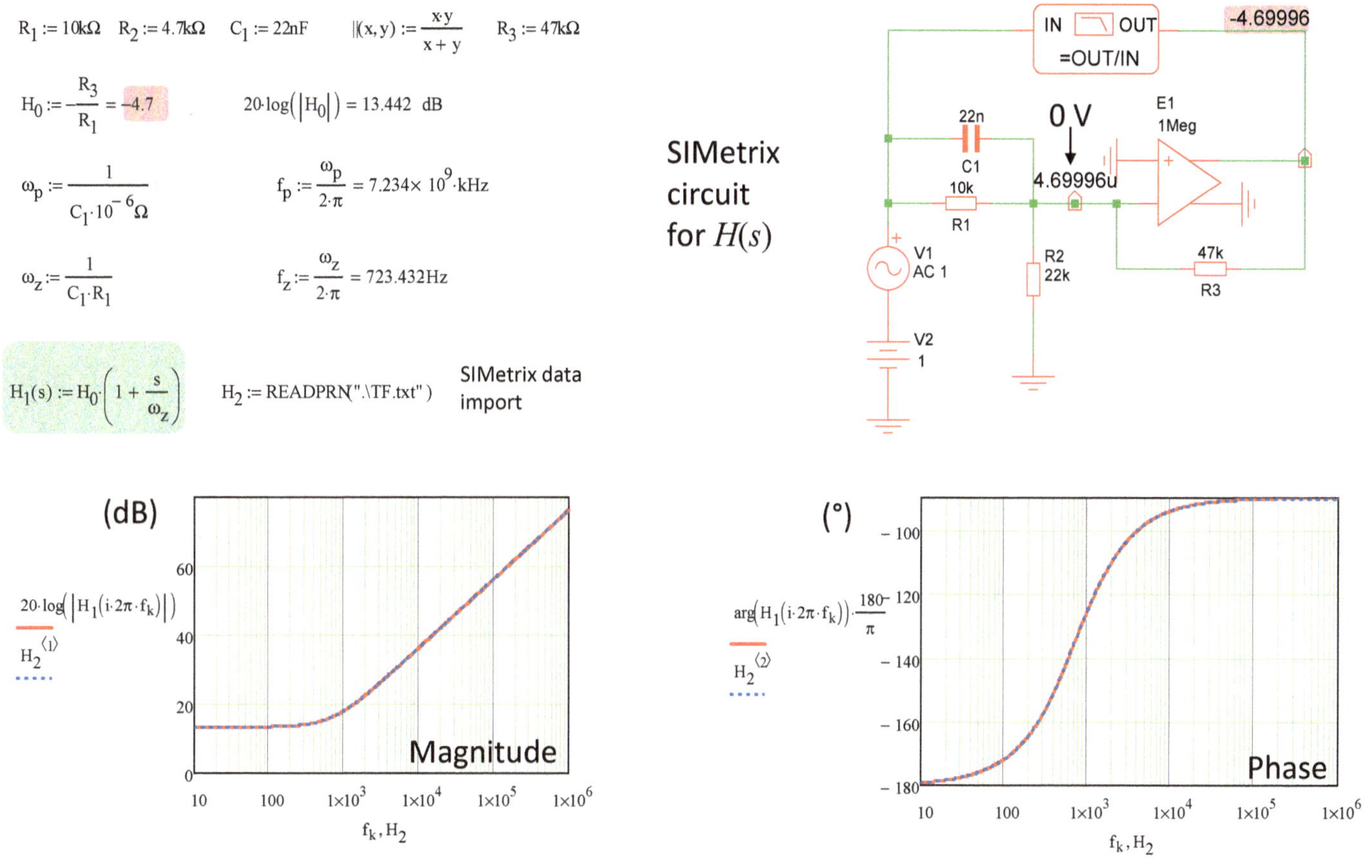

There is a Single Zero in this Circuit via the Virtual Ground Created by the Op-Amp

Figure 5.38

We carry on with the op-amp and now insert a resistor in series with capacitor C_1 as in Figure 5.39.

When the source is zeroed, resistor R_4 sets a time constant which brings a pole to the circuit.

The zero is quickly found by inspection and includes R_4 in series with R_1. This makes the zero naturally lower than the pole and gives a means to separate them for tweaking the phase response.

As you can see, it is boosted between the zero and the pole as Figure 5.40 shows.

Set V_{in} to 0 V and look at the resistance from C_1's terminals. Considering the virtual ground at $V_{(-)}$, R_1 is out of the picture:

$$R = R_4 \rightarrow \omega_p = \frac{1}{C_1 R_4}$$

What could cause the input stimulus from propagating and null the response? If $Z_1(s)$ becomes an transformed open:

$$Z_1(s) = R_1 \parallel \left(\frac{1}{sC_1} + R_4 \right) \rightarrow \infty \quad s_z = -\frac{1}{(R_1 + R_4)C_1} \quad \text{so} \quad \omega_z = \frac{1}{(R_1 + R_4)C_1}$$

$$H(s) = H_0 \frac{1 + \dfrac{s}{\omega_z}}{1 + \dfrac{s}{\omega_p}}$$

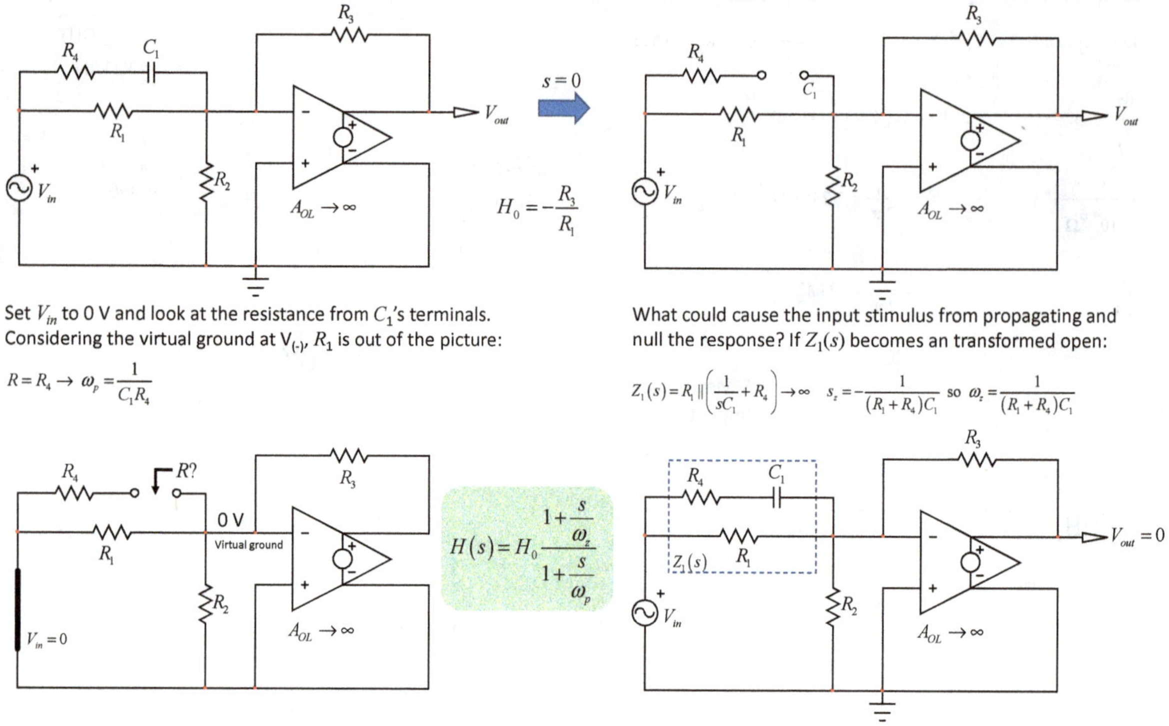

A Resistor is Added in Series with the Capacitor

Figure 5.39

$R_1 := 10\text{k}\Omega \quad R_2 := 4.7\text{k}\Omega \quad C_1 := 22\text{nF} \quad \parallel(x,y) := \dfrac{x \cdot y}{x+y} \quad R_3 := 47\text{k}\Omega \quad R_4 := 150\Omega$

$H_0 := -\dfrac{R_3}{R_1} = -4.7 \qquad 20 \cdot \log\left(|H_0|\right) = 13.442 \ \text{dB}$

$\omega_p := \dfrac{1}{C_1 \cdot R_4} \qquad\qquad f_p := \dfrac{\omega_p}{2 \cdot \pi} = 48.229 \,\text{kHz}$

$\omega_z := \dfrac{1}{C_1 \cdot (R_1 + R_4)} \qquad f_z := \dfrac{\omega_z}{2 \cdot \pi} = 712.74 \,\text{Hz}$

$$H_1(s) := H_0 \cdot \left(\frac{1 + \dfrac{s}{\omega_z}}{1 + \dfrac{s}{\omega_p}} \right)$$

$H_2 := \text{READPRN}(".\backslash\text{TF}.\text{txt}")$ \quad SIMetrix data import

$20 \cdot \log\left(\left|H_1(i \cdot 2\pi \cdot f_k)\right|\right)$

$H_2^{\langle 1 \rangle}$

Magnitude

$\arg\left(H_1(i \cdot 2\pi \cdot f_k)\right) \cdot \dfrac{180}{\pi}$

$H_2^{\langle 2 \rangle}$

Phase

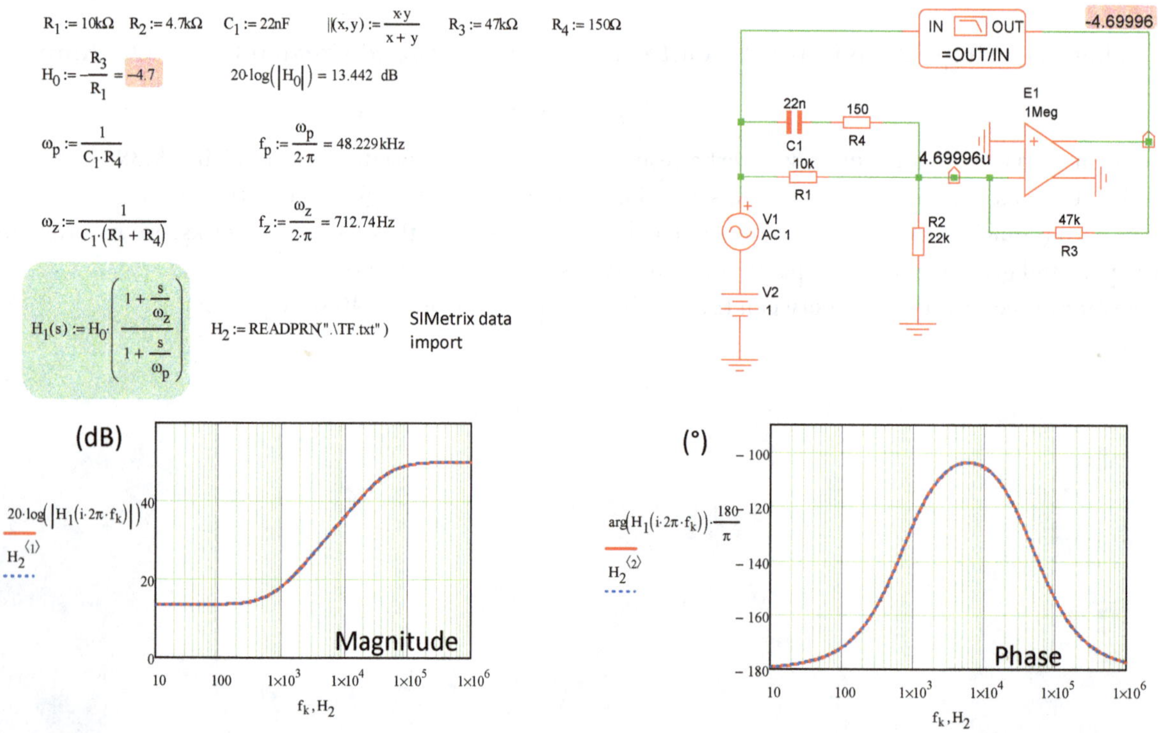

Now, the Response includes a Pole Which Flattens the Magnitude at High Frequency

Figure 5.40

A simple low-pass filter is represented in Figure 5.41. Feedback resistor R_2 closes the local loop in both ac and dc conditions, setting the gain with R_1 for $s = 0$.

The pole depends on R_2 and when its value increases towards infinity—R_2 is removed from the circuit—we have a classic integrator featuring a pole now set at the origin.

Figure 5.42 confirms our results are correct.

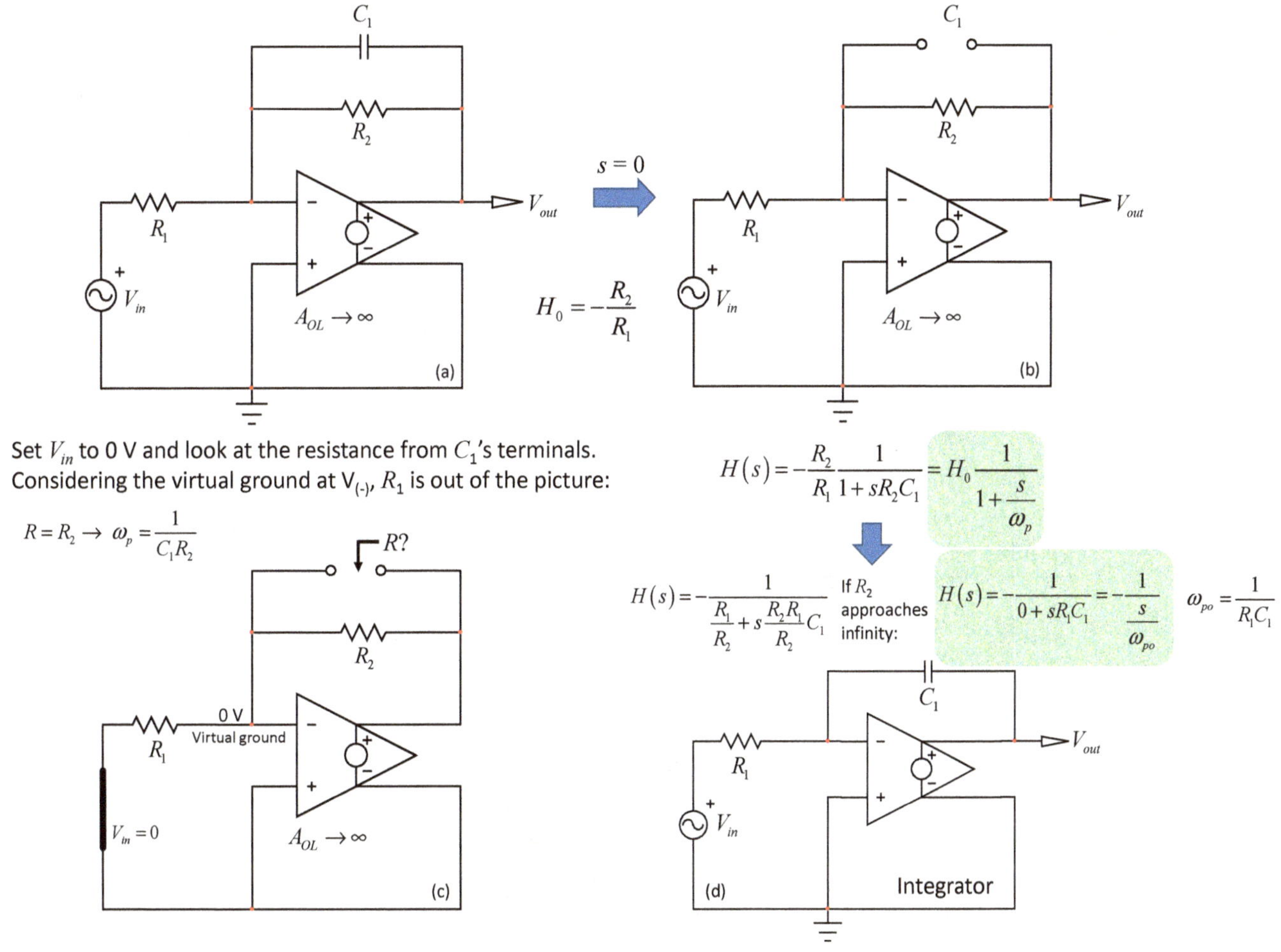

$$s = 0$$

$$H_0 = -\frac{R_2}{R_1}$$

Set V_{in} to 0 V and look at the resistance from C_1's terminals. Considering the virtual ground at $V_{(-)}$, R_1 is out of the picture:

$$R = R_2 \rightarrow \omega_p = \frac{1}{C_1 R_2}$$

$$H(s) = -\frac{R_2}{R_1}\frac{1}{1 + sR_2C_1} = H_0\frac{1}{1 + \dfrac{s}{\omega_p}}$$

$$H(s) = -\frac{1}{\dfrac{R_1}{R_2} + s\dfrac{R_2R_1}{R_2}C_1} \quad \text{If } R_2 \text{ approaches infinity:} \quad H(s) = -\frac{1}{0 + sR_1C_1} = -\frac{1}{\dfrac{s}{\omega_{po}}} \qquad \omega_{po} = \frac{1}{R_1C_1}$$

A Simple Low-Pass filter is Built around an Op-Amp

Figure 5.41

$R_1 := 10\text{k}\Omega \quad R_2 := 47\text{k}\Omega \quad C_1 := 22\text{nF} \qquad \|(x,y) := \dfrac{x \cdot y}{x + y}$

$H_0 := \dfrac{R_2}{R_1} = -4.7 \qquad\qquad 20 \cdot \log\left(\left|H_0\right|\right) = 13.442 \ \text{dB}$

$\omega_p := \dfrac{1}{C_1 \cdot R_2} \qquad\qquad f_p := \dfrac{\omega_p}{2 \cdot \pi} = 0.154 \text{kHz}$

$H_1(s) := H_0 \cdot \left(\dfrac{1}{1 + \dfrac{s}{\omega_p}}\right) \qquad H_2 := \text{READPRN}(".\backslash TF.txt")$ SIMetrix data import

$20 \cdot \log\left(\left|H_1\left(i \cdot 2\pi \cdot f_k\right)\right|\right)$

$H_2^{\langle 1 \rangle}$

Magnitude

$\arg\left(H_1\left(i \cdot 2\pi \cdot f_k\right)\right) \cdot \dfrac{180}{\pi}$

$H_2^{\langle 2 \rangle}$

Phase

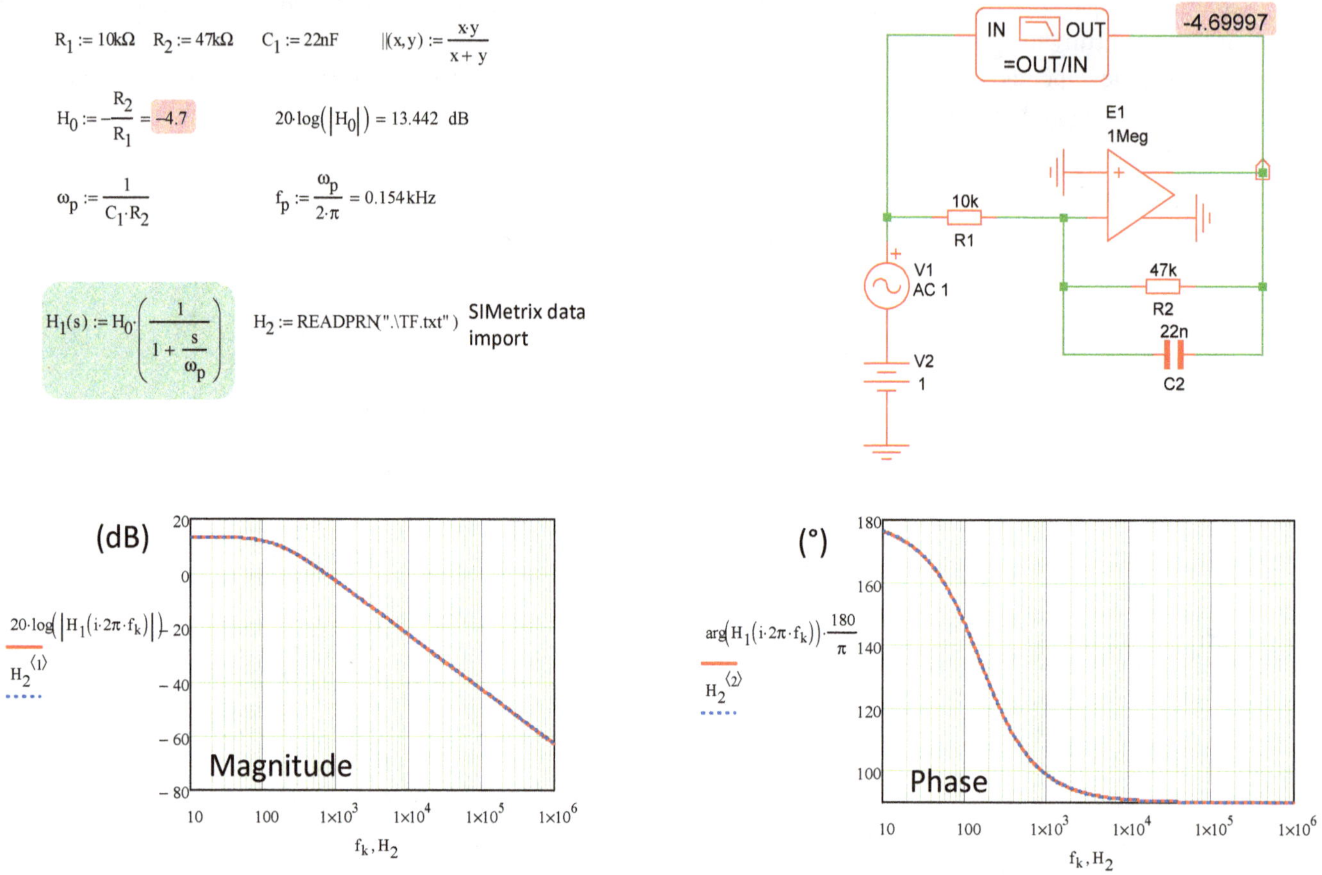

In this Configuration, the Pole is Set by Resistor R_2

Figure 5.42

We can also check the effect of the op-amp open-loop gain A_{OL} on the transfer function of this integrator.

This is proposed in Figure 5.43.

In this circuit, which could be used as a compensator featuring a reference voltage at the non-inverting pin, the resistive divider made of R_1 and R_2 would set the regulated voltage.

$$H_0 = -A_{OL}\frac{R_2}{R_1+R_2}$$

$$V_{out} = \varepsilon \cdot A_{OL}$$

$$\varepsilon = -V_{in}\frac{R_2}{R_1+R_2}$$

$$H(s) = -A_{OL}\frac{R_2}{R_1+R_2}\frac{1}{1+s(R_1\parallel R_2)(1+A_{OL})C_1} = H_0\frac{1}{1+\dfrac{s}{\omega_p}}$$

$$\varepsilon = -I_T(R_1\parallel R_2)$$

$$V_T = -\varepsilon - \varepsilon\cdot A_{OL} = -\varepsilon(1+A_{OL})$$

$$V_T = I_T(R_1\parallel R_2)(1+A_{OL})$$

$$R = \frac{V_T}{I_T} = (R_1\parallel R_2)(1+A_{OL})$$

$$\omega_p = \frac{1}{\left[(R_1\parallel R_2)(1+A_{OL})\right]C_1}$$

If A_{OL} is large

$$H(s) = -\frac{1}{\dfrac{1}{\dfrac{A_{OL}R_2}{R_1+R_2}} + s\left(\dfrac{R_1R_2}{R_1+R_2}\right)\dfrac{(1+A_{OL})}{\dfrac{A_{OL}R_2}{R_1+R_2}}C_1} = -\frac{1}{\underbrace{\dfrac{R_1+R_2}{A_{OL}R_2}}_{\approx 0} + sC_1R_1\underbrace{\left(\dfrac{1+A_{OL}}{A_{OL}}\right)}_{\approx 1}}$$

$$H(s) \approx -\frac{1}{\dfrac{s}{\omega_{po}}} \qquad \omega_{po} = \frac{1}{R_1C_1}$$

Classical integrator in which R_2 plays no ac role.

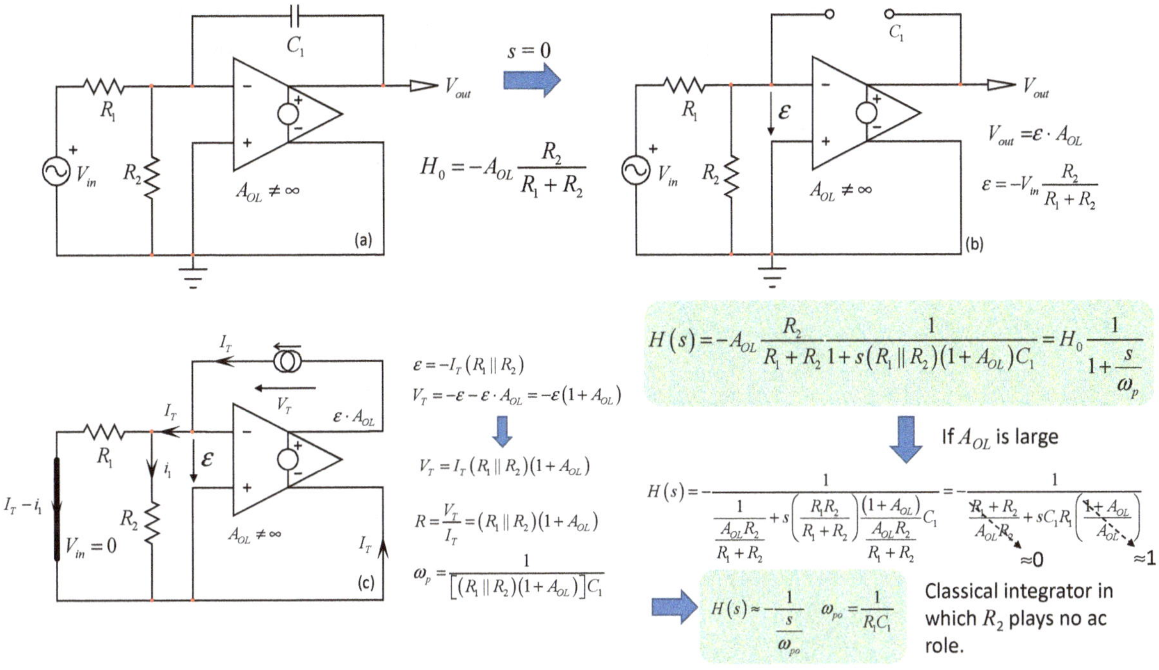

The Op-Amp is Usually Considered Perfect with an Infinite Open-Loop Gain. Let's Examine its Impact on the Transfer Function

Figure 5.43

$$R_1 := 10k\Omega \quad R_2 := 10k\Omega \quad C_1 := 10nF \qquad \parallel(x,y) := \frac{x\cdot y}{x+y} \qquad A_{OL} := 1000$$

$$H_0 := \frac{R_2}{R_1+R_2}\cdot A_{OL} = -500 \qquad 20\cdot\log(|H_0|) = 53.979 \quad \text{dB}$$

$$\omega_p := \frac{1}{C_1\cdot(R_1\parallel R_2)(1+A_{OL})} \qquad f_p := \frac{\omega_p}{2\cdot\pi} = 3.18\,\text{Hz}$$

$$H_1(s) := H_0\cdot\left(\frac{1}{1+\dfrac{s}{\omega_p}}\right) \qquad H_2 := \text{READPRN}(".\backslash TF.txt") \qquad \begin{array}{l}\text{SIMetrix data}\\\text{import}\end{array}$$

Magnitude

Phase

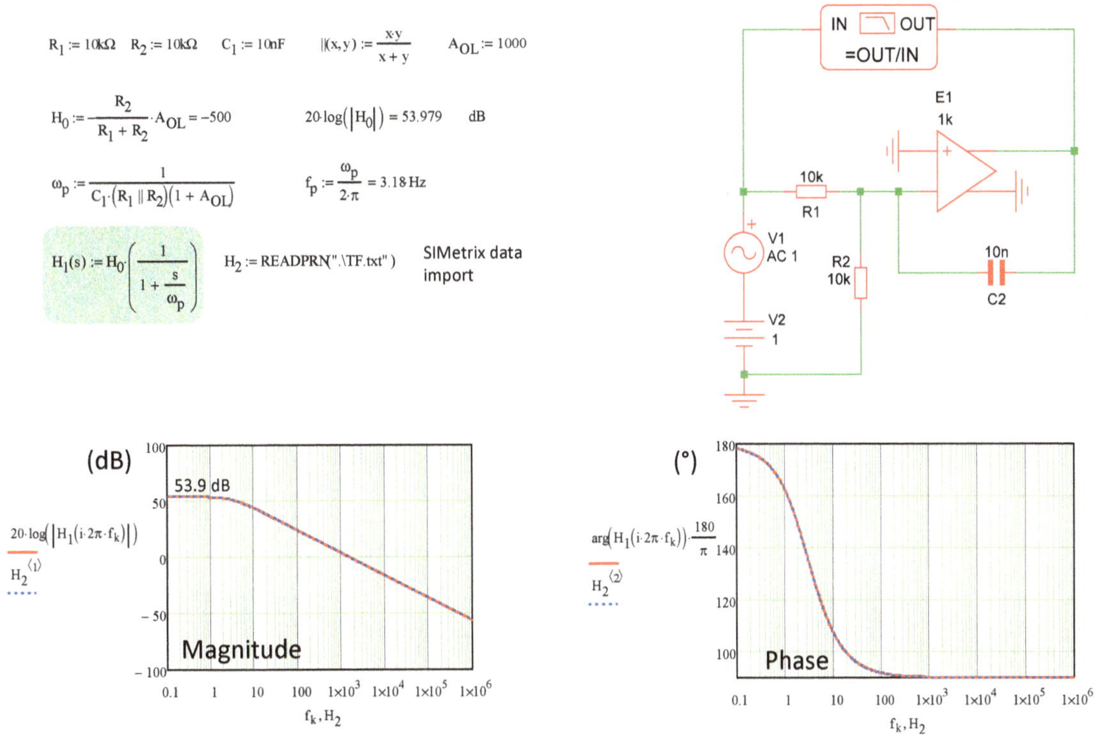

When A_{OL} is Not Infinite, the Low-Side Resistance Plays a Role and Sets the Dc Gain for $s = 0$

Figure 5.44

At dc, when $s = 0$, the op-amp output now depends on the open-loop gain A_{OL} but also on the division ratio. This is because there is no virtual ground—the op-amp runs open loop—and R_2 needs to be considered. The final transfer function is quite complicated and nicely simplifies to that of a perfect integrator when A_{OL} approaches infinity.

We now slightly rearrange the components in Figure 5.45 with a feedback network made of C_1 and R_3 in series.

Set V_{in} to 0 V and look at the resistance from C_1's terminals. Considering the virtual ground at $V_{(-)}$, R_2 is out of the picture:

$R = R_3 + R_4$

$\rightarrow \omega_p = \dfrac{1}{C_1(R_3 + R_4)}$

What could prevent the input stimulus from propagating to form a response? If $Z_1(s)$ becomes a transformed short:

$Z_1(s) = \dfrac{1}{sC_1} + R_3 = 0 \qquad s_z = -\dfrac{1}{R_3 C_1}$ so $\omega_z = \dfrac{1}{R_3 C_1}$

$H(s) = -\dfrac{R_4}{R_1}\dfrac{1 + sR_3C_1}{1 + s(R_3 + R_4)C_1}$

$H(s) = H_0 \dfrac{1}{1 + \dfrac{s}{\omega_p}}$

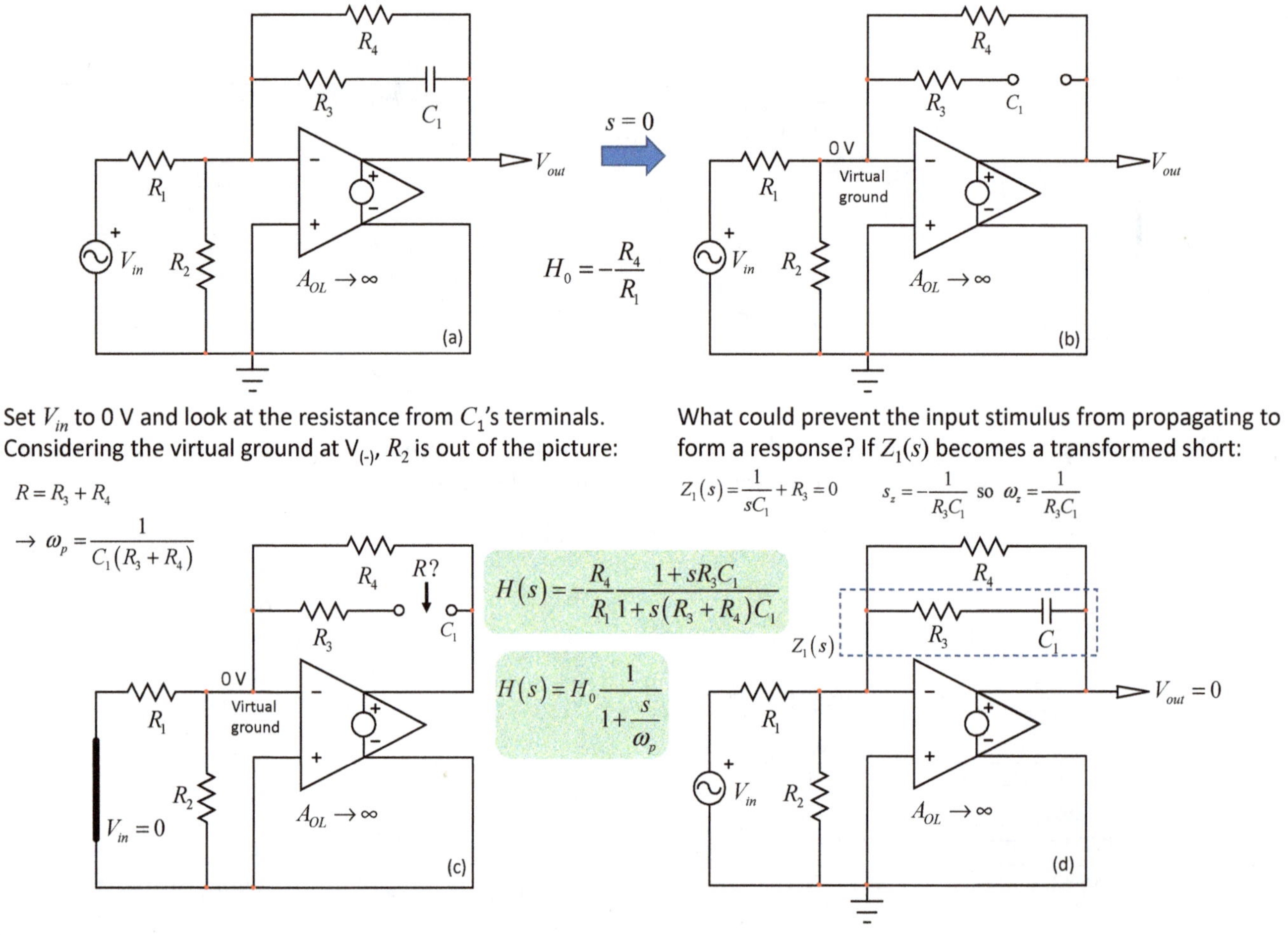

The Feedback Network is now Made of a Capacitor in Series with a Resistor

Figure 5.45

$R_1 := 10k\Omega \quad R_2 := 10k\Omega \quad C_1 := 10nF \quad \|(x,y) := \frac{x \cdot y}{x+y} \quad R_3 := 15k\Omega \quad R_4 := 470k\Omega$

$H_0 := -\frac{R_4}{R_1} = -47 \qquad 20 \cdot \log(|H_0|) = 33.442 \quad dB$

$\omega_p := \frac{1}{C_1 \cdot (R_3 + R_4)} \qquad f_p := \frac{\omega_p}{2 \cdot \pi} = 32.815Hz$

$\omega_z := \frac{1}{C_1 \cdot R_3} \qquad f_z := \frac{\omega_z}{2 \cdot \pi} = 1.061 \times 10^3 \cdot Hz$

$H_1(s) := H_0 \cdot \left(\frac{1 + \frac{s}{\omega_z}}{1 + \frac{s}{\omega_p}} \right) \qquad H_2 := READPRN(".\backslash TF.txt")$ SIMetrix data import

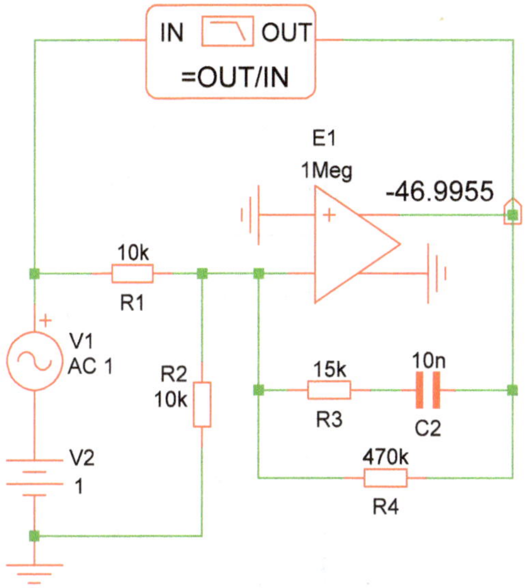

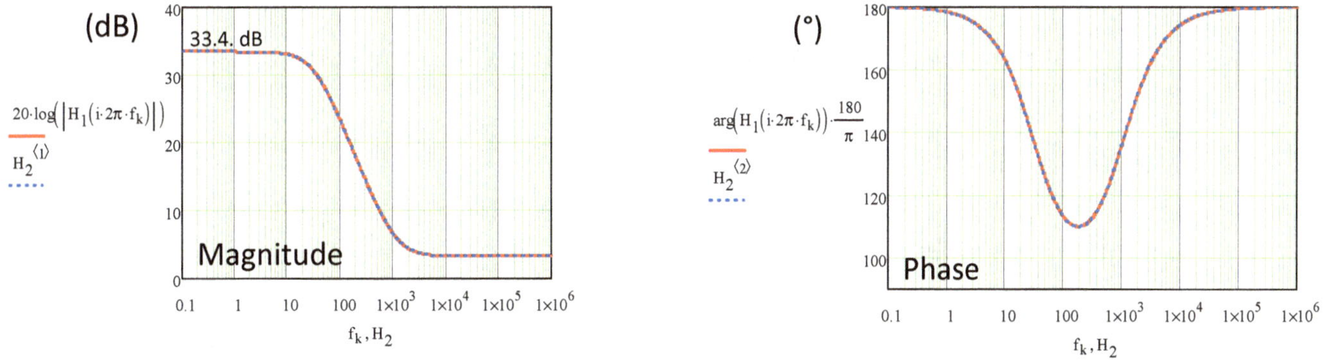

The 1-V Dc Source lets you Determine the Gain H_0 Which is -46.99 as Calculated

Figure 5.46

Because of the virtual ground kept in dc by R_4, the low-side resistor R_2 is excluded from the transfer function.

The zero is easily identified and characterized by inspection (Figure 5.46).

We now placed the resistors a little bit differently and even if this configuration has no particular application as a filter, it remains valid to acquire the skill. A few equations tell us that the dc gain is 1 and we are left with the determination of the resistance driving C_1.

Again, a set of simple expressions shows us the resistive combination which leads to the pole position as shown in Figure 5.47.

(a) $s = 0$

$$i_1 = \frac{V_{in}\dfrac{R_3}{R_1+R_3} + V_{out}\dfrac{R_1}{R_1+R_3} - V_{out}}{R_2} = 0$$

$$V_{out} = V_{in}$$

$$H_0 = 1$$

(b)

$$i_1 = 0$$

Set V_{in} to 0 V and look at the resistance from C_1's terminals.
Install the test generator I_T and determine the voltage V_T.

$$V_A = V_T - I_T R_2 \text{ and } i_1 = \frac{V_A}{R_1} \rightarrow i_1 = \frac{V_T - I_T R_2}{R_1}$$

$$V_T - I_T R_2 = V_T + R_3\left(I_T - \frac{V_T - I_T R_2}{R_1}\right)$$

$$\frac{V_T}{I_T} = \frac{R_1(R_2+R_3) + R_2 R_3}{R_3}$$

$$\tau = RC_1 = \frac{R_1(R_2+R_3) + R_2 R_3}{R_3}C_1$$

$$\omega_p = \frac{R_3}{\left[R_1(R_2+R_3) + R_2 R_3\right]C_1}$$

$$H(s) = \frac{1}{1 + \dfrac{s}{\omega_p}}$$

(c)

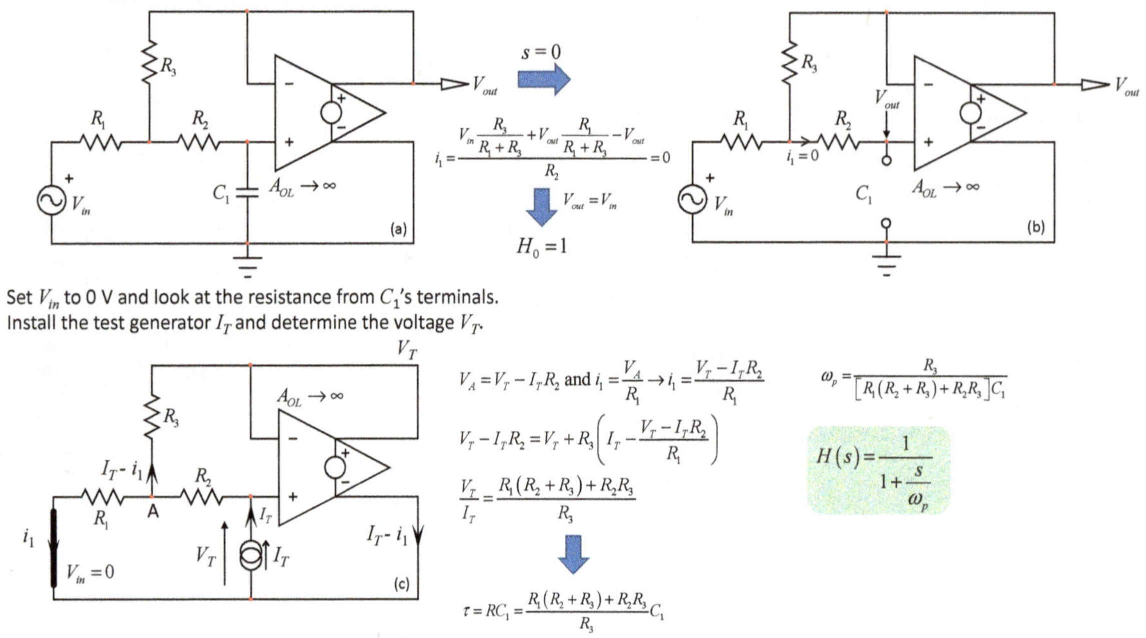

A Resistor Comes from the Output and Biases the Middle Node

Figure 5.47

The ac response in Figure 5.48 confirms our equations are correct.

$R_1 := 10\text{k}\Omega \quad R_2 := 15\text{k}\Omega \quad C_1 := 22\text{nF} \quad \|(x,y) := \dfrac{x \cdot y}{x+y} \quad R_3 := 22\text{k}\Omega$

$H_0 := 1 \qquad 20 \cdot \log(|H_0|) = 0$

$\tau_1 := C_1 \cdot \left(\dfrac{R_1 \cdot R_2 + R_1 \cdot R_3 + R_2 \cdot R_3}{R_3}\right) = 7 \times 10^{-4}\,\text{s}$

$\omega_p := \dfrac{R_3}{C_1 \cdot (R_1 \cdot R_2 + R_1 \cdot R_3 + R_2 \cdot R_3)} \qquad f_p := \dfrac{\omega_p}{2 \cdot \pi} = 227.364\,\text{Hz}$

$H_1(s) := H_0 \cdot \dfrac{1}{1 + \dfrac{s}{\omega_p}} \qquad H_2 := \text{READPRN}(".\backslash\text{TF.txt"})$

SIMetrix data
import

(dB)

$20 \cdot \log(|H_1(i \cdot 2\pi \cdot f_k)|)$

$H_2^{\langle 1 \rangle}$

Magnitude

f_k, H_2

(°)

$\arg(H_1(i \cdot 2\pi \cdot f_k)) \cdot \dfrac{180}{\pi}$

$H_2^{\langle 2 \rangle}$

Phase

f_k, H_2

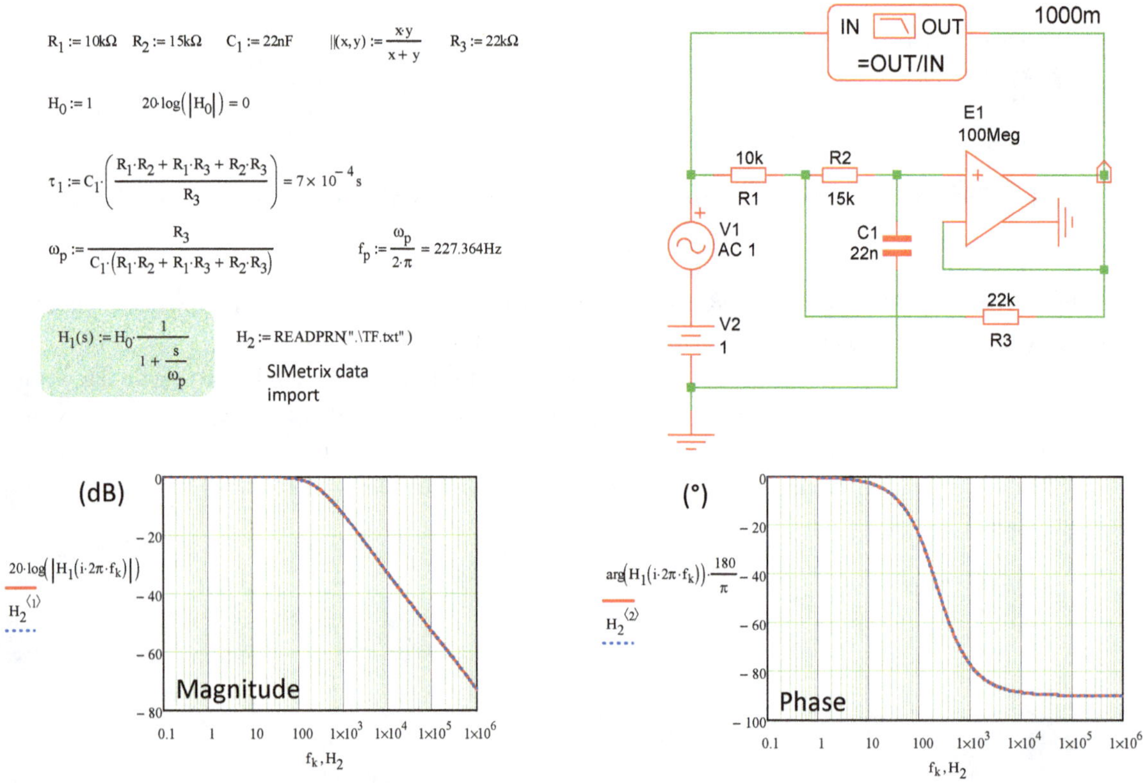

This is the Response of a Classical Low-Pass Filter

Figure 5.48

In the circuit from Figure 5.49, the non-inverting pin of the op-amp is fed from V_{in} via a low-pass filter.

The dc gain of this circuit is unity and quicky determined using superposition.

The pole is also immediate as when V_{in} is zeroed, R_1 is, alone, loading the capacitor.

By placing C_1 in its high-frequency state, there is a response, indicating that this circuit hosts a zero.

The zero can be determined using an NDI after the output has been nulled but applying the generalized transfer function is faster in my opinion. Just determine the high-frequency gain H^1 and a negative zero appears depending on three resistors values.

If R_2 equals R_3, the pole and the zero neutralize in magnitude but considering the negative root in the numerator—this is a right-half-plane zero—the phase lags down to $180°$ as shown in Figure 5.50.

Apply superposition by splitting V_{in} in two branches feeding R_3 and R_1:

$$V_{out} = -V_{in}\frac{R_2}{R_3} + V_{in}\left(\frac{R_2}{R_3}+1\right)$$

Expand

$$H_0 = 1$$

(a)

(b)

Set V_{in} to 0 V and look at the resistance from C_1's terminals. R_1 is grounded and alone across the capacitor.

You can see if you place C_1 in its high-frequency state, the stimulus produces a response: there is a zero in this circuit.

$R = R_1$

$$H^1 = -\frac{R_2}{R_3}$$

$$\tau = R_1 C_1$$

$$H(s) = \frac{H_0 + sH^1\tau}{1+s\tau}$$

$$\omega_p = \frac{1}{R_1 C_1}$$

$$H(s) = \frac{1 - s\frac{R_2}{R_3}R_1 C_1}{1+s\tau}$$

(c)

(d)

Rather than going through a NDI, we can use the generalized transfer function and determine H^1:

$$H(s) = \frac{1-\dfrac{s}{\omega_z}}{1+\dfrac{s}{\omega_p}} \qquad \omega_z = \frac{R_3}{R_2 R_1 C_1} \qquad \text{if } R_2 = R_3$$

$$\omega_p = \frac{1}{R_1 C_1} \qquad \omega_z = \frac{1}{R_1 C_1}$$

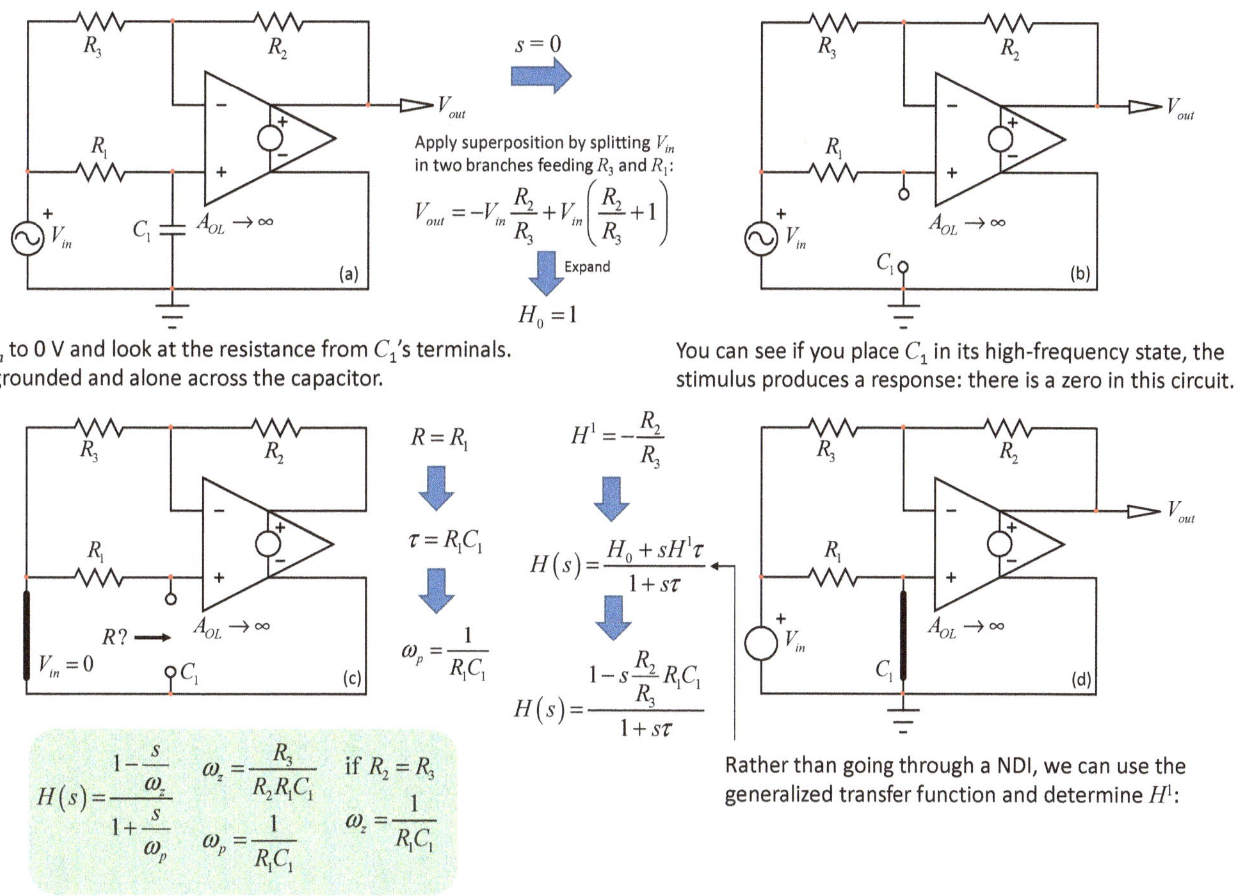

This Arrangement Creates an All-Pass Filter

Figure 5.49

$R_1 := 15k\Omega \quad R_2 := 100k\Omega \quad C_1 := 22nF \qquad \|(x,y) := \frac{x \cdot y}{x+y} \qquad R_3 := 100k\Omega$

$H_0 := 1 \qquad 20 \cdot \log(|H_0|) = 0 \qquad V_{out} = -V_{in}\frac{R_2}{R_3} + V_{in} \cdot \left(1 + \frac{R_2}{R_3}\right) \text{ --> } V_{out} = V_{in}$

$\tau_1 := R_1 \cdot C_1 = 3.3 \times 10^{-4} s$

$\omega_p := \frac{1}{R_1 \cdot C_1} \qquad f_p := \frac{\omega_p}{2 \cdot \pi} = 482.288Hz$

$H_1 := \frac{R_2}{R_3} \qquad H_{20}(s) := \frac{H_0 + s \cdot H_1 \cdot \tau_1}{1 + s \cdot \tau_1} \qquad \omega_z := \frac{R_3}{R_2 \cdot R_1 \cdot C_1} \qquad f_z := \frac{\omega_z}{2 \cdot \pi} = 482.288Hz$

$H_{10}(s) := H_0 \cdot \frac{1 - \frac{s}{\omega_z}}{1 + \frac{s}{\omega_p}} \qquad H_2 := \text{READPRN}(\text{".\textbackslash TF.txt"})$ SIMetrix data import

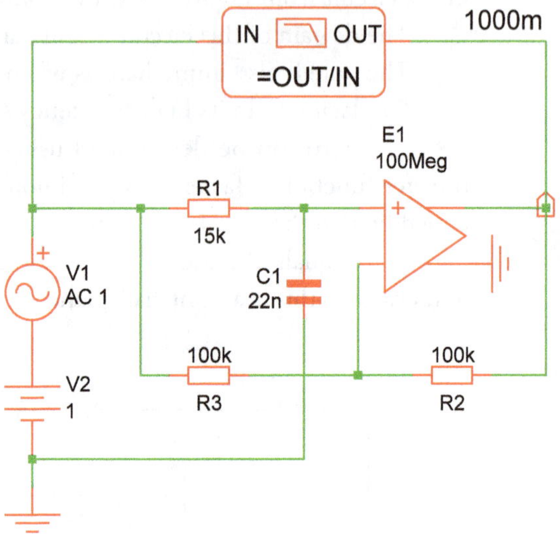

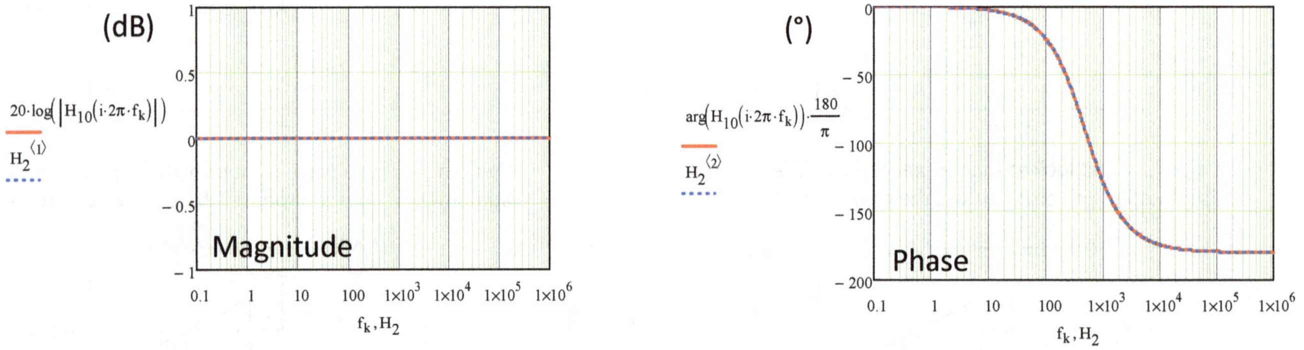

When R_2 and R_3 are Equal, the Pole and Zero are Balanced and the Phase Lags Add

Figure 5.50

Let's increase complexity a little here with an audio shelving filter shown in Figure 5.51.

The potentiometer in the upper left corner lets the user linearly choose which audio band to boost. When the wiper is in the middle position, we will see that the ac response is flat in magnitude and phase. When the wiper is rotated to the left or the right position, the response becomes that of a low- or high-pass 1st-order filter.

First, replace the potentiometer with its equivalent 2-resistor circuit shown in the upper right corner of the figure.

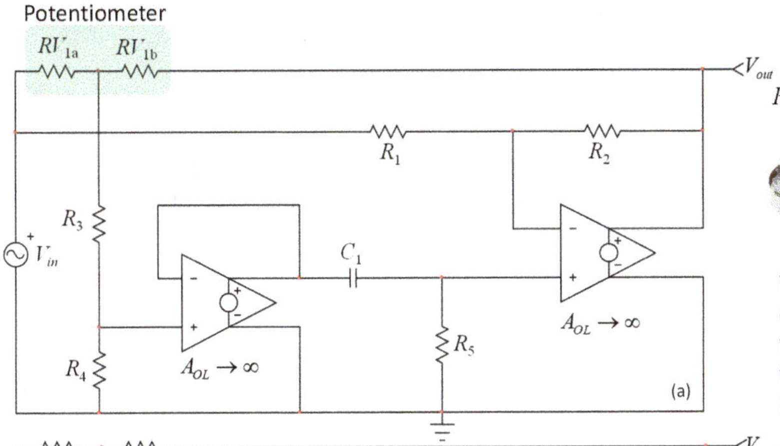

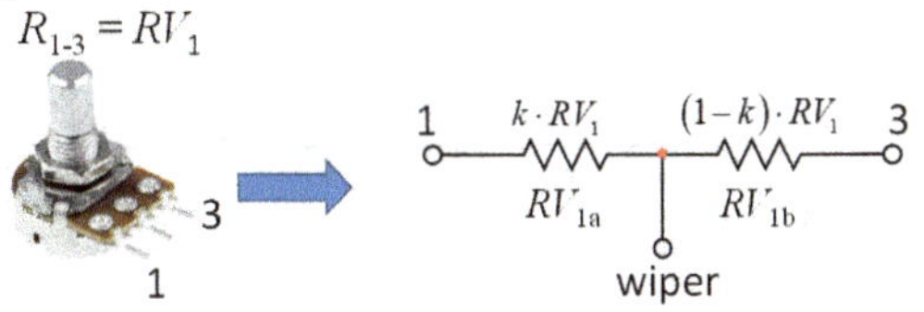

You treat the 3-pin potentiometer with two resistors each affected by a coefficient depending on the wiper position: k = 1, means the shaft is rotated towards 3 while k = 0 implies a shaft rotated towards 1.

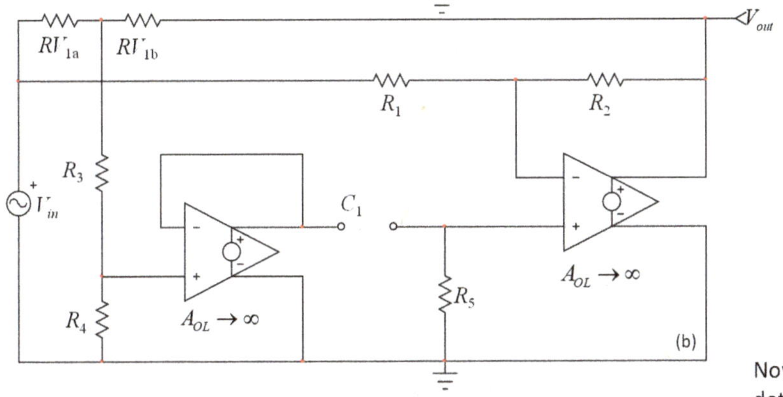

With C_1 open, the second op-amp is a simple inverting configuration and the dc gain H_0 is immediate:

$$H_0 = -\frac{R_2}{R_1}$$

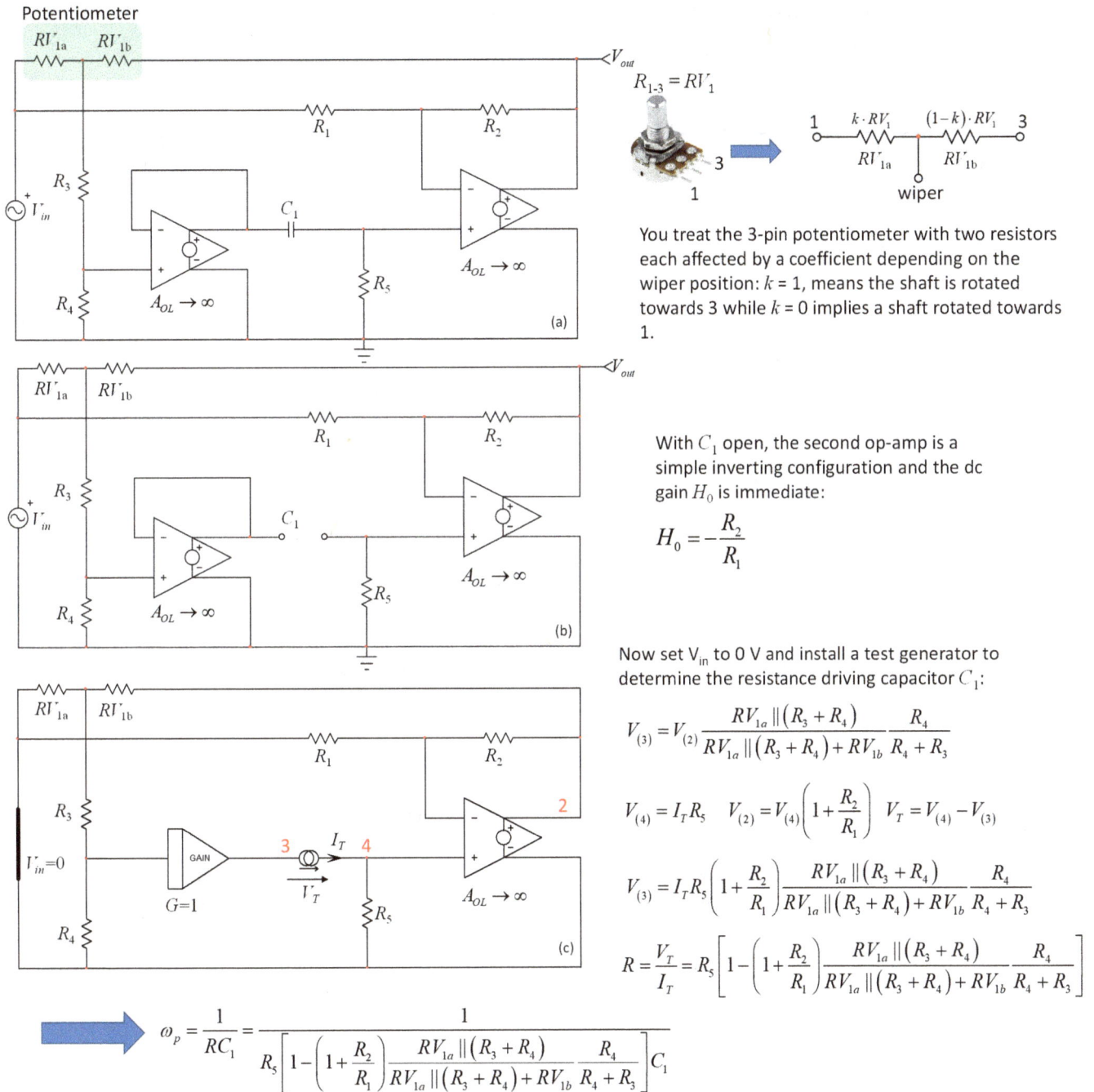

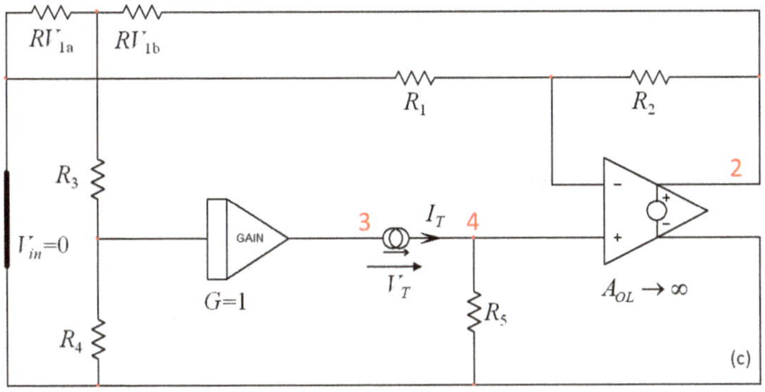

Now set V_{in} to 0 V and install a test generator to determine the resistance driving capacitor C_1:

$$V_{(3)} = V_{(2)}\frac{RV_{1a}\,||\left(R_3+R_4\right)}{RV_{1a}\,||\left(R_3+R_4\right)+RV_{1b}}\frac{R_4}{R_4+R_3}$$

$$V_{(4)} = I_T R_5 \quad V_{(2)} = V_{(4)}\left(1+\frac{R_2}{R_1}\right) \quad V_T = V_{(4)} - V_{(3)}$$

$$V_{(3)} = I_T R_5\left(1+\frac{R_2}{R_1}\right)\frac{RV_{1a}\,||\left(R_3+R_4\right)}{RV_{1a}\,||\left(R_3+R_4\right)+RV_{1b}}\frac{R_4}{R_4+R_3}$$

$$R = \frac{V_T}{I_T} = R_5\left[1-\left(1+\frac{R_2}{R_1}\right)\frac{RV_{1a}\,||\left(R_3+R_4\right)}{RV_{1a}\,||\left(R_3+R_4\right)+RV_{1b}}\frac{R_4}{R_4+R_3}\right]$$

$$\omega_p = \frac{1}{RC_1} = \frac{1}{R_5\left[1-\left(1+\frac{R_2}{R_1}\right)\frac{RV_{1a}\,||\left(R_3+R_4\right)}{RV_{1a}\,||\left(R_3+R_4\right)+RV_{1b}}\frac{R_4}{R_4+R_3}\right]C_1}$$

Transform the Potentiometer into a set of Two Resistors each Affected by a Coefficient k

Figure 5.51

Bring the stimulus in the circuit and solve V_T/I_T for a null in the output:

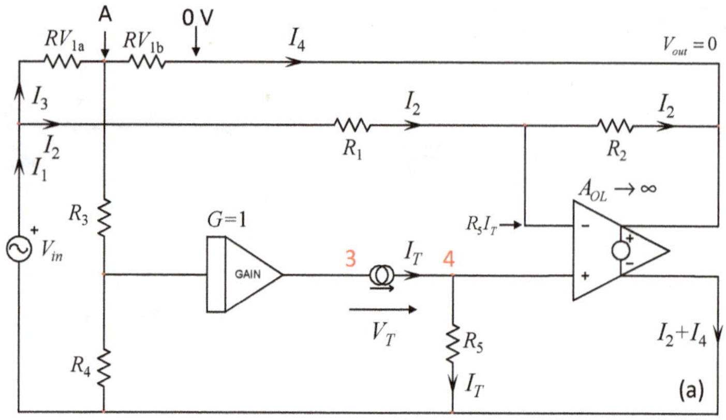

$$V_{(4)} = I_T R_5$$

$$V_{(A)} = V_{in}\frac{RV_{1b}\|(R_3+R_4)}{RV_{1b}\|(R_3+R_4)+RV_{1a}}$$

$$V_{(3)} = V_{in}\frac{RV_{1b}\|(R_3+R_4)}{RV_{1b}\|(R_3+R_4)+RV_{1a}}\frac{R_4}{R_4+R_3}$$

$$I_T R_5 = R_2 I_2 \to I_2 = I_T\frac{R_5}{R_2}$$

$$V_{in} = I_T R_5 + R_1 I_2 \longrightarrow V_{in} = I_T\left(R_5+\frac{R_5}{R_2}R_1\right)$$

$$V_{(3)} = I_T\left(R_5+\frac{R_5}{R_2}R_1\right)\frac{RV_{1b}\|(R_3+R_4)}{RV_{1b}\|(R_3+R_4)+RV_{1a}}\frac{R_4}{R_4+R_3}$$

$$R = \frac{V_T}{I_T} = R_5 - \left(R_5+\frac{R_5}{R_2}R_1\right)\frac{RV_{1b}\|(R_3+R_4)}{RV_{1b}\|(R_3+R_4)+RV_{1a}}\frac{R_4}{R_4+R_3}$$

$$\omega_z = \frac{1}{RC_1} = \frac{1}{\left[R_5 - \left(R_5+\frac{R_5}{R_2}R_1\right)\frac{RV_{1b}\|(R_3+R_4)}{RV_{1b}\|(R_3+R_4)+RV_{1a}}\frac{R_4}{R_4+R_3}\right]C_1}$$

$$H(s) = -\frac{R_2}{R_1}\cdot\frac{1+s\left[R_5-\left(R_5+\frac{R_5}{R_2}R_1\right)\frac{RV_{1b}\|(R_3+R_4)}{RV_{1b}\|(R_3+R_4)+RV_{1a}}\frac{R_4}{R_4+R_3}\right]C_1}{1+s\cdot R_5\left[1-\left(1+\frac{R_2}{R_1}\right)\frac{RV_{1a}\|(R_3+R_4)}{RV_{1a}\|(R_3+R_4)+RV_{1b}}\frac{R_4}{R_4+R_3}\right]C_1} = H_0\frac{1+\frac{s}{\omega_z}}{1+\frac{s}{\omega_p}}$$

Nulling the Output will let us Determine the Position of the Zero

Figure 5.52

We start with the dc gain where the absence of C_1 naturally turns the circuit into a classical inverter involving R_1 and R_2.

The determination of the resistance driving C_1 requires the placement of the test generator I_T as shown in the figure.

A few equations tell us how resistors combine to form the time constant of the circuit.

In a practical circuit, R_5 can be replaced by a variable resistor and gives another means to affect the transfer function cutoff frequency. To determine the zero, inspection won't work and it's best to resort to an NDI as illustrated in Figure 5.52. Here also, if you consider the 0-V bias on the output node and an equal voltage between inverting and non-inverting pins of the right-side op-amp, finding the zero is easy.

The final transfer function shows the presence of a pole and a zero.

The Mathcad® sheet is given in Figure 5.53 and the ac response for different values of k in Figure 5.54.

$R_1 := 22k\Omega \quad R_2 := 22k\Omega \quad R_3 := 7.5k\Omega \quad R_4 := 5.6k\Omega \quad k := 0.7 \quad RV_1 := 10k\Omega$

$RV1_a := k \cdot RV_1 = 7 \cdot k\Omega \quad RV1_b := (1-k) \cdot RV_1 = 3 \cdot k\Omega \quad \|(x,y) := \frac{x \cdot y}{x+y} \quad C_1 := 10nF$

$R_5 := 100k\Omega$

$H_0 := -\frac{R_2}{R_1} \qquad 20\log(|H_0|) = 0 \quad dB$

-------- Determination of the pole -------

$$V_3 = V_2 \cdot \frac{RV1_a \| (R_3+R_4)}{RV1_a \| (R_3+R_4) + RV1_b} \cdot \frac{R_4}{R_4+R_3}$$

$$V_4 = I_T \cdot RV_2$$

$$V_2 = V_4 \cdot \left(1 + \frac{R_2}{R_1}\right)$$

$$V_3 = I_T \cdot RV_2 \cdot \left(1 + \frac{R_2}{R_1}\right) \cdot \left[\frac{RV1_a \| (R_3+R_4)}{RV1_a \| (R_3+R_4) + RV1_b} \cdot \frac{R_4}{R_4+R_3}\right]$$

$$V_T = V_4 - V_3$$

$$I_T \cdot RV_2 \left[1 - \left(1 + \frac{R_2}{R_1}\right) \cdot \left[\frac{RV1_a \| (R_3+R_4)}{RV1_a \| (R_3+R_4) + RV1_b} \cdot \frac{R_4}{R_4+R_3}\right]\right]$$

$$R_{tau1} := R_5 \cdot \left[1 - \left(1 + \frac{R_2}{R_1}\right) \cdot \left[\frac{RV1_a \| (R_3+R_4)}{RV1_a \| (R_3+R_4) + RV1_b} \cdot \frac{R_4}{R_4+R_3}\right]\right] = 48.42105k\Omega$$

$$\tau_1 := C_1 \cdot \left[R_5 \cdot \left[1 - \left(1 + \frac{R_2}{R_1}\right) \cdot \left[\frac{RV1_a \| (R_3+R_4)}{RV1_a \| (R_3+R_4) + RV1_b} \cdot \frac{R_4}{R_4+R_3}\right]\right]\right] = 484.21053\mu s$$

$$\omega_p := \frac{1}{\tau_1} \qquad f_p := \frac{\omega_p}{2 \cdot \pi} = 328.68956Hz$$

$$H_1(s) := H_0 \cdot \frac{1 + \frac{s}{\omega_z}}{1 + \frac{s}{\omega_p}}$$

$H_2 :=$ READPRN(".\TF.txt") SIMetrix data import

-------- Determination of the zero -------

$$V_4 = I_T \cdot R_5$$

$$V_A = V_1 \cdot \frac{RV1_b \| (R_3+R_4)}{RV1_b \| (R_3+R_4) + RV1_a}$$

$$V_3 = V_1 \cdot \frac{RV1_b \| (R_3+R_4)}{RV1_b \| (R_3+R_4) + RV1_a} \cdot \frac{R_4}{R_4+R_3}$$

$$V_1 = I_T \cdot \left(R_5 + \frac{R_5}{R_2} \cdot R_1\right)$$

$$V_3 = \left[I_T \cdot \left(R_5 + \frac{R_5}{R_2} \cdot R_1\right)\right] \cdot \frac{RV1_b \| (R_3+R_4)}{RV1_b \| (R_3+R_4) + RV1_a} \cdot \frac{R_4}{R_4+R_3}$$

$$R_{tau2} := R_5 - \left(R_5 + \frac{R_5}{R_2} \cdot R_1\right) \cdot \frac{RV1_b \| (R_3+R_4)}{RV1_b \| (R_3+R_4) + RV1_a} \cdot \frac{R_4}{R_4+R_3} = 77.89474k\Omega$$

$$\tau_2 := C_1 \cdot R_5 \cdot \left[1 - \left(1 + \frac{R_1}{R_2}\right) \cdot \frac{RV1_b \| (R_3+R_4)}{RV1_b \| (R_3+R_4) + RV1_a} \cdot \frac{R_4}{R_4+R_3}\right] = 778.94737\mu s$$

$$\omega_z := \frac{1}{\tau_2} \qquad f_z := \frac{\omega_z}{2 \cdot \pi} = 204.32054Hz$$

.param RV1=10k
.param k=0.7
.param RV1a={RV1*k}
.param RV1b={RV1*(1-k)}

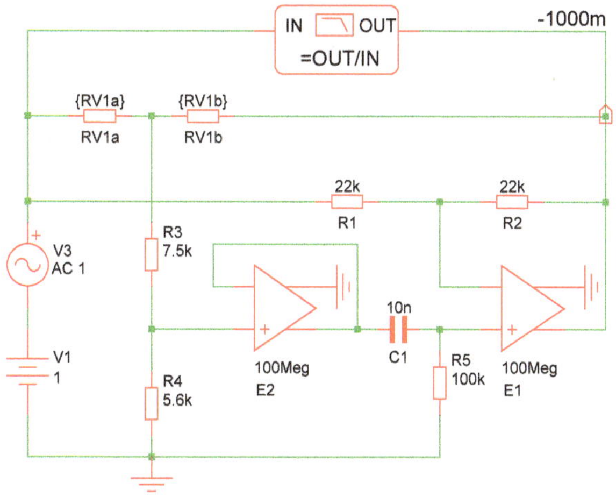

The Transfer Function Features a Pole and a Zero whose Position Depends on the Potentiometer Value

Figure 5.53

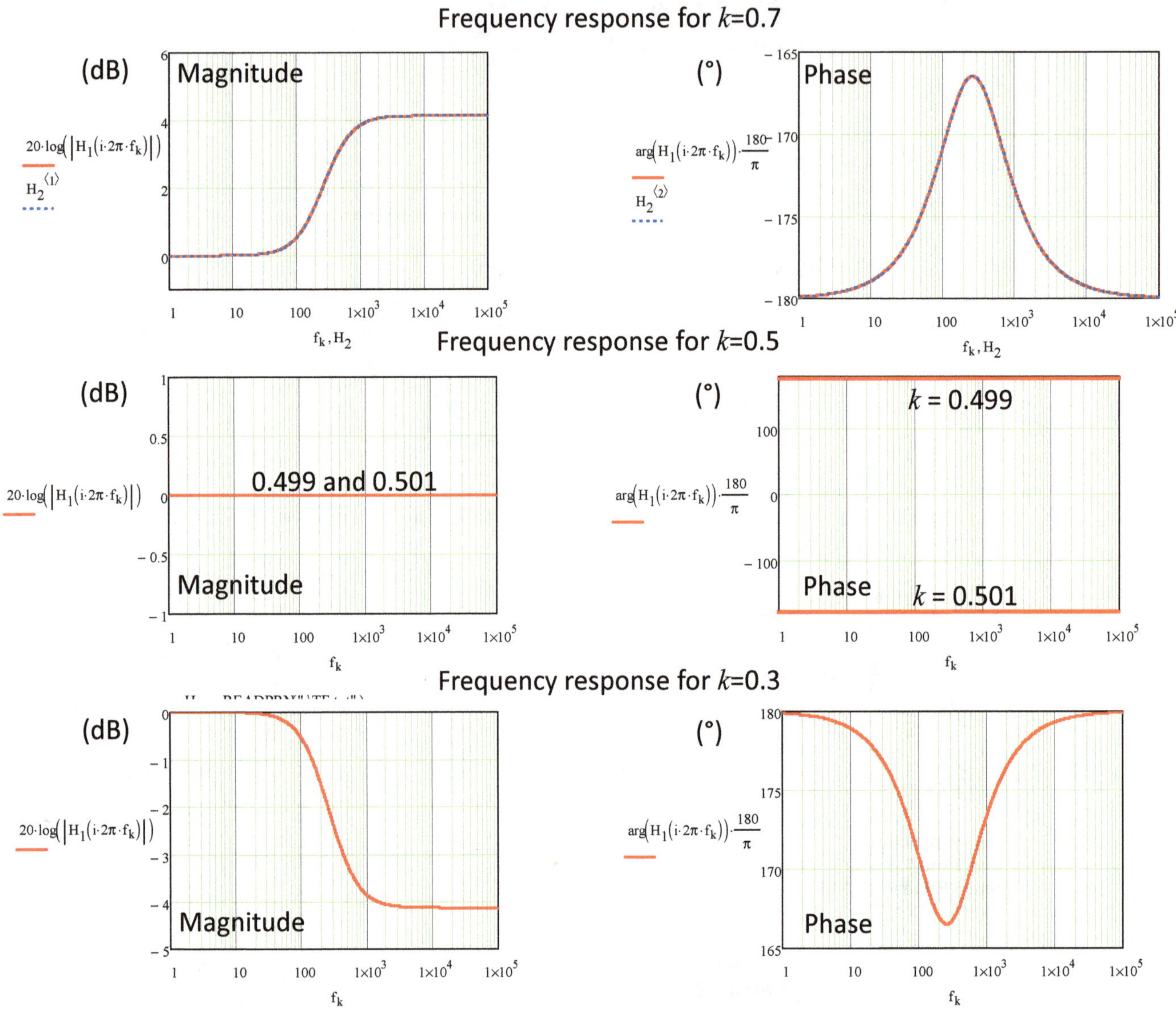

Adjusting the Potentiometer Changes the Ac Response of this Filter

Figure 5.54

In Figure 5.55, the capacitor is not referenced to the ground but sits on the low-side resistor R_3.

The time constants for the pole and zero are quickly derived, respectively by setting V_{in} to zero volts, then nulling V_{out}.

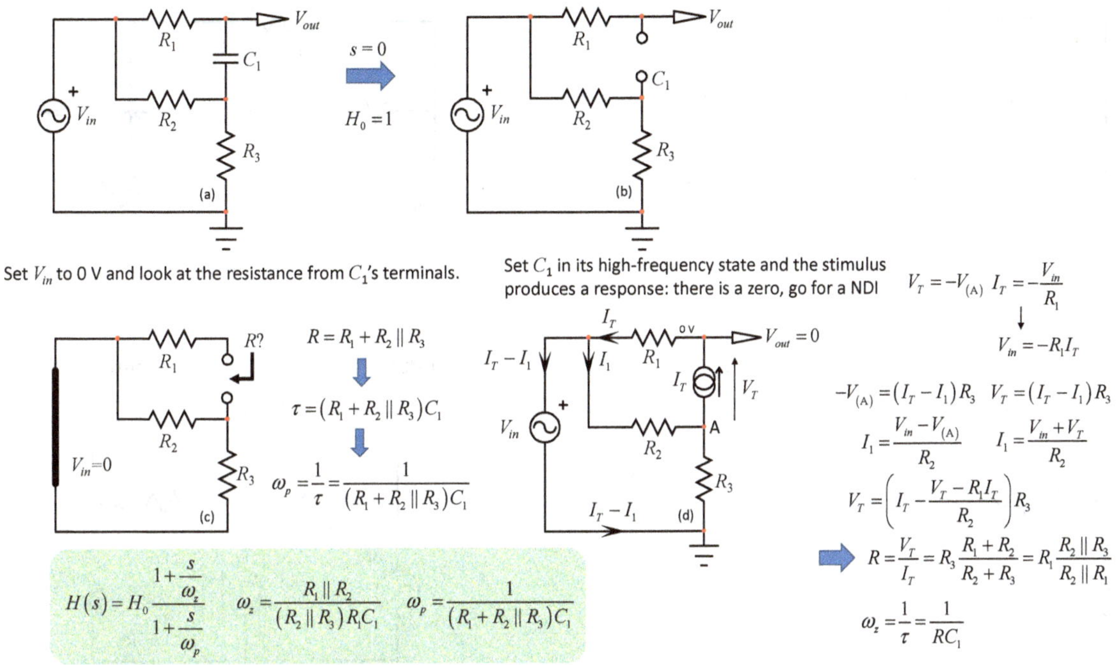

The Capacitor Low-Side Terminal is Connected to the Ground-Referenced Resistor R_3

Figure 5.55

The ac response is given in Figure 5.56 and confirms our analysis.

$\|(x,y):=\frac{x\cdot y}{x+y}$ $R_1:=2k\Omega$ $R_2:=2k\Omega$ $R_3:=100\Omega$ $C_1:=100nF$

$H_0:=1$ $20\cdot\log(H_0)=0$ dB

$\tau_1:=(R_1+R_3\|R_2)\cdot C_1=209.52381\mu s$ $\tau_2:=C_1\cdot\left(R_3\cdot\frac{R_1+R_2}{R_2+R_3}\right)=19.04762\mu s$

$\omega_p:=\frac{1}{\tau_1}=4.77273\times10^3\frac{1}{s}$ $f_p:=\frac{\omega_p}{2\cdot\pi}=0.7596kHz$

$\omega_z:=\frac{1}{\tau_2}=5.25\times10^4\frac{1}{s}$ $f_z:=\frac{\omega_z}{2\cdot\pi}=8.35563kHz$ $f_{z2}:=\frac{R_1\|R_2}{(R_3\|R_2)\cdot R_1\cdot C_1}\cdot\frac{1}{2\cdot\pi}=8.35563kHz$

$H_1(s):=H_0\cdot\frac{1+\frac{s}{\omega_z}}{1+\frac{s}{\omega_p}}$

$H_2:=READPRN(".\backslash TF.txt")$ SIMetrix data import

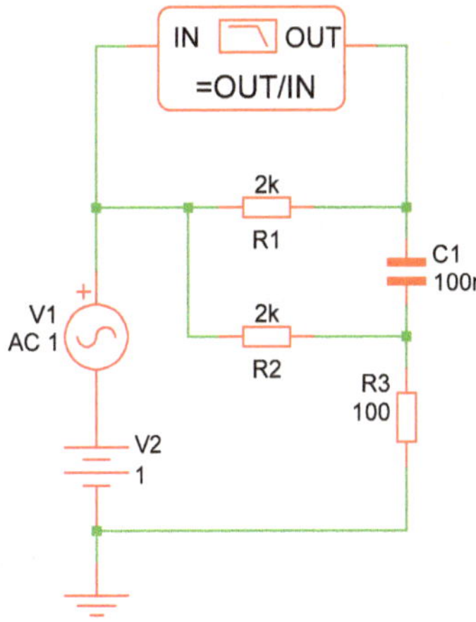

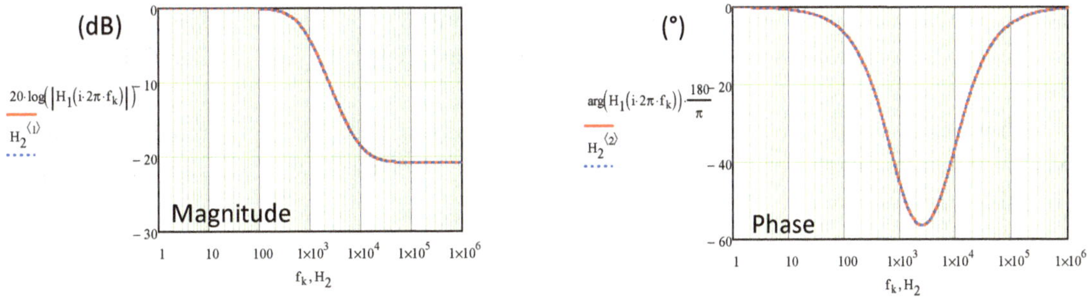

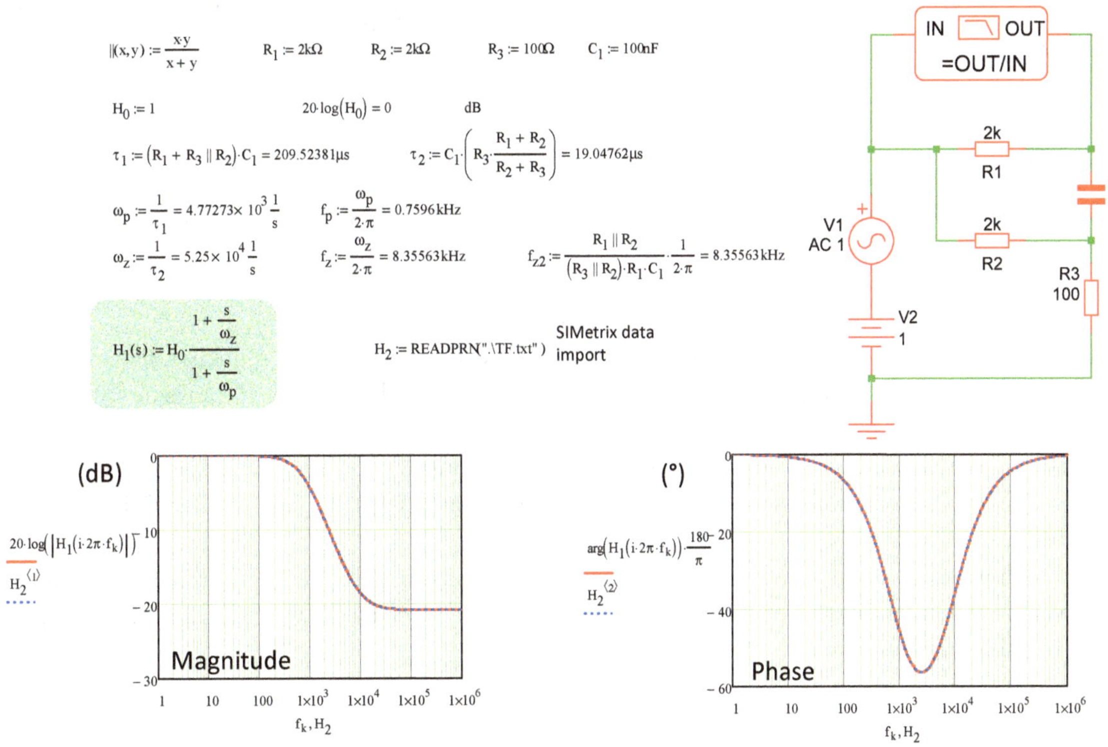

The Response is that of a Classical 1st-order Circuit

Figure 5.56

Set V_{in} to 0 V and look at the resistance from C_1's terminals.

Set C_1 in its high-frequency state and the stimulus produces a response: there is a zero, go for a NDI

$$s = 0$$

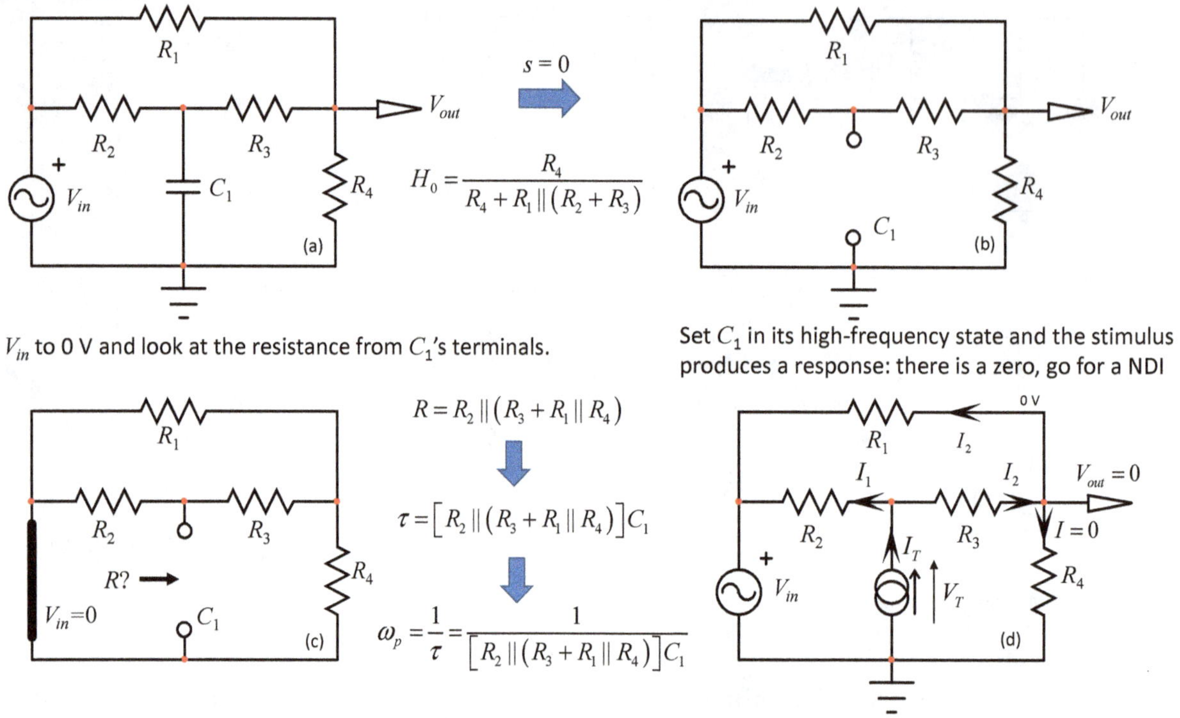

$$H_0 = \frac{R_4}{R_4 + R_1 \| (R_2 + R_3)}$$

$$R = R_2 \| (R_3 + R_1 \| R_4)$$

$$\tau = \left[R_2 \| (R_3 + R_1 \| R_4) \right] C_1$$

$$\omega_p = \frac{1}{\tau} = \frac{1}{\left[R_2 \| (R_3 + R_1 \| R_4) \right] C_1}$$

$$I_1 = \frac{V_T - V_{in}}{R_2} \quad I_2 = \frac{V_T}{R_3} = -\frac{V_{in}}{R_1} \rightarrow V_{in} = -V_T \frac{R_1}{R_3}$$

$$I_T = I_1 + I_2 = \frac{V_T - V_{in}}{R_2} - \frac{V_{in}}{R_1}$$

$$I_T = \frac{V_T (R_1 + R_2 + R_3)}{R_2 R_3} \rightarrow R = \frac{V_T}{I_T} = \frac{R_2 R_3}{R_1 + R_2 + R_3}$$

$$\omega_z = \frac{R_1 + R_2 + R_3}{C_1 R_2 R_3}$$

$$H(s) = \frac{R_4}{R_4 + R_1 \| (R_2 + R_3)} \cdot \frac{1 + sC_1 \frac{R_1 + R_2 + R_3}{R_2 R_3}}{1 + s\left[R_2 \| (R_3 + R_1 \| R_4) \right] C_1} = H_0 \frac{1 + \frac{s}{\omega_z}}{1 + \frac{s}{\omega_p}}$$

The Capacitor is Now in the Middle of the Network with a Resistor from the Input to the Output

Figure 5.57

In Figure 5.57, the capacitor now appears in the middle of the network. Inspection quickly tells you the dc gain and where the pole is located. Again, you can see how quickly the FACTs lead you to the answer without writing a single line of algebra.

For the zero, apply an NDI which requires a few simple equations to determine the associated time constant. You could also resort to the generalized transfer function and determine what the gain is when C_1 is set in its high-frequency state.

This approach would prevent you from running an NDI but leads to a more complicated expression.

This is what is shown in Figure 5.58 where the two zeroes are placed at exactly the same location, but the NDI clearly gives the simplest answer.

$\|(x,y) := \frac{x \cdot y}{x+y}$ $\quad R_1 := 2k\Omega \quad R_2 := 2k\Omega \quad R_3 := 100\Omega \quad C_1 := 100nF \quad R_4 := 250\Omega$

$$H_0 := \frac{R_4}{R_4 + R_1 \| (R_2 + R_3)} = 0.19617 \qquad 20 \cdot \log(H_0) = -14.14725 \text{ dB}$$

$$\tau_1 := [R_2 \| (R_3 + R_1 \| R_4)] \cdot C_1 = 27.7512 \mu s$$

$$\tau_2 := \left(\frac{R_2 \cdot R_3}{R_1 + R_2 + R_3} \right) \cdot C_1 = 4.87805 \mu s$$

$$\omega_p := \frac{1}{\tau_1} = 3.60345 \times 10^4 \frac{1}{s} \qquad f_p := \frac{\omega_p}{2 \cdot \pi} = 5.73507 kHz$$

$$\omega_z := \frac{1}{\tau_2} = 2.05 \times 10^5 \frac{1}{s} \qquad f_z := \frac{\omega_z}{2 \cdot \pi} = 32.62676 kHz$$

Zero obtained With NDI

$$f_{z2} := \frac{\dfrac{R_4}{R_4 + (R_2 + R_3) \| R_1}}{\dfrac{R_4 \| R_3}{R_4 \| R_3 + R_1} \cdot [R_2 \| (R_3 + R_1 \| R_4)] \cdot C_1} \cdot \frac{1}{2 \cdot \pi} = 32.62676 kHz$$

Zero obtained with generalized TF

$$H_1(s) := H_0 \cdot \frac{1 + \frac{s}{\omega_z}}{1 + \frac{s}{\omega_p}}$$

$H_2 :=$ READPRN(".\TF.txt") SIMetrix data import

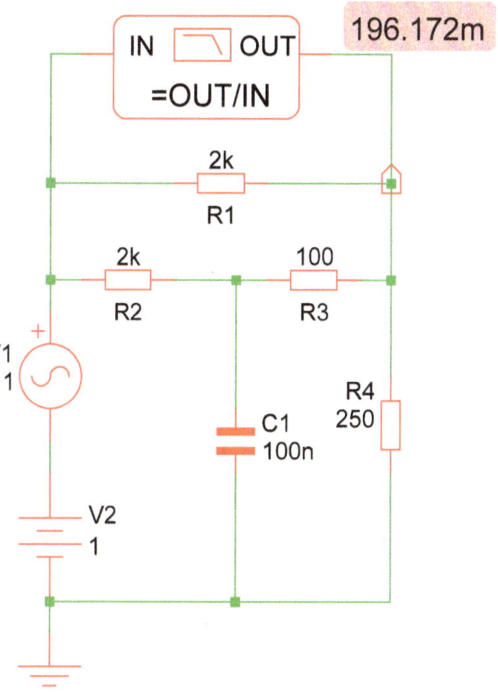

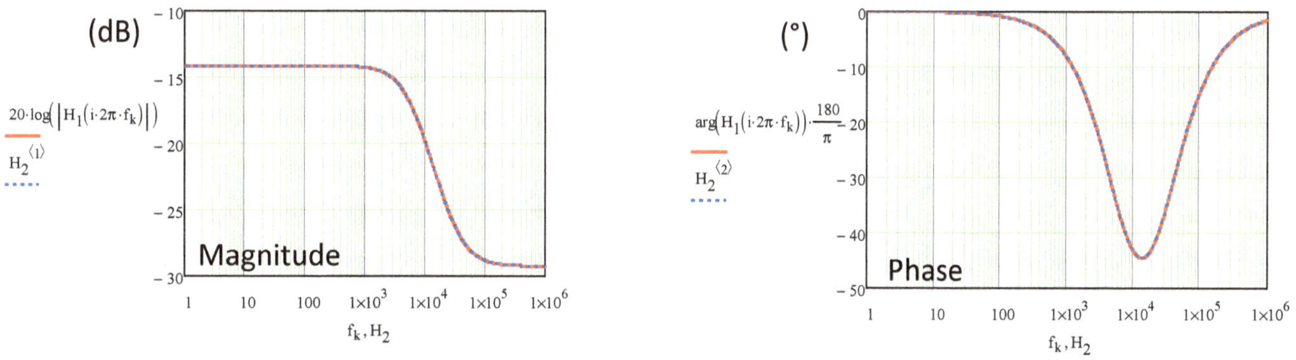

The Ac Response Shows a Pole and a Zero

Figure 5.58

We are going back with an op-amp in Figure 5.59 where a more complicated network feeds the output back to the inverting input.

The dc transfer function H_0 is obtained using superposition but another approach would certainly work.

The idea here is to determine the voltage at the non-inverting pin, voltage which is equal to 0 V in the end considering the virtual ground. For determining the pole, I have installed a test generator I_T and solved a few equations to obtain the time constant. The zero is immediately determined by identifying R_5-C_1 forming a potential transformed short circuit, nulling the output.

Once the transfer function is assembled, its response can be tested versus a SIMetrix® plot as illustrated in Figure 5.60.

Set $s = 0$

Apply superposition by alternatively setting V_{in} to 0 V with V_{out} alive then V_{out} to 0 V and V_{in} active. Determine $V_{(-)}$ in both cases.

(a)

(b)

$$V_{(-)}\Big|_{V_{in}=0} = V_{out}\frac{R_3}{R_3+R_4}\cdot\frac{R_1}{R_2+R_4\parallel R_3+R_1}$$

$$V_{(-)}\Big|_{V_{out}=0} = V_{in}\frac{R_2+R_4\parallel R_3}{R_1+(R_2+R_4\parallel R_3)}$$

Set V_{in} to 0 V

Set V_{out} to 0 V

$$V_{(-)} = V_{(-)}\Big|_{V_{in}=0} + V_{(-)}\Big|_{V_{out}=0}$$
If $A_{OL}\to\infty$ then $\varepsilon\to 0$

(c)

(d)

$$V_{(-)}\Big|_{V_{in}=0} + V_{(-)}\Big|_{V_{out}=0} = 0$$

solve and factor $\dfrac{V_{out}}{V_{in}}$

$$H_0 = -\frac{R_3(R_2+R_4)+R_2R_4}{R_1R_3}$$

To determine the pole, set V_{in} to 0 V and install a test generator I_T. For simplicity, you can temporarily disconnect R_5 and bring it back later.

For the zero, what impedance combination could null the response? Of course, $Z_1(s) = 0$

$$V_T = I_T R_2 + (I_T - i_1)R_4$$

$$i_1 = -\frac{I_T R_2}{R_3}$$

Substitute i_1 in the above

$$\frac{V_T}{I_T} = R_2 + R_4\left(1+\frac{R_2}{R_3}\right)$$

Bring R_5 back

$$R = R_5 + R_2 + R_4\left(1+\frac{R_2}{R_3}\right)$$

$$\omega_p = \frac{1}{\left[R_5 + R_2 + R_4\left(1+\frac{R_2}{R_3}\right)\right]C_1}$$

$$Z_1(s) = R_5 + \frac{1}{sC_1} = 0$$

$$\to s_z = -\frac{1}{R_5C_1}$$

$$\omega_z = \frac{1}{R_5C_1}$$

$Z_1(s) = 0$

(e)

(f)

$$H(s) = H_0\frac{1+\dfrac{s}{\omega_z}}{1+\dfrac{s}{\omega_p}}$$

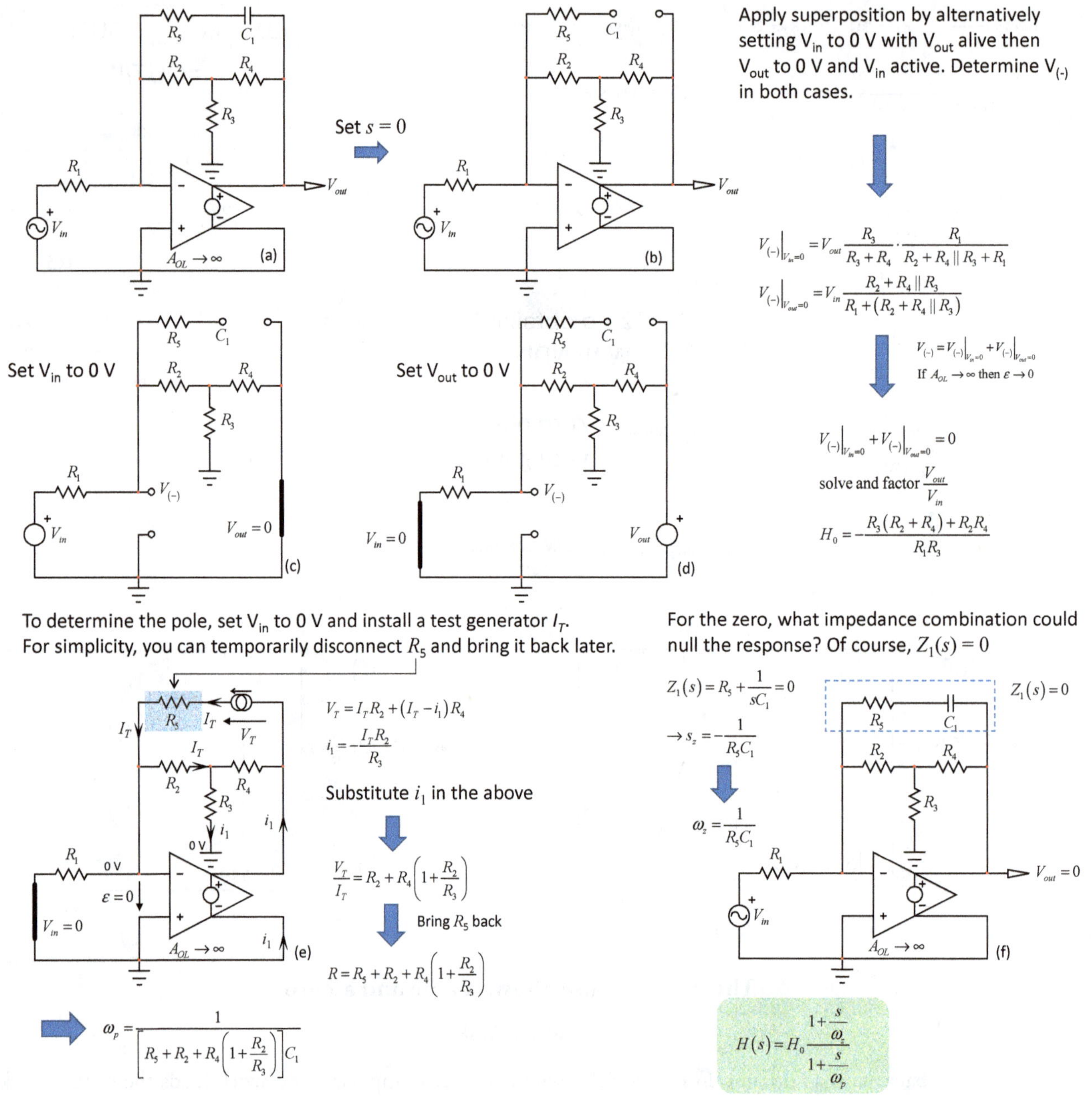

Despite the Unusual Configuration, Determining the Pole and the Zero is Fast

Figure 5.59

$R_1 := 49.9k\Omega \quad R_2 := 200k\Omega \quad R_3 := 1.13k\Omega \quad R_4 := 200k\Omega \quad R_5 := 49.9k\Omega \quad C_1 := 1\mu F$

$\|(x,y) := \frac{x \cdot y}{x+y} \qquad R_{inf} := 10^{23}\Omega$

$V\_1 = V_{out} \cdot \frac{R_3}{R_4+R_3} \cdot \frac{R_1}{R_2 + R_4 \| R_3 + R_1} \qquad V\_2 = V_{in} \cdot \frac{(R_2 + R_4 \| R_3)}{(R_2 + R_4 \| R_3) + R_1}$

$V_{out} \cdot \left(\frac{R_3}{R_4+R_3} \cdot \frac{R_1}{R_2 + R_4 \| R_3 + R_1} \right) + V_{in} \cdot \frac{(R_2 + R_4 \| R_3)}{(R_2 + R_4 \| R_3) + R_1} = 0$

$H_0 := -\frac{R_3 \cdot (R_2 + R_4) + R_2 \cdot R_4}{R_1 \cdot R_3} = -717.399 \qquad 20 \cdot \log(|H_0|) = 57.115 \text{ dB}$

$\omega_p := \frac{1}{C_1 \left[R_4 \cdot \left(1 + \frac{R_2}{R_3} \right) + R_5 + R_2 \right]} \qquad \omega_z := \frac{1}{R_5 \cdot C_1}$

$H_1(s) := H_0 \cdot \frac{1 + \frac{s}{\omega_z}}{1 + \frac{s}{\omega_p}} \qquad f_z := \frac{\omega_z}{2\pi} = 3.189 \text{Hz} \qquad f_p := \frac{\omega_p}{2\pi} = 4.44 \times 10^{-3} \cdot \text{Hz}$

$H_2 :=$ READPRN(".\TF.txt") SIMetrix data import

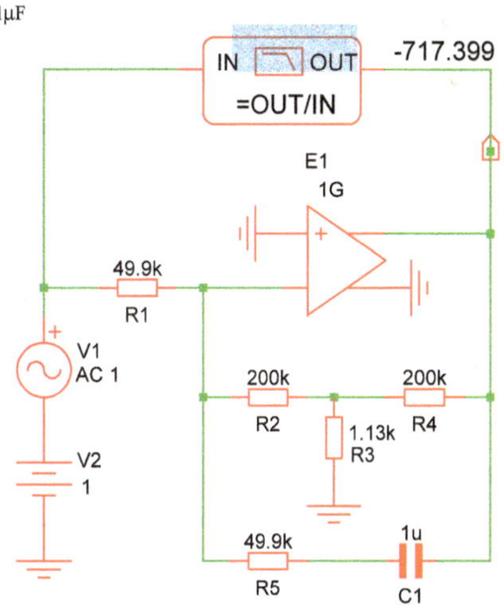

SIMetrix circuit for $H(s)$

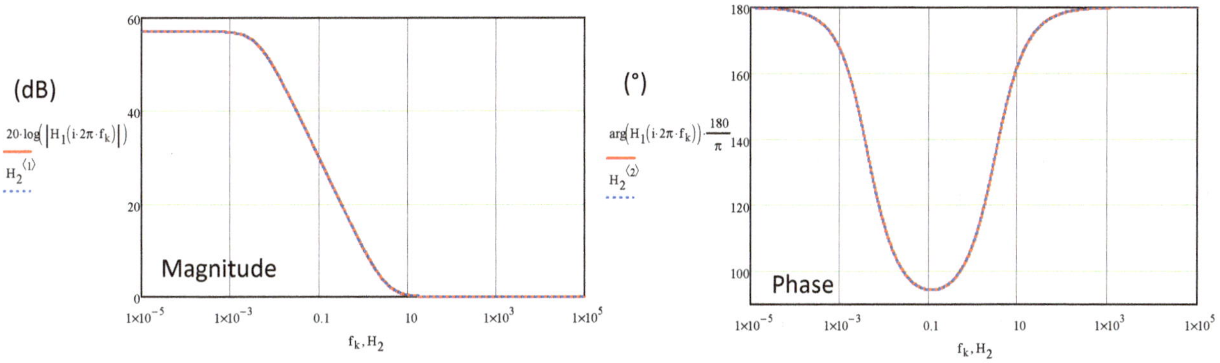

The Ac Response from SIMetrix® Exactly Matches that of Mathcad®

Figure 5.60

In Figure 5.61, we simplified the circuit and replaced resistor R_3 by a capacitor. The determination of the dc gain is easy with C_1 removed. To determine the resistance R driving that capacitor, you can first consider a non-infinite open-loop gain. In this mode, you determine the voltage at the inverting pin and carry on until you have R.

Inspection also works if you take a closer look when considering an infinite gain:

- A_{OL} is infinite then $\varepsilon = 0$
- $i_{R_1} = i_{R_2} = 0 \rightarrow V_T = \varepsilon = 0$
- $R = V_T / I_T = 0 / I_T = 0$

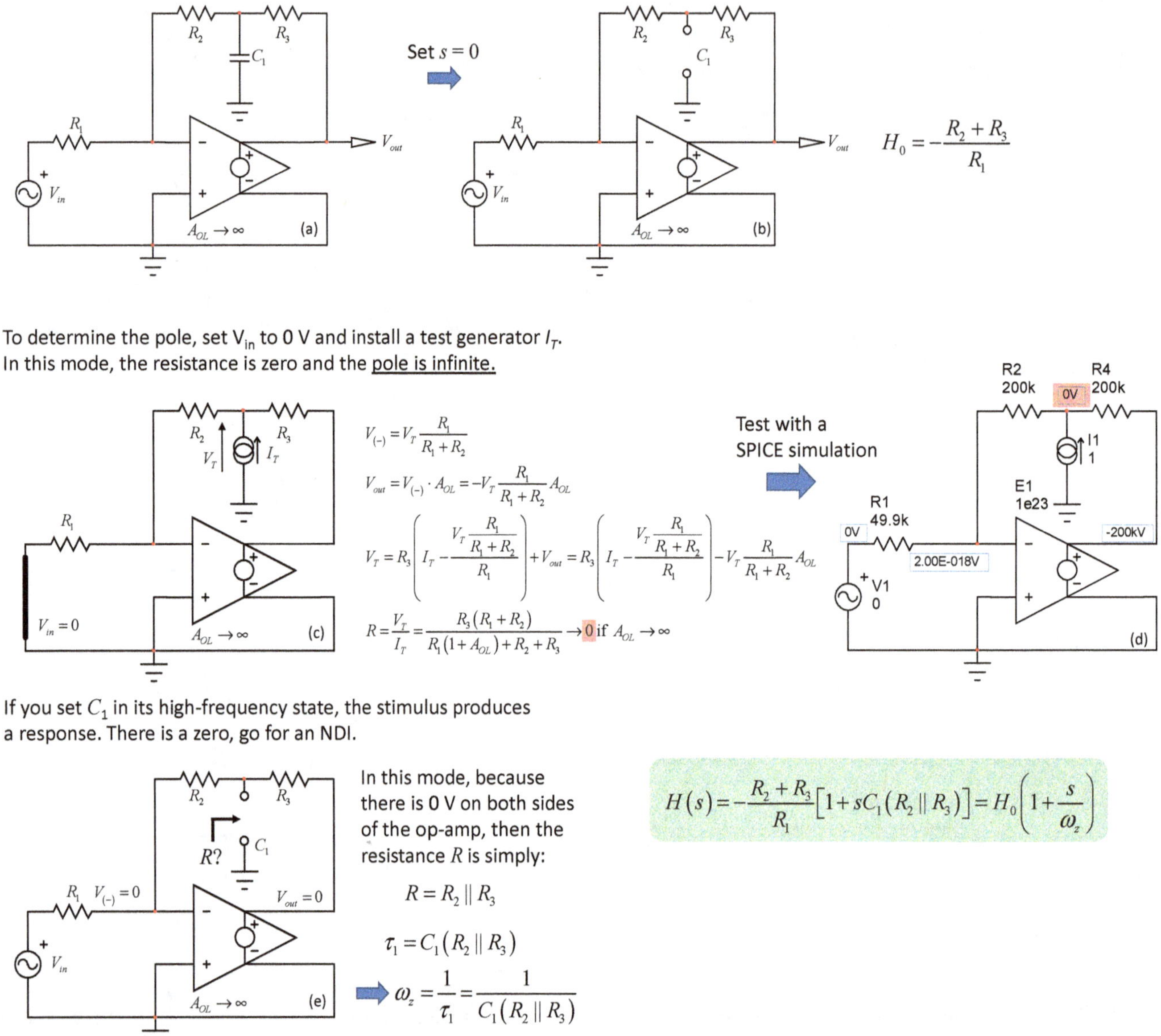
Set $s=0$
V_{out}
$A_{OL}\to\infty$
(a)
(b)
$H_0=-\frac{R_2+R_3}{R_1}$
To determine the pole, set V_{in} to 0 V and install a test generator I_T.
In this mode, the resistance is zero and the pole is infinite.
$V_{(-)}=V_T\frac{R_1}{R_1+R_2}$
$V_{out}=V_{(-)}\cdot A_{OL}=-V_T\frac{R_1}{R_1+R_2}A_{OL}$
$V_T=R_3\left(I_T-\frac{V_T\frac{R_1}{R_1+R_2}}{R_1}\right)+V_{out}=R_3\left(I_T-\frac{V_T\frac{R_1}{R_1+R_2}}{R_1}\right)-V_T\frac{R_1}{R_1+R_2}A_{OL}$
$R=\frac{V_T}{I_T}=\frac{R_3(R_1+R_2)}{R_1(1+A_{OL})+R_2+R_3}\to 0\text{ if }A_{OL}\to\infty$
$V_{in}=0$
(c)
Test with a SPICE simulation
R2
200k
R4
200k
0V
I1
1
E1
1e23
R1
49.9k
0V
-200kV
2.00E-018V
V1
0
(d)
If you set C_1 in its high-frequency state, the stimulus produces a response. There is a zero, go for an NDI.
$R?$
$V_{(-)}=0$
$V_{out}=0$
(e)
In this mode, because there is 0 V on both sides of the op-amp, then the resistance R is simply:
$R=R_2\,||\,R_3$
$\tau_1=C_1(R_2\,||\,R_3)$
$\omega_z=\frac{1}{\tau_1}=\frac{1}{C_1(R_2\,||\,R_3)}$
$H(s)=-\frac{R_2+R_3}{R_1}\left[1+sC_1(R_2\,||\,R_3)\right]=H_0\left(1+\frac{s}{\omega_z}\right)$

A Few Lines of Algebra are Necessary for the Pole, but Nothing too Complicated

Figure 5.61

This resistance becomes a short circuit when the op-amp open-loop gain A_{OL} approaches infinity. This is quickly confirmed by the SPICE simulation shown on the right side.

Having a zeroed resistance means the time constant is also zero and the pole appears at an infinite value: this circuit does not have a pole. The zero is determined inspecting an NDI and, considering a zeroed output, the resistance driving the capacitor is simply the paralleling of R_2 and R_3.

We have all we need and can test the frequency response in Figure 5.62.

$R_1 := 49.9\text{k}\Omega \quad R_2 := 100\text{k}\Omega \quad R_3 := 100\text{k}\Omega \quad C_1 := 10\text{nF}$

$\|(x,y) := \dfrac{x \cdot y}{x + y} \quad R_{inf} := 10^{23}\Omega \quad A_{OL} := 10^{12}$

Pole approaches infinity

$\tau_1 := C_1 \cdot \dfrac{R_1 \cdot R_3 + R_2 \cdot R_3}{R_1 + R_2 + R_3 + A_{OL} \cdot R_1} \qquad \omega_p := \dfrac{1}{\tau_1} = 3.329 \times 10^{14} \dfrac{1}{s}$

$H_0 := -\dfrac{R_2 + R_3}{R_1} = -4.008 \qquad\qquad 20 \cdot \log(|H_0|) = 12.059 \quad dB$

$\omega_z := \dfrac{1}{C_1 \cdot (R_2 \| R_3)}$

SIMetrix data import

$H_1(s) := H_0 \cdot \left(1 + \dfrac{s}{\omega_z}\right) \quad f_z := \dfrac{\omega_z}{2\pi} = 318.31 \text{Hz} \qquad H_2 := \text{READPRN}(".\backslash\text{TF.txt}")$

(dB)

$20 \cdot \log(|H_1(i \cdot 2\pi \cdot f_k)|)$

$H_2^{\langle 1 \rangle}$

Magnitude

f_k, H_2

IN ☐ OUT   -4.00802
=OUT/IN

E1
1G

49.9k
R1

V1
AC 1

100k   100k
R2   R4

C1
10n

V2
1

SIMetrix
circuit
for $H(s)$

(°)

$\arg(H_1(i \cdot 2\pi \cdot f_k)) \cdot \dfrac{180}{\pi}$

$H_2^{\langle 2 \rangle}$

Phase

f_k, H_2

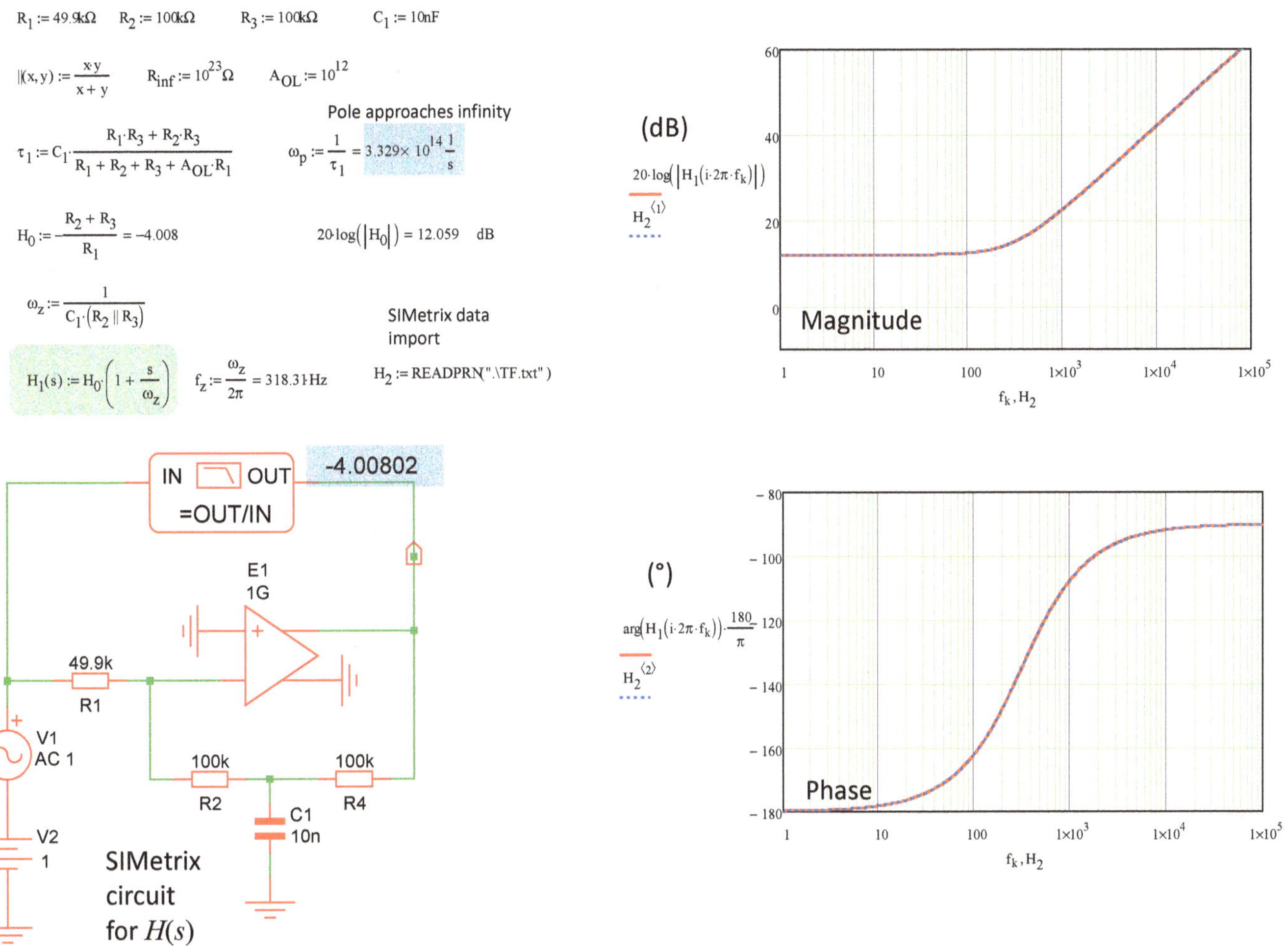

The Ac Response from SIMetrix® and Mathcad® Confirm the Absence of a Pole

Figure 5.62

The circuit proposed in Figure 5.63 is found in x10 oscilloscope probes.

In this network, you want to attenuate the signal by a fixed ratio—10:1 in this example—without affecting the ac response. In other words, you want the flattest magnitude without phase distortion. Considering the two capacitors, it looks like a second-order system. However, when you zero the stimulus, these two capacitors come in parallel and form only one energy-storing element equal to the sum of the capacitors: this is a degenerate case and despite two energy-storing elements, it is still a 1st-order network.

Starting from there, the rest of the analysis is classic and the zero is found by inspection.

You end up with a transfer function featuring a pole/zero pair and a leading term representing the attenuation you want. To compensate the probe, the pole and the zero must be coincident, meaning the ac response becomes flat in magnitude and phase (the pole and the zero neutralize each other). This is what you do when you calibrate the probe with a small trimmer so that the applied 1-kHz square wave goes through perfectly without distortion on its edges. Distortion occurs if the zero and the pole are not coincident—like if the zero occurs before the pole (differentiator) or the pole occurs before the zero (integrator). In this exercise, you can determine the value of capacitor C_2 which would lead to a perfect cancelation of the pole and zero. Solving for C_2 when the two time constants are equal—$\tau_1 = \tau_{1N}$—gives a capacitance of 198 pF.

The ac response is given in Figure 5.64 with a purposely misaligned pole and zero.

Set $s = 0$

$$H_0 = \frac{R_2}{R_1 + R_2}$$

Set V_{in} to 0 V and look through C_1's connections:

When the excitation is zeroed, the network reduces to a 1st-oder system

$$R = R_1 \| R_2$$

$$\tau_1 = (R_1 \| R_2)(C_1 + C_2)$$

$$\omega_p = \frac{1}{\tau_1} = \frac{1}{(R_1 \| R_2)(C_1 + C_2)}$$

If you short-circuit C_1 and apply a stimulus, then there is a response. C_1 contributes a zero. C_2 does not in similar conditions: there is one single zero.

What could block the stimulus propagation for nulling the response?

$$Z_1(s) = R_1 \| \frac{1}{sC_1} = \frac{R_1}{1 + sR_1C_1}$$

$$Z_1(s) \to \infty \text{ if } (1 + sR_1C_1) = 0$$

$$s_z = -\frac{1}{R_1C_1}$$

$$H(s) = \frac{R_2}{R_1 + R_2} \frac{1 + sR_1C_1}{1 + (C_1 + C_2)(R_1 \| R_2)}$$

$$H(s) = H_0 \frac{1 + \dfrac{s}{\omega_z}}{1 + \dfrac{s}{\omega_p}} \quad \omega_z = \frac{1}{R_1C_1} \quad \omega_p = \frac{1}{(C_1 + C_2)(R_1 \| R_2)}$$

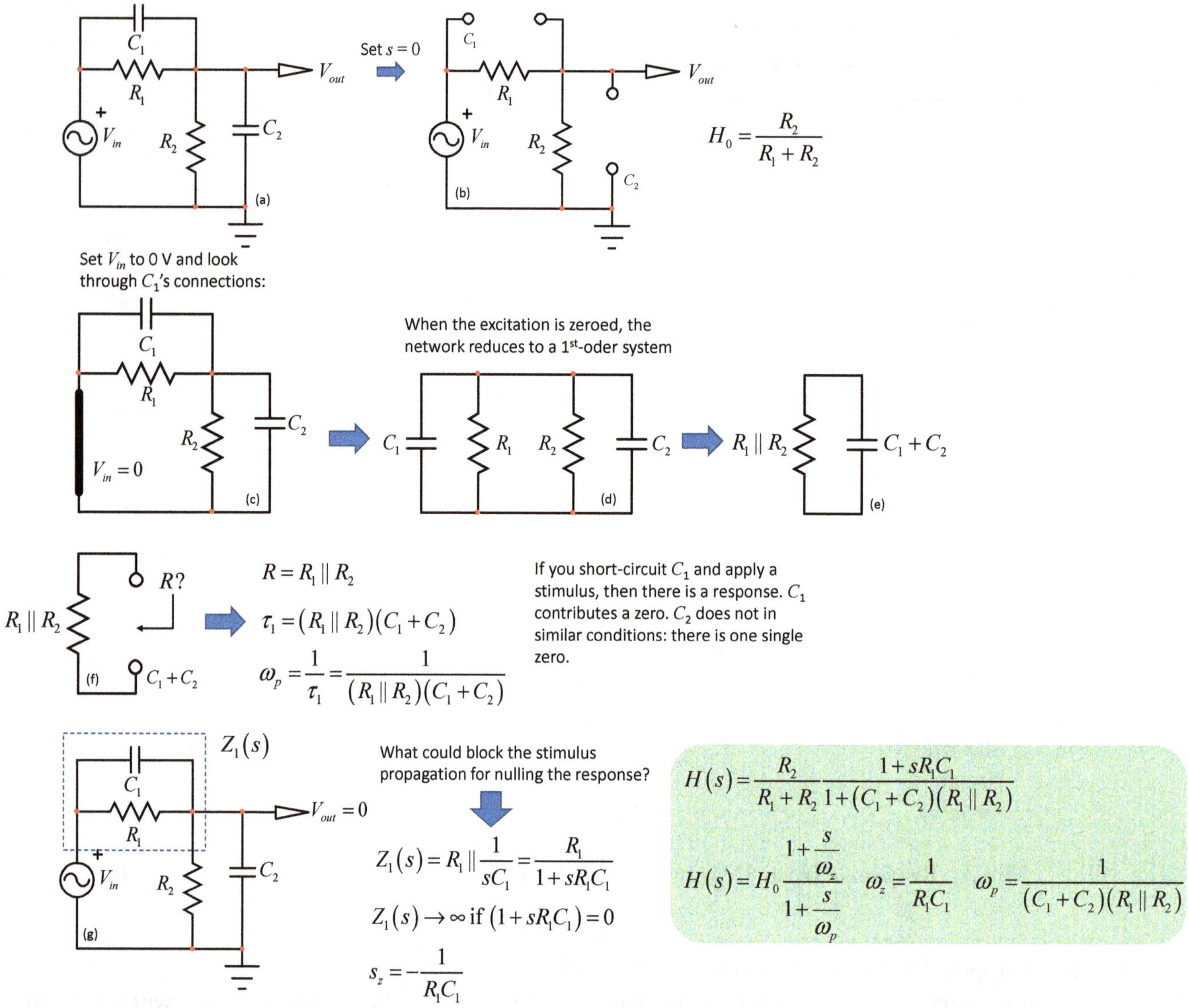

Despite Two Capacitors, this Compensated Probe is a 1st-Order Network

Figure 5.63

$R_1 := 9M\Omega \qquad R_2 := 1M\Omega \qquad C_1 := 22pF \qquad C_2 := 47pF$

$\|(x,y) := \dfrac{x \cdot y}{x + y} \qquad R_{inf} := 10^{23}\Omega \qquad A_{OL} := 10^{12}$

$H_0 := \dfrac{R_2}{R_1 + R_2} = 0.1 \qquad 20 \cdot \log(H_0) = -20 \qquad dB$

$\tau_1 := (C_1 + C_2) \cdot (R_1 \| R_2) = 62.1 \cdot \mu s \qquad \omega_p := \dfrac{1}{\tau_1} \qquad f_p := \dfrac{\omega_p}{2 \cdot \pi} = 2.563kHz$

$\tau_{1N} := R_1 \cdot C_1 = 198\mu s \qquad \omega_z := \dfrac{1}{\tau_{1N}} \qquad f_z := \dfrac{\omega_z}{2 \cdot \pi} = 0.804kHz$

$H_1(s) := H_0 \dfrac{1 + \dfrac{s}{\omega_z}}{1 + \dfrac{s}{\omega_p}} \qquad H_2 := READPRN(".\TF.txt")$

To compensate this divider, equal the two time constants τ_1 and τ_{1N} :

$(C_1 + C_2) \cdot \left(\dfrac{R_1 \cdot R_2}{R_1 + R_2} \right) = R_1 \cdot C_1$

$C_2 = C_1 \dfrac{R_1}{R_2} = 198\ pF$

When C_2 is 198 pF, the pole and the zero are coincident, neutralizing each other. The magnitude and phase are ac flat.

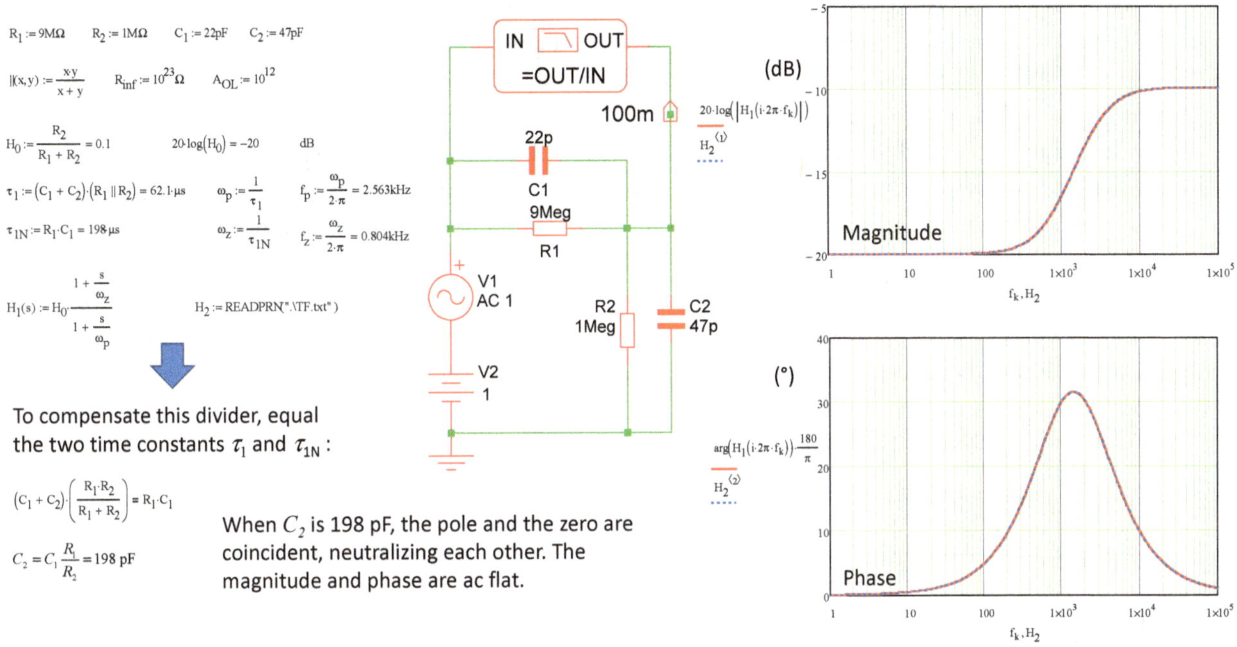

The Two Capacitors are in parallel, a Degenerate Case Reducing the Order to One Despite Two Energy-Storing Elements—the Zero Occurs before the Pole

Figure 5.64

$C_1 = 22\ pF$
$C_2 = 100\ pF$ → $f_p := \dfrac{\omega_p}{2 \cdot \pi} = 1.449kHz$ Pole is higher than zero: differentiator
$f_z := \dfrac{\omega_z}{2 \cdot \pi} = 0.804kHz$

$C_1 = 22\ pF$
$C_2 = 198\ pF$ → $f_p := \dfrac{\omega_p}{2 \cdot \pi} = 0.804kHz$ Pole and zero neutralized
$f_z := \dfrac{\omega_z}{2 \cdot \pi} = 0.804kHz$

$C_1 = 22\ pF$
$C_2 = 250\ pF$ → $f_p := \dfrac{\omega_p}{2 \cdot \pi} = 0.65kHz$ Zero is higher than pole: integrator
$f_z := \dfrac{\omega_z}{2 \cdot \pi} = 0.804kHz$

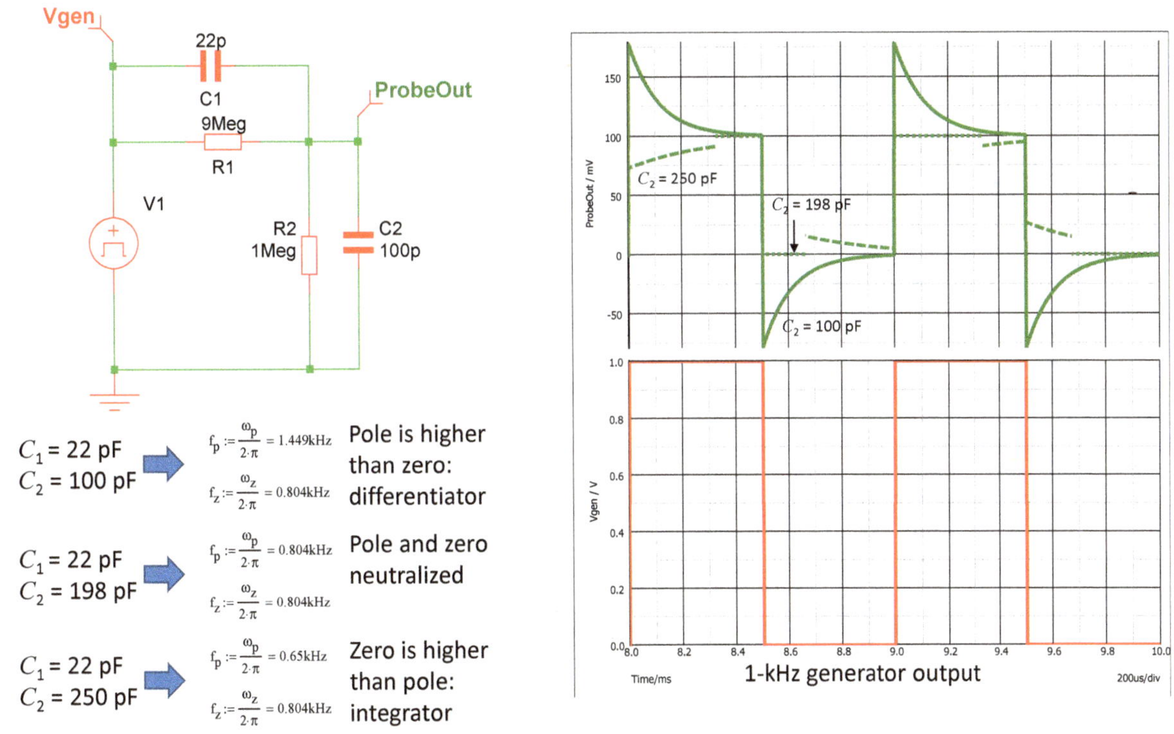

Transient Response for Three C_2 Capacitance Values

Figure 5.65

In Figure 5.65, add a 1-V 1-kHz square-wave generator as oscilloscopes provide on the front panel for calibrating a divide-by-ten probe. The goal is to trim capacitor C_2 so that the pole and the zero neutralize to deliver a perfectly square response. Otherwise, the signal observed on the oscilloscope is distorted and shows integrating or differentiating effects depending on C_2's value.

As confirmed by the dotted line in the middle, a capacitor of 198 pF does the expected job.

Let's have a look now at 1st-order gyrator drawn in Figure 5.66. The goal here is to emulate an inductance with an active circuit built around an op-amp and a capacitor. When determining an impedance, the stimulus is the current source I_T while the response is the voltage V_T collected across the source connecting terminals. The dc input resistance is obtained by inspection, considering the two inputs of the op-amp at 0 V. Then, you turn the stimulus off—the current source is open-circuited—and you determine the resistance R in this mode.

Again, a simple set of equations leads you to the answer, but inspection would work equally well considering the two inputs of the op-amp at the same potential.

Set I_{in} to 0 A and look through C_1's connections. The two op-amp inputs share a common potential.

The zero is obtained by nulling the response. The response is the voltage V_T across the current source. If nulled, it is a degenerate case and the generator can be replaced by a short circuit:

$$V_{(A)} = I_T R_1 - I_T R_2$$
$$V_{(+)} = -I_T R_2$$
$$V_T = V_{(A)} - V_{(+)} = I_T R_1$$
$$R = \frac{V_T}{I_T} = R_1$$
$$\tau_1 = R_1 C_1$$
$$\omega_p = \frac{1}{\tau_1} = \frac{1}{R_1 C_1}$$

$$R = R_2$$
$$\tau_{1N} = R_2 C_1$$
$$\omega_z = \frac{1}{\tau_{1N}} = \frac{1}{R_2 C_1}$$

$$Z_{in}(s) = R_1 \frac{1 + sR_2 C_1}{1 + sR_1 C_1} = R_0 \frac{1 + \dfrac{s}{\omega_z}}{1 + \dfrac{s}{\omega_p}}$$

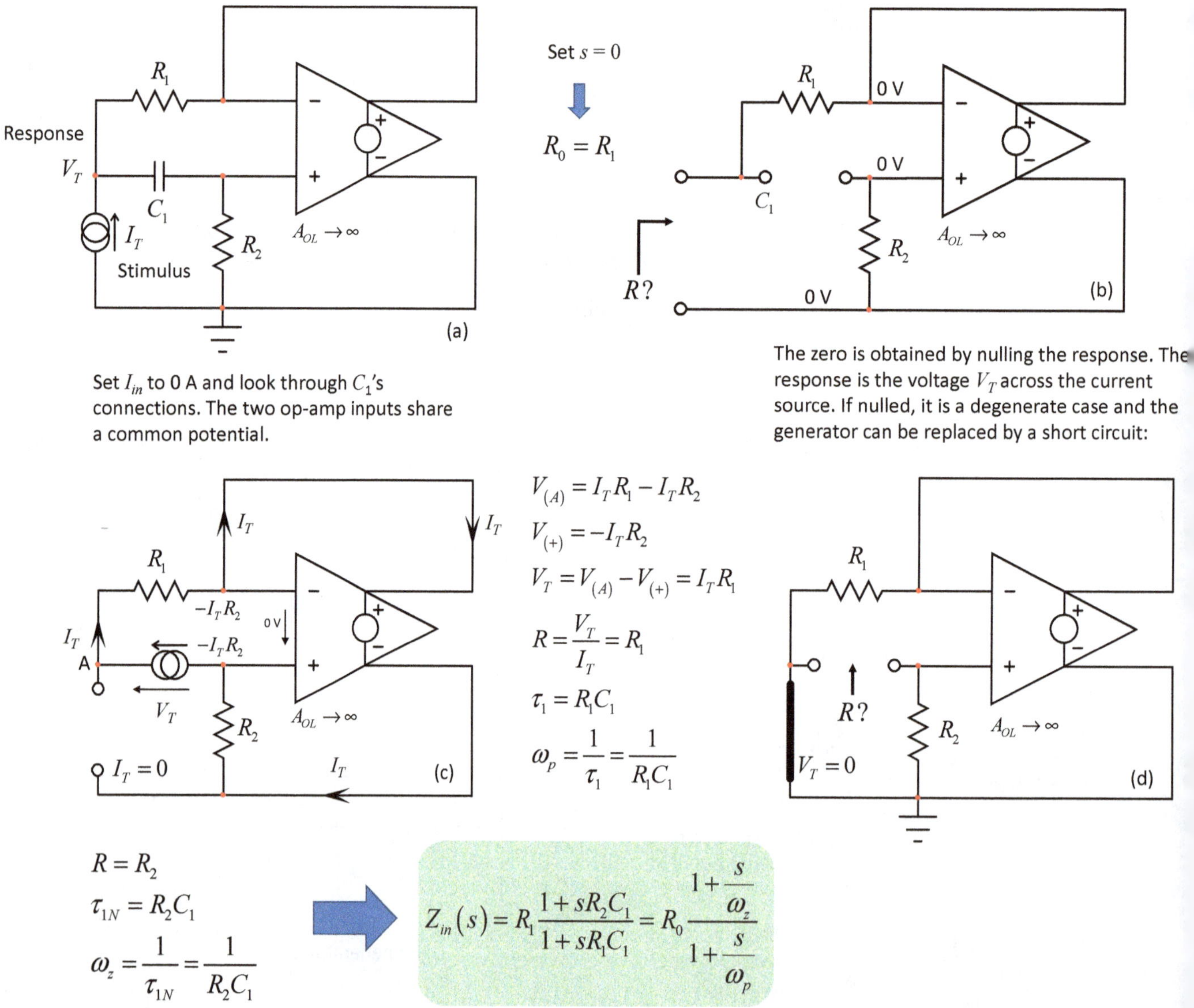

A Gyrator Emulates Inductance with an Op-Amp, a Capacitor and Two Resistors

Figure 5.66

The zero is immediately determined by nulling the response V_T. Zero volts across a current source represents a degenerate case and you can replace the generatorit by a wire in the circuit. The resistance offered by the capacitor's terminals in this mode is simply R_2.

You can now assemble the transfer function and verify its integrity versus a SIMetrix® plot in Figure 5.67.

As you can see, the magnitude and phase responses are perfectly superimposed.

What is interesting now is to determine an approximate passive equivalent circuit to this gyrator. By looking at the magnitude graph in Figure 5.67, we see a flat curve in dc and this is a resistive part. Then the magnitude increases with frequency and it is well an inductive behavior.

Finally, above a certain frequency value, the impedance becomes resistive again. How could we model this frequency response? Figure 5.68 shows the network which will behave similarly in magnitude and phase. If you consider resistor R_1 much smaller than R_2, for instance respectively 10 Ω and 22 kΩ as in our example, then you can match the time constants of the gyrator input impedance expression with that of the *RL* network I drawn.

In this case, the emulated inductance is simply $L_{eq} = R_1R_2C_1$ and its ohmic components are R_1 for the equivalent series resistance (ESR, r_L in the sketch) and R_2 for the high-frequency part.

When you plot the response of the emulated network with its passive equivalent, all curves nicely superimpose.

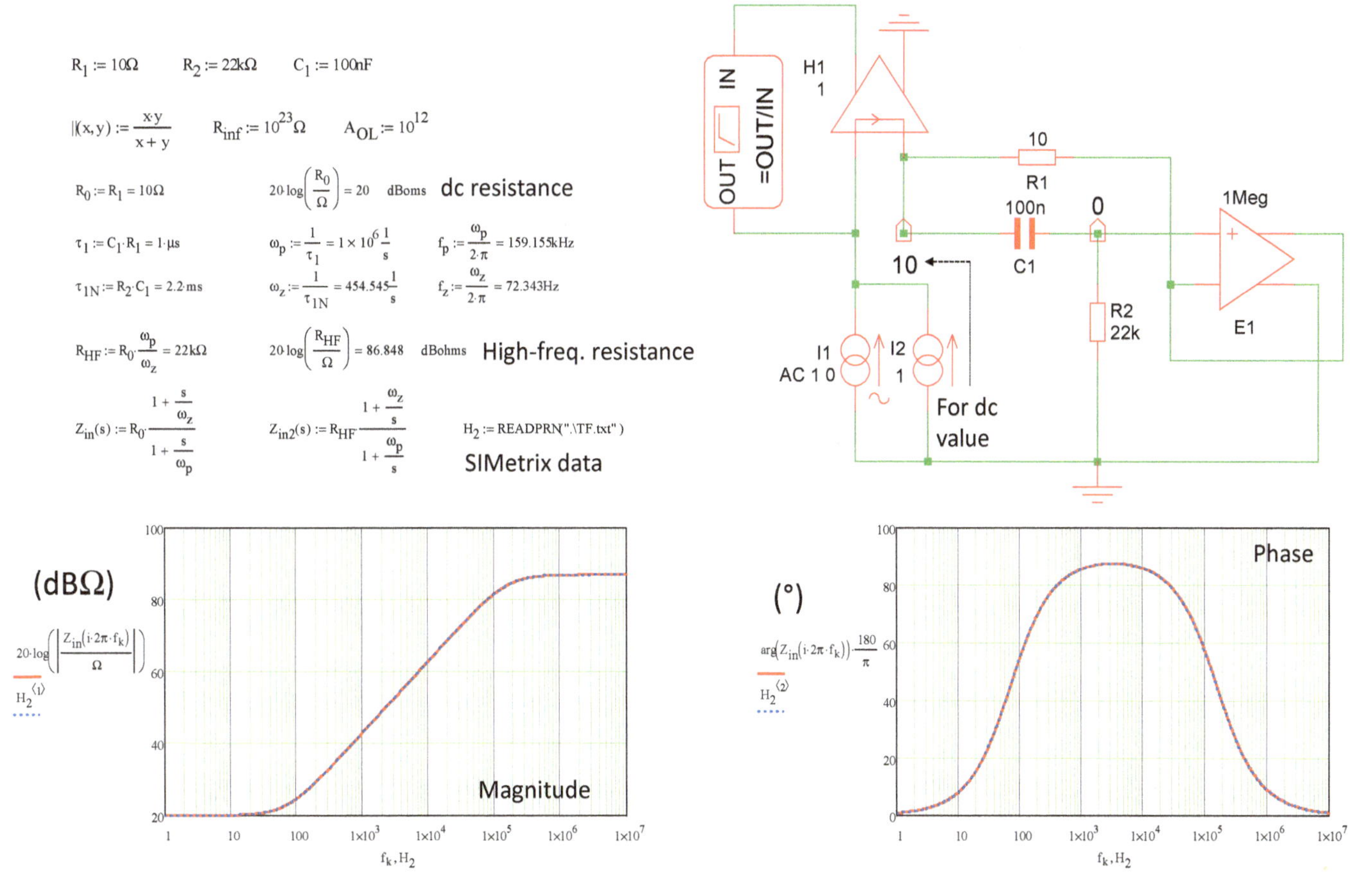

The Transfer Function Shows a Dc Part, a Growing Impedance versus Frequency and a Flat High-Frequency Resistive Portion

Figure 5.67

The magnitude response is that of a lossy inductive circuit like this one:

$Z?$

r_L

R_{HF}

L_{eq}

Its impedance is easily obtained:

$$Z(s) = (r_L \| R_{HF}) \dfrac{1 + s\dfrac{L_{eq}}{r_L}}{1 + s\dfrac{L_{eq}}{r_L + R_{HF}}}$$

Emulated $\quad R_1 \dfrac{1 + sR_2C_1}{1 + sR_1C_1}$

?

Passive Circuit $\quad 1 + s\dfrac{L_{eq}}{r_L}$

$$(r_L \| R_{HF}) \dfrac{}{1 + s\dfrac{L_{eq}}{r_L + R_{HF}}}$$

Equivalent passive circuit for $R_1 \ll R_2$

$$r_L = R_1 \| R_2 \approx R_1 \qquad R_{HF} = R_2$$

$$sR_2C_1 = s\dfrac{L_{eq}}{r_L} \approx s\dfrac{L_{eq}}{R_1} \qquad sR_1C_1 = s\dfrac{L_{eq}}{r_L + R_{HF}} \approx s\dfrac{L_{eq}}{R_2}$$

$$L_{eq} = R_1R_2C_1$$

$f_1 := 10\text{kHz}$

$L_1 := \dfrac{|Z_{in}(i \cdot 2\pi \cdot f_1)|}{2 \cdot \pi \cdot f_1} = 21.957\text{mH} \qquad R_1 \cdot C_1 \cdot R_2 = 22\,\text{mH}$

(dBΩ)

Emulated $\quad 20 \cdot \log\left(\left|\dfrac{Z_{in}(i \cdot 2\pi \cdot f_k)}{\Omega}\right|\right)$

$\quad 20 \cdot \log\left(\left|\dfrac{Z_{in3}(i \cdot 2\pi \cdot f_k)}{\Omega}\right|\right)$

Passive

$L_{eq} = 22\text{ mH}$
@10 kHz

Magnitude

(°)

Phase

Emulated $\quad \arg(Z_{in}(i \cdot 2\pi \cdot f_k)) \cdot \dfrac{180}{\pi}$

$\quad \arg(Z_{in3}(i \cdot 2\pi \cdot f_k)) \cdot \dfrac{180}{\pi}$

Passive

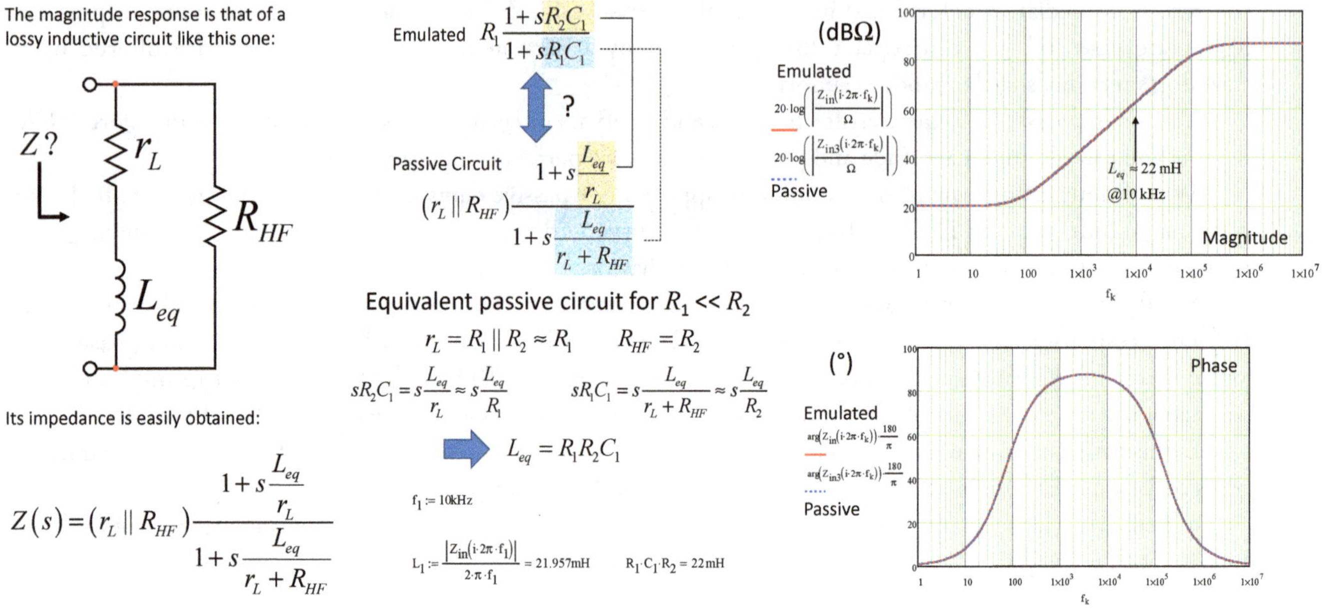

The Equivalent *RL* Network can be Calculated and its Impedance Matches that of the Active Network

Figure 5.68

We will now look at a circuit featuring a bipolar transistor drawn in Figure 5.69.

We want the transfer function linking the input current I_i to the collector current I_c.

A bipolar transistor is a nonlinear component and you have to replace it with its so-called hybrid-π linear model.

The bias resistance R_b remains in the analysis with the dc source V_b replaced by a wire as it is 0 V in ac (same for the V_{cc} rail which is grounded in ac).

For convenience, I transformed the parallel arrangement of R_b and the input stimulus I_i into a Thévenin generator.

With this new circuit, open the capacitor for the dc analysis and determine the gain H_0.

For the pole, set the Thévenin voltage source to 0 V and install a test generator I_T across capacitor C_E's connecting terminals. After a few simple equations, the pole comes out easily. For the zero, we will perform an NDI but in our head this time. Bring the stimulus back and consider the circuit for a nulled response, e.g. $I_c = 0$.

What could bring this collector current to zero?

A nulled base current I_b of course. Considering the presence of the stimulus, what could then justify a zeroed base current? An open emitter, meaning the impedance made of C_E paralleled with R_E approaches infinity for $s = s_z$. Solve the condition for which it happens and you find the zero.

We test the complete transfer function and compare results versus a SIMetrix® plot in Figure 5.70 and they are identical.

What is the transfer function linking I_c to I_i?

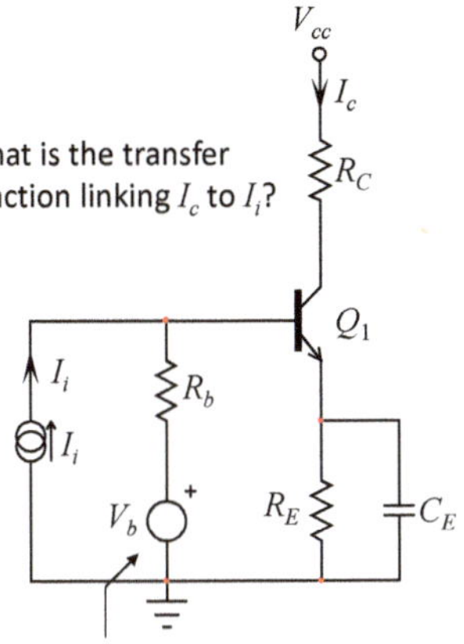

Dc bias source, 0 V in ac

Replace the transistor by its hybrid-pi linear model. V_b is ac 0-V.

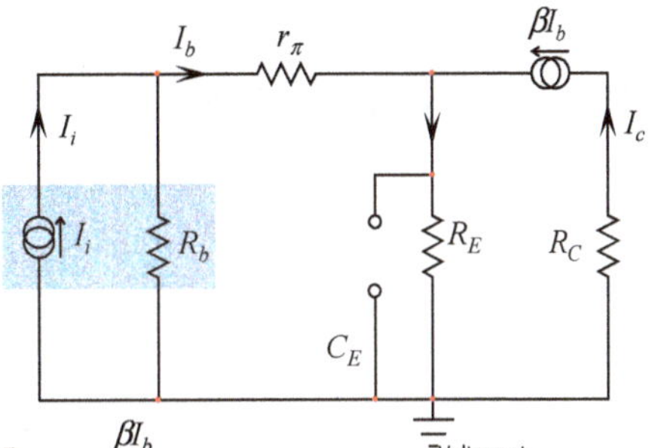

Transform the input source in a Thévenin generator

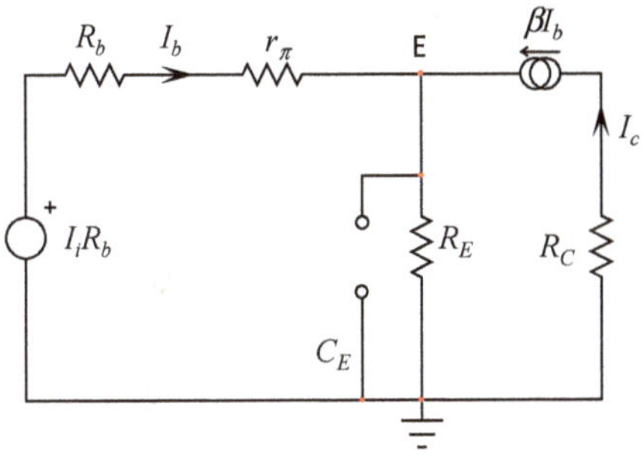

Voltage at node E

$$I_b = \frac{R_b I_i - I_b(\beta+1)R_E}{R_b + r_\pi}$$

$$I_b = \frac{R_b I_i}{R_b + r_\pi + (\beta+1)R_E}$$

$$I_c = \beta I_b = \beta\frac{R_b I_i}{R_b + r_\pi + (\beta+1)R_E}$$

$$H_0 = \frac{I_c}{I_i} = \frac{\beta R_b}{R_b + r_\pi + (\beta+1)R_E}$$

Set the stimulus to 0 V and install a test generator I_T:

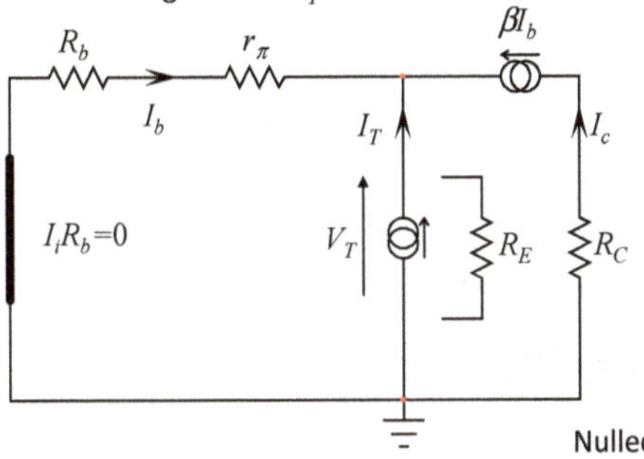

You see that R_E is in parallel with the capacitor and will show as is in the final result. Temporarily disconnect it and bring it back later in parallel with the intermediate result.

$$I_b = -\frac{V_T}{r_\pi + R_b} \qquad I_T = -(i_b + \beta i_b)$$

$$I_T = \frac{V_T(\beta+1)}{R_b + r_\pi}$$

$$\frac{V_T}{I_T} = R = \frac{R_b + r_\pi}{1+\beta}$$

Bring R_E back

$$\tau_1 = \left[\left(\frac{R_b + r_\pi}{1+\beta}\right) \| R_E\right] C_E$$

$$\omega_p = \frac{1}{\tau_1} = \frac{1}{\left[\left(\frac{R_b + r_\pi}{1+\beta}\right) \| R_E\right] C_E}$$

Nulled response

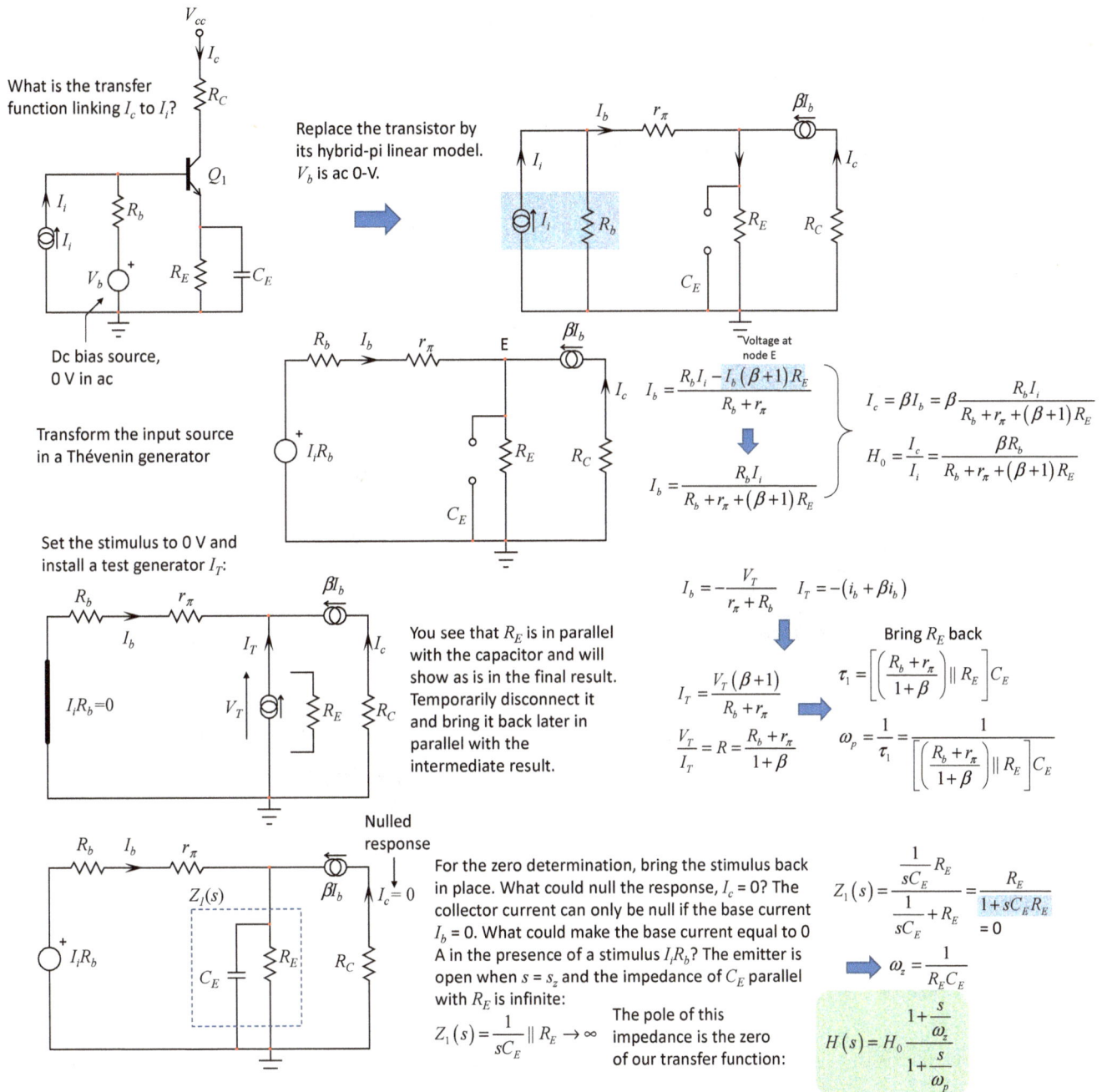

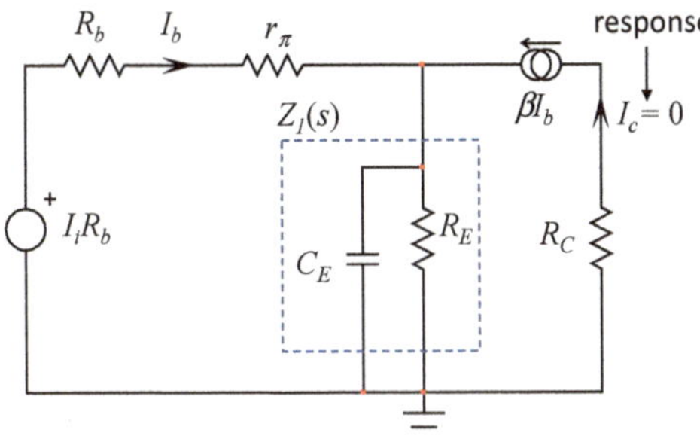

For the zero determination, bring the stimulus back in place. What could null the response, $I_c = 0$? The collector current can only be null if the base current $I_b = 0$. What could make the base current equal to 0 A in the presence of a stimulus I_iR_b? The emitter is open when $s = s_z$ and the impedance of C_E parallel with R_E is infinite:

$$Z_1(s) = \frac{1}{sC_E} \| R_E \to \infty$$

The pole of this impedance is the zero of our transfer function:

$$Z_1(s) = \frac{\frac{1}{sC_E}R_E}{\frac{1}{sC_E} + R_E} = \frac{R_E}{1 + sC_ER_E}$$

$1 + sC_ER_E = 0$

$$\omega_z = \frac{1}{R_EC_E}$$

$$H(s) = H_0\frac{1 + \frac{s}{\omega_z}}{1 + \frac{s}{\omega_p}}$$

Find the Transfer Function Linking the Input Current to the Collector Current

Figure 5.69

$\beta := 100 \quad r_\pi := 1.2k\Omega \quad R_b := 22k\Omega \quad R_E := 470\Omega \quad R_c := 10k\Omega \quad C_E := 470nF \quad \|(x,y) := \frac{x \cdot y}{x+y}$

$$i_{b.} = \frac{R_b \cdot i_i - i_b \cdot (\beta + 1) \cdot R_E}{R_b + r_\pi} \qquad i_b = \frac{R_b \cdot i_i}{R_E + R_b + r_\pi + R_E \cdot \beta}$$

$$i_c = \beta \cdot i_b = \beta \cdot \frac{R_b \cdot i_i}{R_E + R_b + r_\pi + R_E \cdot \beta} \qquad \frac{i_c}{i_i} = \beta \cdot \frac{R_b}{R_E + R_b + r_\pi + R_E \cdot \beta}$$

$$H_0 := \frac{\beta \cdot R_b}{R_E + R_b + r_\pi + R_E \cdot \beta} = 31.131$$

this is the dc gain: $20 \cdot \log(H_0) = 29.864$ dB

Pole determination:

$$i_b = -\frac{V_T}{r_\pi + R_b} \qquad I_T = -(i_b + \beta \cdot i_b) \qquad I_T = \frac{V_T \cdot (1+\beta)}{r_\pi + R_b} \qquad \frac{V_T}{I_T} = \frac{R_b + r_\pi}{1 + \beta}$$

$$R_{eq} := \left(\frac{R_b + r_\pi}{\beta + 1}\right) \| R_E = 154.295\ \Omega \qquad \tau_1 := R_{eq} \cdot C_E = 72.518\ \mu s \qquad \omega_p := \frac{1}{\tau_1} \qquad f_p := \frac{\omega_p}{2 \cdot \pi} = 2.195\ kHz$$

Zero determination:

The zero occurs in a null double condition in which the output is nulled while the stimulus is back in the circuit. A nulled output implies a nulled base current which, in presence of a stimulus biasing the base, an open emitter. The opening of the emitter mesh happens if the impedance made of R_E || C_E becomes infinite a the zero angular frequency. The pole of this impedance is located at:

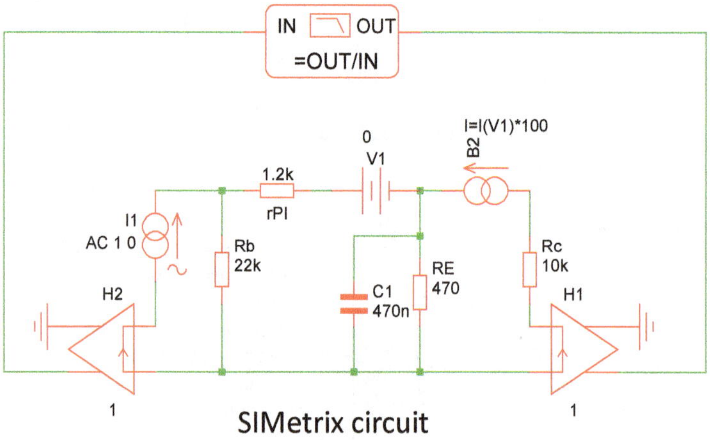

$$\omega_z := \frac{1}{R_E \cdot C_E}$$

$$H(s) = \frac{i_c}{i_i} = \frac{\beta \cdot R_b}{R_E + R_b + r_\pi + R_E \cdot \beta} \cdot \frac{1 + s \cdot R_E \cdot C_E}{1 + s \cdot C_E \cdot \left[\left(\frac{R_b + r_\pi}{1+\beta}\right) \| R_E\right]}$$

$$H_1(s) := H_0 \cdot \frac{1 + \frac{s}{\omega_z}}{1 + \frac{s}{\omega_p}} \qquad H_2 := \text{READPRN}(".\backslash TF.txt")$$

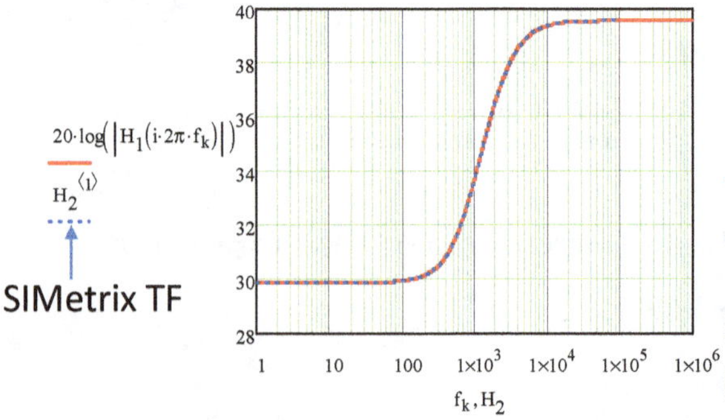

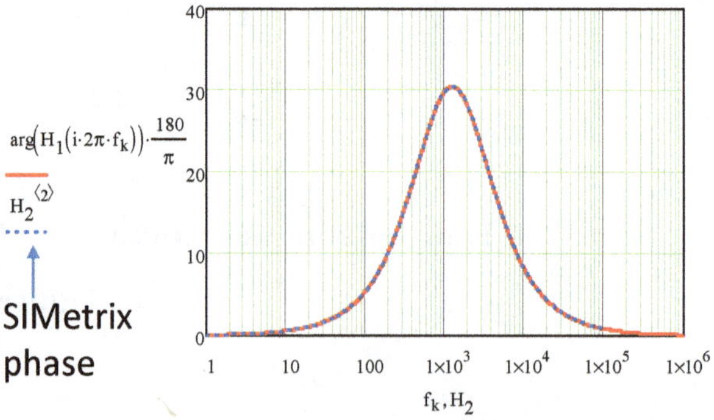

The Equivalent Linear Circuit from SIMetrix® Delivers the Same Curves as those Obtained with Equations

Figure 5.70

5.3 List of Figures and Transfer Functions

For convenient browsing of the transfer functions I derived, below are pictures summarizing the networks studied in this chapter.

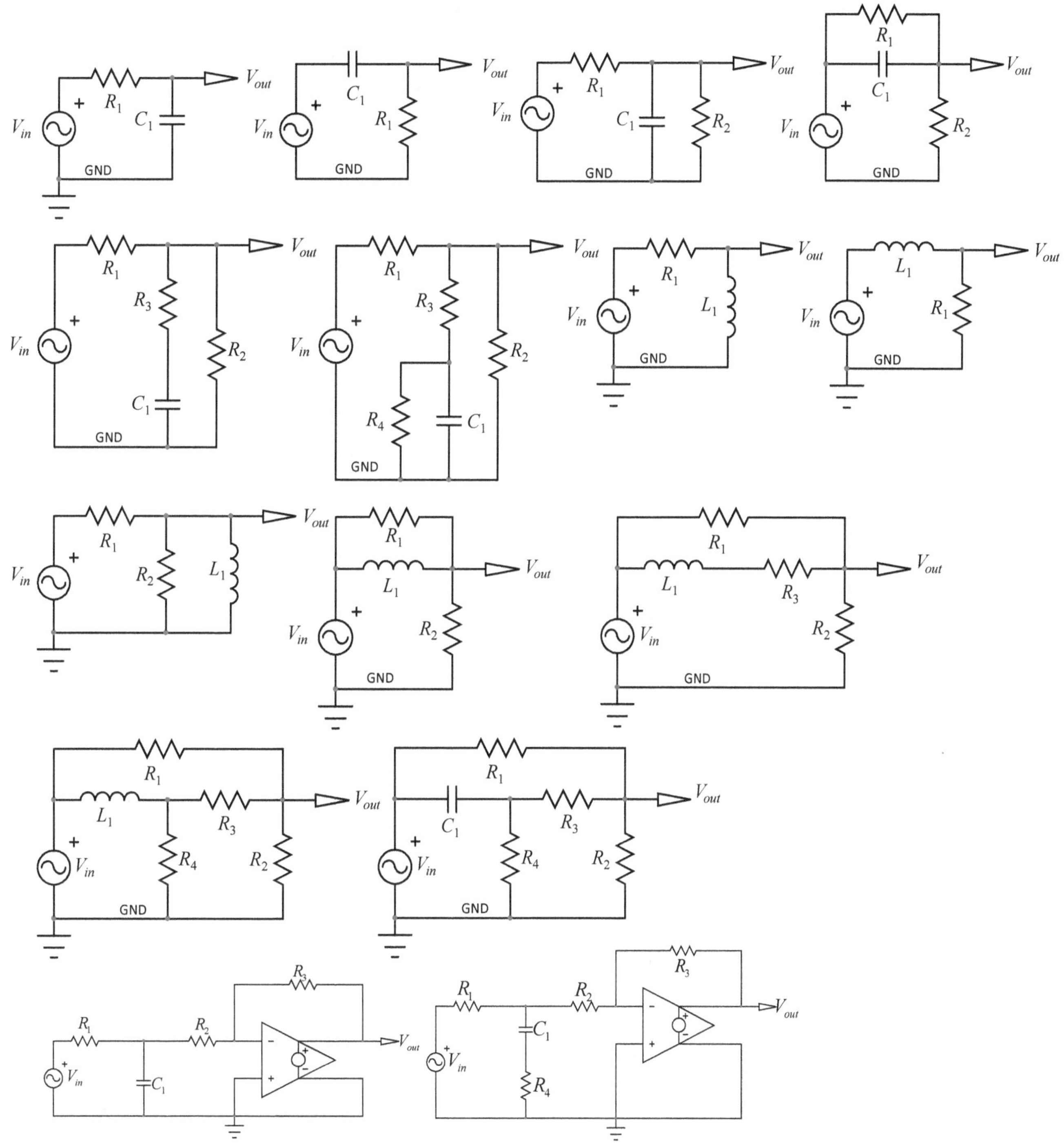

The First Set of 1st-Order Transfer Functions Determined in this Chapter

Figure 5.71

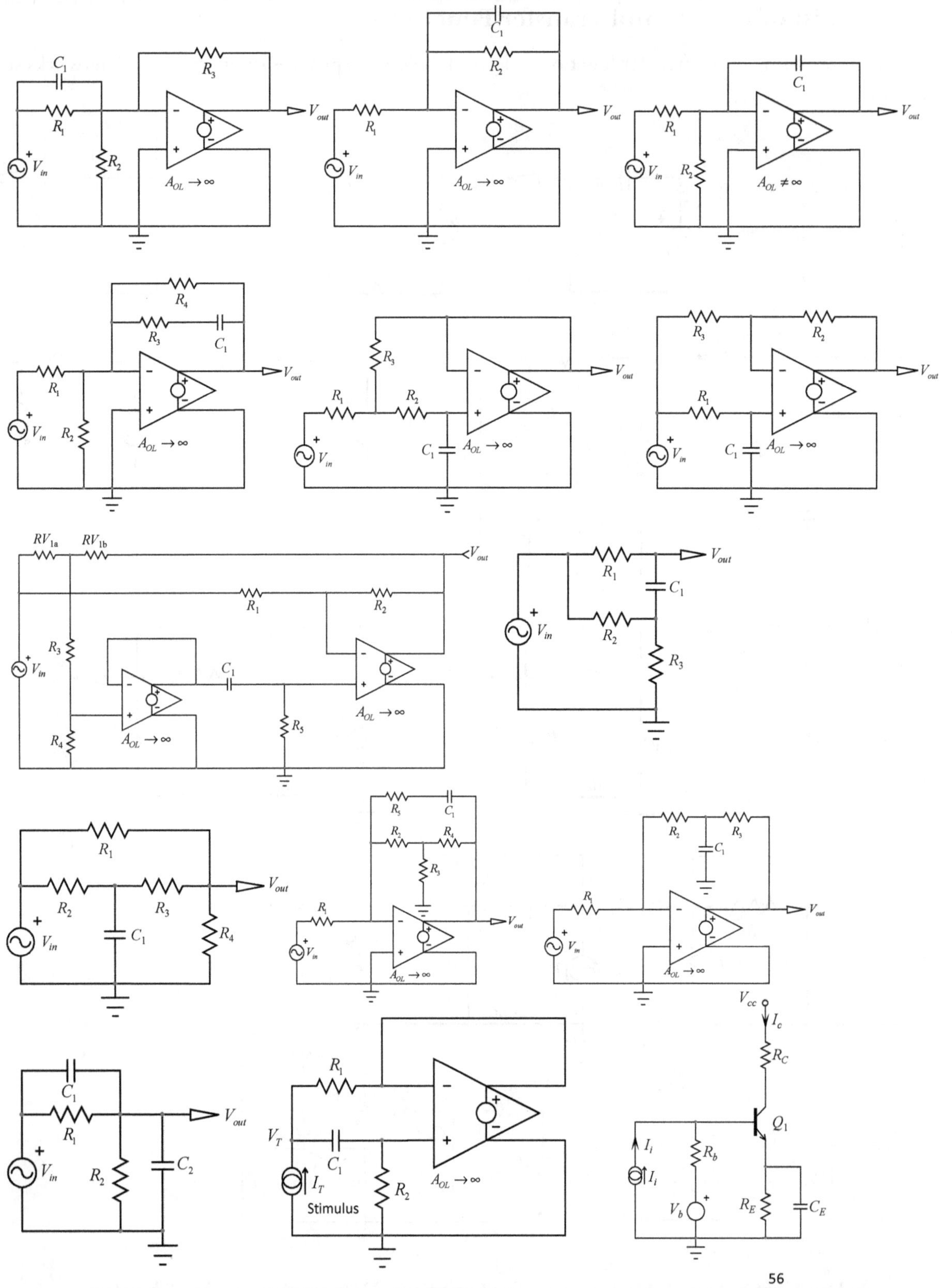

The Second Batch of 1st-Order Transfer Functions Determined in this Chapter

Figure 5.72

Chapter 6: Second-Order Transfer Functions

IN THIS CHAPTER, I will determine transfer functions of networks featuring two energy-storing elements. The principle remains the same: find the time constants with a zeroed excitation to identify the poles and find the zeroes by nulling the output or by resorting to the generalized transfer function.

We start this chapter with a quick summary of the technique.

6.1 Zeroed Excitation and Null Double Injection for 2nd-Order Systems

The denominator of a second-order transfer function obeys the following expression:

$$D(s)=1+b_1 s+b_2 s^2 \tag{6.1}$$

The term b_1 is obtained by summing the time constants associated with each energy-storing element, τ_1 and τ_2:

$$b_1=\tau_1+\tau_2 \tag{6.2}$$

During the exercise, the excitation or the stimulus source is zeroed and you determine the resistance R driving the first energy-storing element while the second is placed in its dc state (an open circuit for a capacitor, a short circuit for an inductor). Then, you repeat the process by finding R driving the second element while the first is now placed in its dc state. With the time constants in hand, you determine b_1.

The second term, b_2, is defined by:

$$b_2=\tau_1\tau_2^1 \tag{6.3}$$

...also equal by redundancy to:

$$b_2=\tau_2\tau_1^2 \tag{6.4}$$

In this approach, reuse one of the two time constants τ_1 or τ_2 and multiply it by another time constant determined as follows:

- τ_2^1 means that energy-storing element labeled 1 is placed in its high-frequency state (a short circuit for a capacitor or an open-circuit for an inductor) and you determine the resistance R driving element 2 in this mode. Then assemble b_2 according to (6.3) by reusing the first time constant τ_1.
- τ_1^2 means that energy-storing element labeled 2 is placed in its high-frequency state. Determine the resistance R driving element 1 in this mode. Then assemble b_2 according to (6.4) by reusing the second time constant τ_2.

Choosing (6.3) or (6.4) depends on the associated network configuration. You will see that, sometimes, one gives a less obvious circuit to analyze (or leads to an indeterminacy) while the other leads to a simpler structure.

In the end, both equations give a similar result but, of course, use the simplest combination.

The numerator of a second-order transfer function will obey the following formula:

$$N(s)=1+a_1s+a_2s^2 \tag{6.5}$$

The term a_1 is determined by summing the time constants associated with each energy-storing element, τ_{1N} and τ_{2N} obtained in a null double injection or NDI:

$$a_1=\tau_{1N}+\tau_{2N} \tag{6.6}$$

Determine the resistance R offered by the connecting terminals of the first energy-storing element while the second is placed in its dc state (an open circuit for a capacitor, a short circuit for an inductor) and null the response.

Then, repeat the process by finding R driving the second element while the first is now placed in its dc state and the response is nulled. With the time constants in hand, determine a_1. This NDI technique requires the installation of a test generator I_T together with the original stimulus that is back in place as documented in the previous chapters.

The second term, a_2, is defined by:

$$a_2=\tau_{1N}\tau_{2N}^{1} \tag{6.7}$$

...also equal by redundancy to:

$$a_2=\tau_{2N}\tau_{1N}^{2} \tag{6.8}$$

In this approach, reuse one of the two time constants already determined for the term a_1, τ_{1N} or τ_{2N}, and multiply it by another time constant determined as follows:

- τ_{2N}^{1} means that energy-storing element labeled 1 is placed in its high-frequency state (a short circuit for a capacitor or an open-circuit for an inductor) and you determine the resistance R driving element 2 while ensuring a null on the output in an NDI configuration. Then assemble a_2 according to (6.7) by reusing the first time constant τ_{1N}.
- τ_{1N}^{2} means that energy-storing element labeled 2 is placed in its high-frequency state and determine the resistance R driving element 1 while ensuring a null on the output in an NDI configuration. Then assemble a_2 according to (6.8) by reusing the first time constant τ_{2N}.

Of course, when inspection works and you can visualize the zeroes in the circuit, the numerator is immediately expressed and you gain a tremendous time. Once the numerator and the denominator are determined, write the transfer function after having determined H_0, the gain obtained for $s=0$:

$$H(s)=H_0\frac{N(s)}{D(s)}=H_0\frac{1+a_1s+a_2s^2}{1+b_1s+b_2s^2} \tag{6.9}$$

With this expression in hand, you can advantageously factor its terms to obtain a normalized 2nd-order polynomial form:

$$H(s)=H_0\frac{1+\dfrac{s}{\omega_{0N}Q_N}+\left(\dfrac{s}{\omega_{0N}}\right)^2}{1+\dfrac{s}{\omega_{0D}Q_D}+\left(\dfrac{s}{\omega_{0D}}\right)^2}=H_0\frac{1+s\dfrac{2\zeta_N}{\omega_{0N}}+\left(\dfrac{s}{\omega_{0N}}\right)^2}{1+s\dfrac{2\zeta_D}{\omega_{0D}}+\left(\dfrac{s}{\omega_{0D}}\right)^2} \tag{6.10}$$

In this expression, the subscripts *N* or *D* respectively refer to the numerator and the denominator elements. *Q* and ζ (zeta) are respectively the quality factor and the damping ratio expressed as combinations of *a* and *b* coefficients.

$$Q_D=\frac{\sqrt{b_2}}{b_1}\quad Q_N=\frac{\sqrt{a_2}}{a_1} \tag{6.11}$$

The resonant frequencies are determined as follows:

$$\omega_{0D}=\frac{1}{\sqrt{b_2}}\quad \omega_{0N}=\frac{1}{\sqrt{a_2}} \tag{6.12}$$

The quality factor and the damping ratio can be used interchangeably:

$$Q=\frac{1}{2\zeta}\leftrightarrow\zeta=\frac{1}{2Q} \tag{6.13}$$

When determining zeroes using an NDI is too complicated or simply because you would like to check your results with an alternative method, use the generalized transfer function extended to a 2nd-order network.

It reuses the time constants already found for the denominator:

$$H(s)=\frac{H_0+s\left(H^1\tau_1+H^2\tau_2\right)+s^2H^{12}\tau_1\tau_2^1}{1+s\left(\tau_1+\tau_2\right)+s^2\tau_1\tau_2^1} \tag{6.14}$$

If H_0 is different than zero, then (6.14) can advantageously be rewritten as:

$$H(s)=H_0\frac{1+s\left(\dfrac{H^1}{H_0}\tau_1+\dfrac{H^2}{H_0}\tau_2\right)+s^2\dfrac{H^{12}}{H_0}\tau_1\tau_2^1}{1+s\left(\tau_1+\tau_2\right)+s^2\tau_1\tau_2^1} \tag{6.15}$$

In these expressions, H^1 and H^2 are high-frequency gains linking the response to the stimulus. H^1 is determined when element 1 is set in its high-frequency state while element 2 remains in dc state.

For H^2, element 2 is placed in its high-frequency state while element 1 remains in dc state.

Finally, H^{12} is a gain obtained with both energy-storing elements set in their high-frequency state. As previously underlined, the generalized transfer function complicates the numerator roots and extra energy may be needed to simplify the final expression.

Nevertheless, the results obtained from (6.9) or (6.15) are identical.

I often use the generalized transfer function when the reference state obtained for $s = 0$ returns zero and makes the use of (6.9) impractical.

6.2 Circuits with two Energy-Storing Elements

This is a classic in textbooks, what is the transfer function of two cascaded RC networks shown in Figure 6.1?

$H_0 = 1$

(a)   (b)

Turn V_{in} off – set it to 0 V

Find R when C_2 is in dc state

$$\tau_1 = R_1 C_1$$

Find R when C_1 is in dc state

$$\tau_2 = (R_1 + R_2)C_2$$

$$b_1 = \tau_1 + \tau_2 = R_1 C_1 + (R_1 + R_2)C_2$$

C_1 is in high-frequency state: determine R

$$\tau_2' = C_2 R_2$$

$$b_2 = \tau_1 \tau_2' = R_1 C_1 C_2 R_2$$

(c)   (d)

$$D(s) = 1 + s(\tau_1 + \tau_2) + s^2 \tau_1 \tau_2' = 1 + s\left[R_1 C_1 + (R_1 + R_2)C_2\right] + s^2 R_1 C_1 R_2 C_2$$

If you set any of the capacitor in their high-frequency state, the stimulus never goes through to form the response: no zero in this circuit.

$$H(s) = \frac{1}{1 + s\left[R_1 C_1 + (R_1 + R_2)C_2\right] + s^2 R_1 C_1 R_2 C_2} = \frac{1}{1 + \dfrac{s}{\omega_0 Q} + \left(\dfrac{s}{\omega_0}\right)^2}$$

$$\left. \begin{array}{l} Q = \dfrac{\sqrt{b_2}}{b_1} = \dfrac{\sqrt{R_1 R_2 C_1 C_2}}{R_1 C_1 + C_2(R_1 + R_2)} \\[2ex] \omega_0 = \dfrac{1}{\sqrt{b_2}} = \dfrac{1}{\sqrt{R_1 R_2 C_1 C_2}} \end{array} \right.$$

When the quality factor Q is much small than 1 or $Q \ll 1$, you can apply the low-Q approximation:

$$\omega_{p_1} = Q\omega_0 = \frac{1}{b_1} = \frac{1}{R_1 C_1 + C_2(R_1 + R_2)}$$

$$\omega_{p_2} = \frac{\omega_0}{Q} = \frac{b_1}{b_2} = \frac{R_1 C_1 + C_2(R_1 + R_2)}{R_1 R_2 C_1 C_2}$$

$$H(s) \approx \frac{1}{\left(1 + \dfrac{s}{\omega_{p_1}}\right)\left(1 + \dfrac{s}{\omega_{p_2}}\right)}$$

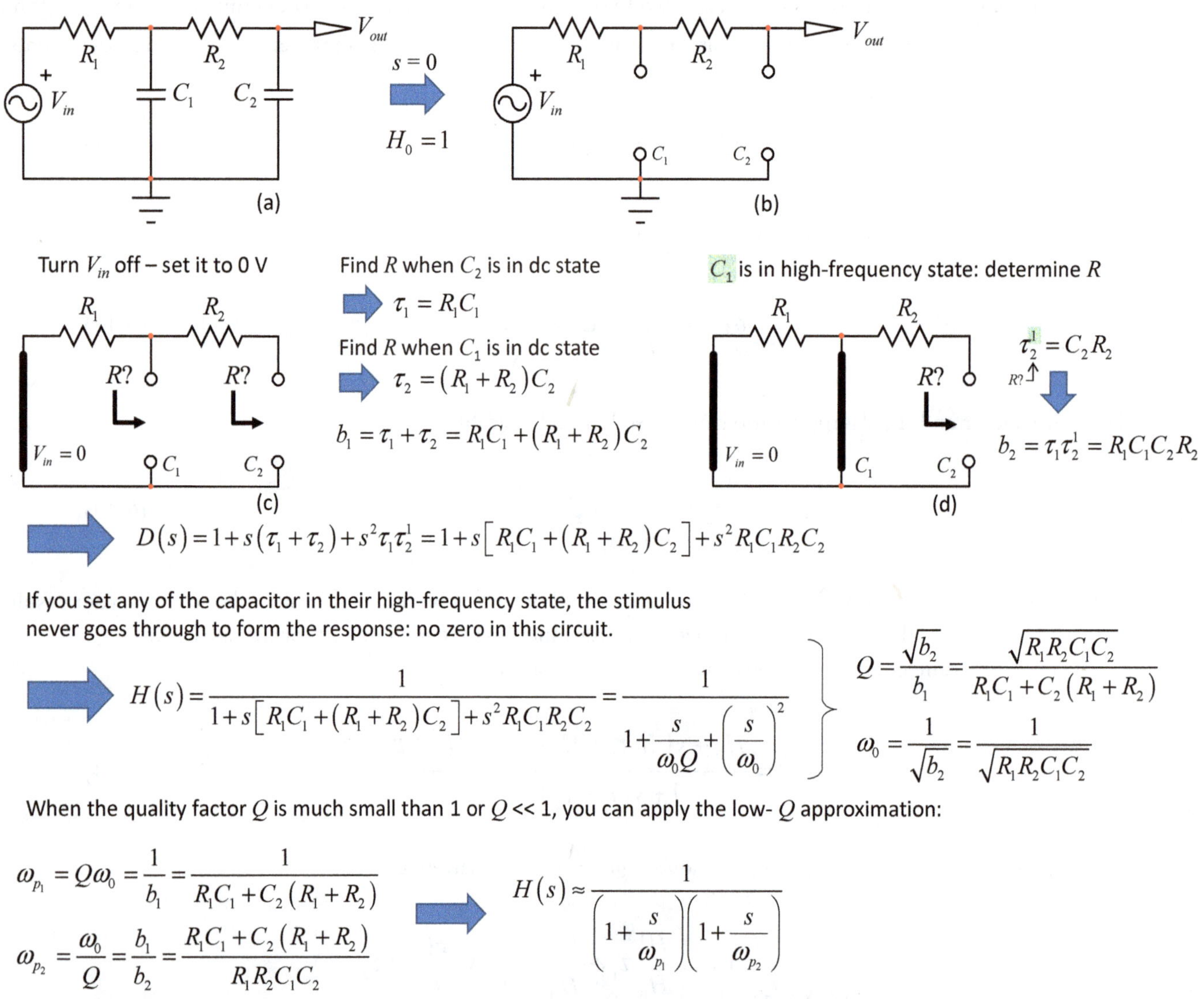

This Simple RC Filter Starts our Analyses with Three Associated Transfer Functions

Figure 6.1

The steps are easy and coefficients come out painlessly and well ordered.

As with any second-order system, you can rework the polynomial under a normalized form and reveal a resonant frequency and a quality factor.

Considering the very low value of the latter, apply the low-Q approximation and rewrite a transfer function featuring a dominant low-frequency pole followed by another, higher-frequency pole.

The ac responses given in Figure 6.2 confirm this approach is correct.

$||(x,y) := \frac{x \cdot y}{x+y}$ $\quad R_{inf} := 10^{23}\Omega$

$C_1 := 0.1\mu F \quad C_2 := 0.15\mu F \quad R_1 := 2.2k\Omega \quad R_2 := 4.7k\Omega$

Transfer function for s=0 $\quad H_0 := 1$

$\tau_1 := C_1 \cdot R_1 = 220 \cdot \mu s$

$\tau_2 := C_2 \cdot (R_1 + R_2) = 1.035 \times 10^3 \cdot \mu s$

$b_1 := \tau_1 + \tau_2 = 1.255 \times 10^3 \cdot \mu s$

$\tau_{12} := C_2 \cdot R_2$

$b_2 := \tau_1 \cdot \tau_{12} = 0.1551 \cdot ms^2$

$D_1(s) := 1 + s \cdot b_1 + s^2 \cdot b_2 \quad Q := \frac{\sqrt{b_2}}{b_1} = 0.31381 \quad \omega_0 := \frac{1}{\sqrt{b_2}} \quad f_0 := \frac{\omega_0}{2 \cdot \pi} = 404.12362Hz$

$\omega_{p1} := \frac{1}{b_1} \quad f_{p1} := \frac{\omega_{p1}}{2\pi} = 126.81669Hz$

$\omega_{p2} := \frac{b_1}{b_2} \quad f_{p2} := \frac{\omega_{p2}}{2\pi} = 1.28781kHz$

$H_1(s) := H_0 \cdot \frac{1}{1 + s \cdot b_1 + s^2 \cdot b_2} \quad H_2(s) := \frac{1}{\left(1 + \frac{s}{\omega_{p1}}\right) \cdot \left(1 + \frac{s}{\omega_{p2}}\right)} \quad H_3(s) := H_0 \cdot \frac{1}{1 + \frac{s}{\omega_0 \cdot Q} + \left(\frac{s}{\omega_0}\right)^2}$

Low-Q approximation $\quad$ Normalized form

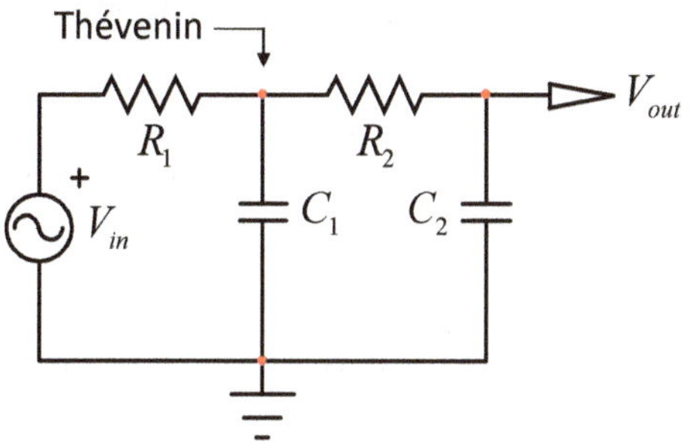

$Z_{th}(s) := R_1 \,||\, \left(\frac{1}{s \cdot C_1}\right)$ Thévenin resistance

$H_{ref}(s) := \frac{\frac{1}{s \cdot C_1}}{\frac{1}{s \cdot C_1} + R_1} \cdot \frac{\frac{1}{s \cdot C_2}}{\frac{1}{s \cdot C_2} + Z_{th}(s) + R_2}$

Brute-force transfer function

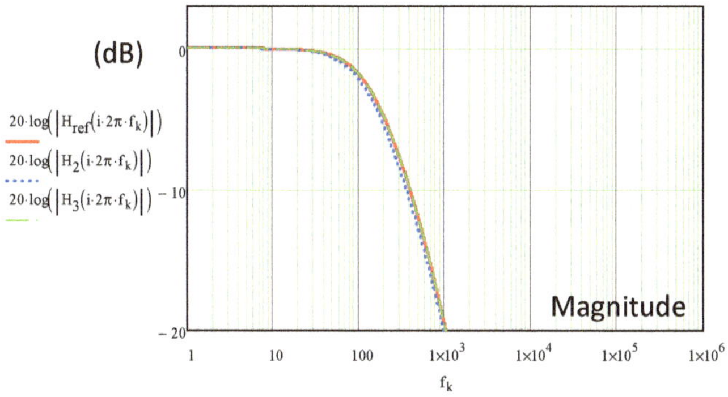

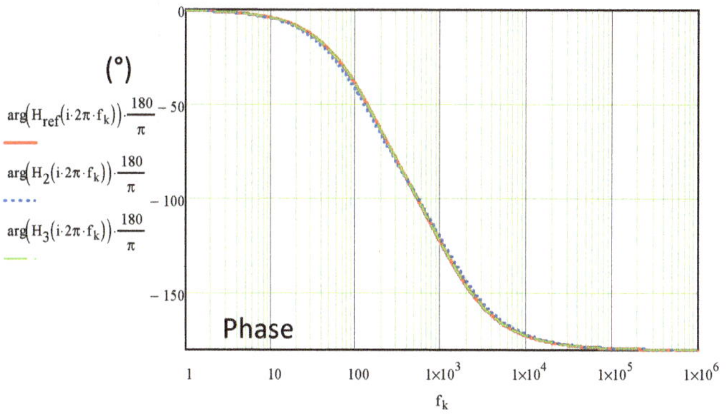

The Ac Responses show Excellent Agreement Between all the Transfer Functions

Figure 6.2

A convenient choice for this type of filter is to select resistors and capacitors of similar values.

In this case, the transfer function simplifies and it is possible to determine the -3-dB cutoff point.

Extract the magnitude from the new expression and solve for f_0 when that magnitude equals 0.707.

This is what is shown in Figure 6.3.

If use capacitors and resistors of equal values:

$$H(s)=\frac{1}{1+3RCs+(sRC)^2} \quad \xrightarrow{\text{Extract magnitude}} \quad |H(s)|=\frac{1}{\sqrt{(1-R^2C^2\omega^2)^2+(3RC\omega)^2}}$$

To determine the -3-dB cutoff point, the magnitude equals 0.707:

$$\frac{1}{\sqrt{2}}=\frac{1}{\sqrt{(1-R^2C^2\omega^2)^2+(3RC\omega)^2}} \quad \Rightarrow \quad f_0=\frac{\sqrt{\frac{\sqrt{53}-7}{2C^2R^2}}}{2\pi}=\frac{0.05956}{RC}$$

$$\|(x,y):=\frac{x\cdot y}{x+y} \qquad R_{inf}:=10^{23}\Omega \qquad f_0:=1\text{kHz}$$

$$R:=1\text{k}\Omega \qquad C:=\frac{0.05956}{R\cdot f_0}=59.56\text{nF}$$

R is fixed and C is determined for f_0

$$C_1:=C \qquad C_2:=C \qquad R_1:=R \qquad R_2:=R$$

$$H_2(s):=H_0\cdot\frac{1}{1+s\cdot 3\cdot R\cdot C+s^2\cdot R^2\cdot C^2}$$

$$\text{Mag}(\omega):=\frac{1}{\sqrt{\left(1-C^2\cdot R^2\cdot\omega^2\right)^2+(3\cdot C\cdot R\cdot\omega)^2}} \qquad \frac{1}{\sqrt{\left(1-C^2\cdot R^2\cdot\omega^2\right)^2+(3\cdot C\cdot R\cdot\omega)^2}}=\frac{1}{\sqrt{2}}$$

$$f_{00}:=\frac{\sqrt{\frac{\sqrt{53}-7}{2\cdot C^2\cdot R^2}}}{2\cdot\pi}=1.00003\times10^3\cdot\text{Hz} \qquad \frac{0.05956}{R\cdot C}=1\times10^3\,\frac{1}{s}$$

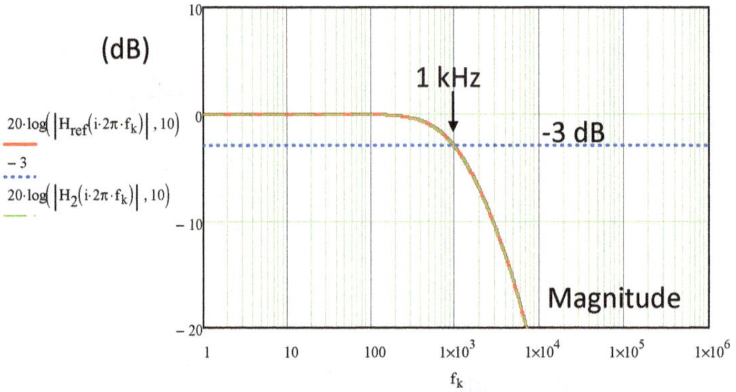

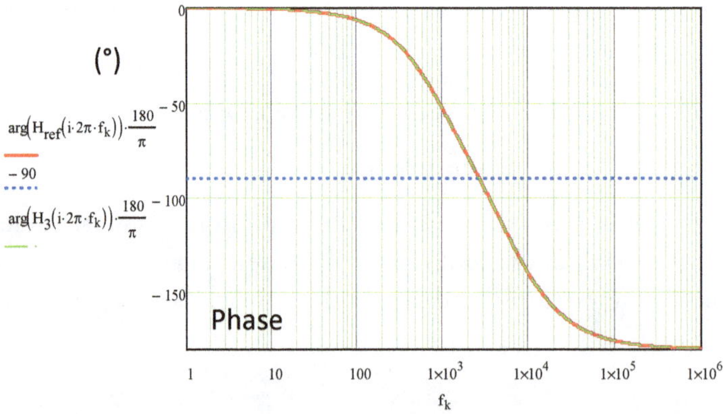

Extracting the Magnitude from the Transfer Function Leads to Finding the Cutoff Point

Figure 6.3

We will now proceed to determine the output impedance of this passive filter as detailed in Figure 6.5.

Install a test generator I_T and the voltage V_T across its terminal is the response.

When I_T is turned off, the circuit returns to the natural state of Figure 6.1 sketch (c).

This is good and we can reuse the denominator already in hand.

(a)

When the excitation is off - $I_T = 0$ - the structure is unchanged as in previous sketches (c) and (d): you can reuse $D(s)$

stimulus $\quad s = 0$

$R_0 = R_1 + R_2$

(b)

$$D(s) = 1 + s\left[R_1C_1 + (R_1 + R_2)C_2\right] + s^2 R_1 C_1 R_2 C_2$$

NDI for the zeroes: degenerate case, response is nulled then replace current source by a wire

$\tau_{1N} = C_1(R_1 \| R_2)$

$\tau_{2N} = C_2 \cdot 0$

$\tau_{2N}^1 = C_2 \cdot 0$

(c)

(d)

Capacitor C_2 cannot contribute a zero because when placed in its high-frequency state, there is no response

$$N(s) = 1 + s\left(\tau_{1N} + \tau_{2N}\right) + s^2 \tau_{1N}\tau_{2N}^1 = 1 + sC_1(R_1 \| R_2)$$

$$Z_{out}(s) = (R_1 + R_2)\frac{1 + sC_1(R_1 \| R_2)}{1 + s\left[R_1C_1 + (R_1 + R_2)C_2\right] + s^2 R_1 C_1 R_2 C_2} = R_0 \frac{1 + \dfrac{s}{\omega_z}}{1 + \dfrac{s}{\omega_0 Q} + \left(\dfrac{s}{\omega_0}\right)^2}$$

$$Z_{out}(s) \approx R_0 \frac{1 + \dfrac{s}{\omega_z}}{\left(1 + \dfrac{s}{\omega_{p_1}}\right)\left(1 + \dfrac{s}{\omega_{p_2}}\right)}$$

$$\omega_{p_1} = Q\omega_0 = \frac{1}{b_1} = \frac{1}{R_1C_1 + C_2(R_1 + R_2)} \qquad \omega_z = \frac{1}{C_1(R_1 \| R_2)}$$

$$\omega_{p_2} = \frac{\omega_0}{Q} = \frac{b_1}{b_2} = \frac{R_1C_1 + C_2(R_1 + R_2)}{R_1 R_2 C_1 C_2}$$

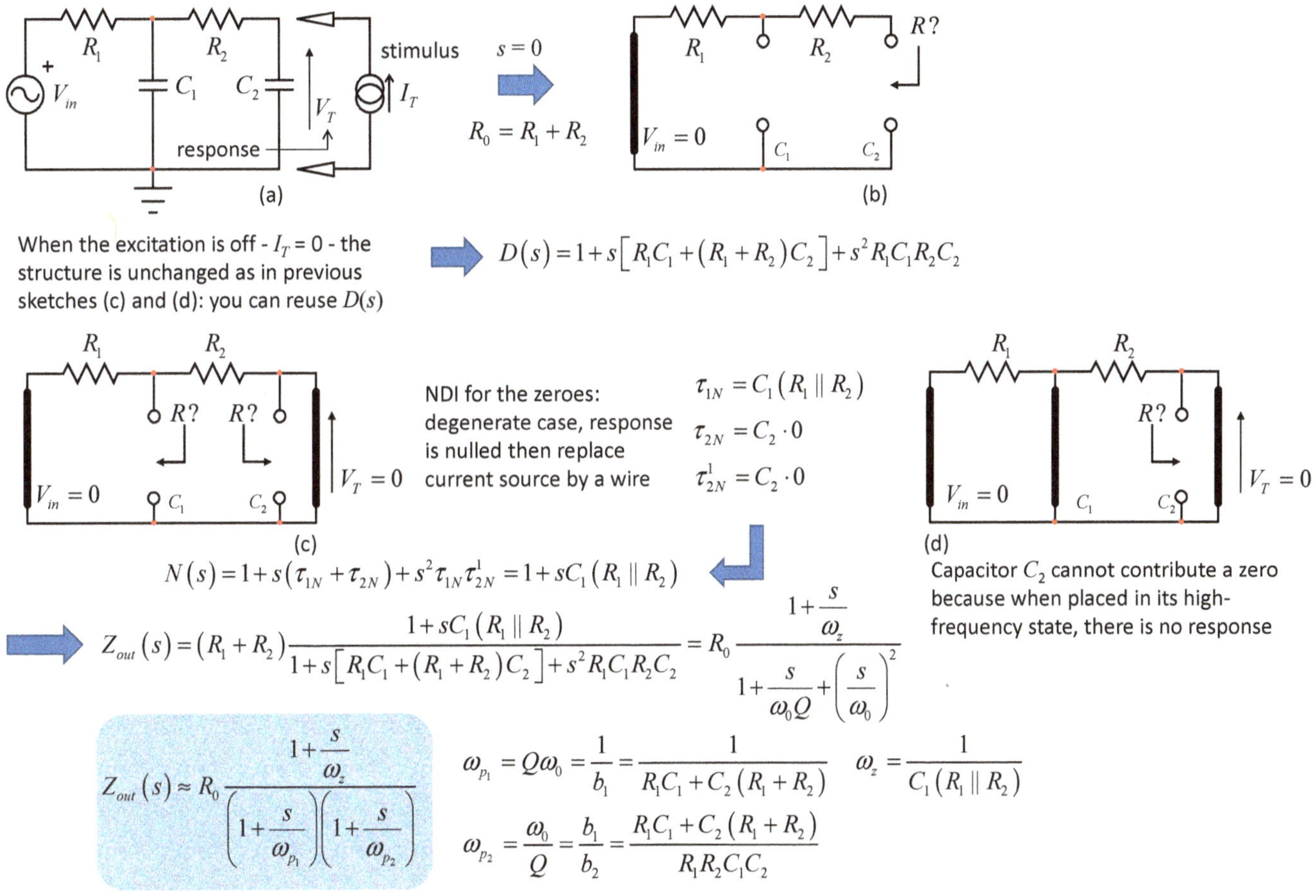

Now determine the Output Impedance of the Cascaded RC Network

Figure 6.4

For the zeroes, we can see that if we alternatively set C_1 and C_2 in their high-frequency state, only C_1 allows a non-zero response V_T whereas C_2 shorts the output terminals: there is a single zero contributed by C_1.

$\|(x, y) := \frac{x \cdot y}{x + y}$ $\quad R_{inf} := 10^{23}\Omega$

$C_1 := 0.1\mu F \quad C_2 := 0.15\mu F \quad R_1 := 2.2k\Omega \quad R_2 := 4.7k\Omega$

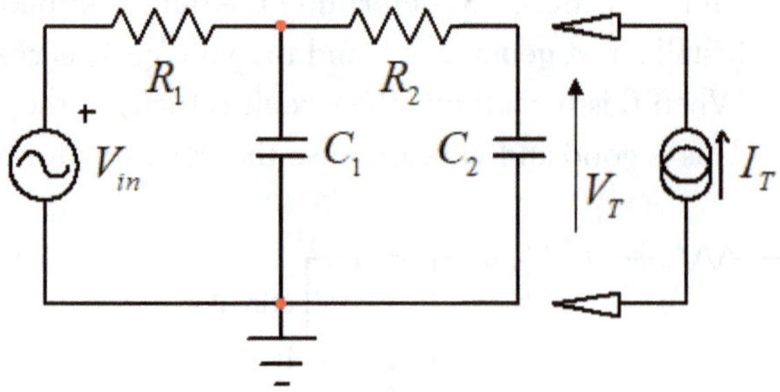

Transfer function for s=0 $\quad R_0 := R_1 + R_2 \quad 20 \cdot \log\left(\frac{R_0}{\Omega}\right) = 76.77698$ dBohms

$\tau_1 := C_1 \cdot R_1 = 220 \cdot \mu s$

$\tau_2 := C_2 \cdot (R_1 + R_2) = 1.035 \times 10^3 \cdot \mu s$

$b_1 := \tau_1 + \tau_2 = 1.255 \times 10^3 \cdot \mu s$

$\tau_{12} := C_2 \cdot R_2$

$b_2 := \tau_1 \cdot \tau_{12} = 0.1551 \cdot ms^2$

$D_1(s) := 1 + s \cdot b_1 + s^2 \cdot b_2 \quad Q := \frac{\sqrt{b_2}}{b_1} = 0.31381 \quad \omega_0 := \frac{1}{\sqrt{b_2}} \quad f_0 := \frac{\omega_0}{2 \cdot \pi} = 404.12362 \cdot Hz$

$Z_1(s) := R_1 \| \left(\frac{1}{s \cdot C_1}\right)$

$Z_{ref}(s) := \left(\frac{1}{s \cdot C_2}\right) \| (R_2 + Z_1(s))$

Brute-force expression

$\omega_{p1} := \frac{1}{b_1} \quad f_{p1} := \frac{\omega_{p1}}{2\pi} = 126.81669 \cdot Hz$

$\omega_{p2} := \frac{b_1}{b_2} \quad f_{p2} := \frac{\omega_{p2}}{2\pi} = 1.28781 \cdot kHz$

$\omega_z := \frac{1}{C_1 \cdot (R_1 \| R_2)} \quad f_z := \frac{\omega_z}{2 \cdot \pi} = 1.06206 kHz$

$Z_1(s) := R_0 \cdot \frac{1 + s \cdot \left[C_1 \cdot (R_1 \| R_2)\right]}{1 + s \cdot b_1 + s^2 \cdot b_2} \quad Z_2(s) := R_0 \cdot \frac{1 + \frac{s}{\omega_z}}{\left(1 + \frac{s}{\omega_{p1}}\right) \cdot \left(1 + \frac{s}{\omega_{p2}}\right)} \quad Z_3(s) := R_0 \cdot \frac{1 + \frac{s}{\omega_z}}{1 + \frac{s}{\omega_0 \cdot Q} + \left(\frac{s}{\omega_0}\right)^2}$

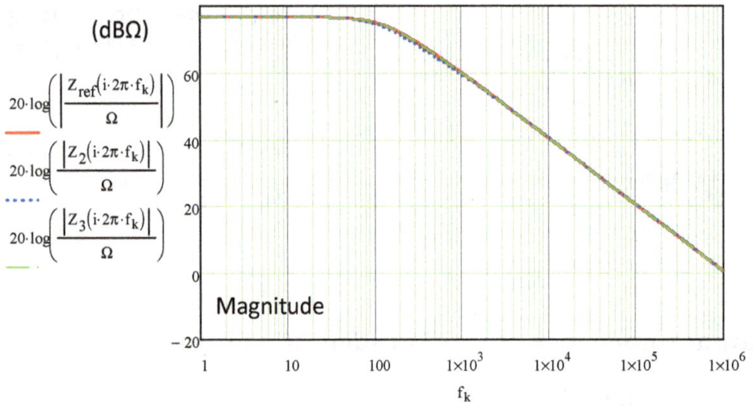

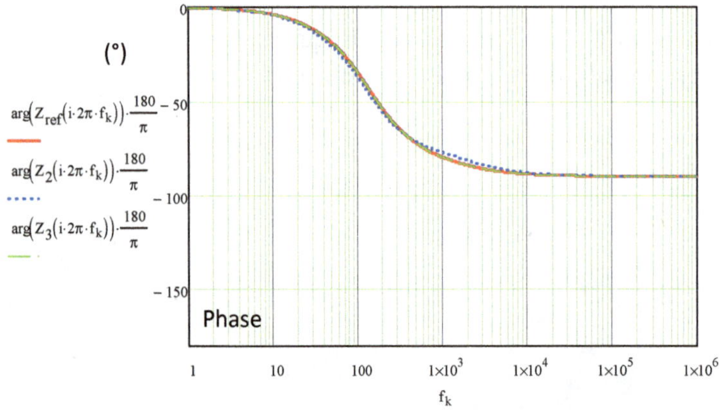

The Ac Responses are all Identical and the Approximation with the Two Cascaded Poles is Good

Figure 6.5

Nevertheless, the NDI circuit is simple with a current source as a zeroed response is similar to replacing it by a wire as shown in sketch (c).

It confirms the single zero brought by C_1. The ac response of this output impedance is given in Figure 6.5.

The input impedance of this 2nd-order network is determined in Figure 6.6.

When s set to zero, the input impedance offers an infinite resistance, but I make it a finite high value R_{inf} for convenient simplifications.

Turn I_T off – set it to 0 A

Find R when C_2 is in dc state
$$\tau_1 = R_{inf} C_1$$
Find R when C_1 is in dc state
$$\tau_2 = R_{inf} C_2$$
$$b_1 = \tau_1 + \tau_2 = R_{inf}\left(C_1 + C_2\right)$$

C_1 is in high-frequency state: determine R
$$\tau_2^1 = C_2 R_2$$
$$b_2 = \tau_1 \tau_2^1 = R_{inf} C_1 C_2 R_2$$

$$D(s) = 1 + s\left(\tau_1 + \tau_2\right) + s^2 \tau_1 \tau_2^1 = 1 + sR_{inf}\left(C_1 + C_2\right) + s^2 R_{inf} C_1 R_2 C_2$$

This is a degenerate case and a 0-V current source can be replaced by a wire. In this mode, you recognize the structure used to determine the poles: we can reuse $D(s)$ of H and Z_{out} as the numerator for Z_{in}!

Same structure as in Z_{out}

$$N(s) = 1 + s\left[R_1 C_1 + \left(R_1 + R_2\right)C_2\right] + s^2 R_1 C_1 R_2 C_2$$

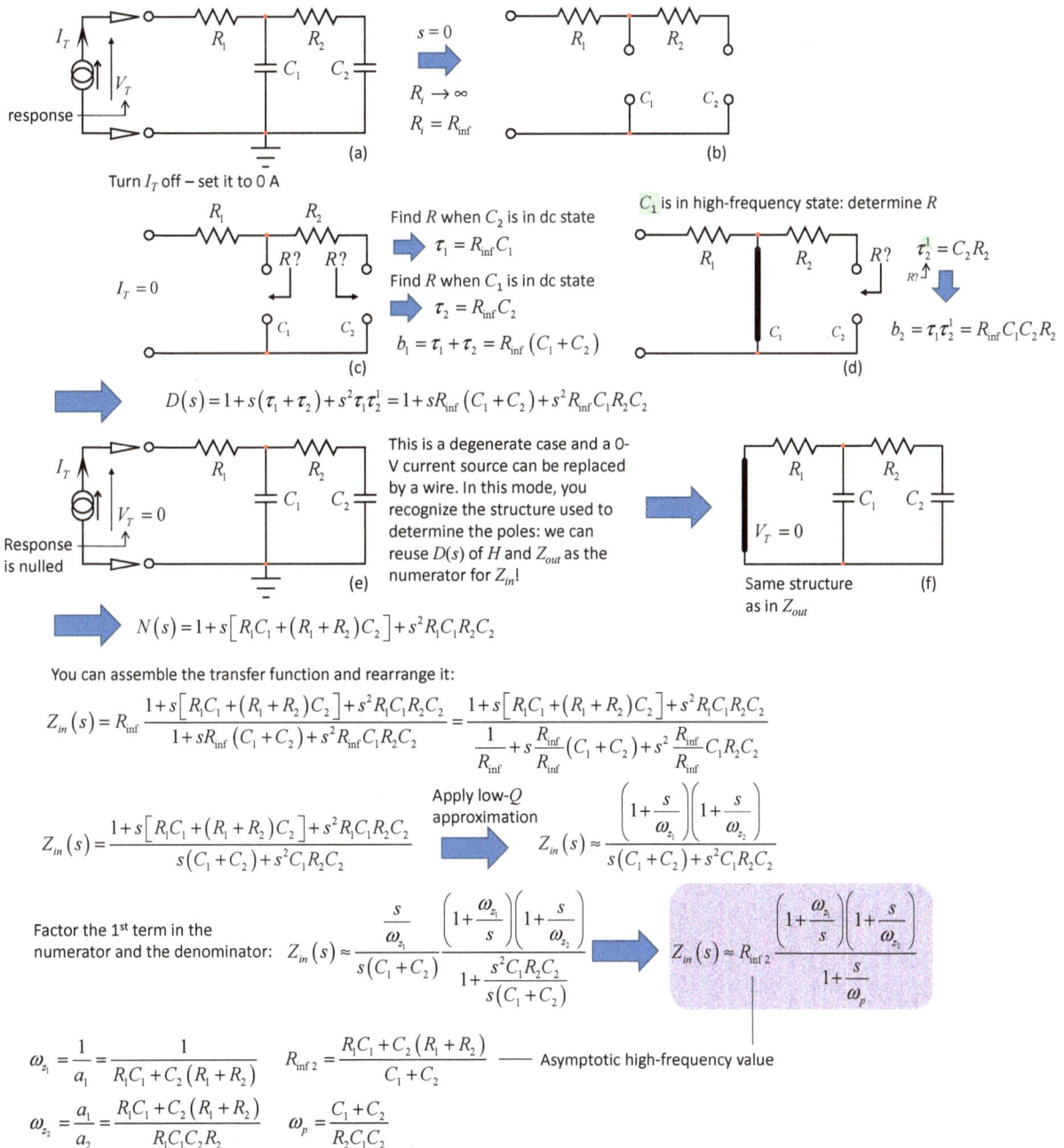

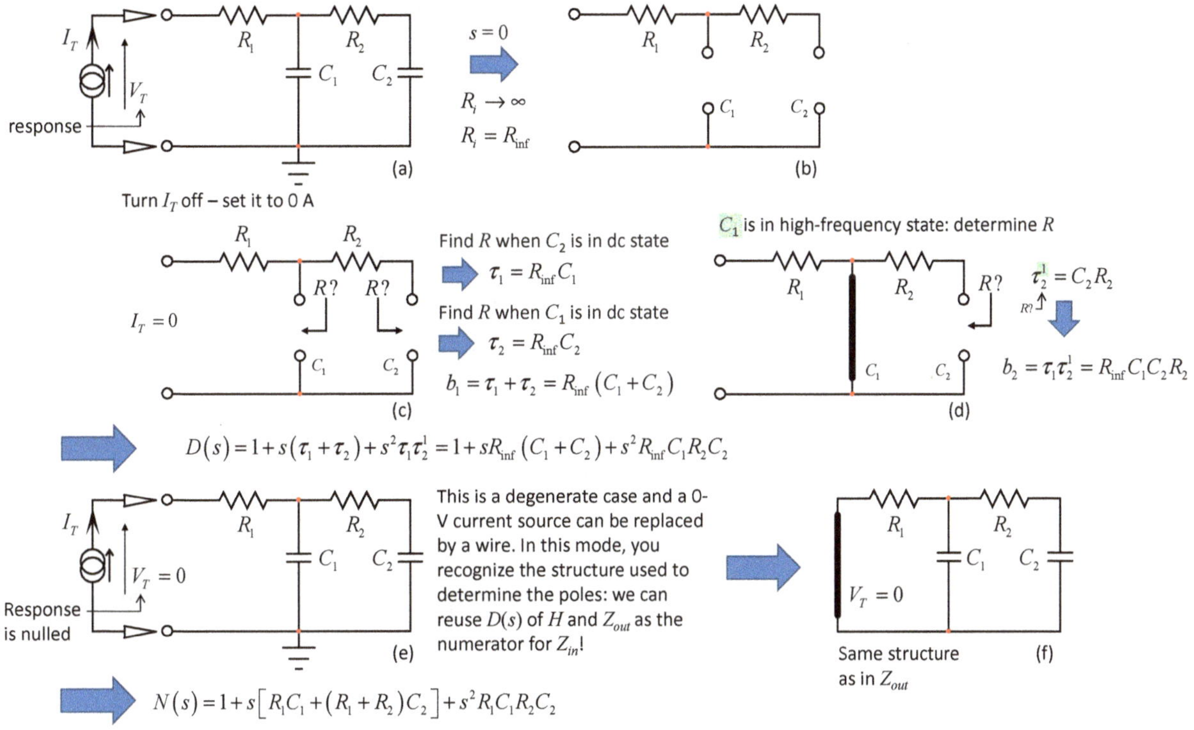

You can assemble the transfer function and rearrange it:

$$Z_{in}(s) = R_{inf}\frac{1 + s\left[R_1 C_1 + \left(R_1 + R_2\right)C_2\right] + s^2 R_1 C_1 R_2 C_2}{1 + sR_{inf}\left(C_1 + C_2\right) + s^2 R_{inf} C_1 R_2 C_2} = \frac{1 + s\left[R_1 C_1 + \left(R_1 + R_2\right)C_2\right] + s^2 R_1 C_1 R_2 C_2}{\dfrac{1}{R_{inf}} + s\dfrac{R_{inf}}{R_{inf}}\left(C_1 + C_2\right) + s^2 \dfrac{R_{inf}}{R_{inf}} C_1 R_2 C_2}$$

Apply low-Q approximation

$$Z_{in}(s) = \frac{1 + s\left[R_1 C_1 + \left(R_1 + R_2\right)C_2\right] + s^2 R_1 C_1 R_2 C_2}{s\left(C_1 + C_2\right) + s^2 C_1 R_2 C_2}$$

$$Z_{in}(s) \approx \frac{\left(1 + \dfrac{s}{\omega_{z_1}}\right)\left(1 + \dfrac{s}{\omega_{z_2}}\right)}{s\left(C_1 + C_2\right) + s^2 C_1 R_2 C_2}$$

Factor the 1st term in the numerator and the denominator:
$$Z_{in}(s) \approx \frac{\dfrac{s}{\omega_{z_1}}\left(1 + \dfrac{\omega_{z_1}}{s}\right)\left(1 + \dfrac{s}{\omega_{z_2}}\right)}{s\left(C_1 + C_2\right)\left(1 + \dfrac{s^2 C_1 R_2 C_2}{s\left(C_1 + C_2\right)}\right)}$$

$$Z_{in}(s) \approx R_{inf\,2}\frac{\left(1 + \dfrac{\omega_{z_1}}{s}\right)\left(1 + \dfrac{s}{\omega_{z_2}}\right)}{1 + \dfrac{s}{\omega_p}}$$

$$\omega_{z_1} = \frac{1}{a_1} = \frac{1}{R_1 C_1 + C_2\left(R_1 + R_2\right)} \qquad R_{inf\,2} = \frac{R_1 C_1 + C_2\left(R_1 + R_2\right)}{C_1 + C_2} \quad \text{——— Asymptotic high-frequency value}$$

$$\omega_{z_2} = \frac{a_1}{a_2} = \frac{R_1 C_1 + C_2\left(R_1 + R_2\right)}{R_1 C_1 C_2 R_2} \qquad \omega_p = \frac{C_1 + C_2}{R_2 C_1 C_2}$$

The Input Impedance Numerator Conveniently Reuses the Denominator of the Output Impedance

Figure 6.6

Both low-frequency time constants are seeing this finite resistance when the excitation is turned off in sketch (c). The denominator is easily determined by combining all these results.

For the zeroes, a nulled response means that V_T across the current source I_T is zero volt. This is the degenerate case that you are now familiar with and you can short the current source as in sketch (f). You recognize the structure already studied in Figure 6.1 (c) and in Figure 6.4 sketch (b). We can thus reuse $D(s)$ and it becomes our numerator for this input impedance. We now assemble all the elements and run simplifications by factoring R_{inf}. Finally, by invoking the low-Q approximation and using an inverted zero, a *low-entropy* expression appears, highlighting the high-frequency asymptote. All the steps are tested and documented in Figure 6.7, confirming our analysis steps.

$\|(x,y) := \frac{x \cdot y}{x+y}$ $\quad R_{inf} := 10^{23}\Omega$

$C_1 := 0.1\mu F \quad C_2 := 0.15\mu F \quad R_1 := 2.2k\Omega \quad R_2 := 4.7k\Omega$

Transfer function for s=0 $\quad R_0 := R_{inf} \quad 20\cdot\log\left(\frac{R_0}{\Omega}\right) = 460$ dBohms

$\tau_1 := R_{inf}\cdot C_1$

$\tau_2 := R_{inf}\cdot C_2$

$b_1 := \tau_1 + \tau_2 = 2.5\times 10^{22}\cdot\mu s$

$\tau_{12} := C_2\cdot R_2$

$b_2 := \tau_1\cdot\tau_{12} = 7.05\times 10^{18}\cdot ms^2$

$D_1(s) := 1 + s\cdot b_1 + s^2\cdot b_2$

$\tau_{1N} := C_1\cdot R_1 = 220\mu s$

$\tau_{2N} := C_2\cdot(R_1 + R_2) = 1.035\times 10^3\cdot\mu s$

$a_1 := \tau_{1N} + \tau_{2N} = 1.255\times 10^3\cdot\mu s$

$\tau_{12N} := C_2\cdot R_2$

$a_2 := \tau_{1N}\cdot\tau_{12N} = 0.1551 ms^2$

$N_1(s) := 1 + s\cdot a_1 + s^2\cdot a_2 \quad Q := \frac{\sqrt{a_2}}{a_1} = 0.31381 \quad \omega_0 := \frac{1}{\sqrt{a_2}} \quad f_0 := \frac{\omega_0}{2\cdot\pi} = 404.12362Hz$

$\omega_{z1} := \frac{1}{a_1} \quad f_{z1} := \frac{\omega_{z1}}{2\pi} = 126.81669Hz$

$\omega_{z2} := \frac{a_1}{a_2} \quad f_{z2} := \frac{\omega_{z2}}{2\pi} = 1.28781kHz$

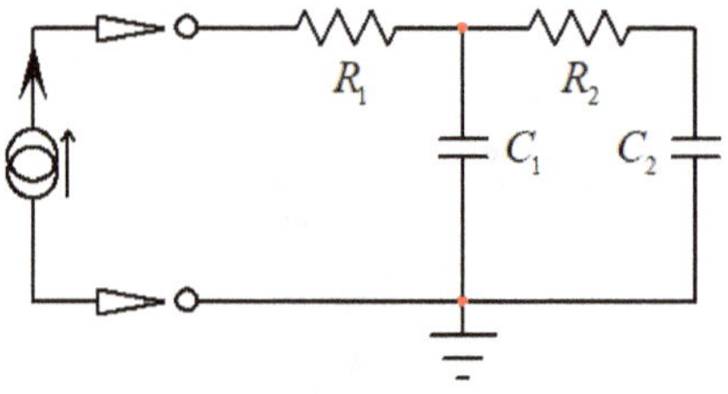

$Z_2(s) := R_2 + \frac{1}{s\cdot C_2}$

$Z_{ref}(s) := R_1 + \left(\frac{1}{s\cdot C_1}\right)\|(Z_2(s))$ — Reference brute-force expression

$Z_{in1}(s) := R_0\cdot\frac{N_1(s)}{D_1(s)}$ $\quad Z_{in2}(s) := \frac{\left[\left(1+\frac{s}{\omega_{z1}}\right)\cdot\left(1+\frac{s}{\omega_{z2}}\right)\right]}{s\cdot(C_1+C_2) + s^2\cdot R_2\cdot C_1\cdot C_2}$ $\quad Z_{in2}(s) := \frac{\left[R_1\cdot C_1 + C_2\cdot(R_1+R_2)\right]}{(C_1+C_2)}\cdot\frac{\left[\left(1+\frac{\omega_{z1}}{s}\right)\cdot\left(1+\frac{s}{\omega_{z2}}\right)\right]}{1+\frac{s\cdot R_2\cdot C_1\cdot C_2}{(C_1+C_2)}}$

$R_{inf2} := \frac{\left[R_1\cdot C_1 + C_2\cdot(R_1+R_2)\right]}{(C_1+C_2)} = 5.02\times 10^3\,\Omega \quad 20\cdot\log\left(\frac{R_{inf2}}{\Omega}\right) = 74.01407$ dBohms $\quad \omega_p := \frac{C_1+C_2}{R_2\cdot C_1\cdot C_2} \quad f_p := \frac{\omega_p}{2\cdot\pi} = 564.37923Hz$

$Z_{in3}(s) := R_{inf2}\cdot\frac{\left[\left(1+\frac{\omega_{z1}}{s}\right)\cdot\left(1+\frac{s}{\omega_{z2}}\right)\right]}{1+\frac{s}{\omega_p}}$

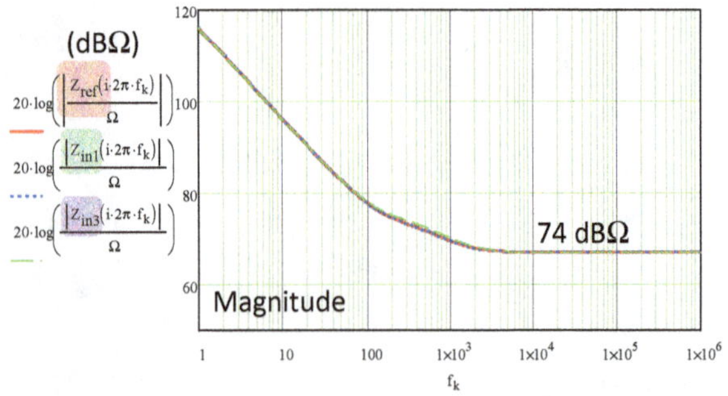

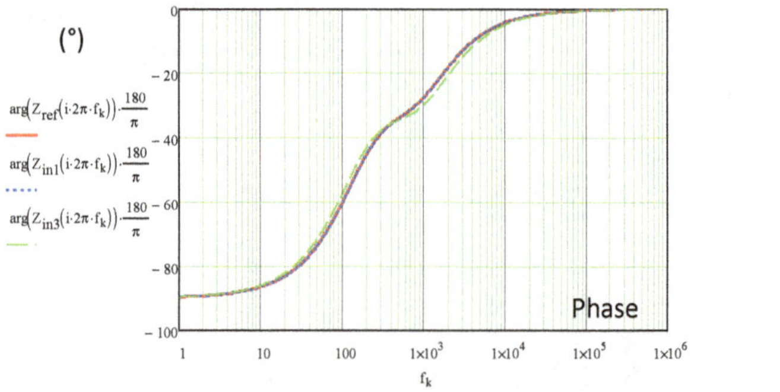

Using an Inverted Zero, it is Possible to Unveil a Leading Term Representative of the High-Frequency Asymptote

Figure 6.7

Let's add a resistance in series with capacitors C_1 and C_2 as in Figure 6.8.

These elements could model the equivalent series resistance or ESR of these capacitors. Just by looking at the circuit, you already know that these added networks contribute a zero to the transfer functions. Considering two energy-storing components, we have two poles and two zeroes. The dc gain is one in the absence of a loading resistor. The determination of the time constants is simple and detailed in the figure.

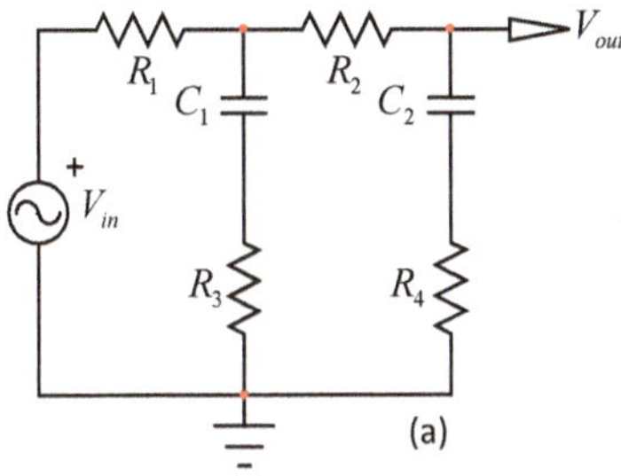

$s=0$ → $H_0=1$

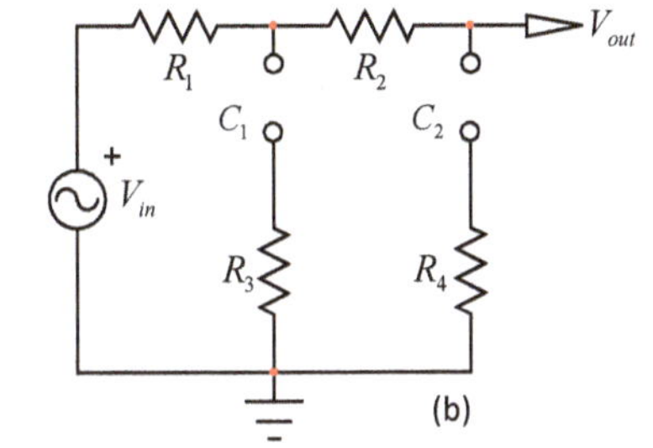

Turn V_{in} off – set it to 0 V

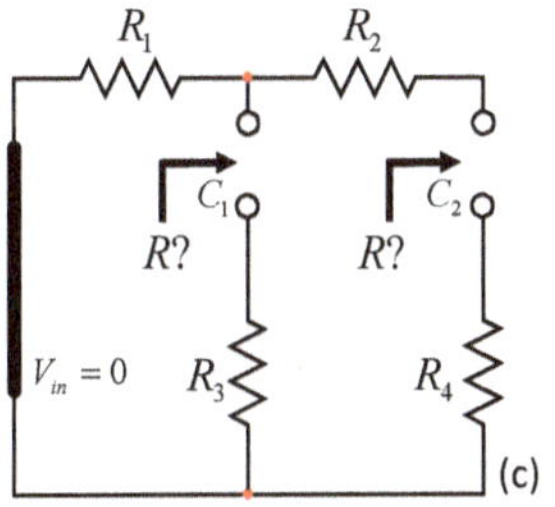

Find R when C_2 is in dc state

$$\tau_1=(R_1+R_3)C_1$$

Find R when C_1 is in dc state

$$\tau_2=(R_1+R_2+R_4)C_2$$

$$b_1=\tau_1+\tau_2=(R_1+R_3)C_1+(R_1+R_2+R_4)C_2$$

C_1 is in high-frequency state: determine R

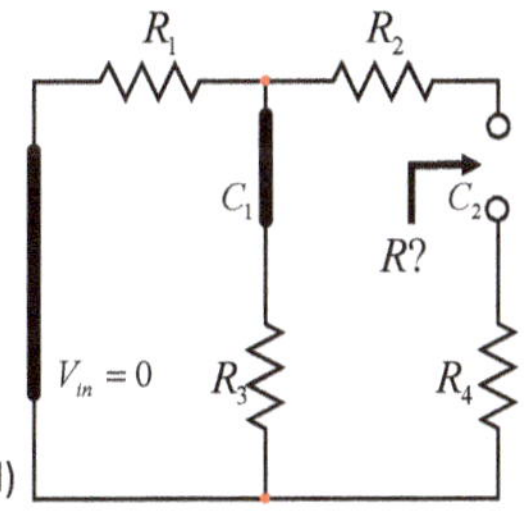

$$\tau_2^1=C_2R_?$$

$$\tau_2^1=(R_4+R_1\,||\,R_3+R_2)C_2$$

$$D(s)=1+s(\tau_1+\tau_2)+s^2\tau_1\tau_2^1=1+s\left[(R_1+R_3)C_1+(R_1+R_2+R_4)C_2\right]+s^2(R_1+R_3)C_1C_2(R_4+R_1\,||\,R_3+R_2)$$

What impedance condition could null the response?

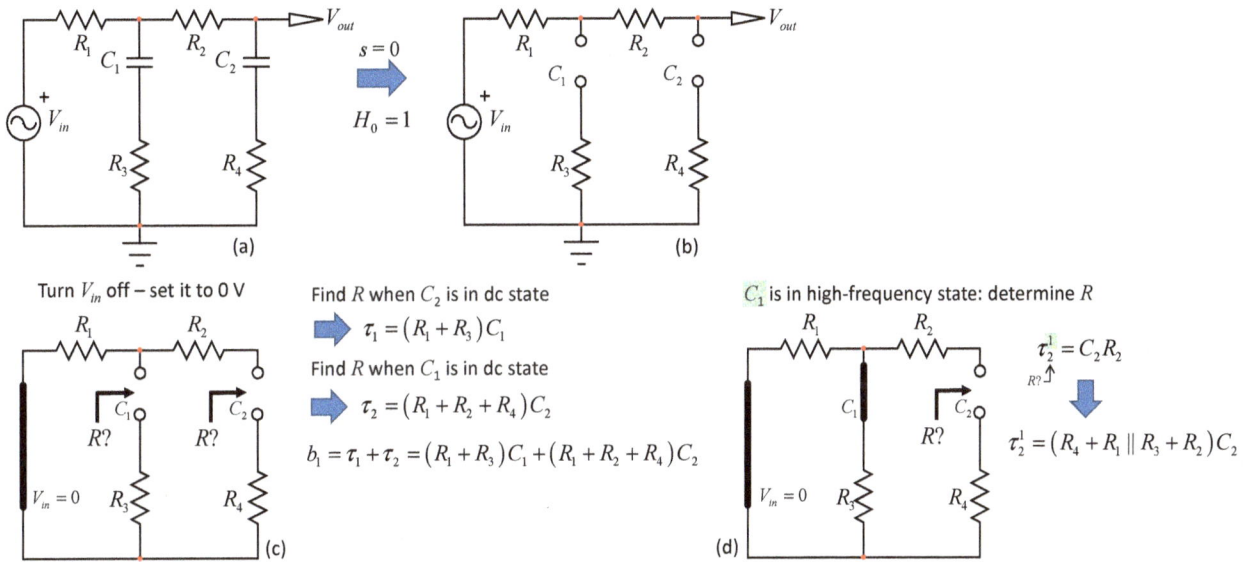

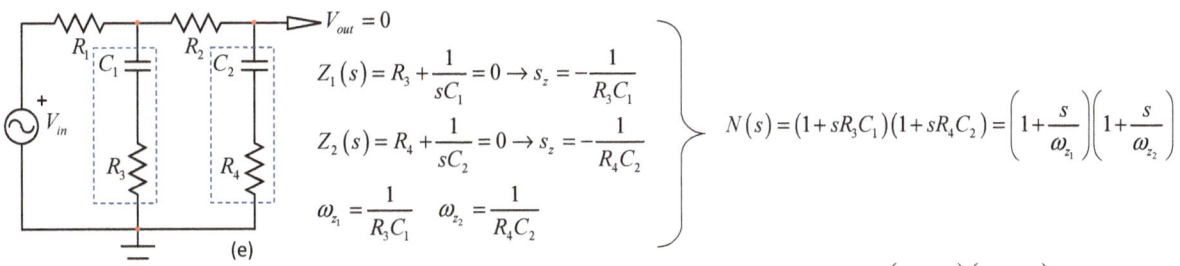

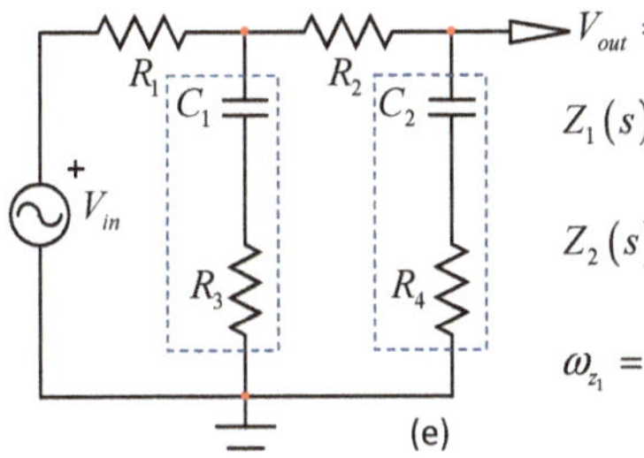

$$Z_1(s)=R_3+\frac{1}{sC_1}=0\rightarrow s_z=-\frac{1}{R_3C_1}$$

$$Z_2(s)=R_4+\frac{1}{sC_2}=0\rightarrow s_z=-\frac{1}{R_4C_2}$$

$$\omega_{z_1}=\frac{1}{R_3C_1}\qquad \omega_{z_2}=\frac{1}{R_4C_2}$$

$$N(s)=(1+sR_3C_1)(1+sR_4C_2)=\left(1+\frac{s}{\omega_{z_1}}\right)\left(1+\frac{s}{\omega_{z_2}}\right)$$

$$H(s)=\frac{(1+sR_3C_1)(1+sR_4C_2)}{1+s\left[(R_1+R_3)C_1+(R_1+R_2+R_4)C_2\right]+s^2(R_1+R_3)C_1C_2(R_4+R_1\,||\,R_3+R_2)}=\frac{\left(1+\frac{s}{\omega_{z_1}}\right)\left(1+\frac{s}{\omega_{z_2}}\right)}{1+\frac{s}{\omega_0Q}+\left(\frac{s}{\omega_0}\right)^2}$$

$$Q=\frac{\sqrt{b_2}}{b_1}=\frac{\sqrt{(R_1+R_3)C_1C_2(R_4+R_1\,||\,R_3+R_2)}}{(R_1+R_3)C_1+(R_1+R_2+R_4)C_2}$$

$$\omega_0=\frac{1}{\sqrt{b_2}}=\frac{1}{\sqrt{(R_1+R_3)C_1C_2(R_4+R_1\,||\,R_3+R_2)}}$$

When the quality factor Q is much small than 1 or Q << 1, you can apply the low-Q approximation:

$$\omega_{p_1}=Q\omega_0=\frac{1}{b_1}=\frac{1}{(R_1+R_3)C_1+(R_1+R_2+R_4)C_2}$$

$$\omega_{p_2}=\frac{\omega_0}{Q}=\frac{b_1}{b_2}=\frac{(R_1+R_3)C_1+(R_1+R_2+R_4)C_2}{(R_1+R_3)C_1C_2(R_4+R_1\,||\,R_3+R_2)}$$

$$H(s)\approx\frac{\left(1+\frac{s}{\omega_{z_1}}\right)\left(1+\frac{s}{\omega_{z_2}}\right)}{\left(1+\frac{s}{\omega_{p_1}}\right)\left(1+\frac{s}{\omega_{p_2}}\right)}$$

Two Resistances are Inserted in Series with Each Capacitor

Figure 6.8

The ac response is given in Figure 6.9 and confirms our analysis: the two poles bring the phase down to -180° while the two zeroes counteract their action and bring the phase back to zero.

$$\|(x,y) := \frac{x \cdot y}{x+y} \qquad R_{inf} := 10^{23}\,\Omega$$

$$C_1 := 0.1\mu F \qquad C_2 := 0.15\mu F \qquad R_1 := 2.2k\Omega \qquad R_2 := 4.7k\Omega \qquad R_3 := 50\Omega \qquad R_4 := 10\Omega$$

Transfer function for s=0 $\qquad H_0 := 1$

$$\tau_1 := C_1 \cdot (R_1 + R_3) = 225\,\mu s$$

$$\tau_2 := C_2 \cdot (R_1 + R_2 + R_4) = 1.0365 \times 10^3 \cdot \mu s$$

$$b_1 := \tau_1 + \tau_2 = 1.2615 \times 10^3 \cdot \mu s$$

$$\tau_{12} := C_2 \cdot (R_4 + R_2 + R_1 \| R_3) = 7.13833 \times 10^{-4}\,s$$

$$b_2 := \tau_1 \cdot \tau_{12} = 0.16061\,ms^2$$

$$D_1(s) := 1 + s \cdot b_1 + s^2 \cdot b_2 \qquad Q := \frac{\sqrt{b_2}}{b_1} = 0.31769 \qquad \omega_0 := \frac{1}{\sqrt{b_2}} \qquad f_0 := \frac{\omega_0}{2 \cdot \pi} = 397.12796 Hz$$

$$\omega_{p1} := \frac{1}{b_1} \qquad f_{p1} := \frac{\omega_{p1}}{2\pi} = 126.16325 Hz$$

$$\omega_{p2} := \frac{b_1}{b_2} \qquad f_{p2} := \frac{\omega_{p2}}{2\pi} = 1.25005 kHz$$

$$\omega_{z1} := \frac{1}{R_3 \cdot C_1} \qquad f_{z1} := \frac{\omega_{z1}}{2\pi} = 31.83099 kHz$$

$$\omega_{z2} := \frac{1}{R_4 \cdot C_2} \qquad f_{z2} := \frac{\omega_{z2}}{2\pi} = 106.1033 kHz$$

$$H_1(s) := H_0 \cdot \frac{\left(1 + \frac{s}{\omega_{z1}}\right)\left(1 + \frac{s}{\omega_{z2}}\right)}{1 + s \cdot b_1 + s^2 \cdot b_2}$$

$$H_2(s) := \frac{\left(1 + \frac{s}{\omega_{z1}}\right)\left(1 + \frac{s}{\omega_{z2}}\right)}{\left(1 + \frac{s}{\omega_{p1}}\right)\left(1 + \frac{s}{\omega_{p2}}\right)}$$

$$H_3(s) := H_0 \cdot \frac{\left(1 + \frac{s}{\omega_{z1}}\right)\left(1 + \frac{s}{\omega_{z2}}\right)}{1 + \frac{s}{\omega_0 \cdot Q} + \left(\frac{s}{\omega_0}\right)^2}$$

$$Z_{th}(s) := R_1 \| \left(\frac{1}{s \cdot C_1} + R_3\right)$$

$$H_{ref}(s) := \frac{\frac{1}{s \cdot C_1} + R_3}{\frac{1}{s \cdot C_1} + R_1 + R_3} \cdot \frac{\frac{1}{s \cdot C_2} + R_4}{\frac{1}{s \cdot C_2} + Z_{th}(s) + R_2 + R_4}$$

Reference brute-force expression

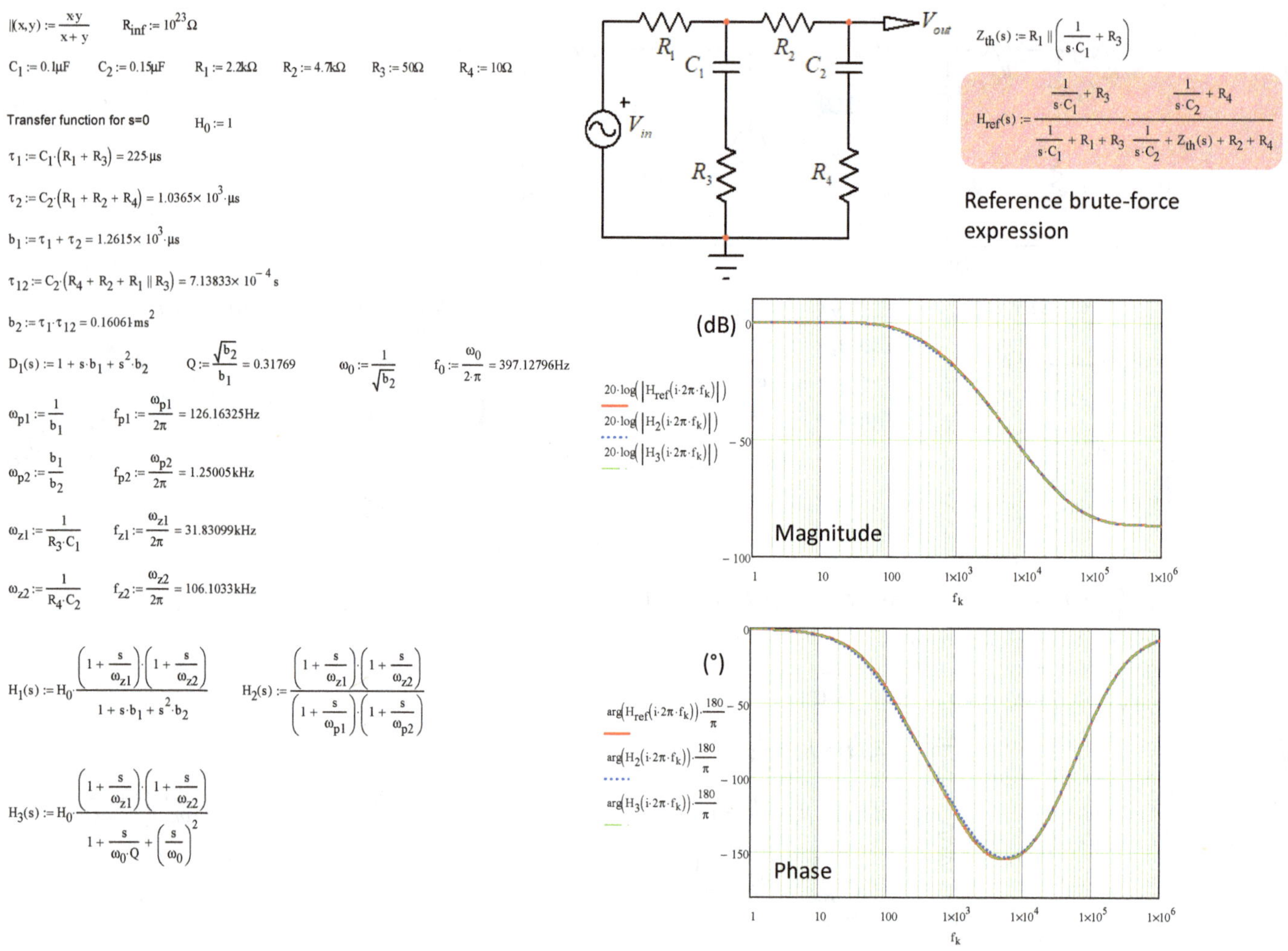

The Ac Response Confirms the Initial Analysis with Two Poles and Two Zeroes

Figure 6.9

Now replace capacitor C_1 by an inductor and see how it combines with the capacitor.

I loaded the circuit this time with resistor R_5 (Figure 6.10). The process remains the same and you can draw more intermediate sketches if you do not immediately see how resistors associate in series-parallel combinations.

By the way, do not develop the paralleled terms as they naturally participate to the *low-entropy* formulation.

In case you would like to assess the effect of one the paralleled terms when either shrinking to a very low value or becoming an infinite resistance, then it is easy to infer the result with the paralleled expressions. On the contrary, if you expand these paralleled terms, they are drowned in the expression and the insight is gone.

The zeroes are found by inspection considering the two transformed short circuits brought by L_1R_3 and C_2R_4.

The denominator is thus immediately expressed without resorting to an NDI which is a tremendous timesaver.

The ac response of the transfer function we have determined using FACTs is given in Figure 6.11 and perfectly matches that obtained with brute-force, by applying the Thévenin theorem.

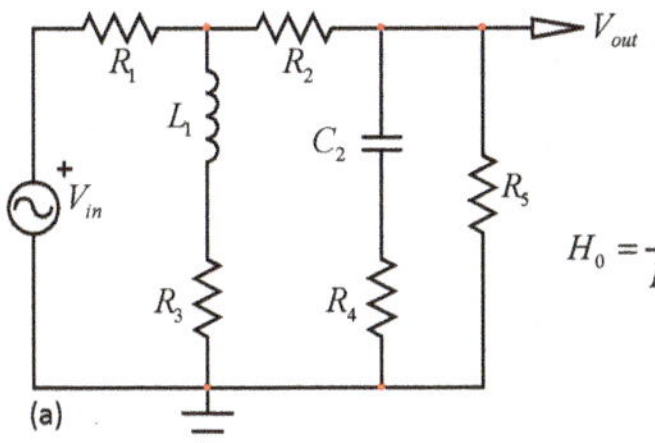

$s=0$

$$H_0=\frac{R_3}{R_3+R_1}\frac{R_5}{R_5+R_2+R_1\parallel R_3}$$

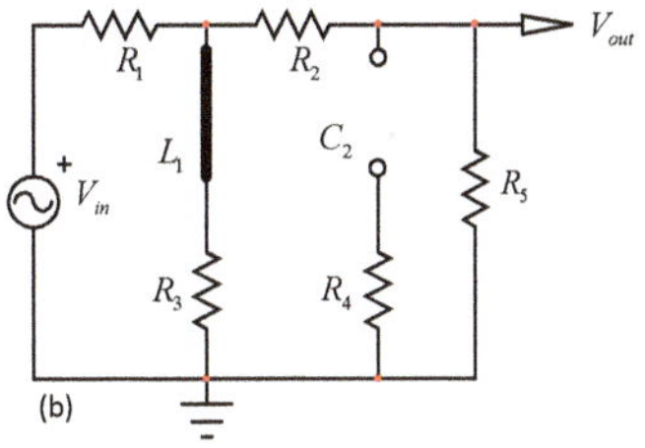

Turn V_{in} off – set it to 0 V

Find R when C_2 is in dc state

$$\tau_1=\frac{L_1}{R_3+R_1\parallel(R_2+R_5)}$$

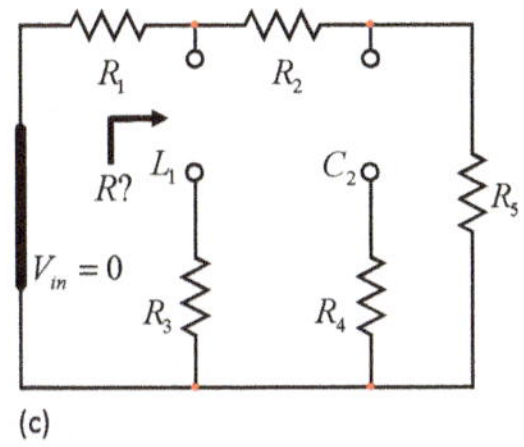

Find R when L_1 is in dc state

$$\tau_2=C_2\left[R_4+(R_2+R_1\parallel R_3)\parallel R_5\right]$$

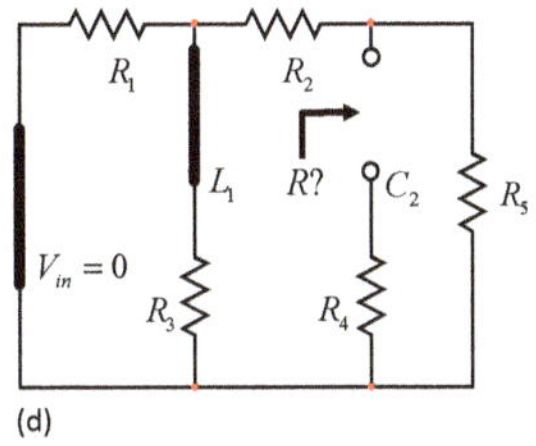

L_1 is in high-frequency state: determine R

$$\tau_2^1=C_2\left[R_4+(R_2+R_1)\parallel R_5\right]$$

$R?$

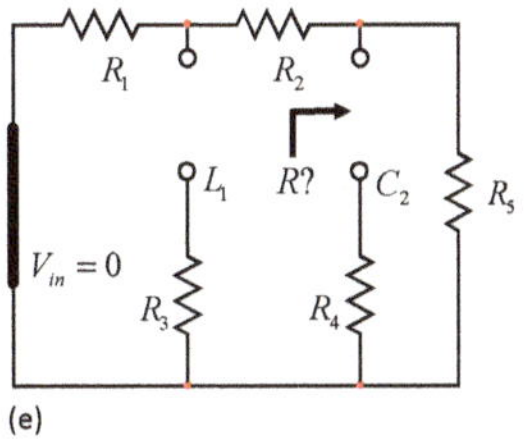

$$D(s)=1+s(\tau_1+\tau_2)+s^2\tau_1\tau_2^1=1+s\left[\frac{L_1}{R_3+R_1\parallel(R_2+R_5)}+C_2\left[R_4+(R_2+R_1\parallel R_3)\parallel R_5\right]\right]+s^2\frac{L_1}{R_3+R_1\parallel(R_2+R_5)}C_2\left[R_4+(R_2+R_1)\parallel R_5\right]$$

What impedance condition could null the response?

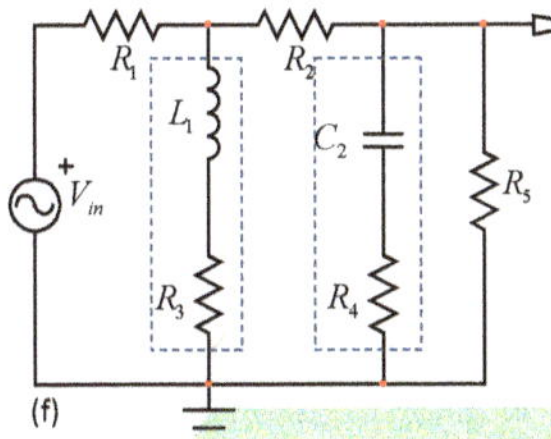

$$Z_1(s)=R_3+sL_1=0\rightarrow s_z=-\frac{R_3}{L_1}$$

$$Z_2(s)=R_4+\frac{1}{sC_2}=0\rightarrow s_z=-\frac{1}{R_4C_2}$$

$$\omega_{z_1}=\frac{R_3}{L_1}\quad\omega_{z_2}=\frac{1}{R_4C_2}$$

$$N(s)=\left(1+s\frac{L_1}{R_3}\right)(1+sR_4C_2)=\left(1+\frac{s}{\omega_{z_1}}\right)\left(1+\frac{s}{\omega_{z_2}}\right)$$

$$H(s)=\frac{R_3}{R_3+R_1}\frac{R_5}{R_5+R_2+R_1\parallel R_3}\frac{\left(1+s\frac{L_1}{R_3}\right)(1+sR_4C_2)}{1+s\left[\frac{L_1}{R_3+R_1\parallel(R_2+R_5)}+C_2\left[R_4+(R_2+R_1\parallel R_3)\parallel R_5\right]\right]+s^2\frac{L_1}{R_3+R_1\parallel(R_2+R_5)}C_2\left[R_4+(R_2+R_1)\parallel R_5\right]}$$

$$H(s)=H_0\frac{\left(1+\frac{s}{\omega_{z_1}}\right)\left(1+\frac{s}{\omega_{z_2}}\right)}{1+\frac{s}{\omega_0Q}+\left(\frac{s}{\omega_0}\right)^2}$$

$$Q=\frac{\sqrt{b_2}}{b_1}=\frac{\sqrt{\frac{L_1}{R_3+R_1\parallel(R_5+R_2)}C_2\left[R_4+(R_2+R_1)\parallel R_5\right]}}{\frac{L_1}{R_3+R_1\parallel(R_5+R_2)}C_2\left[R_4+(R_2+R_1\parallel R_3)\parallel R_5\right]}$$

$$\omega_0=\frac{1}{\sqrt{b_2}}=\frac{1}{\sqrt{\frac{L_1}{R_3+R_1\parallel(R_5+R_2)}C_2\left[R_4+(R_2+R_1)\parallel R_5\right]}}$$

When the quality factor Q is much small than 1 or $Q<<1$, you can apply the low-Q approximation:

$$\omega_{p_1}=Q\omega_0=\frac{1}{b_1}=\frac{1}{\frac{L_1}{R_3+R_1\parallel(R_5+R_2)}+C_2\left[R_4+(R_2+R_1\parallel R_3)\parallel R_5\right]}$$

$$\omega_{p_2}=\frac{\omega_0}{Q}=\frac{b_1}{b_2}=\frac{\frac{L_1}{R_3+R_1\parallel(R_5+R_2)}+C_2\left[R_4+(R_2+R_1\parallel R_3)\parallel R_5\right]}{\frac{L_1}{R_3+R_1\parallel(R_5+R_2)}C_2\left[R_4+(R_2+R_1)\parallel R_5\right]}$$

$$H(s)\approx H_0\frac{\left(1+\frac{s}{\omega_{z_1}}\right)\left(1+\frac{s}{\omega_{z_2}}\right)}{\left(1+\frac{s}{\omega_{p_1}}\right)\left(1+\frac{s}{\omega_{p_2}}\right)}$$

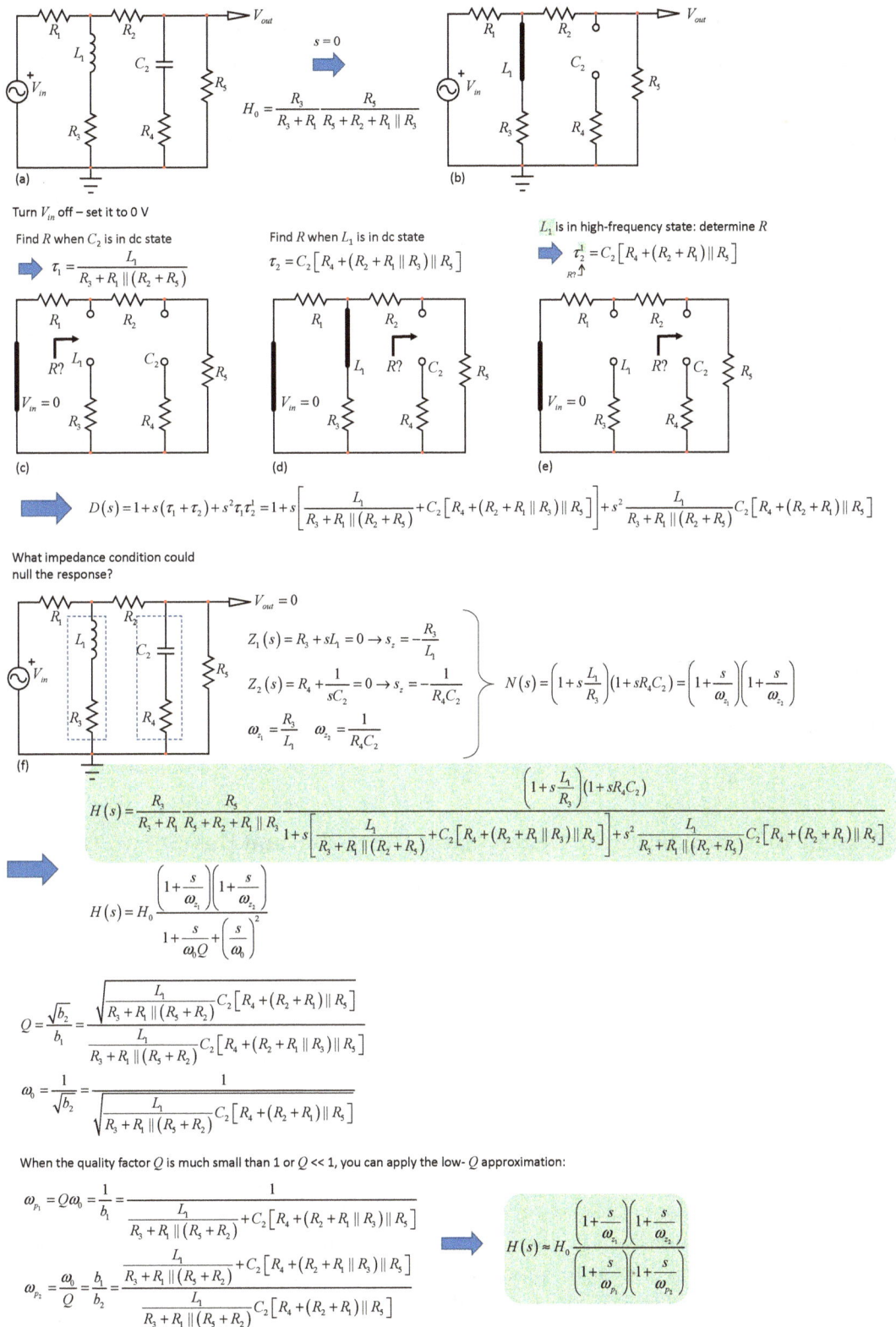

The Inductor Now Part of the Circuit Complicates the Expressions: No Problem with the FACTs!

Figure 6.10

$\|(x,y) := \frac{x \cdot y}{x+y}$ $\quad R_{inf} := 10^{23}\Omega$

$L_1 := 250\mu H \quad C_2 := 0.22\mu F \quad R_1 := 1k\Omega \quad R_2 := 3.2k\Omega \quad R_3 := 50\Omega \quad R_4 := 10\Omega \quad R_5 := 1.5k\Omega$

Transfer function for s=0

$$H_0 := \frac{R_3}{R_3 + R_1} \cdot \frac{R_5}{R_5 + R_2 + R_1 \| R_3} = 0.01505 \qquad 20 \cdot \log(H_0) = -36.45208 \text{ dB}$$

$$\tau_1 := \frac{L_1}{R_3 + R_1 \| (R_5 + R_2)} = 0.28586\mu s$$

$$\tau_2 := C_2 \cdot [R_4 + (R_2 + R_1 \| R_3) \| R_5] = 227.93721\mu s$$

$$b_1 := \tau_1 + \tau_2 = 228.22307\mu s$$

$$\tau_{12} := C_2 \cdot [R_4 + (R_2 + R_1) \| R_5] = 2.45358 \times 10^{-4} s$$

$$b_2 := \tau_1 \cdot \tau_{12} = 7.01374 \times 10^{-5} \cdot ms^2$$

$$D_1(s) := 1 + s \cdot b_1 + s^2 \cdot b_2 \qquad Q := \frac{\sqrt{b_2}}{b_1} = 0.0367 \qquad \omega_0 := \frac{1}{\sqrt{b_2}} \qquad f_0 := \frac{\omega_0}{2 \cdot \pi} = 1.9004 \times 10^4 \cdot Hz$$

$$\omega_{p1} := \frac{1}{b_1} \qquad f_{p1} := \frac{\omega_{p1}}{2\pi} = 697.36571 Hz$$

$$\omega_{p2} := \frac{b_1}{b_2} \qquad f_{p2} := \frac{\omega_{p2}}{2\pi} = 517.88095 kHz$$

$$\omega_{z1} := \frac{R_3}{L_1} \qquad f_{z1} := \frac{\omega_{z1}}{2\pi} = 31.83099 kHz$$

$$\omega_{z2} := \frac{1}{R_4 \cdot C_2} \qquad f_{z2} := \frac{\omega_{z2}}{2\pi} = 72.34316 kHz$$

$$H_1(s) := H_0 \cdot \frac{\left(1 + \frac{s}{\omega_{z1}}\right) \cdot \left(1 + \frac{s}{\omega_{z2}}\right)}{1 + s \cdot b_1 + s^2 \cdot b_2}$$

Full expression

$$H_2(s) := H_0 \cdot \frac{\left(1 + \frac{s}{\omega_{z1}}\right) \cdot \left(1 + \frac{s}{\omega_{z2}}\right)}{\left(1 + \frac{s}{\omega_{p1}}\right) \cdot \left(1 + \frac{s}{\omega_{p2}}\right)}$$

Approximate expression

$$H_3(s) := H_0 \cdot \frac{\left(1 + \frac{s}{\omega_{z1}}\right) \cdot \left(1 + \frac{s}{\omega_{z2}}\right)}{1 + \frac{s}{\omega_0 \cdot Q} + \left(\frac{s}{\omega_0}\right)^2}$$

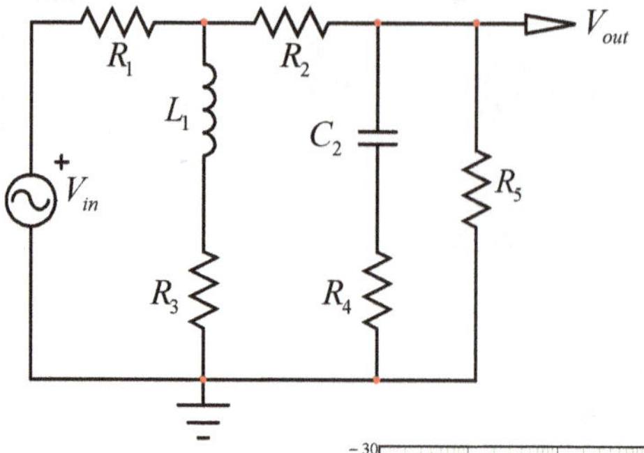

$$Z_{th}(s) := R_1 \| (s \cdot L_1 + R_3)$$

$$H_{ref}(s) := \frac{s \cdot L_1 + R_3}{s \cdot L_1 + R_3 + R_1} \cdot \frac{\left(\frac{1}{s \cdot C_2} + R_4\right) \| R_5}{\left(\frac{1}{s \cdot C_2} + R_4\right) \| R_5 + Z_{th}(s) + R_2}$$

Reference brute-force expression

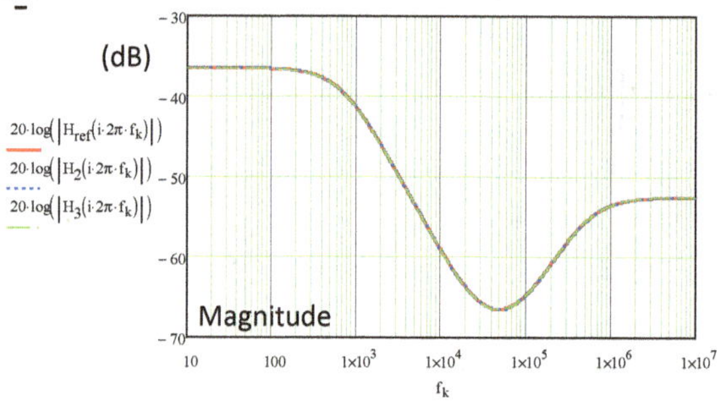

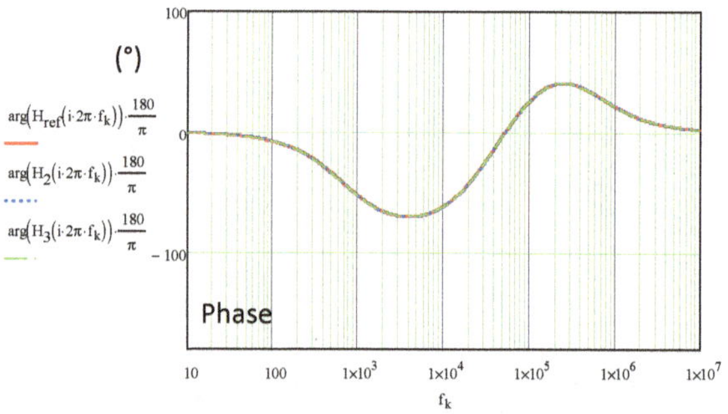

It is Possible to Approximate the Denominator Expression and Unveil Two Distinct Poles

Figure 6.11

What is the output impedance of this network?

To determine it, install a current source I_T sweeping the output as represented in Figure 6.12.

The resistance R_0 is determined at dc and shows many combined resistors.

When the excitation I_T is turned off to determine the poles—you now know it—the circuit is similar to the structure already examined in sketches (c), (d) and (e) of Figure 6.10: we can reuse the denominator $D(s)$ straight away and it represents a major timesaver.

For the zeroes, null the response V_T which corresponds to a degenerate case where the current source is replaced by a short circuit.

Looking at the resistance R driving each of the energy-storing element is then a simple exercise.

Because the numerator is also of second-order, I have highlighted a resonant frequency and a quality factor with a N as subscript to distinguish them for the denominator's components now subscripted with a D. Finally, if Q_N is way smaller than unity, it is possible to approximately factor the numerator with two cascaded zeroes.

This is what I have done in Figure 6.11 where all the ac responses are gathered.

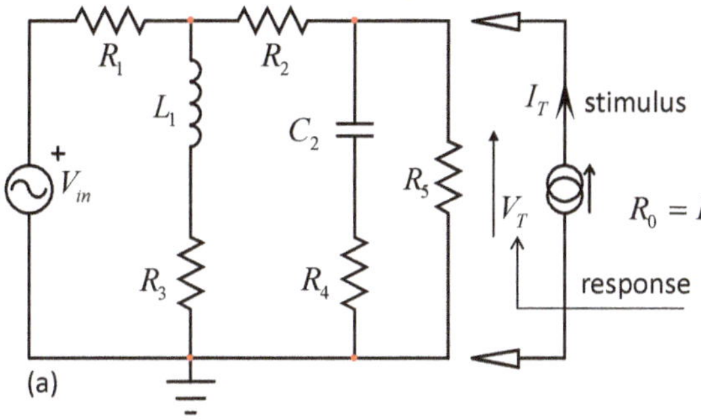

(a)

$$R_0 = R_5 \,||\, (R_2 + R_1 \,||\, R_3)$$

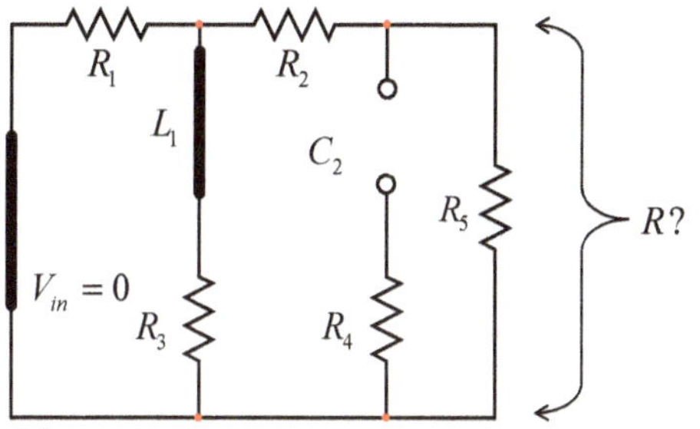

(b)

When the excitation is off - $I_T = 0$ - the structure is unchanged as in previous sketches (c) and (d): you can reuse $D(s)$

$$D(s) = 1 + sb_1 + s^2 b_2$$

$$b_1 = \frac{L_1}{R_3 + R_1 \,||\, (R_2 + R_5)} + C_2\left[R_4 + (R_2 + R_1 \,||\, R_3) \,||\, R_5\right]$$

$$b_2 = \frac{L_1}{R_3 + R_1 \,||\, (R_2 + R_5)} C_2\left[R_4 + (R_2 + R_1) \,||\, R_5\right]$$

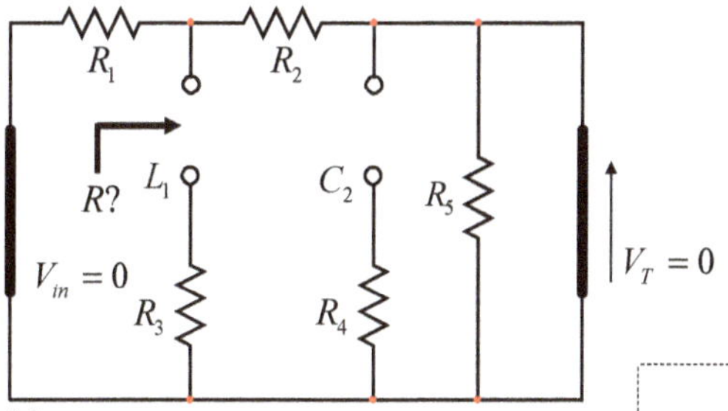

(c)

NDI for the zeroes: degenerate case, response is nulled then replace current source by a wire

$$\tau_{1N} = \frac{L_1}{R_3 + R_1 \,||\, R_2}$$

$$\tau_{2N} = R_4 C_2$$

$$\tau_{2N}^1 = R_4 C_2$$

$R?$

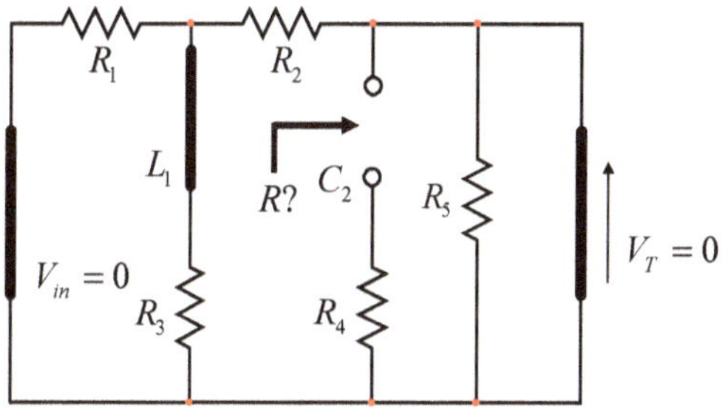

(d)

Set to HF state

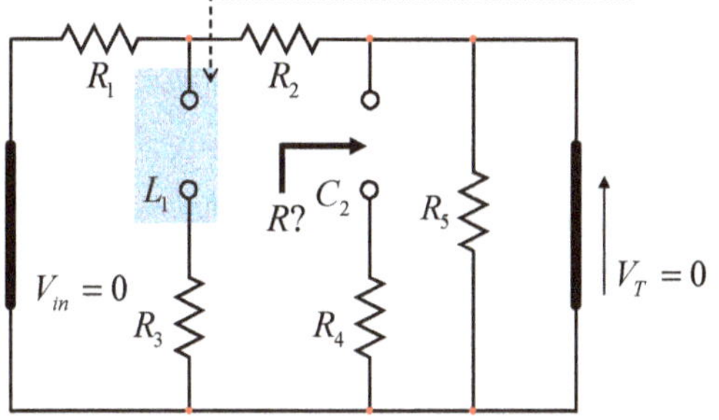

(e)

$$N(s) = 1 + a_1 s + a_2 s^2 = 1 + s(\tau_{1N} + \tau_{2N}) + s^2(\tau_{1N}\tau_{2N}^1)$$

$$a_1 = \frac{L_1}{R_3 + R_1 \,||\, R_2} + R_4 C_2$$

$$a_2 = \frac{L_1}{R_3 + R_1 \,||\, R_2} R_4 C_2$$

$$Z_{out}(s) = \left[R_5 \,||\, (R_2 + R_1 \,||\, R_3)\right] \frac{1 + s\left(\frac{L_1}{R_3 + R_1 \,||\, R_2} + R_4 C_2\right) + s^2\left(\frac{L_1}{R_3 + R_1 \,||\, R_2} R_4 C_2\right)}{1 + s\left(\frac{L_1}{R_3 + R_1 \,||\, (R_2 + R_5)} + C_2\left[R_4 + (R_2 + R_1 \,||\, R_3) \,||\, R_5\right]\right) + s^2 \frac{L_1}{R_3 + R_1 \,||\, (R_2 + R_5)} C_2\left[R_4 + (R_2 + R_1) \,||\, R_5\right]}$$

$$Z_{out}(s) = R_0 \frac{1 + \frac{s}{Q_N \omega_{0N}} + \left(\frac{s}{\omega_{0N}}\right)^2}{1 + \frac{s}{Q_D \omega_{0D}} + \left(\frac{s}{\omega_{0D}}\right)^2}$$

$$R_0 = \left[R_5 \,||\, (R_2 + R_1 \,||\, R_3)\right]$$

$$Q_N = \frac{\sqrt{\frac{L_1}{R_3 + R_1 \,||\, R_2} R_4 C_2}}{\frac{L_1}{R_3 + R_1 \,||\, R_2} + R_4 C_2} \qquad Q_D = \frac{\sqrt{\frac{L_1}{R_3 + R_1 \,||\, (R_2 + R_5)} C_2\left[R_4 + (R_2 + R_1) \,||\, R_5\right]}}{\frac{L_1}{R_3 + R_1 \,||\, (R_2 + R_5)} + C_2\left[R_4 + (R_2 + R_1 \,||\, R_3) \,||\, R_5\right]}$$

$$\omega_{0N} = \frac{1}{\sqrt{\frac{L_1}{R_3 + R_1 \,||\, R_2} R_4 C_2}} \qquad \omega_{0D} = \frac{1}{\sqrt{\frac{L_1}{R_3 + R_1 \,||\, (R_2 + R_5)} C_2\left[R_4 + (R_2 + R_1) \,||\, R_5\right]}}$$

Determine the Output Impedance of this Network by Installing a Test Generator

Figure 6.12

$\|(x,y) := \frac{x \cdot y}{x+y}$ $\quad R_{inf} := 10^{23}\Omega$

$L_1 := 250\mu H \quad C_2 := 0.22\mu F \quad R_1 := 1k\Omega \quad R_2 := 3.2k\Omega \quad R_3 := 50\Omega \quad R_4 := 10\Omega \quad R_5 := 1.5k\Omega$

Transfer function for s=0

$R_0 := R_5 \| (R_2 + R_1 \| R_3) = 1.02608\times 10^3\,\Omega \qquad 20\cdot\log\left(\frac{R_0}{\Omega}\right) = 60.22361 \text{ dB}\Omega$

$\tau_1 := \frac{L_1}{R_3 + R_1 \| (R_5 + R_2)} = 0.28586\mu s$

$\tau_2 := C_2 \cdot [R_4 + (R_2 + R_1 \| R_3) \| R_5] = 227.93721\mu s$

$b_1 := \tau_1 + \tau_2 = 228.22307\mu s$

$\tau_{12} := C_2 \cdot [R_4 + (R_2 + R_1) \| R_5] = 2.45358\times 10^{-4} s$

$b_2 := \tau_1 \cdot \tau_{12} = 7.01374\times 10^{-5} \cdot ms^2$

$D_1(s) := 1 + s\cdot b_1 + s^2 \cdot b_2 \qquad Q_D := \frac{\sqrt{b_2}}{b_1} = 0.0367 \qquad \omega_{0D} := \frac{1}{\sqrt{b_2}} \qquad f_{0D} := \frac{\omega_{0D}}{2\cdot\pi} = 19.00401kHz$

$\tau_{1N} := \frac{L_1}{R_3 + R_1 \| R_2} = 0.30792\mu s$

$\tau_{2N} := R_4 \cdot C_2 = 2.2 \cdot \mu s$

$a_1 := \tau_{1N} + \tau_{2N} = 2.50792\mu s$

$\tau_{12N} := R_4 \cdot C_2 = 2.2 \cdot \mu s$

$a_2 := \tau_{1N} \cdot \tau_{12N} = 6.77419\times 10^{-13} s^2$

$Q_N := \frac{\sqrt{a_2}}{a_1} = 0.32818 \qquad \omega_{0N} := \frac{1}{\sqrt{a_2}} \qquad f_{0N} := \frac{\omega_{0N}}{2\cdot\pi} = 193.37099kHz$

$\omega_{p1} := \frac{1}{b_1} \qquad f_{p1} := \frac{\omega_{p1}}{2\pi} = 697.36571Hz$

$\omega_{p2} := \frac{b_1}{b_2} \qquad f_{p2} := \frac{\omega_{p2}}{2\pi} = 517.88095kHz$

$\omega_{z1} := \frac{1}{a_1} \qquad f_{z1} := \frac{\omega_{z1}}{2\pi} = 63.46099kHz$

$\omega_{z2} := \frac{a_1}{a_2} \qquad f_{z2} := \frac{\omega_{z2}}{2\pi} = 589.21778kHz$

Approximate expression

$$Z_1(s) := R_0 \cdot \frac{1 + a_1 \cdot s + a_2 \cdot s^2}{1 + s\cdot b_1 + s^2 \cdot b_2} \qquad Z_2(s) := R_0 \cdot \frac{\left(1+\frac{s}{\omega_{z1}}\right)\left(1+\frac{s}{\omega_{z2}}\right)}{\left(1+\frac{s}{\omega_{p1}}\right)\left(1+\frac{s}{\omega_{p2}}\right)} \qquad Z_3(s) := R_0 \cdot \frac{1 + \frac{s}{\omega_{0N}\cdot Q_N} + \left(\frac{s}{\omega_{0N}}\right)^2}{1 + \frac{s}{\omega_{0D}\cdot Q_D} + \left(\frac{s}{\omega_{0D}}\right)^2}$$

Full expression

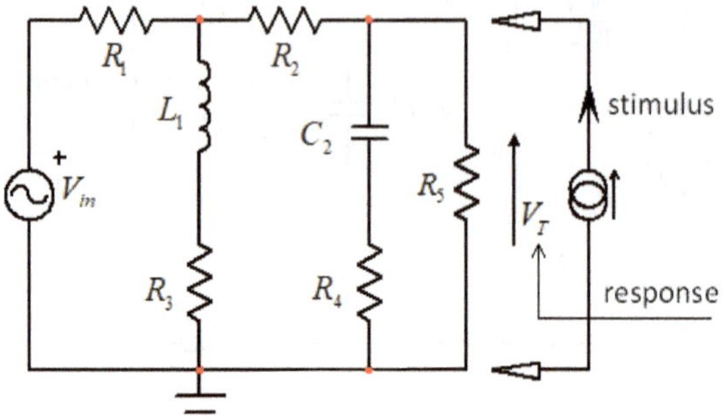

$$Z_{ref}(s) := R_5 \| \left[\left(\frac{1}{s\cdot C_2} + R_4\right) \| [R_2 + (s\cdot L_1 + R_3) \| R_1]\right]$$

Reference brute-force expression

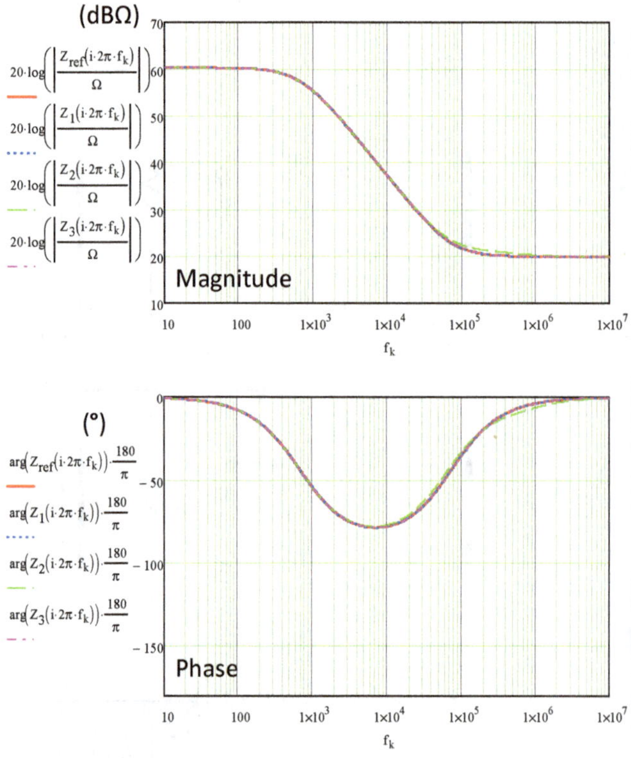

The Plots confirm the Analysis is Correct

Figure 6.13

Now that we have the gain and output impedance, we can extract the input impedance Z_{in} as proposed in Figure 6.14. This is not particularly complicated, but the FACTs are truly the instrument of choice to run the analysis in a swift manner.

As we have seen many times, when determining the three expressions, gain H, output and input impedances, the denominator $D(s)$ determined for H and Z_{out} becomes the numerator of the Z_{in} expression. This is because nulling the response and shorting the stimulus—a degenerate case for the nulled response across the test generator I_T—brings the network back to the configuration already studied for D.

As the time constants are similar, we can reuse them for building the numerator N.

The results are tested in Figure 6.15.

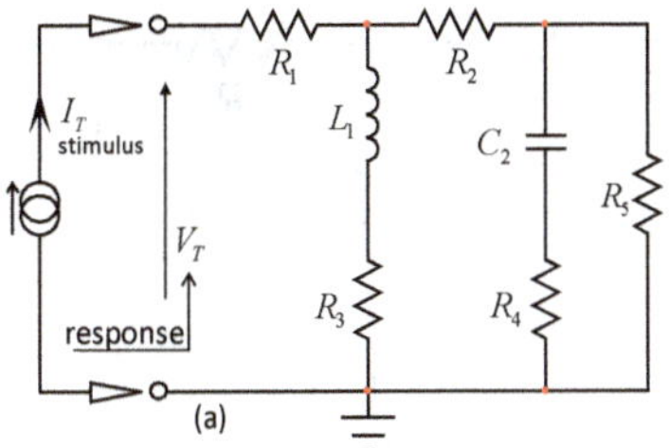

I_T
stimulus
V_T
response
R_1
R_2
L_1
C_2
R_5
R_3
R_4
(a)

$s = 0$

$R_i = R_1 + R_3 \,||\, (R_2 + R_5)$

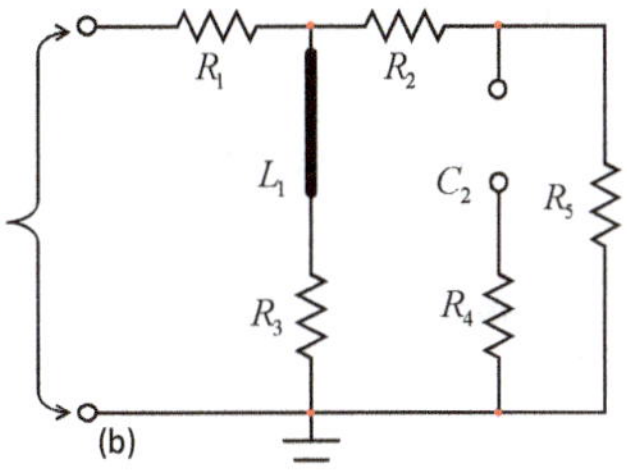

Turn I_T off – set it to 0 A

Find R when C_2 is in dc state

$$\tau_1 = \frac{L_1}{R_3 + R_2 + R_5}$$

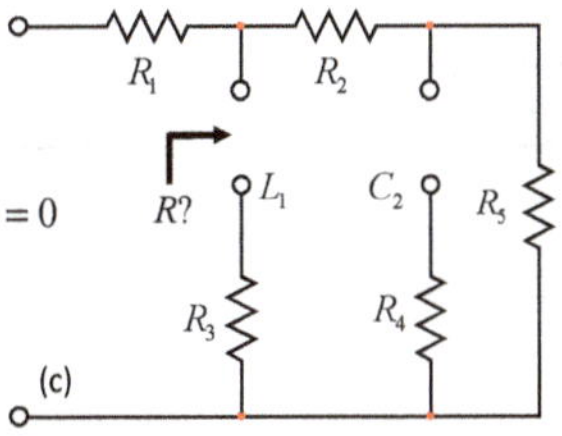

Find R when L_1 is in dc state

$$\tau_2 = C_2\left[R_4 + R_5 \,||\, (R_2 + R_3)\right]$$

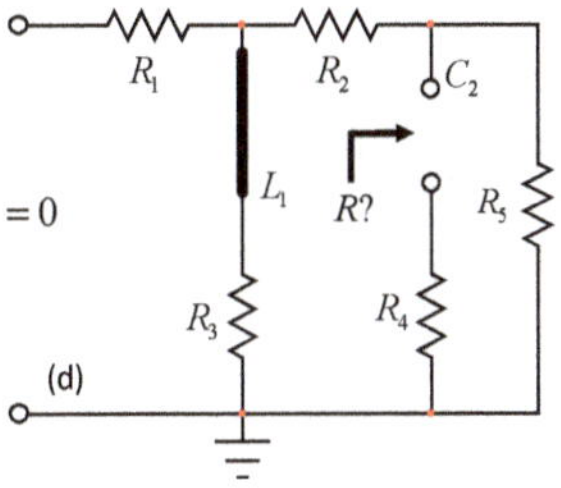

Find R when L_1 is in high-frequency state

$$\tau_2^1 = C_2\left(R_4 + R_5\right)$$

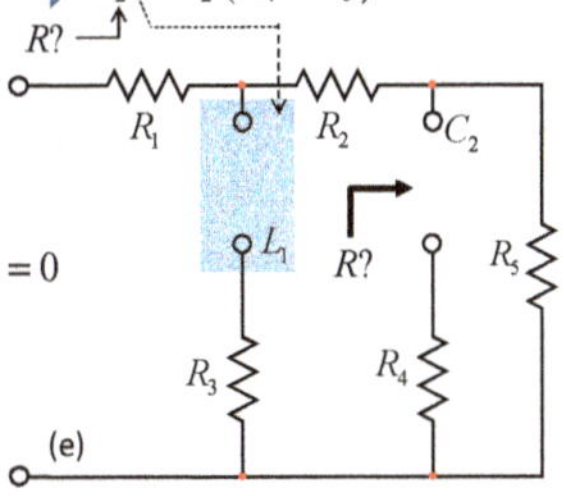

$$b_1 = \frac{L_1}{R_3 + R_2 + R_5} + C_2\left[R_4 + R_5 \,||\, (R_2 + R_3)\right]$$

$$b_2 = \frac{L_1}{R_3 + R_2 + R_5} C_2\left[R_4 + R_5\right]$$

$$D(s) = 1 + sb_1 + s^2 b_2$$

$$D(s) = 1 + s\left(\frac{L_1}{R_3 + R_2 + R_5} + C_2\left[R_4 + R_5 \,||\, (R_2 + R_3)\right]\right) + s^2\left(\frac{L_1}{R_3 + R_2 + R_5} C_2\left[R_4 + R_5\right]\right)$$

$$D(s) = 1 + \frac{s}{\omega_0 Q} + \left(\frac{s}{\omega_0}\right)^2$$

$$Q_D = \frac{\sqrt{b_2}}{b_1} = \frac{\sqrt{\frac{L_1}{R_3 + R_2 + R_5} C_2\left[R_4 + R_5\right]}}{\frac{L_1}{R_3 + R_2 + R_5} + C_2\left[R_4 + R_5 \,||\, (R_2 + R_3)\right]} \qquad \omega_{0D} = \frac{1}{\sqrt{b_2}} = \frac{1}{\sqrt{\frac{L_1}{R_3 + R_2 + R_5} C_2\left[R_4 + R_5\right]}}$$

Null the response V_T across the test generator I_T: degenerate case.

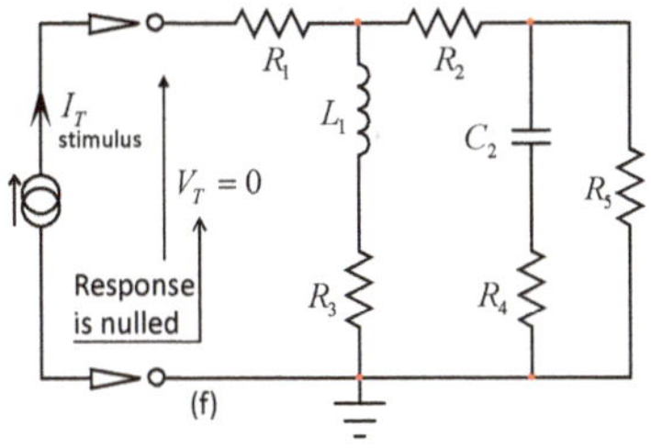

I_T
stimulus
$V_T = 0$
Response is nulled
R_1
R_2
L_1
C_2
R_5
R_3
R_4
(f)

Replace the generator by a wire

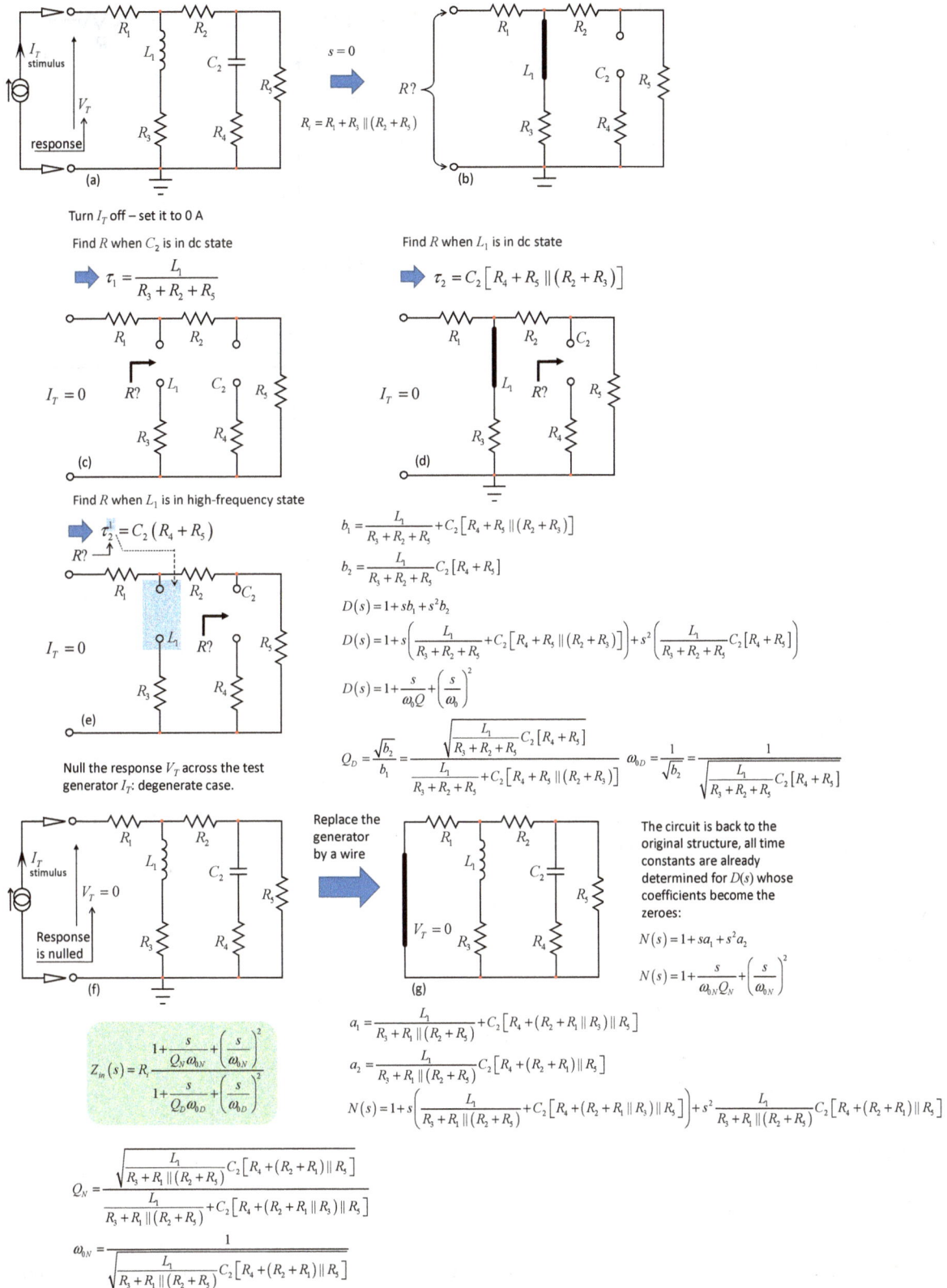

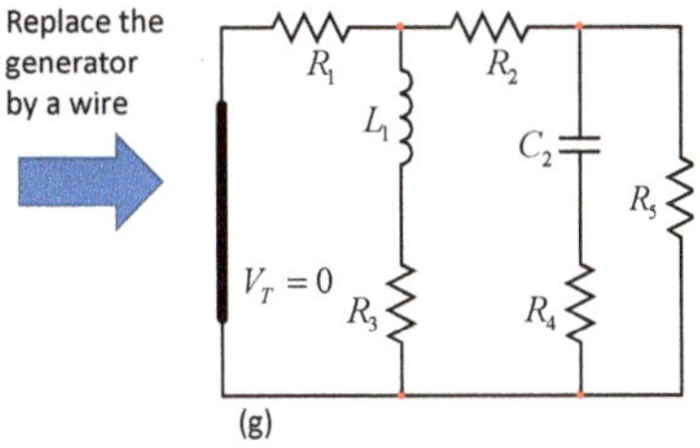

The circuit is back to the original structure, all time constants are already determined for $D(s)$ whose coefficients become the zeroes:

$$N(s) = 1 + sa_1 + s^2 a_2$$

$$N(s) = 1 + \frac{s}{\omega_{0N} Q_N} + \left(\frac{s}{\omega_{0N}}\right)^2$$

$$a_1 = \frac{L_1}{R_3 + R_1 \,||\, (R_2 + R_5)} + C_2\left[R_4 + (R_2 + R_1 \,||\, R_3) \,||\, R_5\right]$$

$$a_2 = \frac{L_1}{R_3 + R_1 \,||\, (R_2 + R_5)} C_2\left[R_4 + (R_2 + R_1) \,||\, R_5\right]$$

$$Z_{in}(s) = R_i \frac{1 + \frac{s}{Q_N \omega_{0N}} + \left(\frac{s}{\omega_{0N}}\right)^2}{1 + \frac{s}{Q_D \omega_{0D}} + \left(\frac{s}{\omega_{0D}}\right)^2}$$

$$N(s) = 1 + s\left(\frac{L_1}{R_3 + R_1 \,||\, (R_2 + R_5)} + C_2\left[R_4 + (R_2 + R_1 \,||\, R_3) \,||\, R_5\right]\right) + s^2 \frac{L_1}{R_3 + R_1 \,||\, (R_2 + R_5)} C_2\left[R_4 + (R_2 + R_1) \,||\, R_5\right]$$

$$Q_N = \frac{\sqrt{\frac{L_1}{R_3 + R_1 \,||\, (R_2 + R_5)} C_2\left[R_4 + (R_2 + R_1) \,||\, R_5\right]}}{\frac{L_1}{R_3 + R_1 \,||\, (R_2 + R_5)} + C_2\left[R_4 + (R_2 + R_1 \,||\, R_3) \,||\, R_5\right]}$$

$$\omega_{0N} = \frac{1}{\sqrt{\frac{L_1}{R_3 + R_1 \,||\, (R_2 + R_5)} C_2\left[R_4 + (R_2 + R_1) \,||\, R_5\right]}}$$

The Test Generator Now Excites the Input Impedance of the Network

Figure 6.14

$$\|(x,y) := \frac{x \cdot y}{x+y} \qquad R_{inf} := 10^{23}\Omega$$

$$L_1 := 100\mu H \quad C_2 := 5\mu F \quad R_1 := 1k\Omega \quad R_2 := 1k\Omega \quad R_3 := 50\Omega \quad R_4 := 10\Omega \quad R_5 := 1.5k\Omega$$

Transfer function for s=0

$$R_i := R_1 + R_3 \| (R_2 + R_5) = 1.04902\times 10^3\,\Omega \qquad 20\cdot\log\left(\frac{R_i}{\Omega}\right) = 60.41567 \text{ dB}$$

$$\tau_1 := \frac{L_1}{R_3 + R_2 + R_5} = 0.39216\mu s$$

$$\tau_2 := C_2\cdot\left[R_4 + (R_2 + R_3) \| R_5\right] = 3.13824\times 10^3\cdot\mu s$$

$$b_1 := \tau_1 + \tau_2 = 3.13863\times 10^3\cdot\mu s$$

$$\tau_{12} := C_2\cdot(R_4 + R_5) = 7.55\times 10^{-3}\,s$$

$$b_2 := \tau_1\cdot\tau_{12} = 2.96078\times 10^{-3}\cdot ms^2$$

$$D_1(s) := 1 + s\cdot b_1 + s^2\cdot b_2 \qquad Q_D := \frac{\sqrt{b_2}}{b_1} = 0.01734 \quad \omega_{0D} := \frac{1}{\sqrt{b_2}} \qquad f_{0D} := \frac{\omega_{0D}}{2\cdot\pi} = 2.92494kHz$$

$$\tau_{1N} := \frac{L_1}{R_3 + R_1 \| (R_5 + R_2)} = 1.30841\mu s$$

$$\tau_{2N} := C_2\cdot\left[R_4 + (R_2 + R_1 \| R_3) \| R_5\right] = 3.13411\times 10^3\cdot\mu s$$

$$a_1 := \tau_{1N} + \tau_{2N} = 3.13542\times 10^3\cdot\mu s$$

$$\tau_{12N} := C_2\cdot\left[R_4 + (R_2 + R_1) \| R_5\right] = 4.33571\times 10^3\cdot\mu s$$

$$a_2 := \tau_{1N}\cdot\tau_{12N} = 5.6729\times 10^{-9}\,s^2$$

$$Q_N := \frac{\sqrt{a_2}}{a_1} = 0.02402 \qquad \omega_{0N} := \frac{1}{\sqrt{a_2}} \qquad f_{0N} := \frac{\omega_{0N}}{2\cdot\pi} = 2.11309kHz$$

$$\omega_{p1} := \frac{1}{b_1} \qquad f_{p1} := \frac{\omega_{p1}}{2\pi} = 50.70845Hz$$

$$\omega_{p2} := \frac{b_1}{b_2} \qquad f_{p2} := \frac{\omega_{p2}}{2\pi} = 168.71478kHz$$

$$\omega_{z1} := \frac{1}{a_1} \qquad f_{z1} := \frac{\omega_{z1}}{2\pi} = 0.05076kHz$$

$$\omega_{z2} := \frac{a_1}{a_2} \qquad f_{z2} := \frac{\omega_{z2}}{2\pi} = 87.96523kHz$$

Approximate expression

$$Z_1(s) := R_i\cdot\frac{1 + a_1\cdot s + a_2\cdot s^2}{1 + s\cdot b_1 + s^2\cdot b_2} \qquad Z_2(s) := R_i\cdot\frac{\left(1+\frac{s}{\omega_{z1}}\right)\cdot\left(1+\frac{s}{\omega_{z2}}\right)}{\left(1+\frac{s}{\omega_{p1}}\right)\cdot\left(1+\frac{s}{\omega_{p2}}\right)} \qquad Z_3(s) := R_i\cdot\frac{1+\frac{s}{\omega_{0N}\cdot Q_N}+\left(\frac{s}{\omega_{0N}}\right)^2}{1+\frac{s}{\omega_{0D}\cdot Q_D}+\left(\frac{s}{\omega_{0D}}\right)^2}$$

Full expression

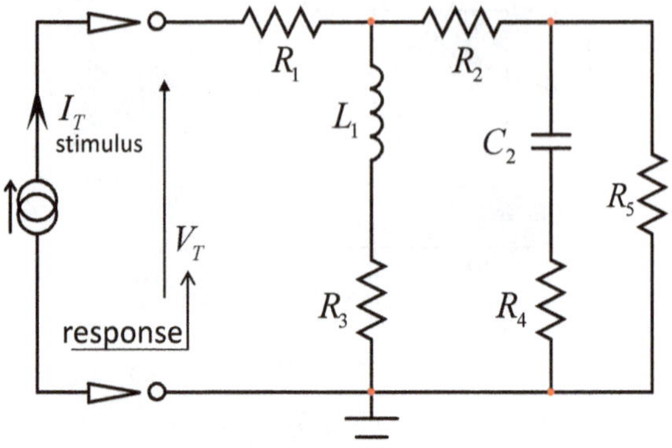

$$Z_{ref}(s) := R_1 + (s\cdot L_1 + R_3) \| \left[R_2 + \left(\frac{1}{s\cdot C_2} + R_4\right) \| R_5\right]$$

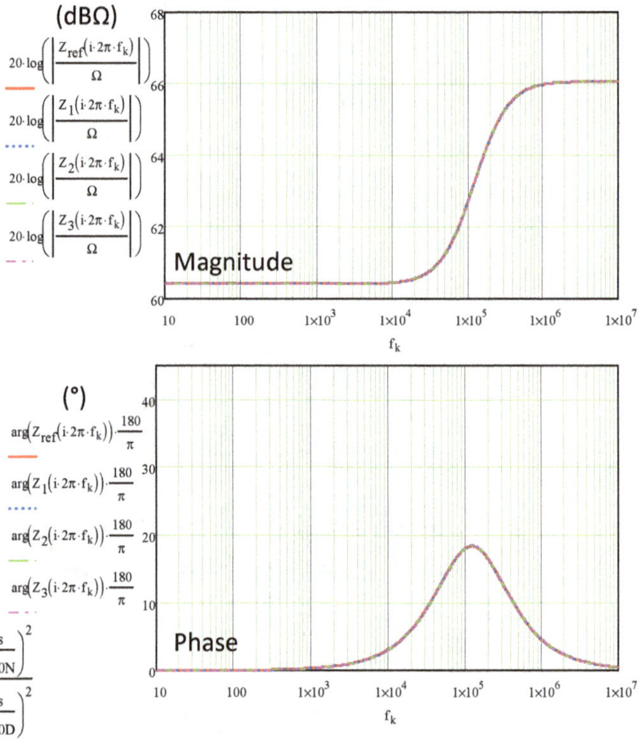

The Ac Response Shows an Excellent Correlation between the Full-Blown and Rearranged Expressions

Figure 6.15

Let's now have a look at a classic *LC* filter as shown in Figure 6.16.

The energy-storing elements are affected by their equivalent series resistor, respectively r_C and r_L for the capacitor and the inductor. This is a classic circuit you can find as a front-end EMI filter. Its transfer function is obtained quickly with the FACTs.

The ac response is given in Figure 6.17 and there is perfect agreement between the FACTs and the brute-force expression. The response from the low-*Q*-approximated transfer function is different and does not hold as *Q* is greater than one with the adopted values. When the circuit is well damped, implying large ohmic losses on the inductor or the capacitor, then the low-*Q* approximation holds, and modeling the transfer function with two distinct cascaded poles gives excellent results as shown in Figure 6.18.

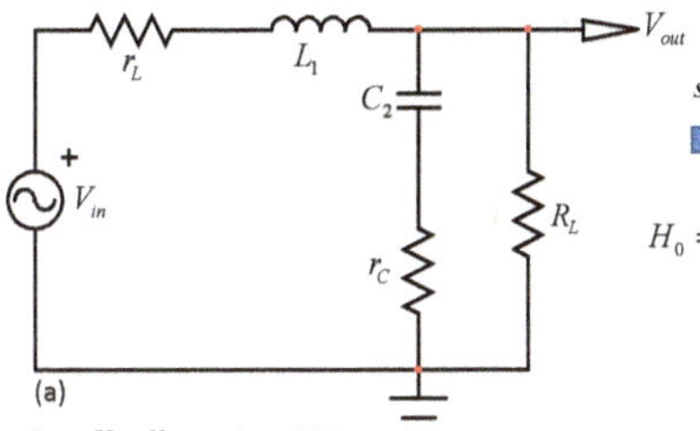

(a)

$s = 0$

$$H_0 = \frac{R_L}{R_L + r_L}$$

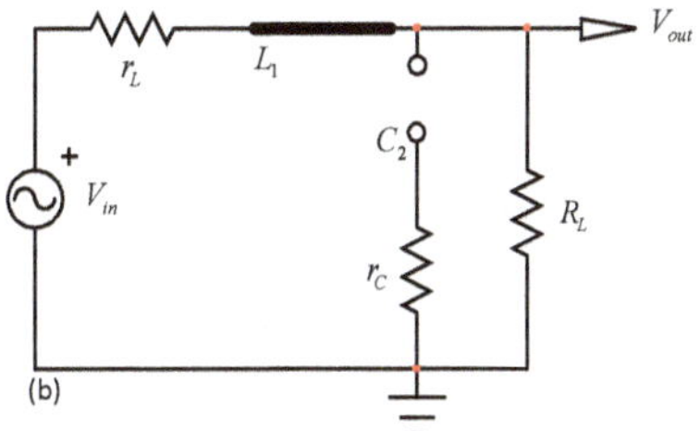

(b)

Turn V_{in} off – set it to 0 V

Find R when C_2 is in dc state

$$\tau_1 = \frac{L_1}{r_L + R_L}$$

Find R when L_1 is in dc state

$$\tau_2 = C_2\left[r_C + r_L \parallel R_L\right]$$

L_1 is in high-frequency state: determine R

$$\tau_2^1 = C_2\left(r_C + R_L\right)$$

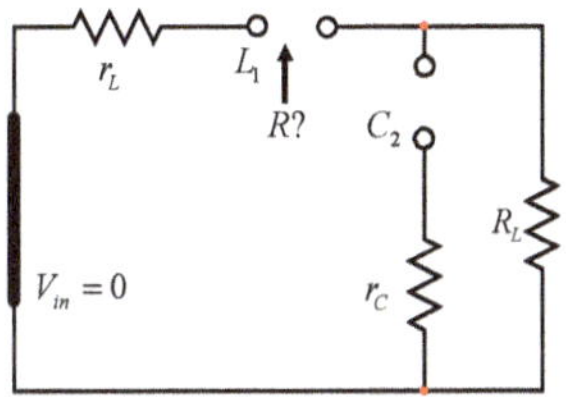

(c)

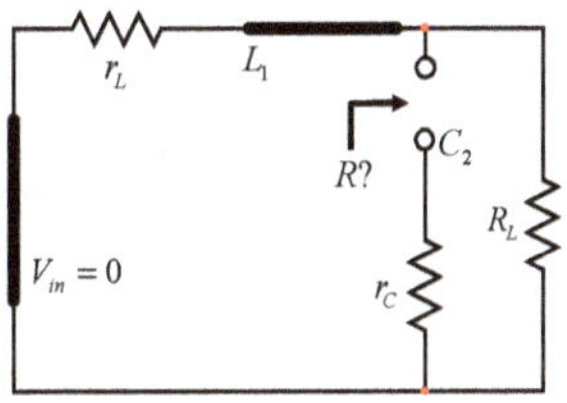

(d)

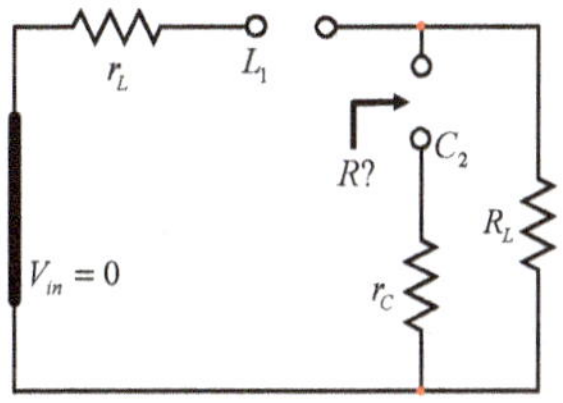

(e)

$$D(s) = 1 + s(\tau_1 + \tau_2) + s^2\tau_1\tau_2^1 = 1 + s\left[\frac{L_1}{r_L + R_L} + C_2\left(r_C + r_L \parallel R_L\right)\right] + s^2\frac{L_1}{r_L + R_L}C_2\left(r_C + R_L\right)$$

What impedance condition could null the response?

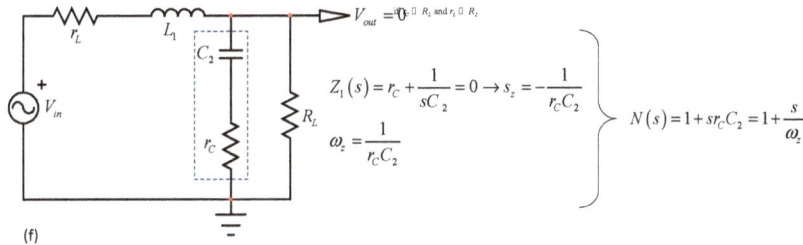

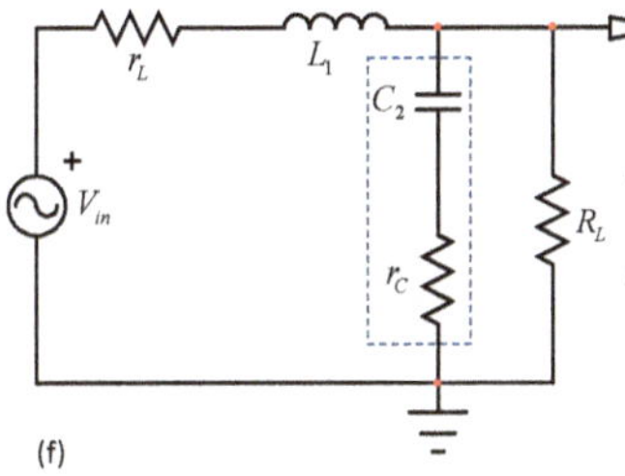

(f)

$$Z_1(s) = r_C + \frac{1}{sC_2} = 0 \rightarrow s_z = -\frac{1}{r_C C_2}$$

$$\omega_z = \frac{1}{r_C C_2}$$

$$N(s) = 1 + s r_C C_2 = 1 + \frac{s}{\omega_z}$$

$$H(s) = \frac{R_L}{r_L + R_L}\frac{1 + s r_C C_2}{1 + s\left[\frac{L_1}{r_L + R_L} + C_2\left(r_C + r_L \parallel R_L\right)\right] + s^2\frac{L_1}{r_L + R_L}C_2\left(r_C + R_L\right)}$$

$$H(s) = H_0\frac{1 + \frac{s}{\omega_z}}{1 + \frac{s}{\omega_0 Q} + \left(\frac{s}{\omega_0}\right)^2}$$

$$H(s) = \frac{1}{1 + s\frac{L_1}{R_L} + s^2 L_1 C_2}$$

$$Q = \frac{\sqrt{b_2}}{b_1} = \frac{\sqrt{\frac{L_1}{r_L + R_L}C_2\left(r_C + R_L\right)}}{\frac{L_1}{r_L + R_L} + C_2\left(r_C + r_L \parallel R_L\right)} \approx \frac{\sqrt{L_1 C_2}}{\frac{L_1}{R_L} + C_2\left(r_C + r_L\right)} \quad \text{If } r_C \ll R_L \text{ and } r_L \ll R_L$$

$$\omega_0 = \frac{1}{\sqrt{b_2}} = \frac{1}{\sqrt{\frac{L_1}{r_L + R_L}C_2\left(r_C + R_L\right)}} \approx \frac{1}{\sqrt{L_1 C_2}} \quad \text{If } r_C \ll R_L \text{ and } r_L \ll R_L$$

When the quality factor Q is much smaller than 1 or $Q \ll 1$, you can apply the low- Q approximation:

$$\omega_{p_1} = Q\omega_0 = \frac{1}{b_1} = \frac{1}{\frac{L_1}{r_L + R_L} + C_2\left(r_C + r_L \parallel R_L\right)}$$

$$\omega_{p_2} = \frac{\omega_0}{Q} = \frac{b_1}{b_2} = \frac{\frac{L_1}{r_L + R_L} + C_2\left(r_C + r_L \parallel R_L\right)}{\frac{L_1}{r_L + R_L}C_2\left(r_C + R_L\right)}$$

$$H(s) \approx H_0\frac{1 + \frac{s}{\omega_z}}{\left(1 + \frac{s}{\omega_{p_1}}\right)\left(1 + \frac{s}{\omega_{p_2}}\right)}$$

This is a Classic *LC* Filter Featuring Parasitic Elements

Figure 6.16

$\|(x,y) := \frac{x \cdot y}{x+y}$ $\quad R_{inf} := 10^{23}\Omega$

$r_L := 0.1\Omega \quad r_C := 0.01\Omega \quad C_2 := 100\mu F \quad L_1 := 47\mu H \quad R_L := 5\Omega$

Transfer function for s=0 $\quad H_0 := \frac{R_L}{R_L + r_L} = 0.98039 \quad 20 \cdot \log(H_0) = -0.172$ dB

$\tau_1 := \frac{L_1}{r_L + R_L} = 9.21569\mu s$

$\tau_2 := C_2 \cdot (r_C + r_L \| R_L) = 10.80392\mu s$

$b_1 := \tau_1 + \tau_2 = 20.01961\mu s$

$\tau_{12} := C_2 \cdot (r_C + R_L) = 5.01 \times 10^{-4} s$

$b_2 := \tau_1 \cdot \tau_{12} = 4.61706 \times 10^{-3} \cdot ms^2$

$D_1(s) := 1 + s \cdot b_1 + s^2 \cdot b_2$

$Q := \frac{\sqrt{b_2}}{b_1} = 3.39412 \quad Q_a := \frac{\sqrt{L_1 \cdot C_2}}{\frac{L_1}{R_L} + C_2 \cdot (r_C + r_L)} = 3.36062 \quad \zeta := \frac{1}{2 \cdot Q} = 0.14731$

$\omega_0 := \frac{1}{\sqrt{b_2}} = 1.47169 \times 10^4 \frac{1}{s} \quad \omega_{0a} := \frac{1}{\sqrt{L_1 \cdot C_2}} = 1.45865 \times 10^4 \frac{1}{s}$

$f_0 := \frac{\omega_0}{2 \cdot \pi} = 2.34227 \times 10^3 \frac{1}{s} \quad \omega_d := \omega_0 \sqrt{1 - \zeta^2} \quad f_d := \frac{\omega_d}{2 \cdot \pi} = 2.31672 kHz$

$\omega_{p1} := \frac{1}{b_1} \quad f_{p1} := \frac{\omega_{p1}}{2\pi} = 7.94995 kHz$

$\omega_{p2} := \frac{b_1}{b_2} \quad f_{p2} := \frac{\omega_{p2}}{2\pi} = 0.6901 kHz$

$\omega_z := \frac{1}{r_C \cdot C_2} \quad f_z := \frac{\omega_z}{2 \cdot \pi} = 159.15494 kHz$

Low-*Q* approximation — Low-entropy version

$H_1(s) := H_0 \cdot \frac{1 + \frac{s}{\omega_z}}{1 + s \cdot b_1 + s^2 \cdot b_2} \quad H_2(s) := H_0 \cdot \frac{1 + \frac{s}{\omega_z}}{\left(1 + \frac{s}{\omega_{p1}}\right) \cdot \left(1 + \frac{s}{\omega_{p2}}\right)} \quad H_3(s) := H_0 \cdot \frac{1 + \frac{s}{\omega_z}}{1 + \frac{s}{\omega_0 \cdot Q} + \left(\frac{s}{\omega_0}\right)^2}$

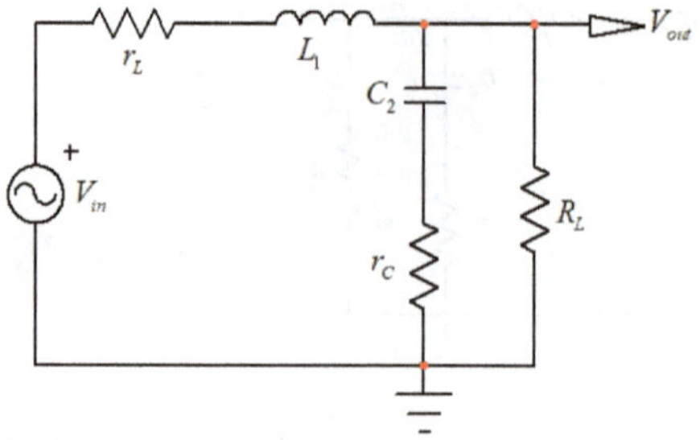

$Z_1(s) := R_L \| \left(\frac{1}{s \cdot C_2} + r_C\right)$

$H_{ref}(s) := \frac{Z_1(s)}{Z_1(s) + r_L + s \cdot L_1}$

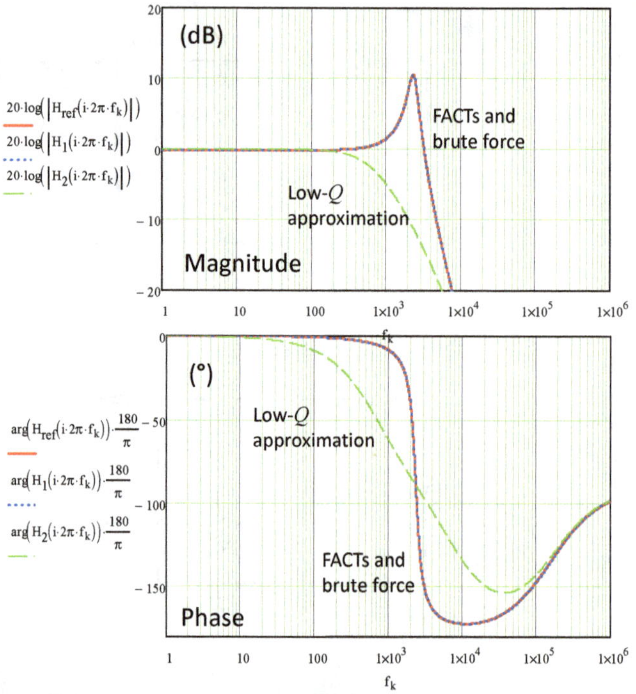

The Response of this *LC* Filter Exhibits Peaking when *Q* is High

Figure 6.17

$r_L := 1\Omega \quad r_C := 10\Omega \quad C_2 := 100\mu F \quad L_1 := 47\mu H \quad R_L := 5\Omega$

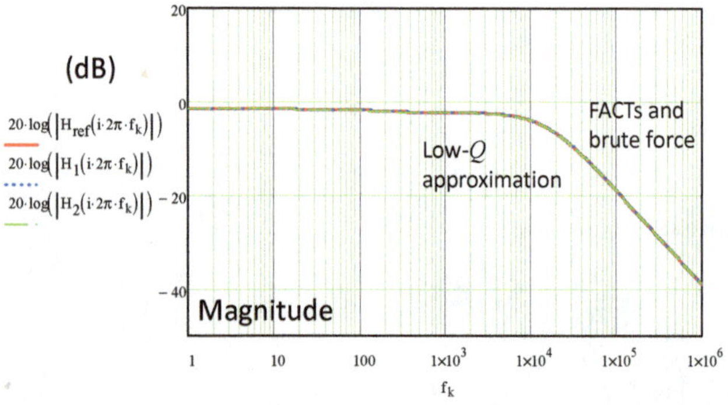

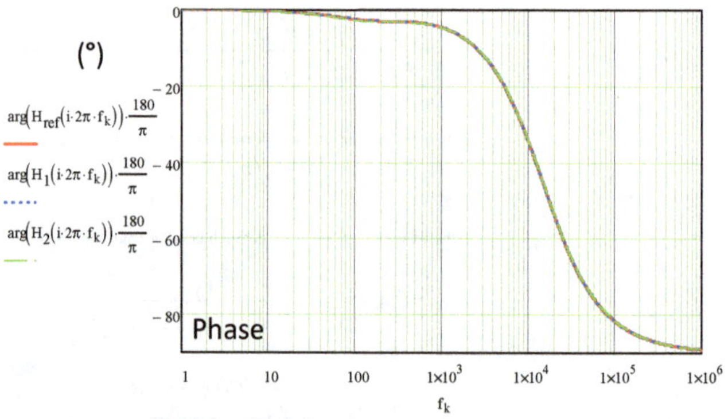

When the Quality Factor is Less than One (0.099), Implying a Well-Damped Circuit, the Low-*Q* Approximation Holds Well

Figure 6.18

As this circuit can be used as a front-end EMI filter, let's have a look at the attenuation it brings. The typical circuit to consider is that of Figure 6.19. A clean dc (or rectified ac) source feeds a noisy switching converter and you want to limit the high-frequency ripple flowing back into the source. You expect the inductor to block this ripple, confining the vast majority of it in the capacitor while the source supplies dc current with a little ripple in it.

The design process consists of determining the fundamental of the noisy current and determining what attenuation is needed to pass the input ripple specifications with margins. When the desired attenuation is found, position the filter cutoff frequency.

To calculate the element values, you need a transfer function linking I_1 (the stimulus and signature) to I_2, the response current you want to minimize at a given frequency. Start by shorting the battery and opening the excitation source I_2 to determine the time constants. It is important to include ohmic losses in the inductor and the capacitor as they enormously affect the attenuation.

The updated circuit and its analysis appear in Figure 6.20 while the frequency response compared to a SIMetrix® plot is proposed in Figure 6.21.

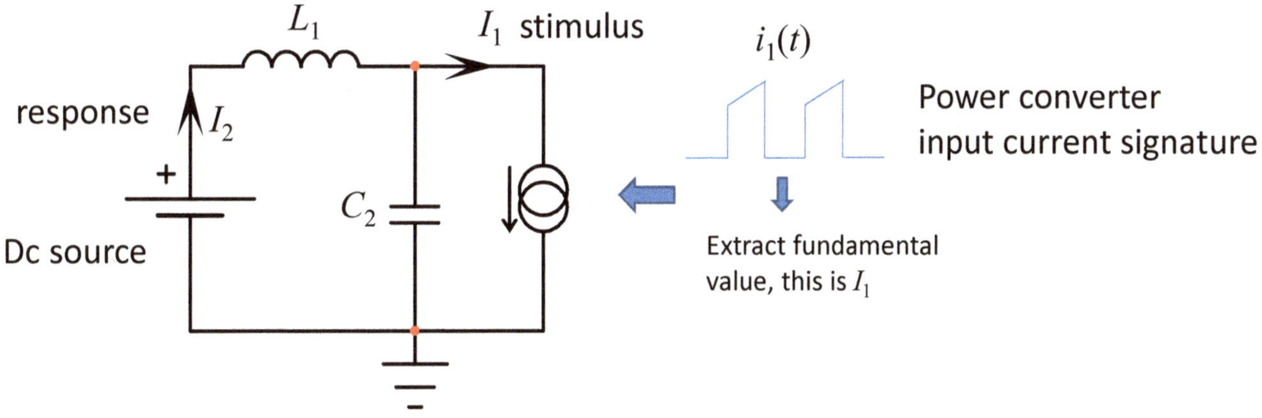

When a Battery Feeds a Noisy Source, Expect the *LC* Filter to Reduce the High-Frequency Ripple Polluting the Source

Figure 6.19

The result is a 2nd-order transfer function. We are interested in its high-frequency attenuation to calibrate the filter design. For design purposes, it is desirable to simplify the expression by concentrating on the high-frequency response when the s^2 term dominates the denominator. The obtained expression thus gives a convenient and easy way to position the filter frequency for obtaining the desired filter attenuation.

Assume you have determined that your converter absorbs a current whose fundamental value is 1 A peak. The specification states a maximum ripple of 15 mA peak at 10 kHz. In this case, the required attenuation is 15m/1 = 0.015 or -36.5 dB. Using the expression given in the bottom of Figure 6.20, we find that the cutoff frequency should be set to $\sqrt{0.015} \cdot 10\ \text{kHz} = 1.2\ \text{kHz}$.

Now select an inductor and a capacitor to meet the cutoff frequency of 1.2 kHz.

As mentioned, parasitics will affect the final attenuation and design margin is necessary to account for their presence. The plot in Figure 6.22 confirms our calculations. This is a first step to design a filter and the second-order network can be supplemented with more parasitics like the inter-turn capacitance for the inductor and the equivalent series inductance for the capacitor.

Both will affect the filter performance.

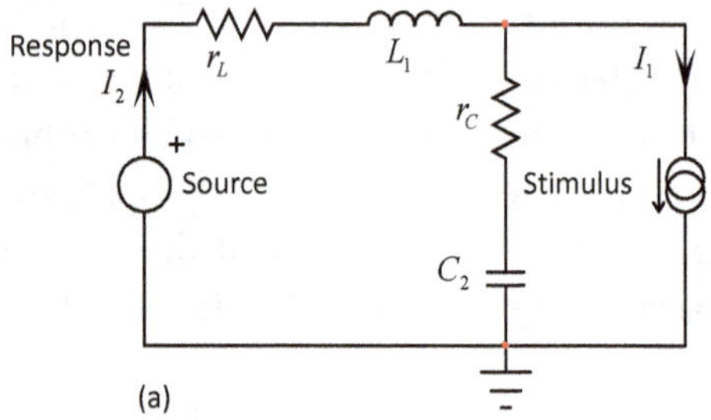
Response
I_2
r_L
L_1
r_C
I_1
Source
Stimulus
C_2
(a)

$s = 0$

$$H_0 = \frac{I_2}{I_1} = 1$$

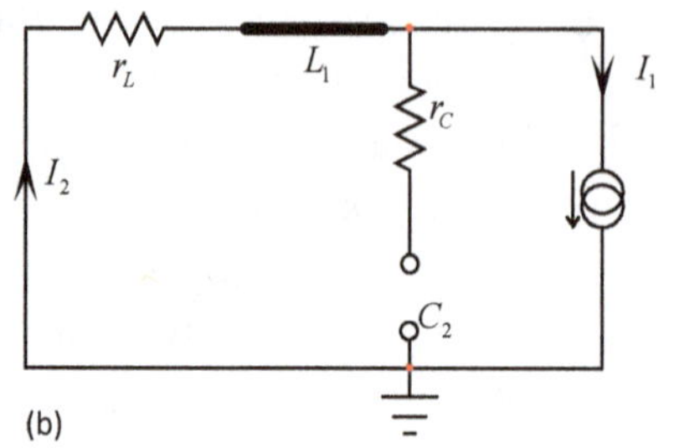

Turn I_1 off – set it to 0 A

Find R when C_2 is in dc state

$$\tau_1 = \frac{L_1}{R_{\text{inf}}}$$

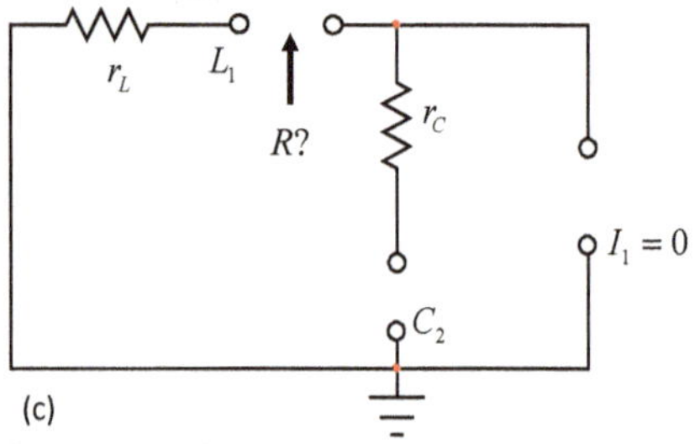

Find R when L_1 is in dc state

$$\tau_2 = C_2 (r_C + r_L)$$

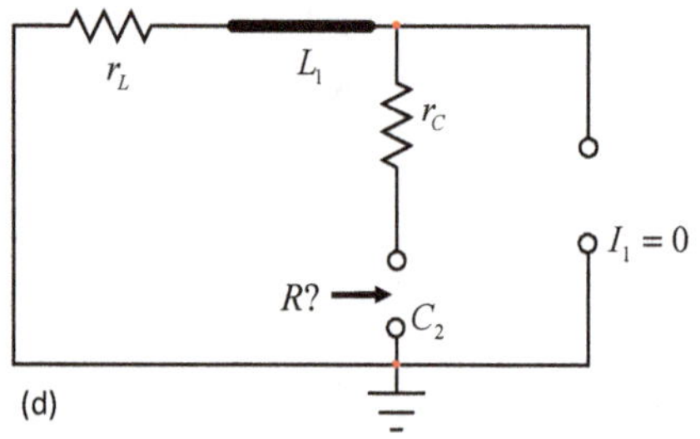

L_1 is in high-frequency state: determine R

$$\tau_2^1 = C_2 R_{\text{inf}}$$

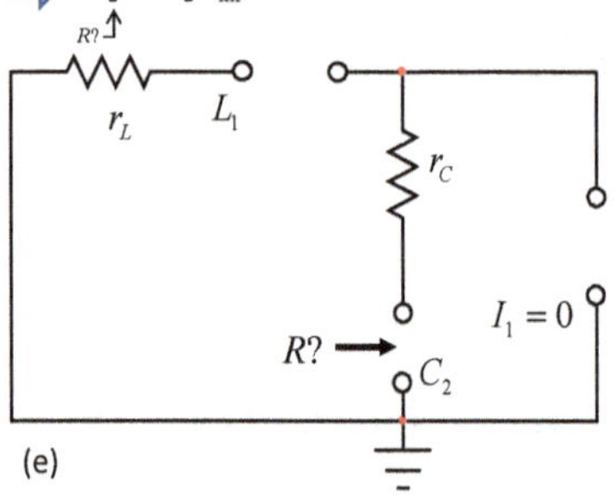

$$D(s) = 1 + s(\tau_1 + \tau_2) + s^2 \tau_1 \tau_2^1$$

$$= 1 + s\left[\frac{L_1}{R_{\text{inf}}} + C_2 (r_C + r_L)\right] + s^2 \frac{L_1}{R_{\text{inf}}} C_2 R_{\text{inf}}$$

$$= 1 + sC_2 (r_C + r_L) + s^2 L_1 C_2$$

You can simplify by R_{inf} which is finite value.

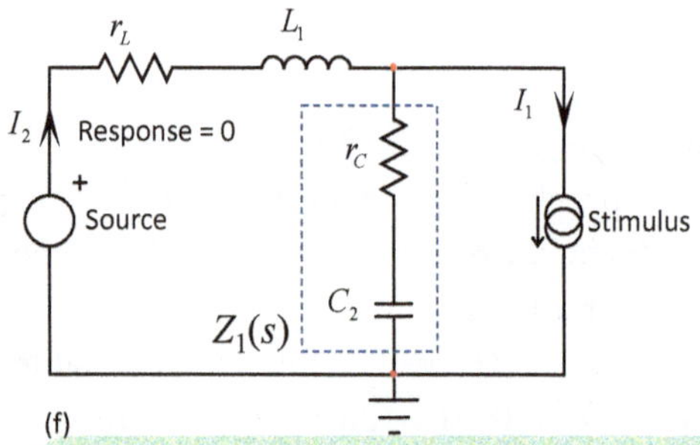
r_L
L_1
I_2
Response = 0
I_1
r_C
Source
Stimulus
C_2
$Z_1(s)$
(f)

Bring the stimulus I_1 back and check what condition could null the response I_2? If Z_1 becomes a transformed short, the response is nulled.

$$Z_1(s) = r_C + \frac{1}{sC_2} = 0 \rightarrow s_z = -\frac{1}{r_C C_2}$$

$$\omega_z = \frac{1}{r_C C_2}$$

$$N(s) = 1 + \frac{s}{\omega_z}$$

$$H(s) = \frac{1 + s r_C C_2}{1 + sC_2 (r_C + r_L) + s^2 L_1 C_2} = \frac{1 + \frac{s}{\omega_z}}{1 + \left(\frac{s}{\omega_0 Q}\right) + \left(\frac{s}{\omega_0}\right)^2}$$

$$Q = \frac{\sqrt{L_1 C_2}}{C_2 (r_L + r_C)} \qquad \omega_0 = \frac{1}{\sqrt{L_1 C_2}}$$

In high-frequencies, beyond resonant frequency, the transfer function can be simplified as:

$$H(s) \approx \frac{1}{\left(\frac{s}{\omega_0}\right)^2} \rightarrow |H(\omega)| \approx \left(\frac{\omega_0}{\omega}\right)^2$$

If A_{filter} is the attenuation needed at ω then position the cutoff frequency ω_0 at:

$$\omega_0 = \sqrt{A_{filter}} \cdot \omega$$

The Circuit Must Include Equivalent Series Resistors Which Significantly Affect Attenuation

Figure 6.20

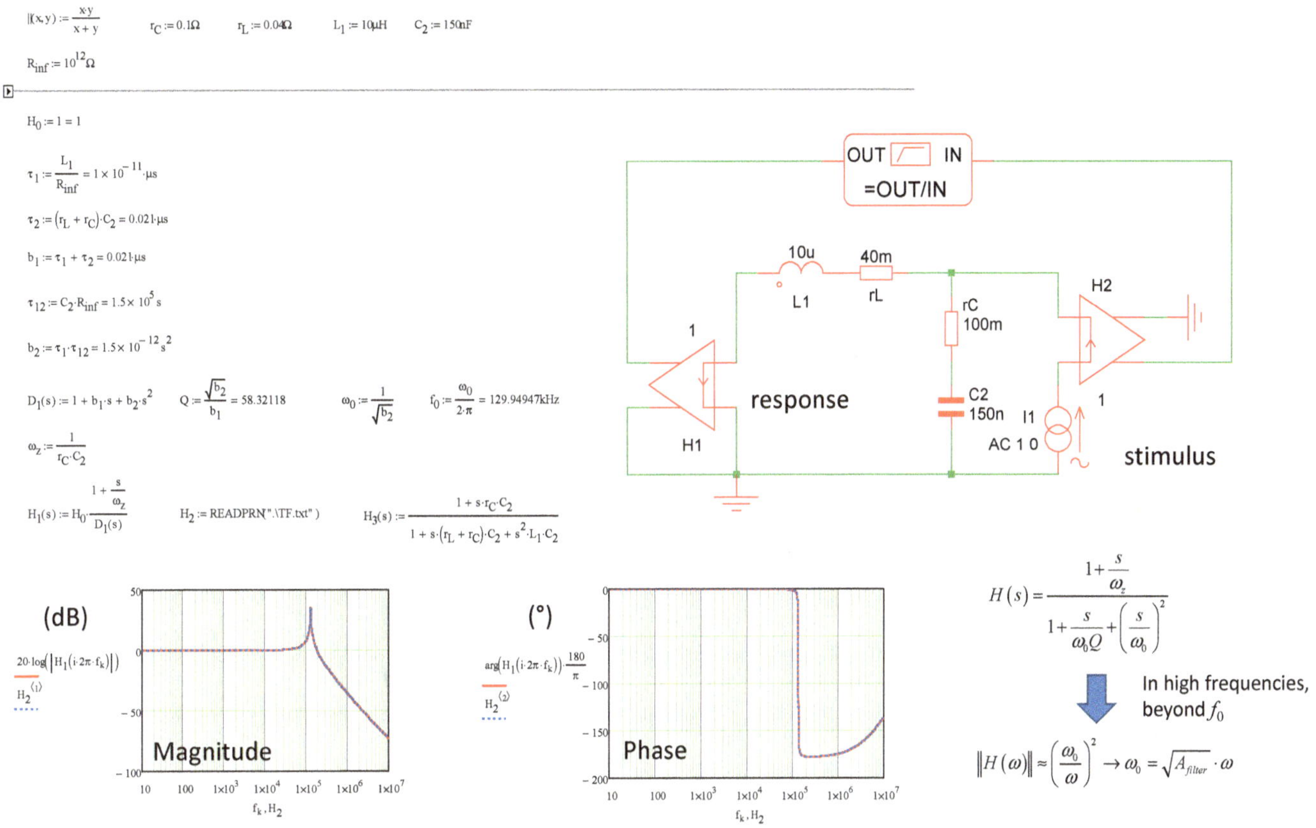
(dB)
Magnitude
(°)
Phase
response
stimulus

$\|(x,y) := \dfrac{x \cdot y}{x + y}$ $r_C := 0.1\Omega$ $r_L := 0.04\Omega$ $L_1 := 10\mu H$ $C_2 := 150nF$

$R_{inf} := 10^{12}\Omega$

$H_0 := 1 = 1$

$\tau_1 := \dfrac{L_1}{R_{inf}} = 1 \times 10^{-11} \cdot \mu s$

$\tau_2 := (r_L + r_C) \cdot C_2 = 0.021 \cdot \mu s$

$b_1 := \tau_1 + \tau_2 = 0.021 \cdot \mu s$

$\tau_{12} := C_2 \cdot R_{inf} = 1.5 \times 10^5 \, s$

$b_2 := \tau_1 \cdot \tau_{12} = 1.5 \times 10^{-12} \, s^2$

$D_1(s) := 1 + b_1 \cdot s + b_2 \cdot s^2$ $Q := \dfrac{\sqrt{b_2}}{b_1} = 58.32118$ $\omega_0 := \dfrac{1}{\sqrt{b_2}}$ $f_0 := \dfrac{\omega_0}{2 \cdot \pi} = 129.94947 kHz$

$\omega_z := \dfrac{1}{r_C \cdot C_2}$

$H_1(s) := H_0 \cdot \dfrac{1 + \dfrac{s}{\omega_z}}{D_1(s)}$ $H_2 := READPRN(" .\backslash TF.txt")$ $H_3(s) := \dfrac{1 + s \cdot r_C \cdot C_2}{1 + s \cdot (r_L + r_C) \cdot C_2 + s^2 \cdot L_1 \cdot C_2}$

(diagram: OUT / IN = OUT/IN; L1 10u, rL 40m, rC 100m, C2 150n, H2, response, H1, I1 AC 1 0, stimulus)

(dB) Magnitude plot: $20 \cdot \log(|H_1(i \cdot 2 \cdot \pi \cdot f_k)|)$, $H_2^{\langle 1 \rangle}$; x-axis f_k, H_2

(°) Phase plot: $arg(H_1(i \cdot 2 \cdot \pi \cdot f_k)) \cdot \dfrac{180}{\pi}$, $H_2^{\langle 2 \rangle}$; x-axis f_k, H_2

$H(s) = \dfrac{1 + \dfrac{s}{\omega_z}}{1 + \dfrac{s}{\omega_0 Q} + \left(\dfrac{s}{\omega_0}\right)^2}$

In high frequencies, beyond f_0

The Final Expression Must be Simplified to Serve our Design Purposes

Figure 6.21

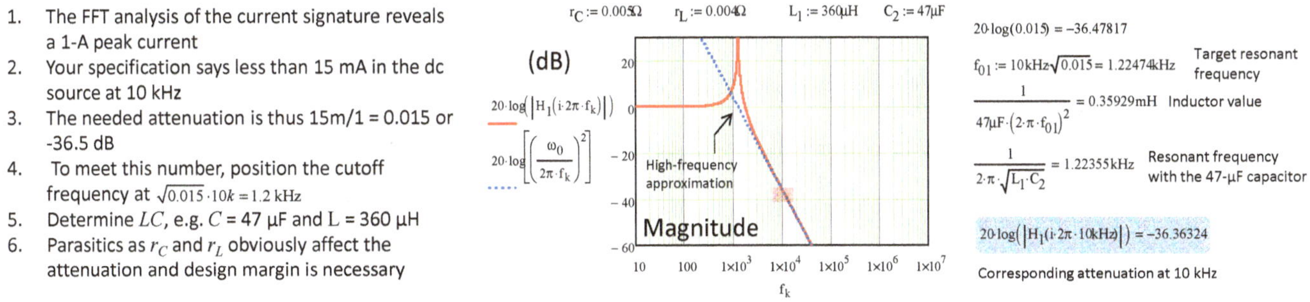
1. The FFT analysis of the current signature reveals a 1-A peak current
2. Your specification says less than 15 mA in the dc source at 10 kHz
3. The needed attenuation is thus 15m/1 = 0.015 or -36.5 dB
4. To meet this number, position the cutoff frequency at $\sqrt{0.015}\cdot 10k = 1.2$ kHz
5. Determine LC, e.g. C = 47 µF and L = 360 µH
6. Parasitics as r_C and r_L obviously affect the attenuation and design margin is necessary
(dB)
High-frequency approximation
Magnitude
Target resonant frequency
Inductor value
Resonant frequency with the 47-µF capacitor
Corresponding attenuation at 10 kHz

In this Example, the Cutoff Frequency is Set to 1.2 kHz and Brings an Attenuation of 36.4 dB at 10 kHz

Figure 6.22

Another important parameter for this EMI filter is its output impedance.

It is well-known that the negative incremental resistance of a switching converter can interact with the output impedance of the front-end filter and bring instabilities.

For this reason, it is important to determine the output impedance of the filter and study its peaking. The determination of the output impedance follows a similar path as illustrated in Figure 6.23. To avoid an indeterminacy in cases where the resistance seen from the connecting terminals of L_1 or C_2 is infinite, I actually replace the infinite value by a high-valued finite term R_{inf}.

This way, I can later simplify expressions in which the term appears in the numerator and the denominator. This is a very useful trick whose counterpart is R_s, the smallest possible resistance you can use instead of a 0-Ω wire.

(a)

(b)

Turn the excitation off - $I_T = 0$ - and determine the time constants:

Find R when C_2 is in dc state

$$\tau_1 = \frac{L_1}{R_{inf}}$$

Find R when L_1 is in dc state

$$\tau_2 = C_2 (r_C + r_L)$$

L_1 is in high-frequency state: determine R

$$\tau_2^1 = C_2 R_{inf}$$

(c)   (d)   (e)

$$D(s) = 1 + s(\tau_1 + \tau_2) + s^2 \tau_1 \tau_2^1 = 1 + s\left[\frac{L_1}{R_{inf}} + C_2(r_C + r_L)\right] + \frac{L_1}{R_{inf}} C_2 R_{inf} = 1 + s(r_C + r_L)C_2 + s^2 L_1 C_2$$

Null the response V_T across the test generator I_T: what impedance combination could zero the response?

The response is zeroed when one of the impedance becomes a transformed short.

$$Z_1(s) = r_L + sL_1 = 0$$

$$s_{z_1} = -\frac{r_L}{L_1} \rightarrow \omega_{z_1} = \frac{r_L}{L_1}$$

$$Z_2(s) = r_C + \frac{1}{sC_2} = 0$$

$$s_{z_1} = -\frac{1}{r_C C_2} \rightarrow \omega_{z_2} = \frac{1}{r_C C_2}$$

(f)   (g)

$$Z_{out}(s) = r_L \frac{\left(1 + s\frac{r_L}{L_1}\right)(1 + sr_C C_2)}{1 + s(r_C + r_L)C_2 + s^2 L_1 C_2} = R_0 \frac{\left(1 + \frac{s}{\omega_{z_1}}\right)\left(1 + \frac{s}{\omega_{z_2}}\right)}{1 + \left(\frac{s}{\omega_0 Q}\right) + \left(\frac{s}{\omega_0}\right)^2}$$

$$\omega_0 = \frac{1}{L_1 C_2}$$

$$Q = \frac{1}{r_L + r_C}\sqrt{\frac{L_1}{C_2}}$$

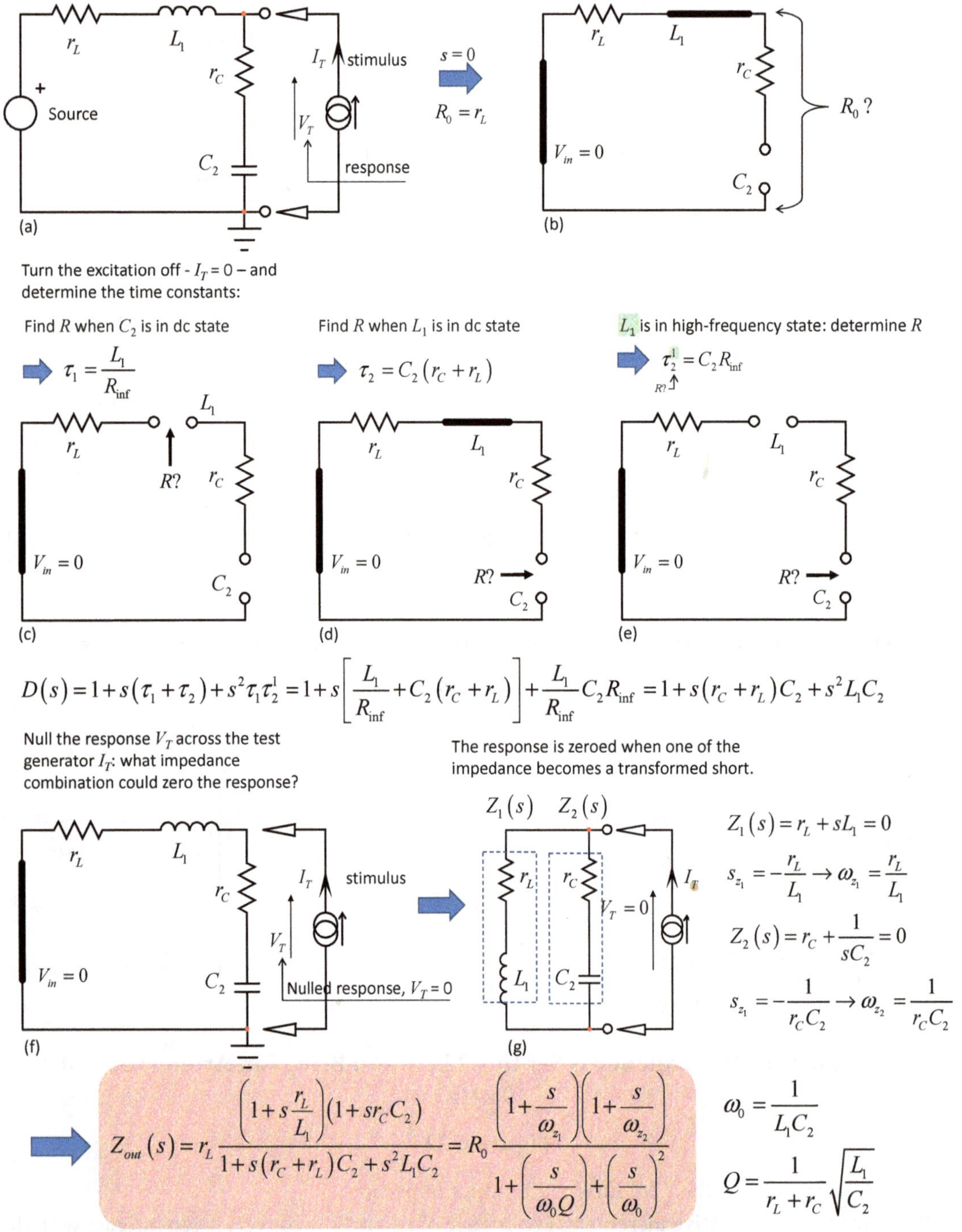

It is Important to Evaluate the Peaking of the EMI Filter Output Impedance

Figure 6.23

The ac response is given in Figure 6.24 and confirms peaking at the resonant frequency.

Determine the peak impedance value by differentiating the magnitude with respect to frequency. The resulting expression is shown at the bottom of the figure and confirms a peak above 50 dBΩ. If you now connect a downstream dc-dc switching converter to this filter, you must study its input impedance to prevent any overlap with the front-end output impedance.

For instance, assume an input impedance of 45 dBΩ for the switching converter. When connected to the filter we designed in this example, oscillations or instabilities would likely be revealed at power on: damping is necessary as explained in [1].

$$\|(x,y) := \frac{x \cdot y}{x + y} \qquad r_C := 0.1\Omega \qquad r_L := 0.04\Omega \qquad L_1 := 10\mu H \qquad C_2 := 150nF$$

$$R_{inf} := 10^{12}\Omega$$

$$R_0 := r_L \qquad 20 \cdot \log\left(\frac{R_0}{\Omega}\right) = -27.9588 \quad dBohms$$

$$\tau_1 := \frac{L_1}{R_{inf}} = 1 \times 10^{-11} \cdot \mu s$$

$$\tau_2 := (r_L + r_C) \cdot C_2 = 0.021 \cdot \mu s$$

$$b_1 := \tau_1 + \tau_2 = 0.021 \cdot \mu s$$

$$\tau_{12} := C_2 \cdot R_{inf} = 1.5 \times 10^{5} s$$

$$b_2 := \tau_1 \cdot \tau_{12} = 1.5 \times 10^{-12} s^2$$

$$D_1(s) := 1 + b_1 \cdot s + b_2 \cdot s^2 \qquad Q := \frac{\sqrt{b_2}}{b_1} = 58.32118 \qquad Q_a := \frac{1}{r_L + r_C}\sqrt{\frac{L_1}{C_2}} = 58.32118$$

$$\omega_0 := \frac{1}{\sqrt{b_2}} \qquad f_0 := \frac{\omega_0}{2 \cdot \pi} = 129.94947 kHz$$

$$\omega_{z1} := \frac{r_L}{L_1} \qquad \omega_{z2} := \frac{1}{r_C \cdot C_2} \qquad f_{z1} := \frac{\omega_{z1}}{2 \cdot \pi} = 636.61977 Hz \qquad f_{z2} := \frac{\omega_{z2}}{2 \cdot \pi} = 10.61033 MHz$$

$$Z_{out}(s) := R_0 \cdot \frac{\left(1 + \frac{s}{\omega_{z1}}\right)\left(1 + \frac{s}{\omega_{z2}}\right)}{D_1(s)} \qquad H_2 := READPRN("\TF.txt") \qquad H_3(s) := \frac{1 + s \cdot r_C \cdot C_2}{1 + s \cdot (r_L + r_C) \cdot C_2 + s^2 \cdot L_1 \cdot C_2}$$

$$Z_{peak} := \frac{R_0 \cdot Q}{\omega_{z1} \cdot \omega_{z2}}\sqrt{\left(\omega_0^2 + \omega_{z1}^2\right) \cdot \left(\omega_0^2 + \omega_{z2}^2\right)} \qquad 20 \cdot \log\left(\frac{Z_{peak}}{\Omega}\right) = 53.55637 \quad dBohms$$

$$20 \cdot \log\left(\left|\frac{Z_{out}(i \cdot \omega_0)}{\Omega}\right|\right) = 53.55637 \quad dBohms$$

$$20 \cdot \log\left(\left|\frac{Z_{out}(i \cdot 2\pi \cdot f_k)}{\Omega}\right|\right)$$

$$\arg\left(Z_{out}(i \cdot 2\pi \cdot f_k)\right) \cdot \frac{180}{\pi}$$

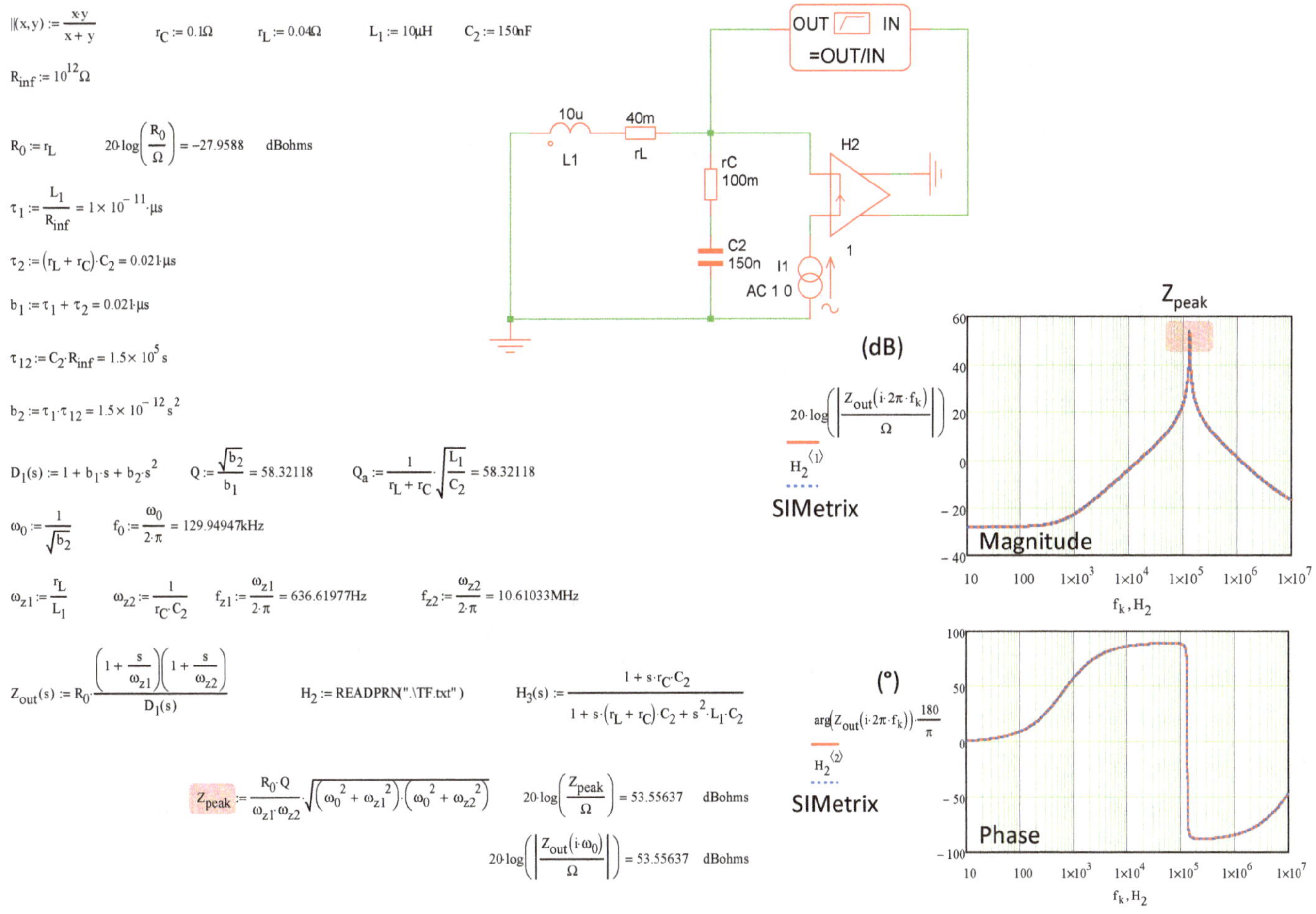

The Magnitude Peaks at Resonance and Must be Studied before Connecting a Downstream Dc-Dc Switching Converter

Figure 6.24

We now look at an active filter such as the one shown in Figure 6.25.

The op-amp is considered perfect and both inputs share a common potential.

I replaced the op-amp symbol with a simple voltage-controlled source whose output equals the voltage at the non-inverting pin.

The dc gain is instantly obtained and the rest of the time constants comes easily.

In sketch (f), I reshuffled the combination into a simpler arrangement for which the answer is straightforward: this is a typical example where redundancy helps us proceed faster.

(a) Turn V_{in} off – set it to 0 V

Find R when C_1 is in dc state

$\tau_2 = C_2(R_1 + R_2)$

$A_{OL} \to \infty$

Voltage follower
$V_{out} = V_{(+)}$
Replace op-amp by simple unity-gain
voltage-controlled source

(b) $s = 0$
$H_0 = 1$

Determining the resistance by inspection is not
that obvious. Install a test generator I_T and
determine the voltage across its terminals.

(c)

(d)

$V_T = R_1 I_T - R_1 I_T = 0$

$R = \dfrac{V_T}{I_T} = \dfrac{0}{I_T} = 0$

$\tau_1 = 0 \cdot C_1$

(e)

C_1 in HF state

(f)

Now set C_1 in its high-frequency state and determine τ_2^1

Not obvious by inspection, try τ_1^2

(g)

There are no zeroes in this circuit: if you place
C_1 or C_2 in their high-frequency state, no response.

$\tau_1^2 \leftarrow R?$

$R = R_1 \| R_2$
$\tau_1^2 = RC_1 = (R_1 \| R_2)C_1$

$D(s) = 1 + s(\tau_1 + \tau_2) + s^2 \tau_1 \tau_2^1 = 1 + s(\tau_1 + \tau_2) + s^2 \tau_2 \tau_1^2$

$D(s) = 1 + s[0 \cdot C_1 + C_2(R_1 + R_2)] + s^2 C_2(R_1 + R_2)C_1(R_1 \| R_2)$

$D(s) = 1 + sC_2(R_1 + R_2) + s^2 C_1 C_2 R_1 R_2$

$H(s) = \dfrac{1}{1 + sC_2(R_1 + R_2) + s^2 C_1 C_2 R_1 R_2}$

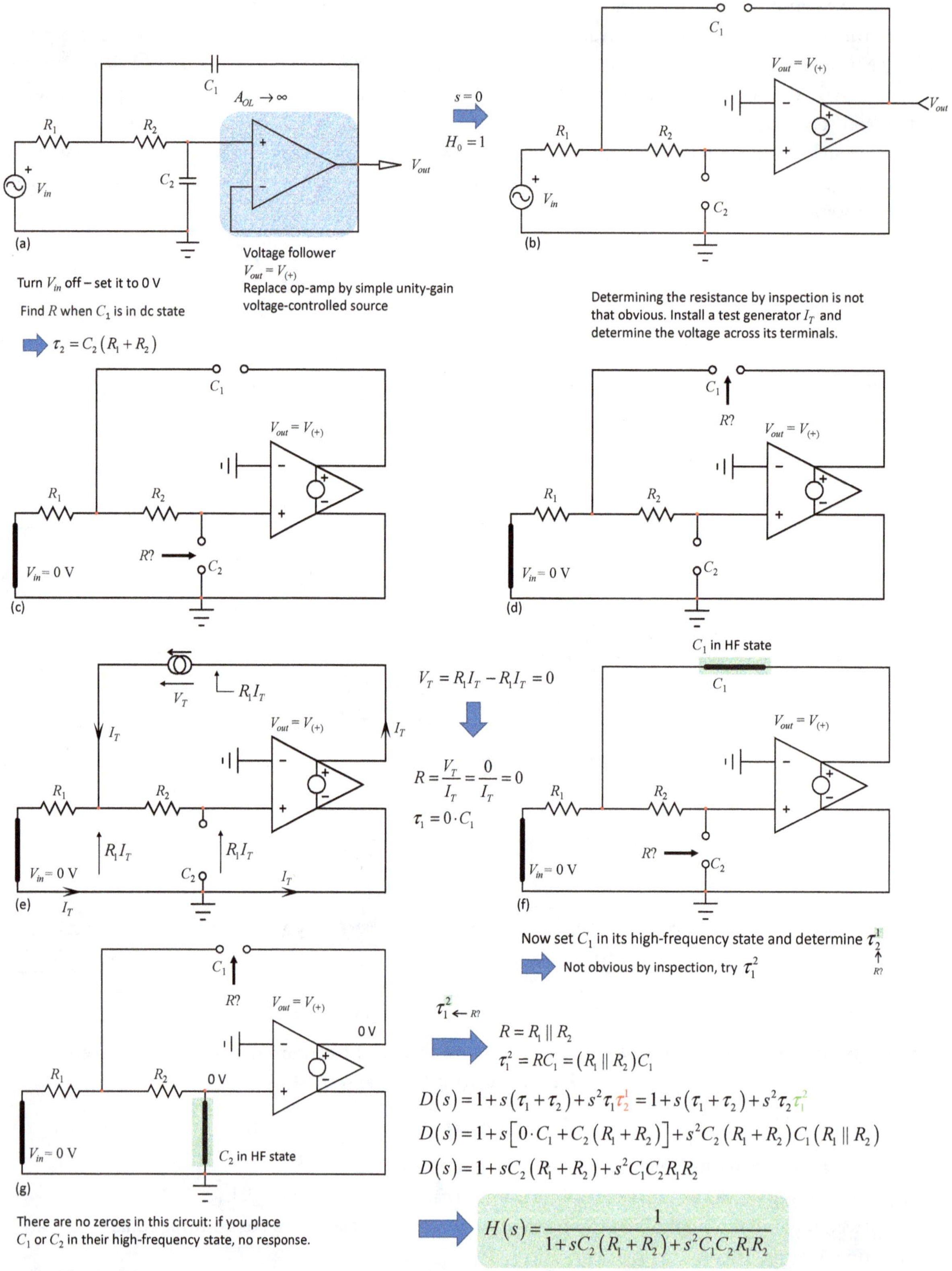

This Second-Order Sallen-Key Filter can be Analyzed with the FACTs

Figure 6.25

The ac response of the filter is given in Figure 6.26 and confirms the analysis is correct.

$\|(x,y) := \dfrac{x \cdot y}{x + y}$ $R_1 := 1.3k\Omega$ $R_2 := 1.3k\Omega$ $C_1 := 22nF$ $C_2 := 47nF$

$H_0 := 1$

$\tau_1 := 0 \cdot C_1 = 0 \cdot \mu s$

$\tau_2 := (R_1 + R_2) \cdot C_2 = 122.2 \mu s$

$b_1 := \tau_1 + \tau_2 = 122.2 \mu s$

$\tau_{21} := (R_1 \| R_2) \cdot C_1 = 1.43 \times 10^{-5} s$

$b_2 := \tau_2 \cdot \tau_{21} = 1.74746 \times 10^{-9} s^2$

$D_1(s) := 1 + b_1 \cdot s + b_2 \cdot s^2$ $Q := \dfrac{\sqrt{b_2}}{b_1} = 0.34208$ $\omega_0 := \dfrac{1}{\sqrt{b_2}}$ $f_0 := \dfrac{\omega_0}{2 \cdot \pi} = 3.80729 kHz$

$H_1(s) := H_0 \cdot \dfrac{1}{D_1(s)}$ $H_2 := \mathrm{READPRN}(".\backslash TF.txt")$ $H_3(s) := \dfrac{1}{1 + s \cdot C_2 \cdot (R_1 + R_2) + s^2 \cdot R_1 \cdot R_2 \cdot C_1 \cdot C_2}$

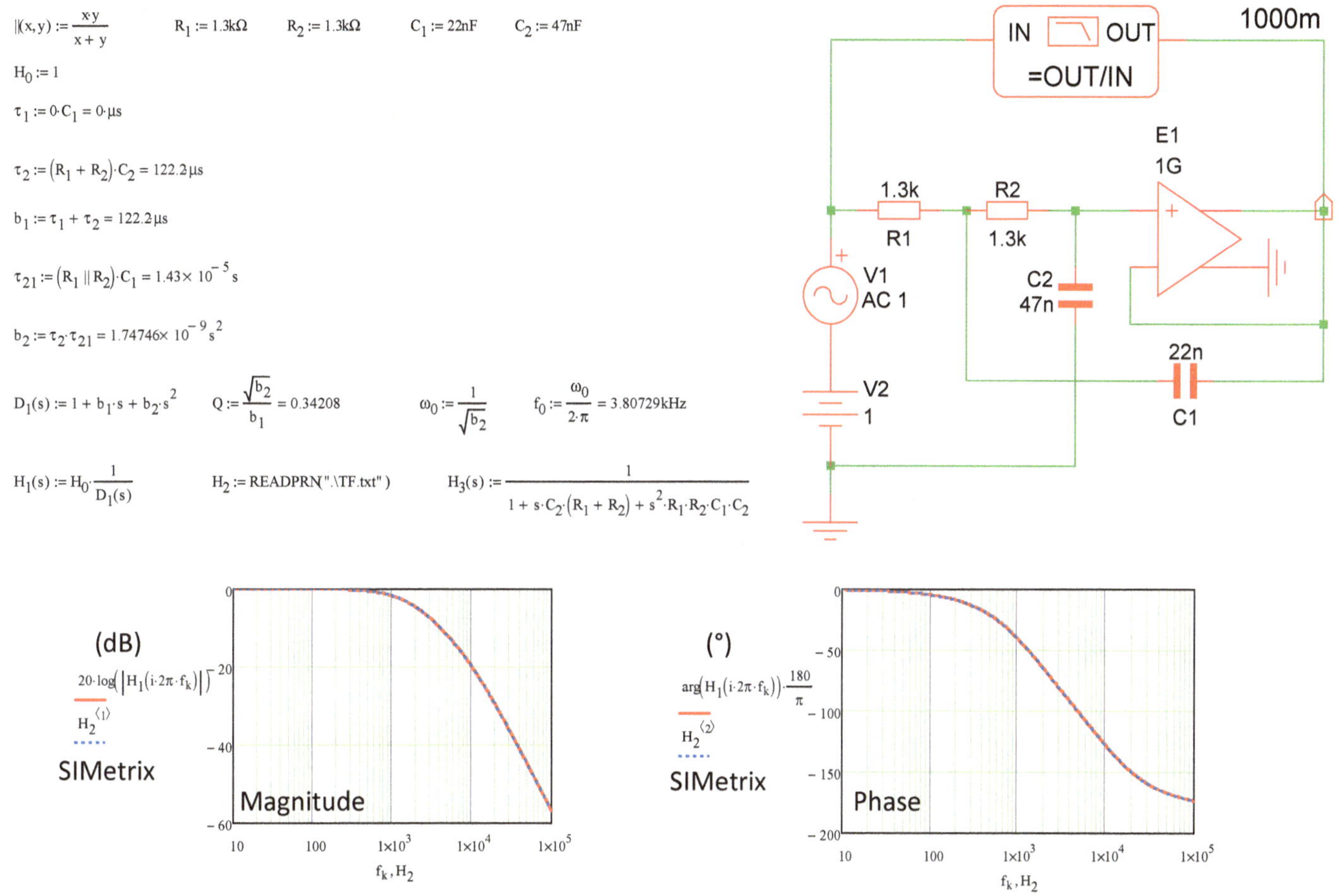

The Transfer Function is Quickly Obtained After a Few Steps

Figure 6.26

We now look at a classical *RLC* filter shown in Figure 6.27.

The presence of L_2 tells you immediately that the gain at dc is zero (L_2 in dc short circuits the response) but C_1 also implies a zeroed gain as *s* approaches infinity: this is a bandpass filter.

It is important to rearrange the transfer function so that the mid-band gain and the resonant frequency are clearly expressed.

From this expression, you will set the component values to meet your design goal; something a raw expression would not let you do easily.

The ac response is given in Figure 6.28.

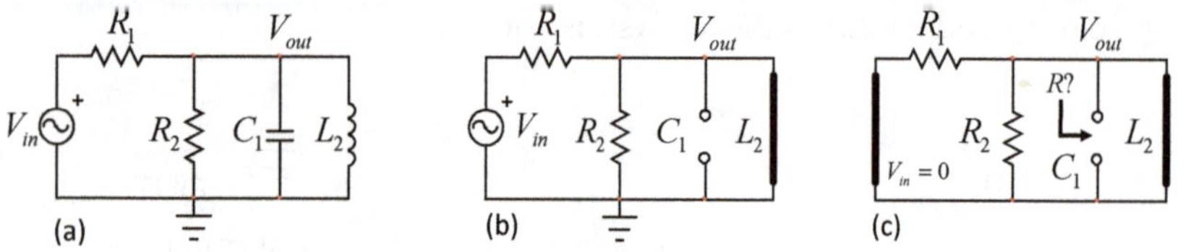

Set C_1 in its dc state

Open the capacitor and short the inductor

$s=0 \Rightarrow H_0=0$

Set V_{in} to 0 V and short L_2

$\Rightarrow \tau_1 = 0 \cdot C_1$

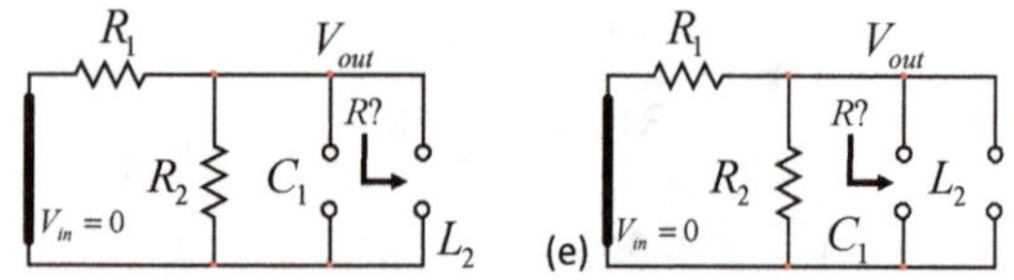

$\Rightarrow \tau_2 = L_2/(R_1 \| R_2)$

(d)

Set L_2 in its high-frequency state and determine R:

$\Rightarrow \tau_1^2 = C_1(R_1 \| R_2)$

$$D(s)=1+s(\tau_1+\tau_2)+s^2\tau_1\tau_2^1 = 1+s(\tau_1+\tau_2)+s^2\tau_2\tau_1^2$$

$$D(s)=1+s\left[0\cdot C_1+\frac{L_2}{R_1 \| R_2}\right]+s^2\frac{L_2}{R_1 \| R_2}C_1(R_1 \| R_2)$$

$$D(s)=1+s\frac{L_2}{R_1 \| R_2}+s^2C_1L_2$$

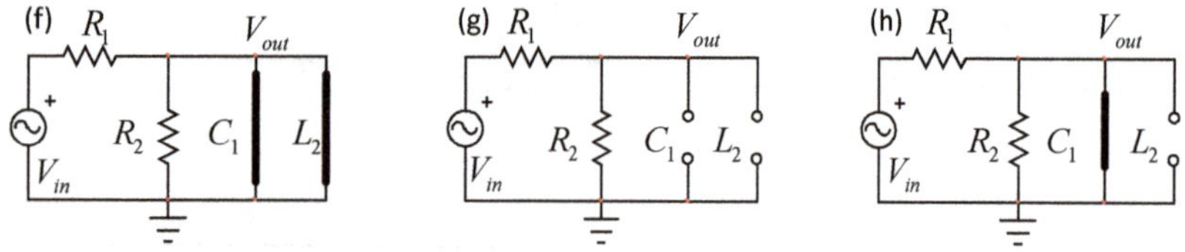

To obtain the zeroes, we will determine high-frequency gains with C_1 in its HF state and L_2 in its dc state:

$\Rightarrow H^1=0$

C_1 in its DC state and L_2 in its high-frequency state:

$\Rightarrow H^2=\frac{R_2}{R_1+R_2}$

C_1 and L_2 are in their high-frequency state:

$\Rightarrow H^{12}=0$

$$N(s)=H_0+s(H^1\tau_1+H^2\tau_2)+s^2H^{21}\tau_2\tau_1^2$$

$$N(s)=0+s\left(0\cdot\tau_1+\frac{R_2}{R_1+R_2}\frac{L_2}{R_1 \| R_2}\right)+s^2 0\cdot\tau_2\tau_1^2$$

$$N(s)=s\frac{L_2}{R_1}=\frac{s}{\omega_z}$$

$$H(s)=\frac{s\frac{L_2}{R_1}}{1+s\frac{L_2}{R_1 \| R_2}+s^2C_1L_2}=\frac{\frac{s}{\omega_z}}{1+\frac{s}{\omega_0 Q}+\left(\frac{s}{\omega_0}\right)^2}$$

$$H(s)=\frac{s}{\omega_z}\frac{1}{\frac{s}{\omega_0 Q}\left(\frac{\omega_0 Q}{s}+1+\frac{s^2}{\omega_0^2}\frac{\omega_0 Q}{s}\right)}=H_{MB}\frac{1}{1+Q\left(\frac{\omega_0}{s}+\frac{s}{\omega_0}\right)}$$

$$\omega_z=\frac{R_1}{L_2} \quad \omega_o=\frac{1}{\sqrt{L_2C_1}} \quad Q=(R_1 \| R_2)\sqrt{\frac{C_1}{L_2}} \quad H_{MB}=\frac{Q\omega_0}{\omega_z}=\frac{R_2}{R_1+R_2}$$

Mid-band gain

This *RLC* Filter is a Classic and can be Analyzed Painlessly with the FACTs

Figure 6.27

$L_2 := 1\text{mH} \quad C_1 := 22\text{nF} \quad R_1 := 1\text{k}\Omega \quad R_2 := 1\text{k}\Omega \quad \|(x,y) := \frac{x \cdot y}{x+y}$

$H_0 := 0$

$\tau_1 := 0 \cdot C_1 = 0 \cdot \mu s \qquad \tau_2 := \frac{L_2}{R_1 \| R_2} = 2 \cdot \mu s$

$b_1 := \tau_1 + \tau_2 = 2 \cdot \mu s$

$\tau_{21} := C_1 \cdot (R_1 \| R_2) = 11 \cdot \mu s$

$b_2 := \tau_2 \cdot \tau_{21} = 22 \cdot \mu s^2 \qquad L_2 \cdot C_1 = 22 \cdot \mu s^2 \qquad \omega_z := \frac{R_1}{L_2}$

$H_1 := 0 \qquad H_2 := \frac{R_2}{R_2 + R_1} \qquad H_{12} := 0 \qquad \omega_0 := \frac{1}{\sqrt{L_2 \cdot C_1}} \qquad Q := (R_1 \| R_2) \cdot \sqrt{\frac{C_1}{L_2}}$

$H_1(s) := \frac{H_0 + H_2 \cdot \tau_2 \cdot s}{1 + b_1 \cdot s + b_2 \cdot s^2} \qquad H_3(s) := \frac{\frac{L_2}{R_1} \cdot s}{1 + b_1 \cdot s + b_2 \cdot s^2} \qquad H_4(s) := \frac{\frac{s}{\omega_z}}{1 + \frac{s}{\omega_0 \cdot Q} + \left(\frac{s}{\omega_0}\right)^2}$

$H_5(s) := \frac{\frac{s}{\omega_z}}{1 + s \cdot \frac{L_2}{R_1 \| R_2} + s^2 \cdot L_2 \cdot C_1} \qquad H_{mb} := \frac{\omega_0 \cdot Q}{\omega_z} = 0.5 \qquad 20 \cdot \log(H_{mb}) = -6.021$

$H_{mb} = \frac{R_2}{R_1 + R_2}$

$H_6(s) := H_{mb} \cdot \frac{1}{1 + Q \cdot \left(\frac{s}{\omega_0} + \frac{\omega_0}{s}\right)}$

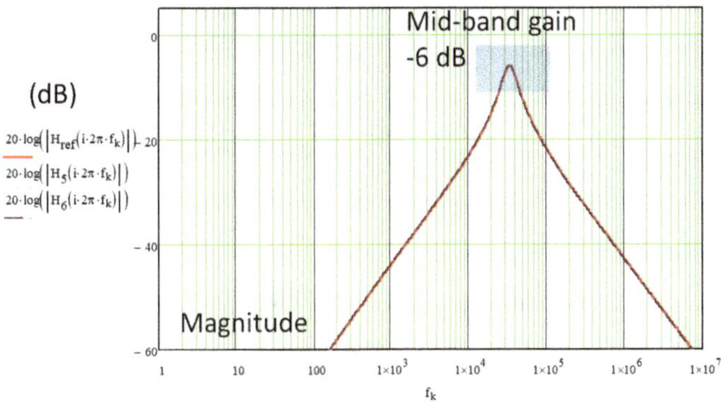

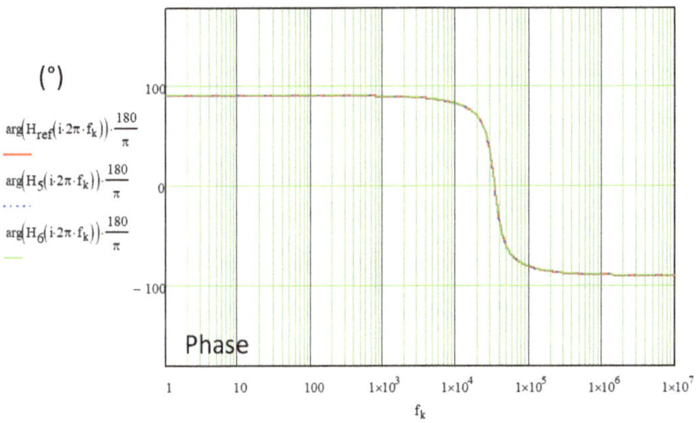

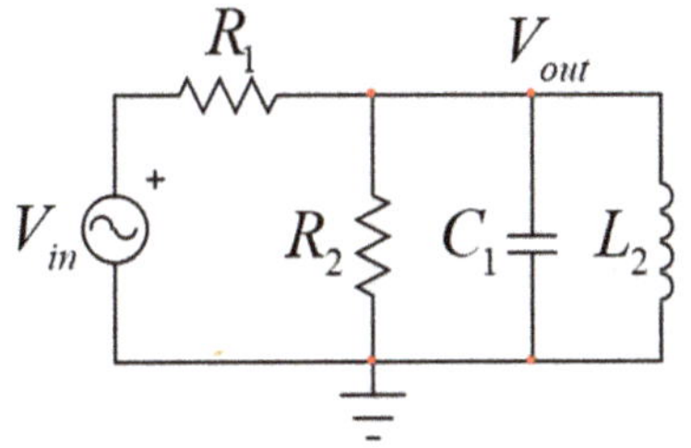

$H_{ref}(s) := \frac{(s \cdot L_2) \| \left(\frac{1}{s \cdot C_1}\right) \| R_2}{R_1 + (s \cdot L_2) \| \left(\frac{1}{s \cdot C_1}\right) \| R_2}$

The Final Transfer Function is Rearranged to Reveal Design Parameters

Figure 6.28

We will now look at a slightly different arrangement where the inductor and the capacitor are connected in series.

In Figure 6.29, a series resistor R_s models the ohmic losses while a parallel resistor R_p sets the gain at dc when C_1 is open.

As usual, start with the dc gain and proceed with the zeroing of the stimulus V_{in}.

Then alternately determine the resistance R driving each energy-storing element to form the denominator $D(s)$.

Finally, simple expressions define the high-frequency gains and the numerator is obtained by reusing the natural time constants. After simplification, set both D and N in a normalized polynomial form and find two different quality factors Q_D and Q_N.

The ac response is given in Figure 6.30 and shows no difference between all possible arrangements and the reference brute-force transfer function $H_{ref}(s)$.

Open the capacitor and short the inductor

Set V_{in} to 0 V and short L_2

Open C_1 and look a R driving L_2

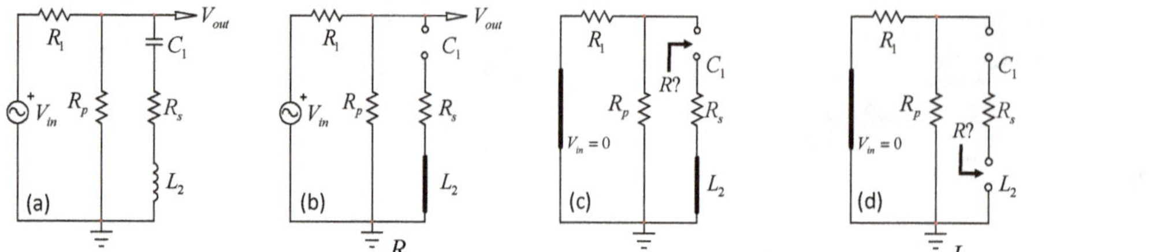

$$s=0 \Rightarrow H_0=\frac{R_p}{R_p+R_1} \qquad \tau_1=RC_1=\left(R_s+R_p \parallel R_1\right)C_1 \qquad \tau_2=\frac{L_2}{R_{\text{inf}}}$$

$$D(s)=1+s\left(\tau_1+\tau_2\right)+s^2\tau_1\tau_2^1$$

$$D(s)=1+s\left[\left(R_s+R_p \parallel R_1\right)C_1+\frac{L_2}{R_{\text{inf}}}\right]+s^2\left(R_s+R_p \parallel R_1\right)C_1\frac{L_2}{R_s+R_p \parallel R_1}$$

$$D(s)=1+s\left(R_s+R_p \parallel R_1\right)C_1+s^2C_1L_2$$

$$D(s)=1+\frac{s}{Q_D\omega_0}+\left(\frac{s}{\omega_0}\right)^2$$

$$\omega_0=\frac{1}{\sqrt{L_2C_1}}$$

$$Q_D=\frac{1}{R_s+R_1 \parallel R_p}\sqrt{\frac{L_2}{C_1}}$$

Now set C_1 in its high-frequency state and determine τ_2^1

Determine the zeroes by calculating high-frequency gains

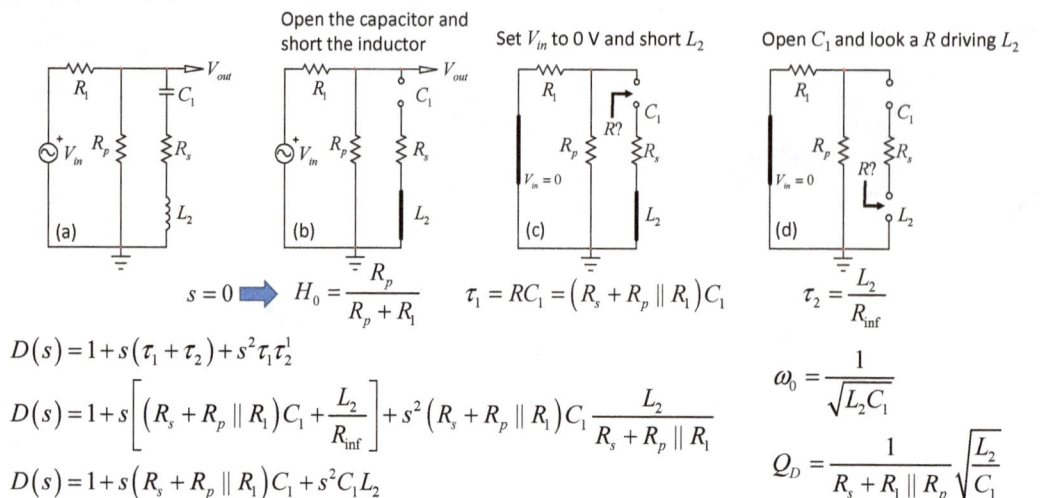

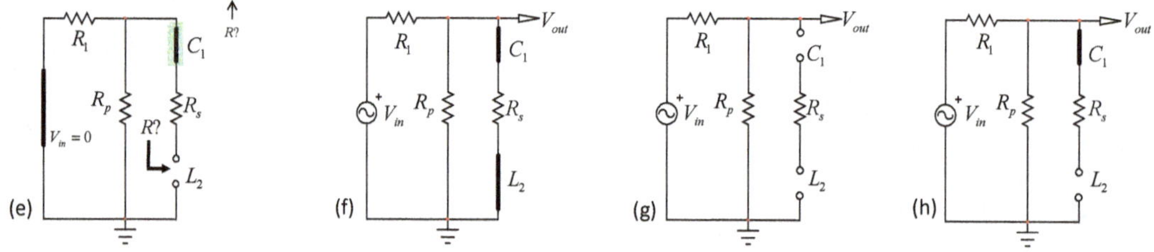

$$\tau_2^1=\frac{L_2}{R_s+R_p \parallel R_1} \qquad H^1=\frac{R_s \parallel R_p}{R_s \parallel R_p+R_1} \qquad H^2=\frac{R_p}{R_p+R_1} \qquad H^{12}=\frac{R_p}{R_p+R_1}$$

H^1: C_1 is in HF state, L_2 is in its dc state

H^2: C_1 is in dc state, L_2 is in its HF state

H^{12}: C_1 is in HF state, L_2 is in its HF state

$$N(s)=H_0+s\left(H^1\tau_1+H^2\tau_2\right)+s^2H^{12}\tau_1\tau_2^1$$

$$N(s)=H_0\left[1+s\left(\frac{H^1}{H_0}\tau_1+\frac{H^2}{H_0}\tau_2\right)+s^2\frac{H^{12}}{H_{.0}}\tau_1\tau_2^1\right]$$

$$N(s)=\frac{R_p}{R_p+R_1}\left[1+s\left(\frac{\frac{R_s \parallel R_p}{R_1+R_s \parallel R_p}}{\frac{R_p}{R_p+R_1}}\left(R_s+R_1 \parallel R_p\right)C_1+\frac{\frac{R_p}{R_p+R_1}}{\frac{R_p}{R_p+R_1}}\cdot\frac{L_2}{R_{\text{inf}}}\right)+s^2\frac{\frac{R_p}{R_p+R_1}}{\frac{R_p}{R_p+R_1}}\left(R_s+R_1 \parallel R_p\right)C_1\frac{L_2}{R_s+R_1 \parallel R_p}\right]$$

$$N(s)=1+sR_sC_1+s^2L_2C_1$$

$$N(s)=1+\left(\frac{s}{\omega_0Q_N}\right)+\left(\frac{s}{\omega_0}\right)^2 \qquad Q_N=\frac{1}{R_s}\sqrt{\frac{L_2}{C_1}}$$

$$\Rightarrow \quad H(s)=H_0\frac{1+\left(\frac{s}{\omega_0Q_N}\right)+\left(\frac{s}{\omega_0}\right)^2}{1+\left(\frac{s}{\omega_0Q_D}\right)+\left(\frac{s}{\omega_0}\right)^2}$$

The Series Resonant *LC* Network Introduces a Pair of Zeroes Located at ω_0

Figure 6.29

$L_2 := 1\text{mH} \quad C_1 := 22\text{nF} \quad R_1 := 1\text{k}\Omega \quad R_p := 1\text{k}\Omega \quad R_s := 1\Omega$

$\|(x,y) := \dfrac{x \cdot y}{x+y} \quad R_{inf} := 10^{23}\Omega$

$H_0 := \dfrac{R_p}{R_1 + R_p}$

$\tau_1 := (R_s + R_1 \| R_p) \cdot C_1 = 11.022\mu s \qquad \tau_2 := \dfrac{L_2}{R_{inf}} = 0 \cdot \mu s$

$b_1 := \tau_1 + \tau_2 = 11.022\mu s$

$\tau_{12} := \dfrac{L_2}{(R_s + R_1 \| R_p)} = 1.996\mu s$

$b_2 := \tau_1 \cdot \tau_{12} = 22 \cdot \mu s^2 \qquad L_2 \cdot C_1 = 22 \cdot \mu s^2$

$H_1 := \dfrac{R_s \| R_p}{R_1 + R_s \| R_p} \qquad H_2 := \dfrac{R_p}{R_p + R_1} \qquad H_{12} := \dfrac{R_p}{R_p + R_1}$

$H_{10}(s) := \dfrac{H_0 + (H_1 \cdot \tau_1 + H_2 \cdot \tau_2) \cdot s + s^2 \cdot H_{12} \cdot \tau_1 \cdot \tau_{12}}{1 + b_1 \cdot s + b_2 \cdot s^2} \qquad \omega_0 := \dfrac{1}{\sqrt{L_2 \cdot C_1}} \qquad Q_D := \dfrac{1}{(R_s + R_1 \| R_p)} \cdot \sqrt{\dfrac{L_2}{C_1}}$

$H_5(s) := H_0 \cdot \dfrac{1 + (R_s \cdot C_1) \cdot s + s^2 \cdot L_2 \cdot C_1}{1 + \dfrac{s}{\omega_0 \cdot Q_D} + \left(\dfrac{s}{\omega_0}\right)^2} \qquad H_6(s) := H_0 \cdot \dfrac{1 + \dfrac{s}{\omega_0 \cdot Q_N} + \left(\dfrac{s}{\omega_0}\right)^2}{1 + \dfrac{s}{\omega_0 \cdot Q_D} + \left(\dfrac{s}{\omega_0}\right)^2} \qquad Q_N := \dfrac{1}{R_s} \cdot \sqrt{\dfrac{L_2}{C_1}}$

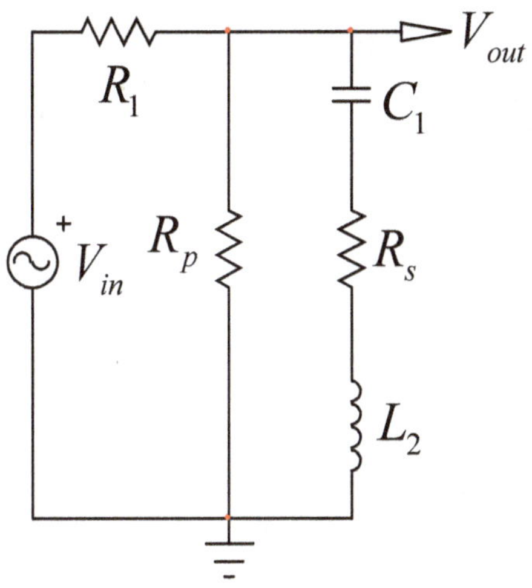

$H_{ref}(s) := \dfrac{\left(s \cdot L_2 + R_s + \dfrac{1}{s \cdot C_1}\right) \| R_p}{R_1 + \left(s \cdot L_2 + R_s + \dfrac{1}{s \cdot C_1}\right) \| R_p}$

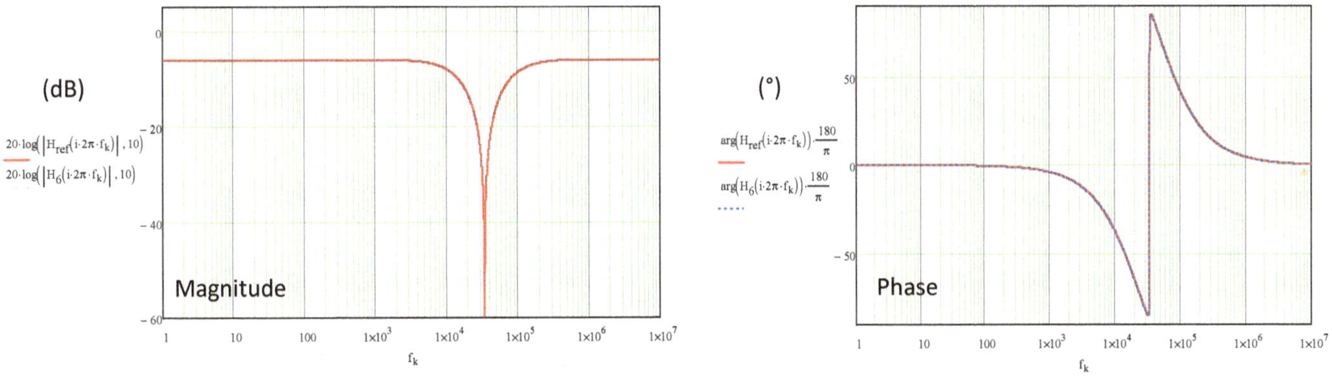

The Ac Response Shows a Notch and Confirms the FACTs led to the Correct Answer

Figure 6.30

The components are now arranged in a different way as shown in Figure 6.31.

The output voltage is collected across capacitor C_2. The presence of this element entirely shunts the response at high frequency while the inductor location sets the dc gain at zero: it is another bandpass filter.

The ac response is given in Figure 6.32.

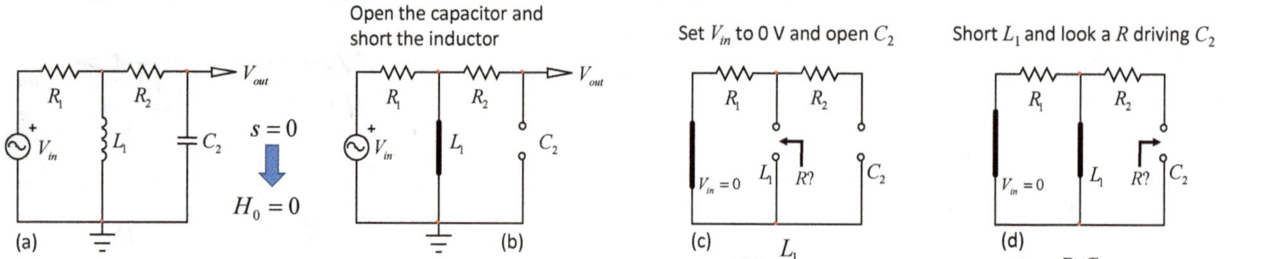
Open the capacitor and short the inductor
Set V_{in} to 0 V and open C_2
Short L_1 and look a R driving C_2
V_{out}
R_1
R_2
L_1
C_2
V_{in}
$s=0$
$H_0=0$
$V_{in}=0$
$R?$
(a)
(b)
(c)
(d)

$$\tau_1=\frac{L_1}{R_1}$$

$$\tau_2=R_2C_2$$

$$D(s)=1+s(\tau_1+\tau_2)+s^2\tau_1\tau_2^1$$

$$D(s)=1+s\left[\frac{L_1}{R_1}+R_2C_2\right]+s^2\frac{L_1}{R_1}(R_1+R_2)C_2$$

$$D(s)=1+\frac{s}{Q\omega_0}+\left(\frac{s}{\omega_0}\right)^2$$

$$\omega_0=\sqrt{\frac{R_1}{R_1+R_2}\frac{1}{L_1C_2}}$$

$$Q=\frac{\sqrt{C_2L_1R_1(R_1+R_2)}}{L_1+R_1R_2C_2}$$

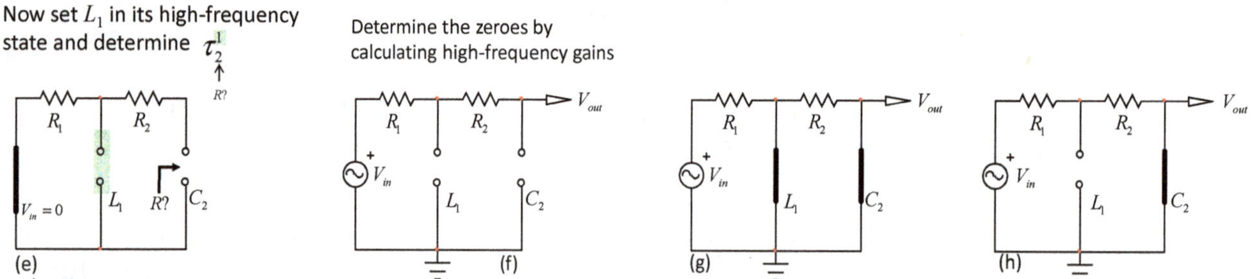
Now set L_1 in its high-frequency state and determine τ_2^1
$R?$
Determine the zeroes by calculating high-frequency gains
V_{out}
R_1
R_2
L_1
C_2
V_{in}
$V_{in}=0$
(e)
(f)
(g)
(h)

$$\tau_2^1=(R_1+R_2)C_2$$

$$H^1=1$$

C_2 is in HF state
L_1 is in its dc state

$$H^2=0$$

C_2 is in HF state
L_1 is in its dc state

$$H^{12}=0$$

C_2 is in HF state
L_1 is in its HF state

$$N(s)=H_0+s(H^1\tau_1+H^2\tau_2)+s^2H^{12}\tau_1\tau_2^1$$

$$N(s)=0+s(\tau_1+0\cdot\tau_2)+s^2\cdot 0\cdot\tau_1\tau_2^1$$

$$N(s)=s\frac{L_1}{R_1}$$

$$N(s)=\frac{s}{\omega_z}\qquad \omega_z=\frac{R_1}{L_1}$$

$$H(s)=\frac{\frac{s}{\omega_z}}{1+\frac{s}{Q\omega_0}+\left(\frac{s}{\omega_0}\right)^2}$$

$$H(s)=\frac{s}{\omega_z}\frac{1}{\frac{s}{\omega_0 Q}\left(\frac{\omega_0 Q}{s}+1+\frac{s^2}{\omega_0^2}\frac{\omega_0 Q}{s}\right)}=H_{MB}\frac{1}{1+Q\left(\frac{\omega_0}{s}+\frac{s}{\omega_0}\right)}$$

$$H_{MB}=\frac{Q\omega_0}{\omega_z}=\frac{L_1}{R_1\left(\frac{L_1}{R_1}+R_2C_2\right)}$$ Mid-band gain

This is Another Bandpass Filter with a Zero Gain at Dc

Figure 6.31

$L_1 := 1\text{mH} \quad C_2 := 22\text{nF} \quad R_1 := 1\text{k}\Omega \quad R_2 := 100\Omega \quad \|(x,y) := \frac{x \cdot y}{x+y}$

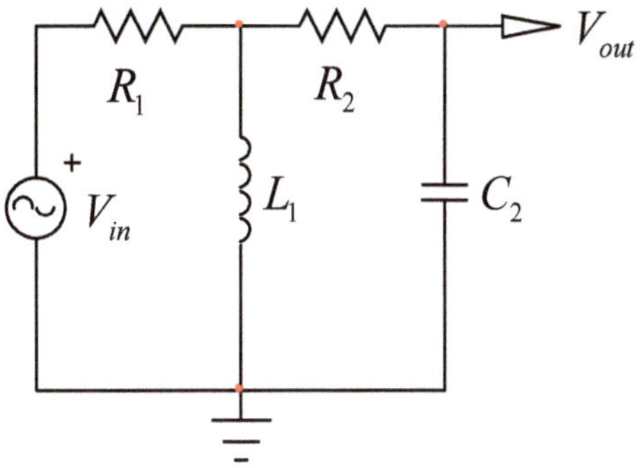

$H_0 := 0$

$\tau_1 := \frac{L_1}{R_1} = 1 \cdot \mu s \qquad \tau_2 := R_2 \cdot C_2 = 2.2 \cdot \mu s$

$b_1 := \tau_1 + \tau_2 = 3.2 \mu s$

$\tau_{12} := C_2 \cdot (R_1 + R_2) = 24.2 \mu s$

$b_2 := \tau_1 \cdot \tau_{12} = 24.2 \mu s^2$

$\omega_0 := \frac{1}{\sqrt{L_1 \cdot C_2 \cdot \frac{R_1 + R_2}{R_1}}} \qquad Q := \frac{\sqrt{b_2}}{b_1} = 1.537 \qquad \frac{\sqrt{C_2 \cdot L_1 \cdot R_1 \cdot (R_1 + R_2)}}{L_1 + R_1 \cdot R_2 \cdot C_2} = 1.537$

$H_1 := 1 \quad H_2 := 0 \quad H_{12} := 0$

$V_{th}(s) := \frac{s \cdot L_1}{R_1 + s \cdot L_1} \qquad R_{th}(s) := (s \cdot L_1) \| R_1$

$H_{2a}(s) := \frac{H_0 + (H_1 \cdot \tau_1 + H_2 \cdot \tau_2) \cdot s + s^2 \cdot H_{12} \cdot \tau_1 \cdot \tau_{12}}{1 + b_1 \cdot s + b_2 \cdot s^2} \qquad \omega_z := \frac{R_1}{L_1} \qquad H_{mb} := \frac{L_1}{R_1 \cdot \left(\frac{L_1}{R_1} + R_2 \cdot C_2 \right)} = 0.313$

$H_{ref}(s) := \frac{s \cdot L_1}{R_1 + s \cdot L_1} \cdot \frac{\frac{1}{s \cdot C_2}}{\frac{1}{s \cdot C_2} + R_{th}(s) + R_2}$

$H_4(s) := \frac{\frac{s}{\omega_z}}{1 + \frac{s}{\omega_0 \cdot Q} + \left(\frac{s}{\omega_0} \right)^2} \qquad H_5(s) := H_{mb} \cdot \frac{1}{1 + Q \cdot \left(\frac{s}{\omega_0} + \frac{\omega_0}{s} \right)} \qquad 20 \cdot \log(H_{mb}) = -10.103$

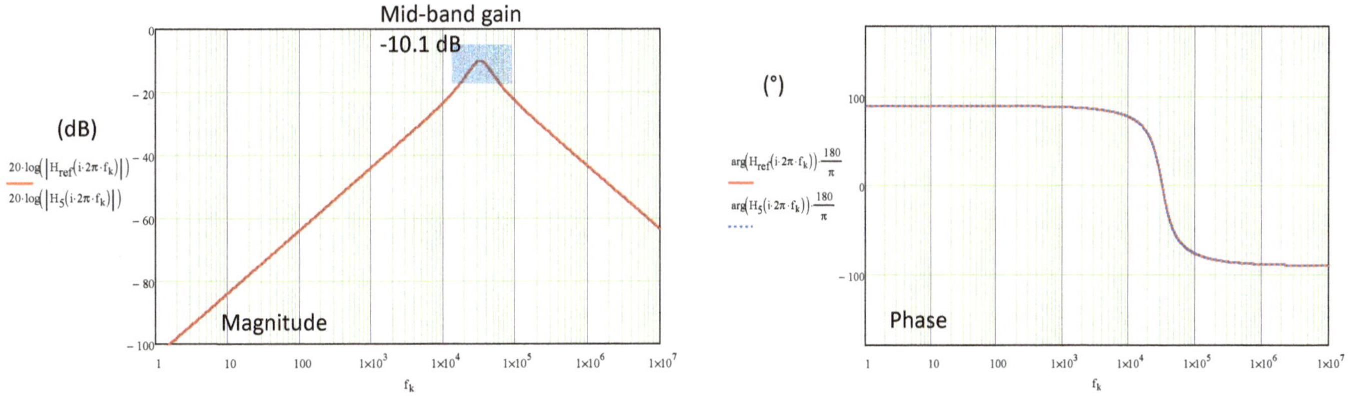

The Ac Response Obtained by the FACTs matches the Brute-Force Analysis

Figure 6.32

We will now change the probe location and consider a response delivered across L_1 rather than C_2.

By inspection, realize that changing the response observation point does not affect the natural time constants of the circuit: the denominator $D(s)$ remains unchanged and can be reused.

However, the combination of R_2 and C_2 now forms an additional zero and an update to the numerator is necessary. There you go, you have the new transfer function without restarting from scratch and the updated expression appears in Figure 6.33.

You still have the peak at resonance, but the gain lands at a value set by the resistive divider made of R_1 and R_2 when s approaches infinity.

$L_1 := 1\text{mH} \quad C_2 := 22\text{nF} \quad R_1 := 1\text{k}\Omega \quad R_2 := 100\Omega \quad \|(x,y) := \frac{x \cdot y}{x+y}$

$H_0 := 0$

$\tau_1 := \frac{L_1}{R_1} = 1\cdot\mu s \qquad \tau_2 := R_2 \cdot C_2 = 2.2\cdot\mu s$

$b_1 := \tau_1 + \tau_2 = 3.2\cdot\mu s$

$\tau_{12} := C_2 \cdot (R_1 + R_2) = 24.2\,\mu s$

$b_2 := \tau_1 \cdot \tau_{12} = 24.2\,\mu s^2$

$\omega_0 := \frac{1}{\sqrt{L_1 \cdot C_2 \cdot \frac{R_1 + R_2}{R_1}}} \qquad Q := \frac{\sqrt{b_2}}{b_1} = 1.537 \qquad \frac{\sqrt{C_2 \cdot L_1 \cdot R_1 \cdot (R_1 + R_2)}}{L_1 + R_1 \cdot R_2 \cdot C_2} = 1.537$

$H_1 := 1 \qquad H_2 := 0 \qquad H_{12} := \frac{R_2}{R_1 + R_2}$

$H_{2a}(s) := \frac{H_0 + (H_1 \cdot \tau_1 + H_2 \cdot \tau_2)\cdot s + s^2 \cdot H_{12} \cdot \tau_1 \cdot \tau_{12}}{1 + b_1 \cdot s + b_2 \cdot s^2} \qquad \omega_{z1} := \frac{R_1}{L_1} \qquad \omega_{z2} := \frac{1}{R_2 \cdot C_2}$

$H_{mb} := \frac{L_1}{R_1 \cdot \left(\frac{L_1}{R_1} + R_2 \cdot C_2\right)} = 0.313 \qquad 20\log(H_{mb}) = -10.103$

$H_4(s) := \frac{\left(\frac{s}{\omega_{z1}}\right)\cdot\left(1 + \frac{s}{\omega_{z2}}\right)}{1 + \frac{s}{\omega_0 \cdot Q} + \left(\frac{s}{\omega_0}\right)^2} \qquad H_5(s) := H_{mb} \cdot \frac{1 + \frac{s}{\omega_{z2}}}{1 + Q \cdot \left(\frac{s}{\omega_0} + \frac{\omega_0}{s}\right)}$

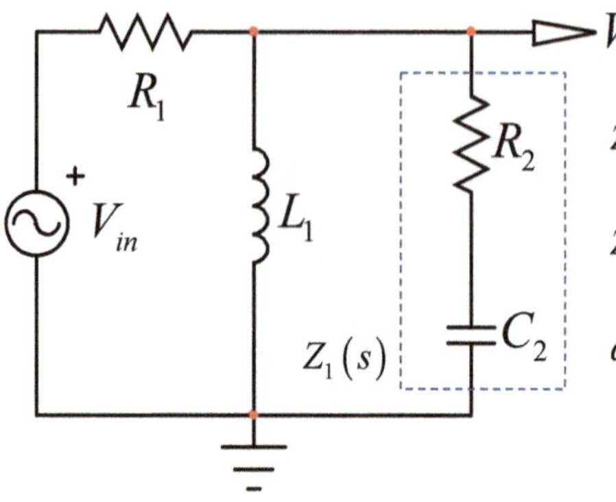

$$Z_1(s) = R_2 + \frac{1}{sC_2}$$

$$Z_1(s) = 0 \rightarrow s_z = -\frac{1}{R_2C_2}$$

$$\omega_z = \frac{1}{R_2C_2}$$

$V_{th}(s) := \frac{s \cdot L_1}{R_1 + s \cdot L_1} \qquad R_{th}(s) := (s \cdot L_1) \| R_1$

$H_{ref}(s) := \frac{s \cdot L_1}{R_1 + s \cdot L_1} \cdot \frac{\frac{1}{s \cdot C_2} + R_2}{\frac{1}{s \cdot C_2} + R_{th}(s) + R_2}$

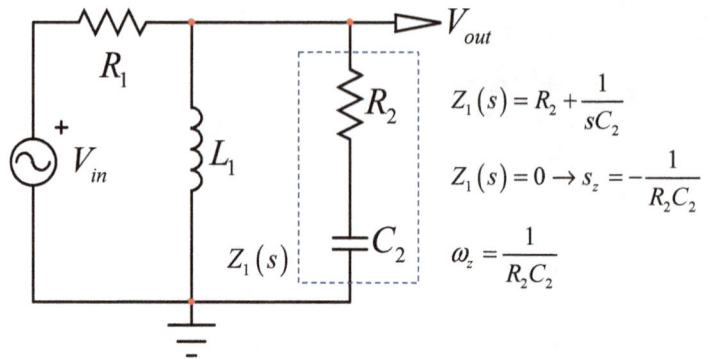

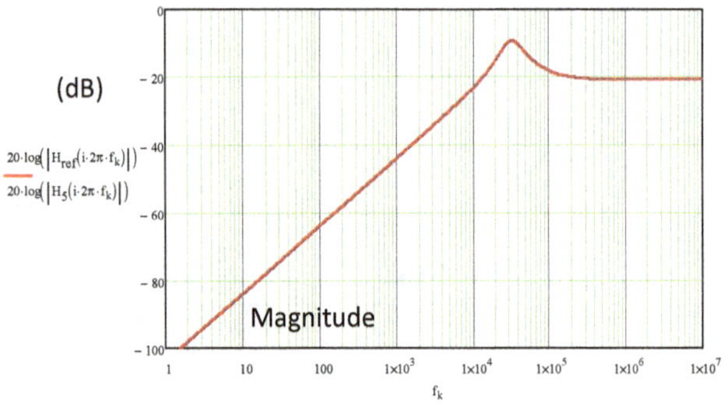

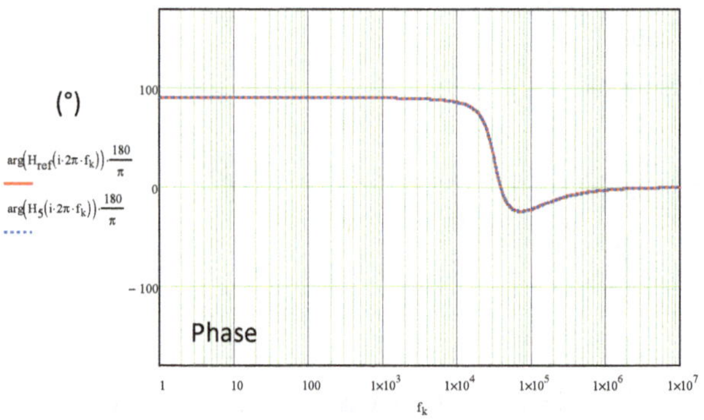

Changing the Probe Position to Sense the Response does not Change the Denominator

Figure 6.33

We now look at a classic passive bandpass filter made of resistors and capacitors as shown in Figure 6.34.

The procedure does not change and there is no trap in this simple arrangement.

The zero at the origin is determined using the high-frequency gains and, as only one remains, H^1, the numerator greatly simplifies to one coefficient only.

The result is formatted in a compact way to highlight the mid-band gain as illustrated in Figure 6.35.

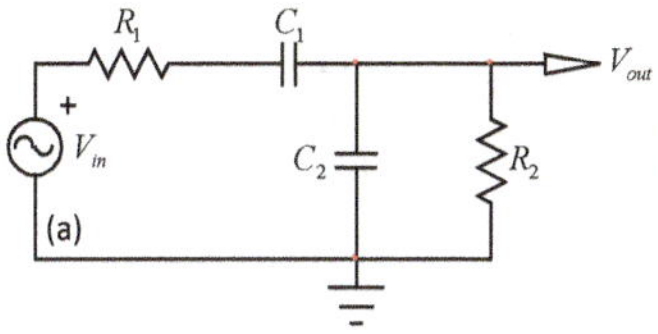

Open the capacitors $s = 0$

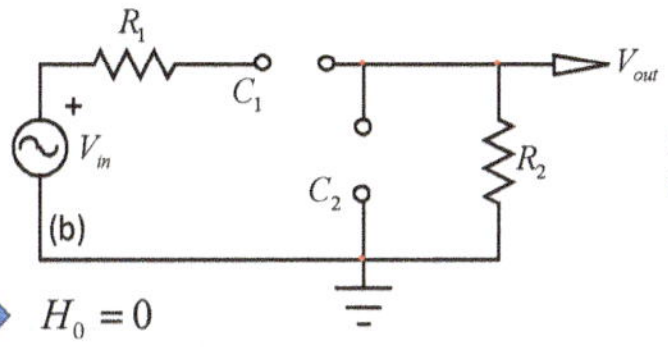

$H_0 = 0$

Set V_{in} to 0 V

Open C_1 and look a R driving C_2
Open C_2 and look a R driving C_1

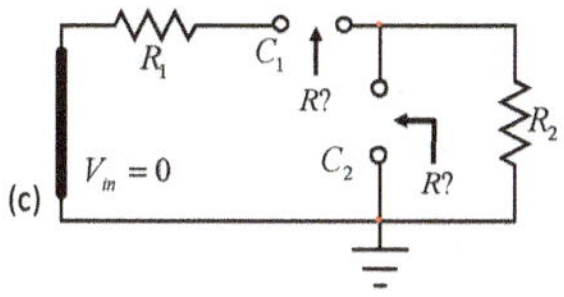

$$\tau_1 = C_1(R_1 + R_2)$$
$$\tau_2 = R_2C_2$$

Now set C_2 in its high-frequency state and determine τ_1^2

$$\tau_1^2 = C_1R_1$$

$$D(s) = 1 + s(\tau_1 + \tau_2) + s^2\tau_1\tau_2^1 = 1 + s\left[C_1(R_1 + R_2) + R_2C_2\right] + s^2R_2C_2R_1C_1$$

$$D(s) = 1 + \frac{s}{Q\omega_0} + \left(\frac{s}{\omega_0}\right)^2$$

$$\omega_0 = \frac{1}{\sqrt{R_1R_2C_1C_2}}$$

$$Q = \frac{\sqrt{b_2}}{b_1} = \frac{\sqrt{R_1R_2C_1C_2}}{C_2R_2\left(\frac{C_1(R_1 + R_2)}{C_2R_2} + 1\right)}$$

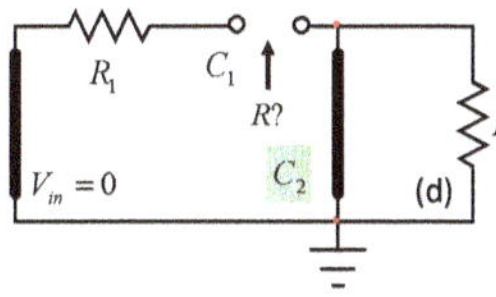

C_2 is in HF state
C_1 is in its HF state

$$H^{12} = 0$$

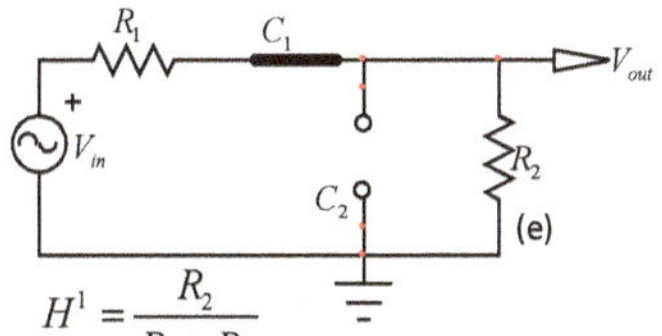

$$H^1 = \frac{R_2}{R_1 + R_2}$$

C_1 is in HF state
C_2 is in its dc state

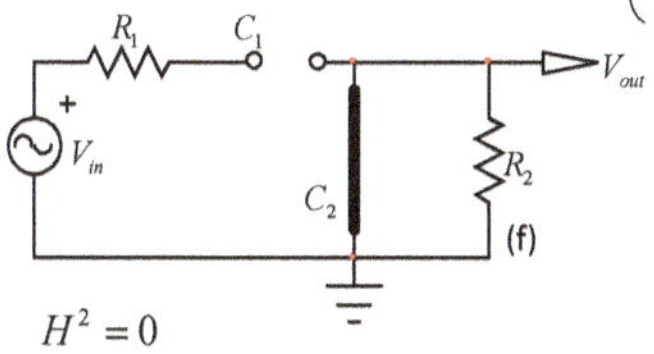

$$H^2 = 0$$

C_2 is in HF state
C_1 is in its dc state

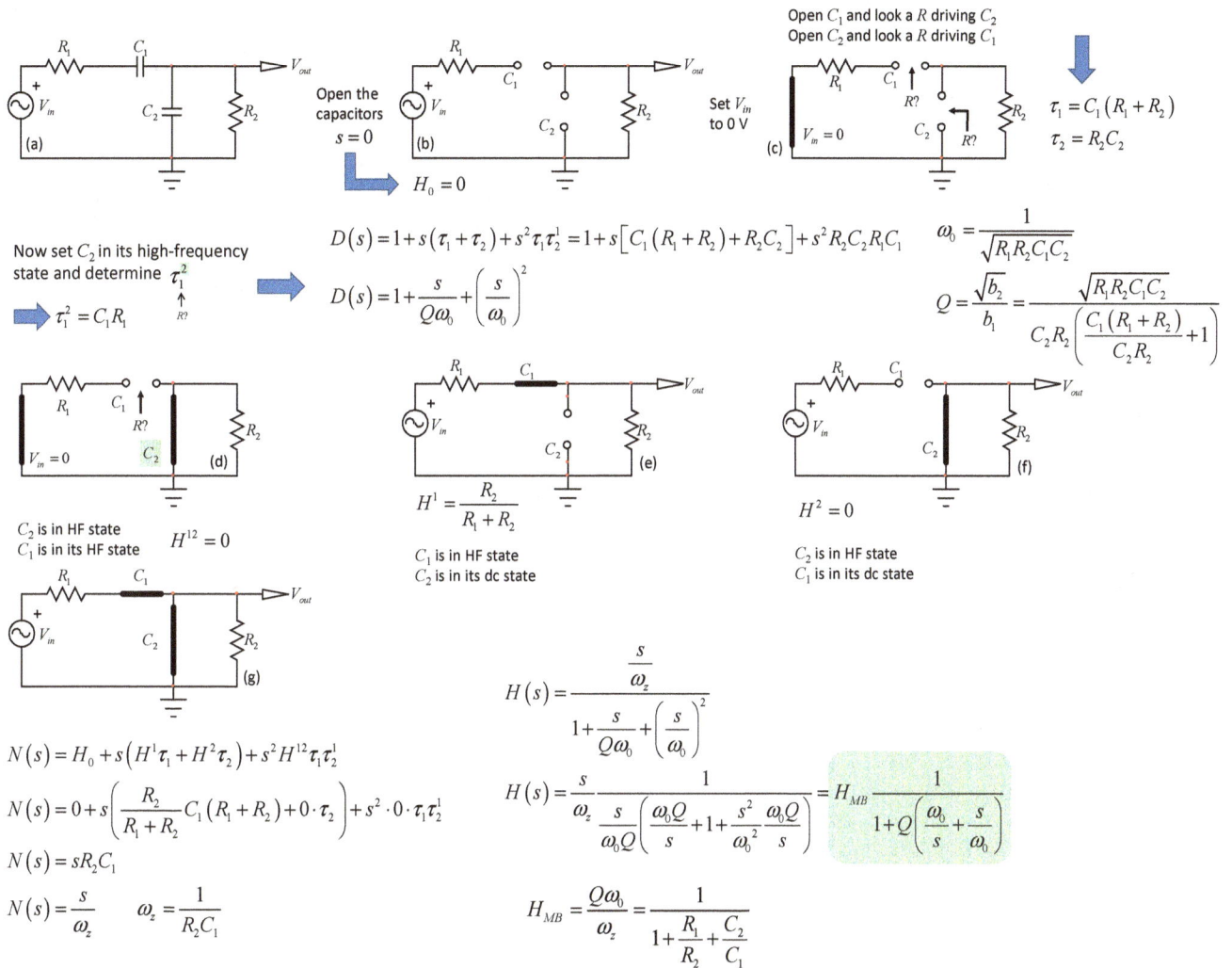

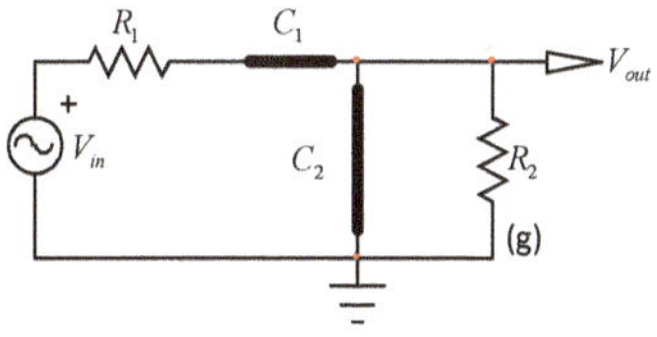

$$N(s) = H_0 + s\left(H^1\tau_1 + H^2\tau_2\right) + s^2H^{12}\tau_1\tau_2^1$$

$$N(s) = 0 + s\left(\frac{R_2}{R_1 + R_2}C_1(R_1 + R_2) + 0\cdot\tau_2\right) + s^2\cdot 0\cdot\tau_1\tau_2^1$$

$$N(s) = sR_2C_1$$

$$N(s) = \frac{s}{\omega_z} \qquad \omega_z = \frac{1}{R_2C_1}$$

$$H(s) = \frac{\frac{s}{\omega_z}}{1 + \frac{s}{Q\omega_0} + \left(\frac{s}{\omega_0}\right)^2}$$

$$H(s) = \frac{s}{\omega_z}\frac{1}{\frac{s}{\omega_0 Q}\left(\frac{\omega_0 Q}{s} + 1 + \frac{s^2}{\omega_0^2}\frac{\omega_0 Q}{s}\right)} = H_{MB}\frac{1}{1 + Q\left(\frac{\omega_0}{s} + \frac{s}{\omega_0}\right)}$$

$$H_{MB} = \frac{Q\omega_0}{\omega_z} = \frac{1}{1 + \frac{R_1}{R_2} + \frac{C_2}{C_1}}$$

The Quality Factor of this Bandpass Filter made of Resistors and Capacitors is Quite Low

Figure 6.34

$\|(x,y) := \frac{x \cdot y}{x+y}$ $\quad C_1 := 0.01\mu F$ $\quad C_2 := 0.022\mu F$ $\quad R_1 := 22k\Omega$ $\quad R_2 := 10k\Omega$

Transfer function for s=0 $\quad H_0 := 0$

$\tau_1 := C_1 \cdot (R_1 + R_2) = 320 \cdot \mu s$

$\tau_2 := C_2 \cdot R_2 = 220 \cdot \mu s$

$b_1 := \tau_1 + \tau_2 = 540 \cdot \mu s$

$\tau_{21} := C_1 \cdot R_1$

$b_2 := \tau_2 \cdot \tau_{21} = 4.84 \times 10^{-8} s^2$

$D_1(s) := 1 + s \cdot b_1 + s^2 \cdot b_2$

$Q := \frac{\sqrt{b_2}}{b_1} = 0.40741$ This is Q $\quad \omega_0 := \frac{1}{\sqrt{b_2}} = 4.54545 \times 10^3 \frac{1}{s}$

$\frac{\sqrt{C_1 \cdot R_1 \cdot C_2 \cdot R_2}}{C_2 \cdot R_2 \cdot \left[\frac{C_1 \cdot (R_1 + R_2)}{C_2 \cdot R_2} + 1\right]} = 0.40741$ $\quad \omega_0 := \frac{1}{\sqrt{C_1 \cdot C_2 \cdot R_1 \cdot R_2}} = 4.54545 \times 10^3 \frac{1}{s}$

Calculate the hi-frequency gains, H1, H2 and H12

When C1 is short circuit: $H_1 := \frac{R_2}{R_1 + R_2} = 0.3125$ $\quad 20 \cdot \log(H_1) = -10.103$ dB

When C2 is a short circuit: $H_2 := 0$ Same for H12 when both caps. are shorted.

$N(s) = H_0 + s \cdot (H_1 \cdot \tau_1 + H_2 \cdot \tau_2) + s^2 \cdot H_1 \cdot H_{12} \cdot \tau_1 \cdot \tau_{12} = s \cdot H_1 \cdot \tau_1$ $\quad \omega_z := \frac{1}{R_2 \cdot C_1}$

$H_2(s) := \frac{s \cdot R_2 \cdot C_1}{1 + s \cdot [C_1 \cdot (R_1 + R_2) + C_2 \cdot R_2] + s^2 \cdot C_2 \cdot R_2 \cdot C_1 \cdot R_1}$ $\quad H_{mb} := \frac{\omega_0 \cdot Q}{\omega_z} = 0.18519$

$H_3(s) := H_{mb} \cdot \frac{1}{1 + Q \cdot \left(\frac{s}{\omega_0} + \frac{\omega_0}{s}\right)}$ $\quad \frac{1}{\frac{R_1}{R_2} + 1 + \frac{C_2}{C_1}} = 0.18519$ $\quad 20 \cdot \log(H_{mb}) = -14.64788$ dB

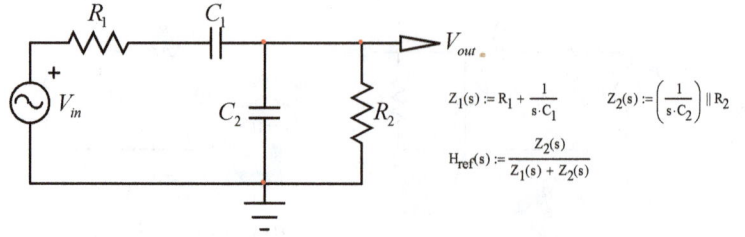

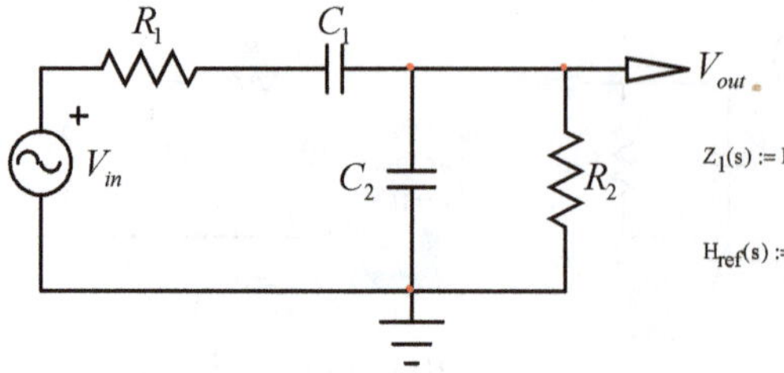

$Z_1(s) := R_1 + \frac{1}{s \cdot C_1}$ $\quad Z_2(s) := \left(\frac{1}{s \cdot C_2}\right) \| R_2$

$H_{ref}(s) := \frac{Z_2(s)}{Z_1(s) + Z_2(s)}$

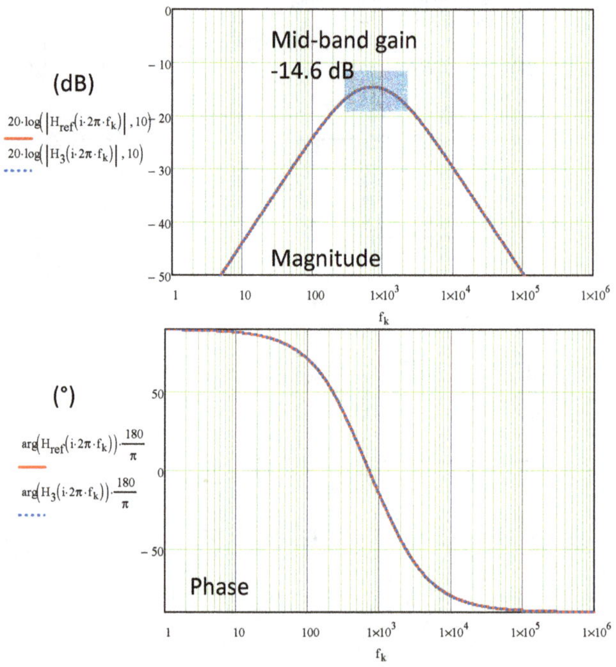

The Response is not Selective Considering the Small Quality Coefficient

Figure 6.35

We can also connect the components a bit differently as shown in Figure 6.36 and determine the new transfer function in this configuration.

Open C_1 and look a R driving C_2
Open C_2 and look a R driving C_1

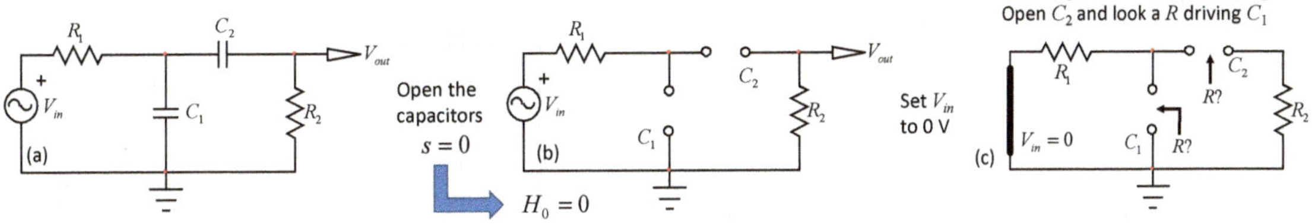

(a)
Open the capacitors $s = 0$
(b)
$H_0 = 0$
Set V_{in} to 0 V
(c)
$V_{in} = 0$
R?

$\tau_1 = R_1C_1$

$\tau_2 = (R_1 + R_2)C_2$

Now set C_1 in its high-frequency state and determine τ_2^1

$\tau_2^1 = C_2R_2$

$$D(s) = 1 + s(\tau_1 + \tau_2) + s^2\tau_1\tau_2^1 = 1 + s\left[R_1C_1 + C_2(R_1 + R_2)\right] + s^2R_1C_1R_2C_2$$

$$D(s) = 1 + \frac{s}{Q\omega_0} + \left(\frac{s}{\omega_0}\right)^2$$

$$\omega_0 = \frac{1}{\sqrt{R_1R_2C_1C_2}}$$

$$Q = \frac{\sqrt{b_2}}{b_1} = \frac{\sqrt{R_1R_2C_1C_2}}{C_2(R_1 + R_2) + R_1C_1}$$

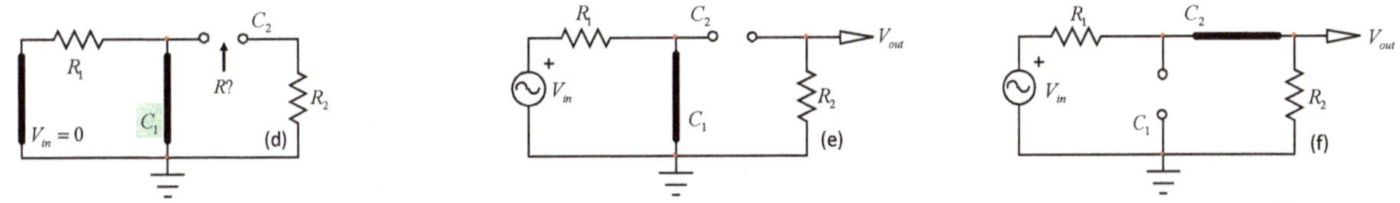

C_2 is in its HF state
C_1 is in its HF state
$H^{12} = 0$

C_1 is in HF state
C_2 is in its dc state
$H^1 = 0$

C_2 is in HF state
C_1 is in its dc state
$H^2 = \frac{R_2}{R_1 + R_2}$

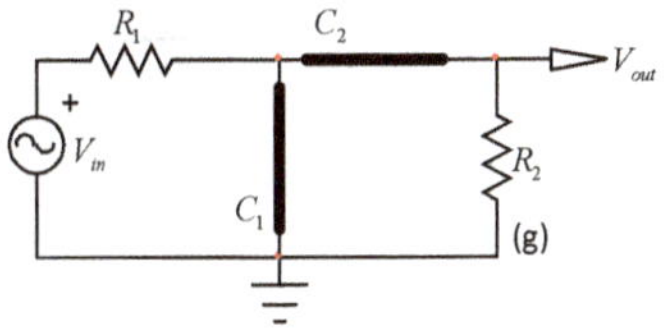

$$N(s) = H_0 + s(H^1\tau_1 + H^2\tau_2) + s^2H^{12}\tau_1\tau_2^1$$

$$N(s) = 0 + s\left(R_1C_1 \cdot 0 + \frac{R_2}{R_1 + R_2}\cdot\tau_2\right) + s^2 \cdot 0 \cdot \tau_1\tau_2^1$$

$$N(s) = sR_2C_2$$

$$N(s) = \frac{s}{\omega_z} \qquad \omega_z = \frac{1}{R_2C_2}$$

$$H(s) = \frac{\frac{s}{\omega_z}}{1 + \frac{s}{Q\omega_0} + \left(\frac{s}{\omega_0}\right)^2}$$

$$H(s) = \frac{s}{\omega_z}\frac{1}{\frac{s}{\omega_0Q}\left(\frac{\omega_0Q}{s} + 1 + \frac{s^2}{\omega_0^2}\frac{\omega_0Q}{s}\right)} = H_{MB}\frac{1}{1 + Q\left(\frac{\omega_0}{s} + \frac{s}{\omega_0}\right)}$$

$$H_{MB} = \frac{Q\omega_0}{\omega_z} = \frac{1}{1 + \frac{R_1}{R_2} + \frac{R_1C_1}{R_2C_2}}$$

This is still a Band-Pass filter despite the Slightly Different Arrangement

Figure 6.36

$C_1 := 0.01\mu F$ $\quad C_2 := 0.022\mu F$ $\quad R_1 := 20k\Omega$ $\quad R_2 := 10k\Omega$ $\quad \|(x,y) := \frac{x \cdot y}{x+y}$

Transfer function for s=0 $\quad H_0 := 0$

$\tau_1 := C_1 \cdot R_1 = 200\,\mu s$

$\tau_2 := C_2 \cdot (R_2 + R_1) = 660\,\mu s$

$b_1 := \tau_1 + \tau_2 = 860\,\mu s$

$\tau_{12} := C_2 \cdot R_2$

$b_2 := \tau_1 \cdot \tau_{12} = 4.4 \times 10^{-8}\,s^2$

$D_1(s) := 1 + s \cdot b_1 + s^2 \cdot b_2$ $\quad Q := \frac{\sqrt{b_2}}{b_1} = 0.24391$ $\quad \omega_0 := \frac{1}{\sqrt{b_2}}$

This is Q $\frac{\sqrt{C_1 \cdot R_1 \cdot C_2 \cdot R_2}}{C_2 \cdot (R_1 + R_2) + C_1 \cdot R_1} = 0.24391$ $\quad \omega_0 := \frac{1}{\sqrt{C_1 \cdot C_2 \cdot R_1 \cdot R_2}}$

Calculate the hi-frequency gains, H1, H2 and H12

When C1 is short circuit: $H_1 := 0$ Same for H12 when both caps. are shorted.

When C2 is a short circuit: $H_2 := \frac{R_2}{R_1 + R_2} = 0.33333$

$N(s) = H_0 + s \cdot (H_1 \cdot \tau_1 + H_2 \cdot \tau_2) + s^2 \cdot H_1 \cdot H_{12} \cdot \tau_1 \cdot \tau_{12} = s \cdot H_1 \cdot \tau_1$

$H_2(s) := \frac{s \cdot R_2 \cdot C_2}{1 + s \cdot [C_2 \cdot (R_1 + R_2) + C_1 \cdot R_1] + s^2 \cdot C_2 \cdot R_2 \cdot C_1 \cdot R_1}$ $\quad \omega_z := \frac{1}{R_2 \cdot C_2}$

$D_3(s) := 1 + \frac{s}{\omega_0 \cdot Q} + \left(\frac{s}{\omega_0}\right)^2$ $\quad H_{00} := \frac{\omega_0 \cdot Q}{\omega_z} = 0.25581$ $\quad 20 \log(H_{00}) = -11.84152$

$\frac{1}{\frac{C_1 \cdot R_1}{C_2 \cdot R_2} + \frac{R_1}{R_2} + 1} = 0.25581$

$H_3(s) := H_{00} \cdot \frac{1}{1 + Q \cdot \left(\frac{s}{\omega_0} + \frac{\omega_0}{s}\right)}$ Final transfer function in a low-entropy form

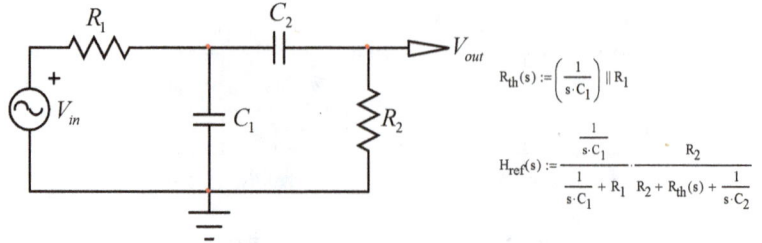

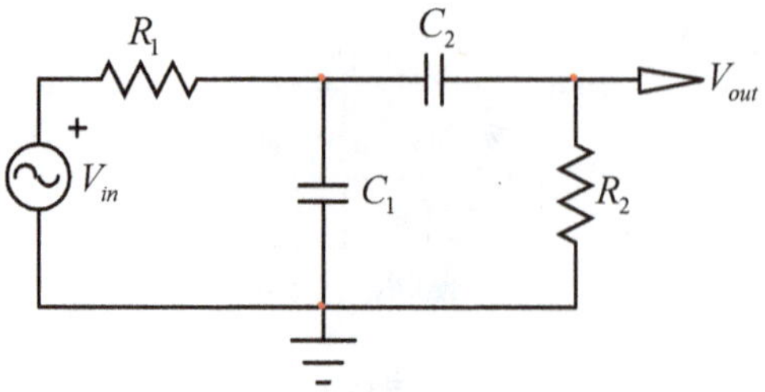

$R_{th}(s) := \left(\frac{1}{s \cdot C_1}\right) \| R_1$

$H_{ref}(s) := \frac{\frac{1}{s \cdot C_1}}{\frac{1}{s \cdot C_1} + R_1} \cdot \frac{R_2}{R_2 + R_{th}(s) + \frac{1}{s \cdot C_2}}$

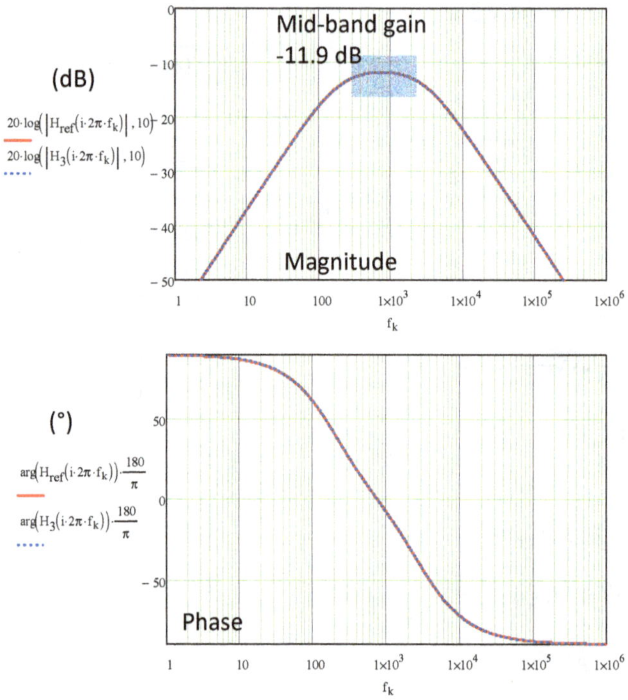

The Ac Response Confirms a Band-Pass Filter Exhibiting a 11.9-dB Attenuation at its Peak

Figure 6.37

We now cascade two *RC* networks loaded by a resistor.

Compared to the first example we solved at the beginning of this chapter; the situation complicates but is surmountable.

Figure 6.38 details the adopted steps.

There are no zeroes in this circuit because should C_1 or C_2 set in its high-frequency state, the response is always zero.

The ac response is given in Figure 6.39 with various expressions.

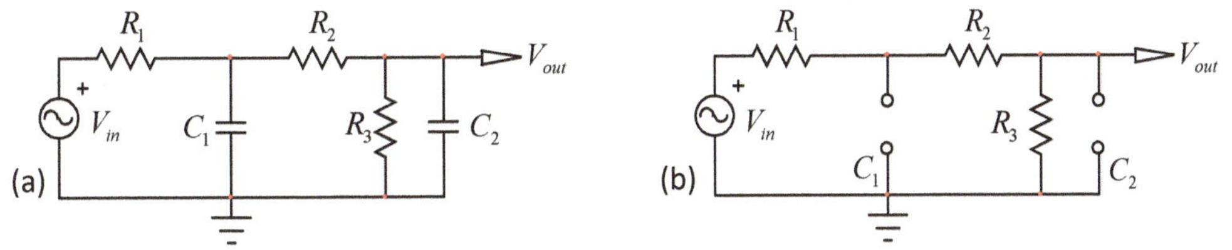

$s=0$

Open the capacitors

$$H_0=\frac{R_3}{R_1+R_2+R_3}$$

Open C_1 and look a R driving C_2
Open C_2 and look a R driving C_1

Now set C_1 in its high-frequency state and determine $\tau_{2\leftarrow R?}^{1}$

Set V_{in} to 0 V

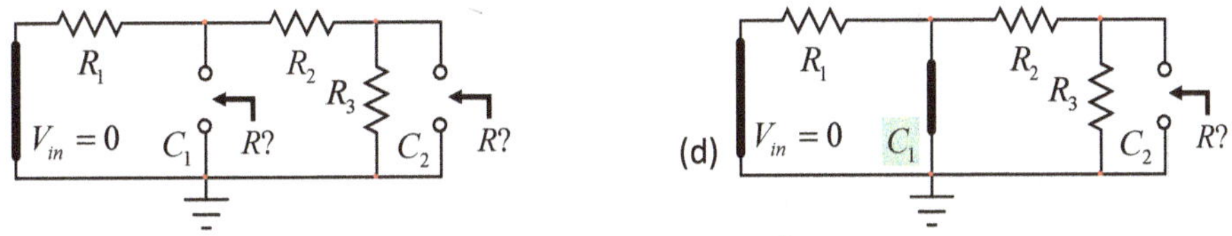

$$\tau_2^1=C_2\left(R_2\,\|\,R_3\right)$$

$$\tau_1=C_1\left[R_1\,\|\left(R_2+R_3\right)\right]$$

$$\tau_2=C_2\left[R_3\,\|\left(R_1+R_2\right)\right]$$

$$D(s)=1+b_1s+b_2s^2=1+s\left(\tau_1+\tau_2\right)+s^2\tau_1\tau_2^1$$

$$D(s)=1+s\Big[C_1\left[R_1\,\|\left(R_2+R_3\right)\right]+C_2\left[R_3\,\|\left(R_1+R_2\right)\right]\Big]+s^2C_1\left[R_1\,\|\left(R_2+R_3\right)\right]C_2\left(R_2\,\|\,R_3\right)$$

$$D(s)=1+\frac{s}{Q\omega_0}+\left(\frac{s}{\omega_0}\right)^2 \Rightarrow H(s)=H_0\frac{1}{1+\frac{s}{Q\omega_0}+\left(\frac{s}{\omega_0}\right)^2}\qquad \omega_0=\frac{1}{\sqrt{b_2}}\qquad Q=\frac{\sqrt{b_2}}{b_1}$$

If $Q \ll 1$, the low- Q approximation holds:

$$\omega_{p_1}=Q\omega_0=\frac{1}{C_1\left[R_1\,\|\left(R_2+R_3\right)\right]+C_2\left[R_3\,\|\left(R_1+R_2\right)\right]}$$

$$\omega_{p_2}=\frac{\omega_0}{Q}=\frac{C_1\left[R_1\,\|\left(R_2+R_3\right)\right]+C_2\left[R_3\,\|\left(R_1+R_2\right)\right]}{C_1\left[R_1\,\|\left(R_2+R_3\right)\right]C_2\left(R_2\,\|\,R_3\right)}$$

$$\Rightarrow H(s)\approx H_0\frac{1}{\left(1+\frac{s}{\omega_{p_1}}\right)\left(1+\frac{s}{\omega_{p_2}}\right)}$$

The Cascaded *RC* Networks explored as a First Example are Now Loaded by a Resistor

Figure 6.38

$\parallel(x,y) := \frac{x \cdot y}{x+y}$ $\quad R_{inf} := 10^{23}\Omega$

$C_1 := 0.01\mu F \quad C_2 := 0.022\mu F \quad R_1 := 22k\Omega \quad R_2 := 10k\Omega \quad R_3 := 12k\Omega$

Transfer function for s=0 $\quad H_0 := \frac{R_3}{R_3 + R_1 + R_2} = 0.27273$

$\tau_1 := C_1 \cdot [R_1 \parallel (R_2 + R_3)] = 110 \cdot \mu s$

$\tau_2 := C_2 \cdot [R_3 \parallel (R_1 + R_2)] = 192 \cdot \mu s$

$b_1 := \tau_1 + \tau_2 = 302 \cdot \mu s \quad b_{1a} := C_1 \cdot [R_1 \parallel (R_2 + R_3)] + C_2 \cdot [R_3 \parallel (R_1 + R_2)] = 302 \mu s$

$\tau_{12} := C_2 \cdot (R_2 \parallel R_3)$

$b_2 := \tau_1 \cdot \tau_{12} = 1.32 \times 10^{-8} s^2 \quad b_{2a} := C_1 \cdot [R_1 \parallel (R_2 + R_3)] \cdot [C_2 \cdot (R_2 \parallel R_3)] = 1.32 \times 10^{-8} s^2$

$D_1(s) := 1 + s \cdot b_1 + s^2 \cdot b_2 \quad Q := \frac{\sqrt{b_2}}{b_1} = 0.38043 \quad \omega_0 := \frac{1}{\sqrt{b_2}}$

$\omega_{p1} := \frac{1}{b_1} \quad f_{p1} := \frac{\omega_{p1}}{2\pi} = 527.00312Hz \quad \frac{1}{2 \cdot \pi \cdot [C_1 \cdot [R_1 \parallel (R_2 + R_3)] + C_2 \cdot [R_3 \parallel (R_1 + R_2)]]} = 527.00312Hz$

$\omega_{p2} := \frac{b_1}{b_2} \quad f_{p2} := \frac{\omega_{p2}}{2\pi} = 3.64127kHz \quad \frac{C_1 \cdot [R_1 \parallel (R_2 + R_3)] + C_2 \cdot [R_3 \parallel (R_1 + R_2)]}{2 \cdot \pi \cdot [C_1 \cdot [R_1 \parallel (R_2 + R_3)]] \cdot [C_2 \cdot (R_2 \parallel R_3)]} = 3.64127kHz$

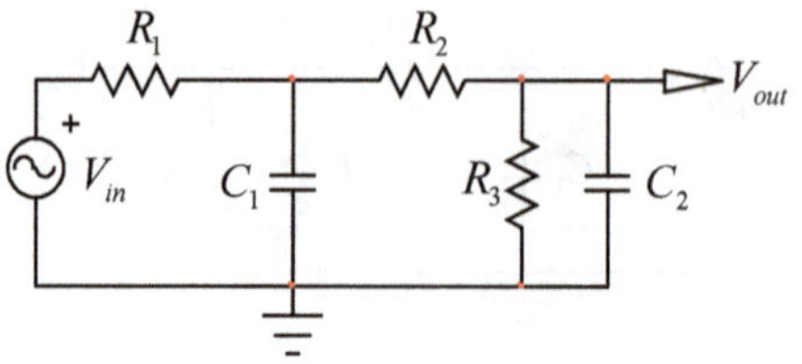

$R_{th}(s) := \left(\frac{1}{s \cdot C_1}\right) \parallel R_1 \quad Z_1(s) := \left(\frac{1}{s \cdot C_2}\right) \parallel R_3$

$H_{ref}(s) := \frac{\frac{1}{s \cdot C_1}}{\frac{1}{s \cdot C_1} + R_1} \cdot \frac{Z_1(s)}{Z_1(s) + R_{th}(s) + R_2}$

Brute-force approach

$H_1(s) := H_0 \cdot \frac{1}{1 + s \cdot b_1 + s^2 \cdot b_2} \quad H_2(s) := H_0 \cdot \frac{1}{\left(1 + \frac{s}{\omega_{p1}}\right)\left(1 + \frac{s}{\omega_{p2}}\right)}$

$H_3(s) := H_0 \cdot \frac{1}{1 + s \cdot [C_1 \cdot [R_1 \parallel (R_2 + R_3)] + C_2 \cdot [R_3 \parallel (R_1 + R_2)]] + s^2 \cdot [C_1 \cdot [R_1 \parallel (R_2 + R_3)]] \cdot [C_2 \cdot (R_2 \parallel R_3)]}$

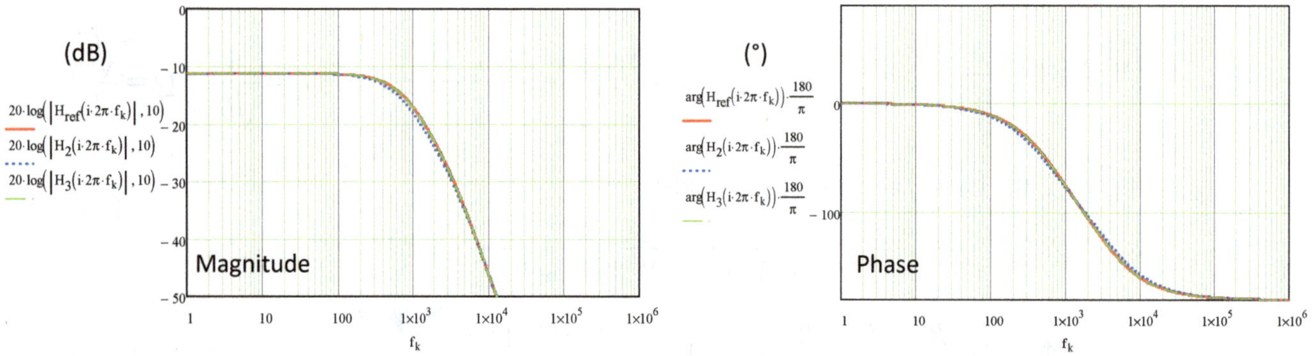

The Transfer Function can be Rearranged in Cascaded Poles when the Quality Factor is much Smaller than 1

Figure 6.39

Let's now explore the impedance of a parallel *RLC* network—which is also a classic.

A few sketches are necessary to extract the time constants for this impedance, but nothing as complicated as the circuit shown in Figure 6.40.

The zeroes are obtained by nulling the response across the test generator—which is equivalent to replacing it by a wire in the sketch. This simplifies the analysis a lot for impedance determinations.

The ac response appears in Figure 6.41 and confirms the analysis is correct.

The compact form already revealed in the previous examples predicts a magnitude peak at 100 Ω or 40 dBΩ.

What is the resistance R_0 **for** $s = 0$?

$R_0 = 0$ (a)   (b)

Zero the stimulus Determine R for C_1

(c)

$R = R_s$
$\tau_1 = R_s C_1$

Zero the stimulus Determine R for L_2

(d)

$R = R_1$

$\tau_2 = \dfrac{L_2}{R_1}$

Determine R **for** L_2 **with** C_1 **in HF state** τ_2^1

(e)

$R = R_s$

$\tau_2^1 = \dfrac{L_2}{R_s}$

Determine R **for** C_1 **with** L_2 **in HF state** τ_1^2

(f)

$R = R_1$

$\tau_1^2 = R_1 C_1$

$D(s) = 1 + s(\tau_1 + \tau_2) + s^2 \tau_1 \tau_2^1 = 1 + s b_1 + s^2 b_2$

$D(s) = 1 + s\left(R_s C_1 + \dfrac{L_2}{R_1}\right) + s^2 R_s C_1 \dfrac{L_2}{R_s}$ or $D(s) = 1 + s\left(R_s C_1 + \dfrac{L_2}{R_1}\right) + s^2 \dfrac{L_2}{R_1} R_1 C_1$

$D(s) = 1 + s\dfrac{L_2}{R_1} + s^2 L_2 C_1$ $Q = \dfrac{\sqrt{b_2}}{b_1} = R_1 \sqrt{\dfrac{C_1}{L_2}}$

$D(s) = 1 + \dfrac{s}{\omega_0 Q} + \left(\dfrac{s}{\omega_0}\right)^2$ $\omega_0 = \dfrac{1}{\sqrt{b_2}} = \dfrac{1}{\sqrt{L_2 C_1}}$

C_1 is in its HF state
L_2 is in its dc state $H^1 = 0$

C_1 is in its dc state
L_2 is in its HF state $H^2 = R_1$

C_1 is in its HF state
L_2 is in its HF state $H^{12} = 0$

(g)   (h)   (i)

$N(s) = R_0 + s\left(H^1 \tau_1 + H^2 \tau_2\right) + s^2 H^{12} \tau_1 \tau_2^1$

$N(s) = 0 + s\left(0 \cdot \tau_1 + R_1 \cdot \tau_2\right) + s^2 \cdot 0 \cdot \tau_1 \tau_2^1$

$N(s) = sR_1 \dfrac{L_2}{R_1} = sL_2$

$Z_{in}(s) = \dfrac{sL_2}{1 + s\dfrac{L_2}{R_1} + s^2 L_2 C_1} = \dfrac{sL_2}{s\dfrac{L_2}{R_1}} \dfrac{1}{1 + \dfrac{R_1}{sL_2} + sR_1 C_1} = R_1 \dfrac{1}{1 + Q\left(\dfrac{s}{\omega_0} + \dfrac{\omega_0}{s}\right)}$

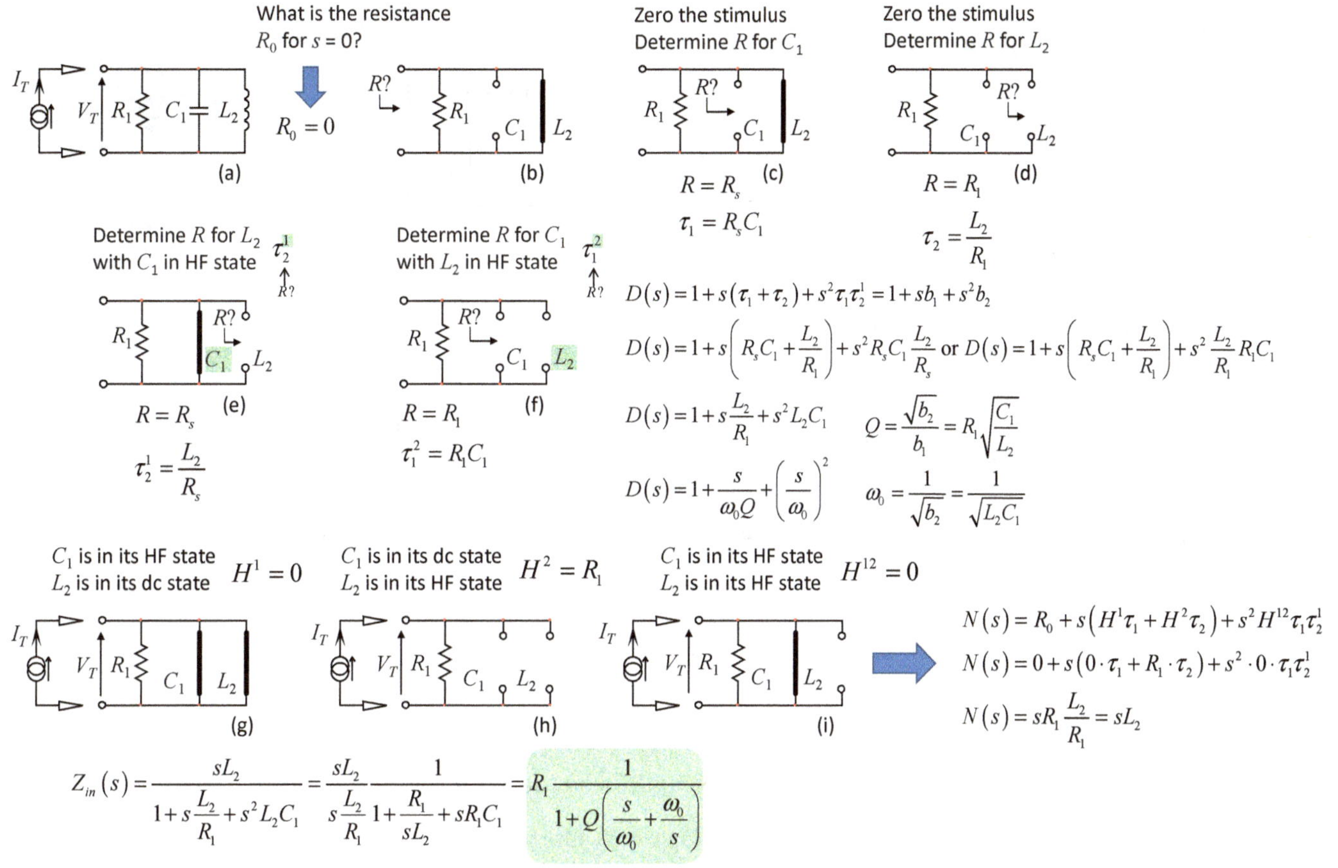

A Test Generator I_T **is Installed to Determine the Input Impedance of this Series** *RLC* **Network**

Figure 6.40

$C_1 := 1\mu F$ $\|(x,y) := \dfrac{x \cdot y}{x+y}$ $R_1 := 100\Omega$ $L_2 := 100\mu H$ $R_s := 10^{-23}\Omega$

$R_0 := 0$

$\tau_1 := C_1 \cdot R_s = 0\mu s$ $\tau_2 := \dfrac{L_2}{R_1} = 1 \cdot \mu s$

$b_1 := \tau_1 + \tau_2 = 1 \cdot \mu s$

$\tau_{12} := \dfrac{L_2}{R_s} = 1 \times 10^{22} \cdot ms$ $\tau_{21} := R_1 \cdot C_1$

$b_2 := \tau_1 \cdot \tau_{12} = 1 \times 10^{-10} s^2$ $b_{2a} := \tau_2 \cdot \tau_{21} = 1 \times 10^{-10} s^2$

$H_{ref}(s) := \left(\dfrac{1}{s \cdot C_1}\right) \| (s \cdot L_2) \| R_1$

$D_1(s) := 1 + b_1 \cdot s + b_2 \cdot s^2$ $Q := \dfrac{\sqrt{b_2}}{b_1} = 10$ $\omega_0 := \dfrac{1}{\sqrt{b_2}}$ $f_0 := \dfrac{\omega_0}{2 \cdot \pi} = 15.91549 kHz$

$H_2 := R_1$ $R_1 \sqrt{\dfrac{C_1}{L_2}} = 10$ $\omega_z := \dfrac{R_1}{L_2}$

$H_1(s) := \dfrac{H_2 \cdot \tau_2 \cdot s}{D_1(s)}$ $H_3(s) := \dfrac{s \cdot L_2}{1 + s \cdot \dfrac{L_2}{R_1} + s^2 \cdot L_2 \cdot C_1}$ $\left|H_1(\omega_0 \cdot i)\right| = 100\Omega$

$H_6(s) := R_1 \cdot \dfrac{\dfrac{s}{\omega_z}}{1 + \dfrac{s}{\omega_0 \cdot Q} + \left(\dfrac{s}{\omega_0}\right)^2}$ $H_4(s) := R_1 \cdot \dfrac{1}{1 + Q \cdot \left(\dfrac{s}{\omega_0} + \dfrac{\omega_0}{s}\right)}$

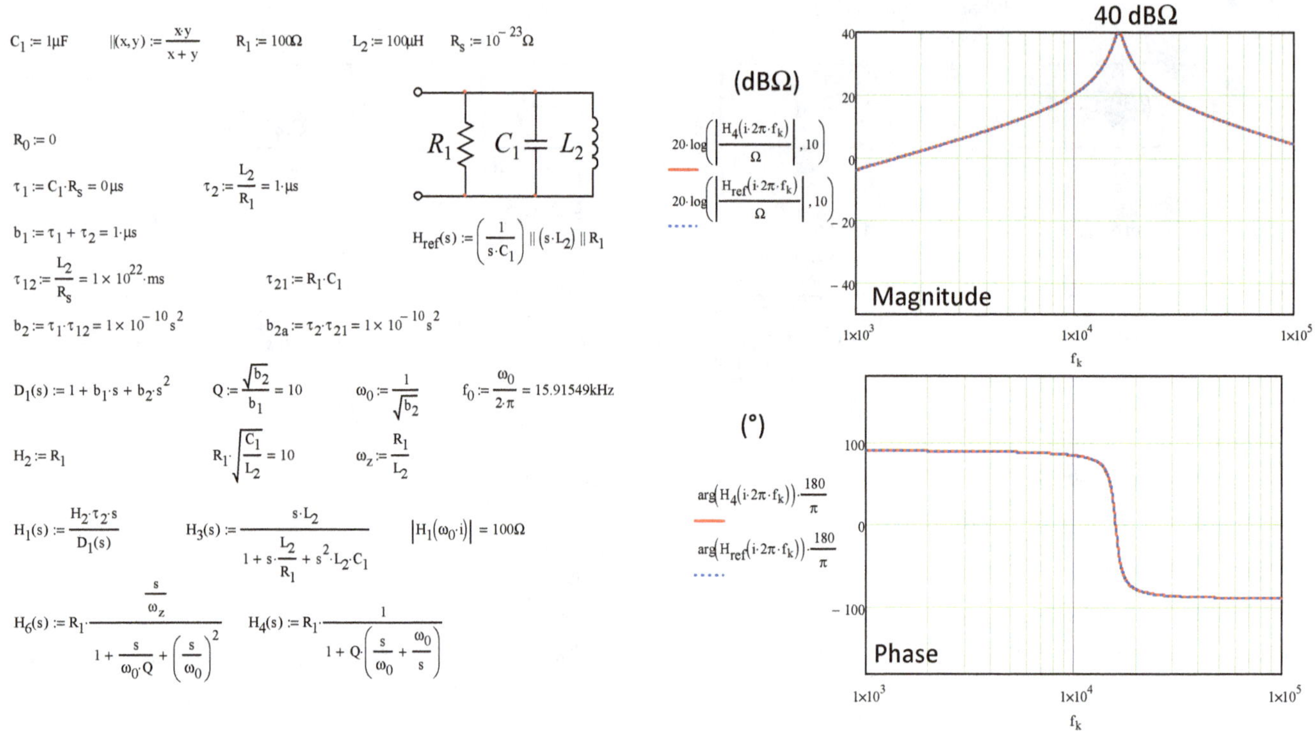

The Impedance Peaks to resistor R_1's Value which is 100 Ω or 40 dBΩ

Figure 6.41

We now look at Figure 6.42 where we still have an inductor and a capacitor forming a second-order network.

Start by shorting the inductor and opening the capacitor to determine the dc gain which is 1 in this mode.

The rest comes easily with the determination of the time constants in different conditions.

Finally, if the quality factor Q is sufficiently low, then the denominator can be expressed as two cascaded poles.

This is shown in Figure 6.43.

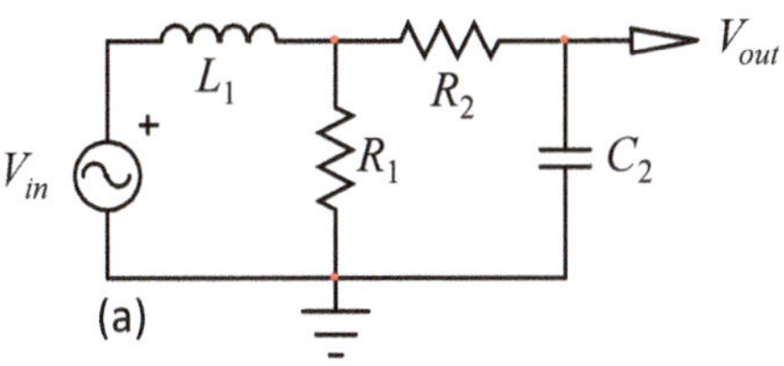

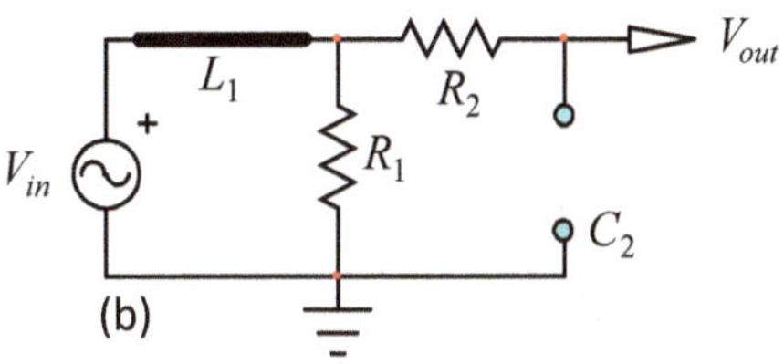

$s = 0$

Open the capacitor – short the inductor

$$H_0 = 1$$

Zero the stimulus
Determine R for L_1

$$\tau_1 = \frac{L_1}{R_1}$$

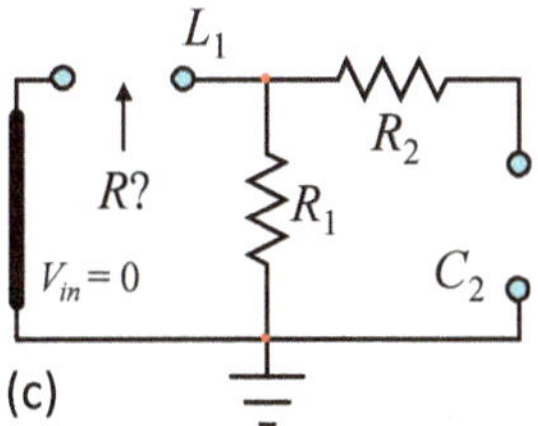

Zero the stimulus
Determine R for C_2

$$\tau_2 = C_2 R_2$$

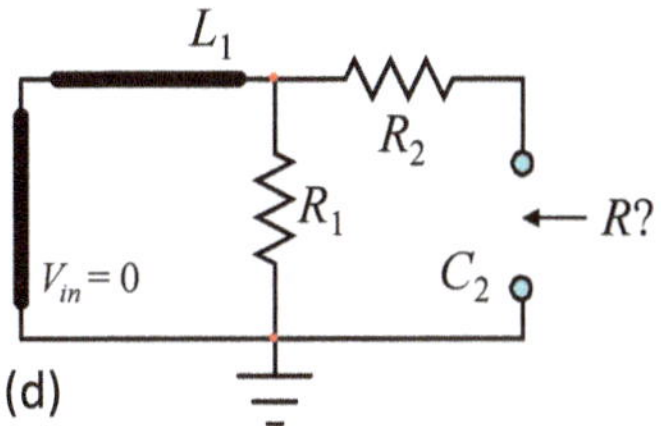

Determine R for C_2
with L_1 in HF state

$$\tau_2^1 = C_2 (R_1 + R_2)$$

$R?$

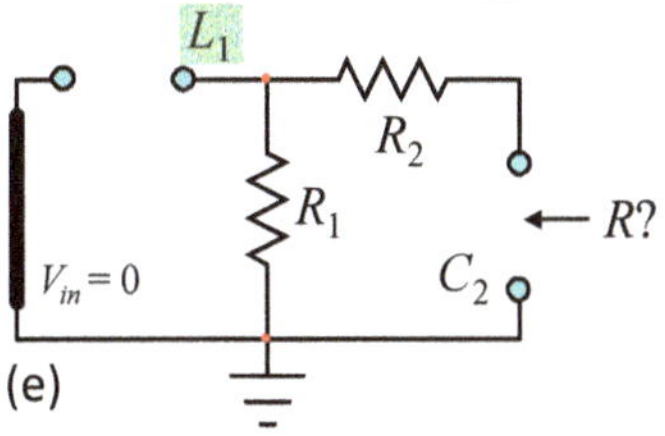

$$D(s) = 1 + s(\tau_1 + \tau_2) + s^2 \tau_1 \tau_2^1 = 1 + s b_1 + s^2 b_2$$

$$D(s) = 1 + s\left(\frac{L_1}{R_1} + R_2 C_2\right) + s^2 \frac{L_1}{R_1} C_2 (R_1 + R_2)$$

$$D(s) = 1 + \frac{s}{\omega_0 Q} + \left(\frac{s}{\omega_0}\right)^2 \Rightarrow H(s) = H_0 \frac{1}{1 + \frac{s}{Q\omega_0} + \left(\frac{s}{\omega_0}\right)^2}$$

$$Q = \frac{\sqrt{b_2}}{b_1} = \frac{\sqrt{\frac{L_1}{R_1} C_2 (R_1 + R_2)}}{\frac{L_1}{R_1} + C_2 R_2} \qquad \omega_0 = \frac{1}{\sqrt{b_2}} = \frac{1}{\sqrt{\frac{L_1}{R_1} C_2 (R_1 + R_2)}}$$

If $Q << 1$, the low- Q approximation holds:

$$\omega_{p_1} = Q\omega_0 = \frac{1}{\frac{L_1}{R_1} + R_2 C_2}$$

$$\omega_{p_2} = \frac{\omega_0}{Q} = \frac{L_1 + C_2 R_1 R_2}{C_2 L_1 (R_1 + R_2)}$$

$$\Rightarrow H(s) \approx H_0 \frac{1}{\left(1 + \frac{s}{\omega_{p_1}}\right)\left(1 + \frac{s}{\omega_{p_2}}\right)}$$

Here, an *LC* Filter is Cascaded with an *RC* Network

Figure 6.42

$R_2 := 10k\Omega \quad C_2 := 10nF \quad L_1 := 470\mu H \quad R_1 := 1k\Omega \quad \|(x,y) := \frac{xy}{x+y}$

$H_0 := 1 \quad 20 \cdot \log(H_0) = 0 \quad dB$

$\tau_1 := \frac{L_1}{R_1} = 0.47 \cdot \mu s$

$\tau_2 := R_2 \cdot C_2 = 100 \cdot \mu s$

$b_1 := \tau_1 + \tau_2 = 100.47 \mu s$

$\tau_{12} := C_2 \cdot (R_1 + R_2) = 110 \cdot \mu s$

$b_2 := \tau_1 \cdot \tau_{12} = 51.7 \cdot \mu s^2$

$H_1(s) := H_0 \cdot \frac{1}{1 + b_1 \cdot s + b_2 \cdot s^2}$

$Q := \frac{\sqrt{b_2}}{b_1} = 0.072 \quad \omega_0 := \frac{1}{\sqrt{b_2}} \quad f_0 := \frac{\omega_0}{2 \cdot \pi} = 22.135 kHz$

$\omega_{p1} := \frac{1}{\frac{L_1}{R_1} + R_2 \cdot C_2} \quad f_{p1} := \frac{\omega_{p1}}{2 \cdot \pi} = 1.584 kHz$

$\omega_{p2} := \frac{L_1 + C_2 \cdot R_1 \cdot R_2}{C_2 \cdot L_1 \cdot (R_1 + R_2)} \quad f_{p2} := \frac{\omega_{p2}}{2 \cdot \pi} = 309.29 kHz$

$H_3(s) := H_0 \cdot \frac{1}{1 + \frac{s}{\omega_0 \cdot Q} + \left(\frac{s}{\omega_0}\right)^2} \quad H_4(s) := \frac{1}{\left(1 + \frac{s}{\omega_{p1}}\right) \cdot \left(1 + \frac{s}{\omega_{p2}}\right)}$

Low-Q approximation

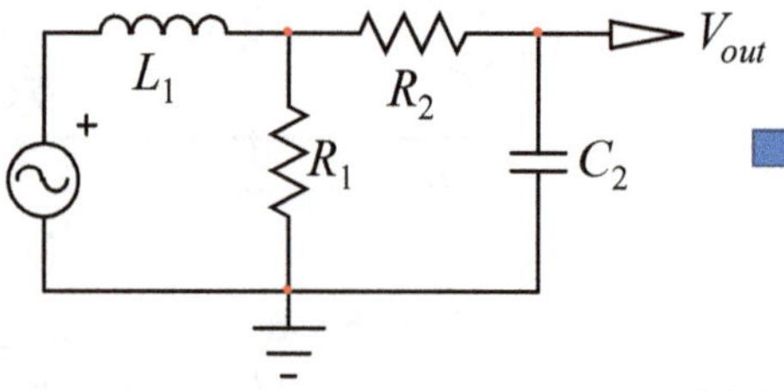

$H_{ref}(s) := \frac{R_1}{R_1 + s \cdot L_1} \cdot \frac{\frac{1}{s \cdot C_2}}{R_2 + (s \cdot L_1) \| (R_1) + \frac{1}{s \cdot C_2}}$

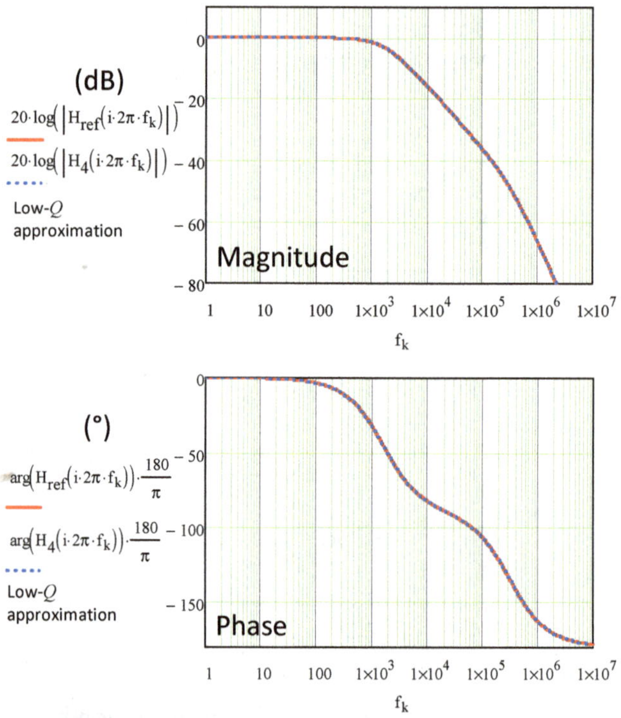

If the Quality Factor Q is Sufficiently Low, the Transfer Function can be Approximated as Two Cascaded Poles

Figure 6.43

In Figure 6.44, you can see the inductor now connected to ground, introducing a zero at the origin since its dc state is a short circuit.

The network thus combines a high-pass filter with a low-pass filter involving capacitor C_2.

There is one zero in this circuit because if L_1 is set in its high-frequency state (while C_2 is in its dc state), then you see the stimulus going through to generate a response.

It is the single zero as the other high-frequency gains return zero.

The ac response is given in Figure 6.45.

Open the capacitor – short the inductor

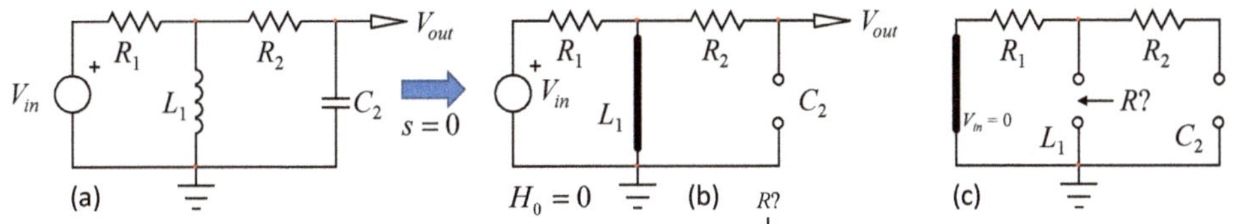

$H_0 = 0$

Zero the stimulus
Determine R for L_1

$\tau_1 = \frac{L_1}{R_1}$

Zero the stimulus
Determine R for C_2 $\quad \tau_2 = R_2C_2$

Determine R for L_1 with C_2 in HF state $\quad \tau_1^2 = \frac{L_1}{R_1 \| R_2}$

$$D(s) = 1 + s(\tau_1 + \tau_2) + s^2\tau_2\tau_1^2 = 1 + sb_1 + s^2b_2$$

$$D(s) = 1 + s\left(\frac{L_1}{R_1} + R_2C_2\right) + s^2R_2C_2\frac{L_1}{R_1 \| R_2}$$

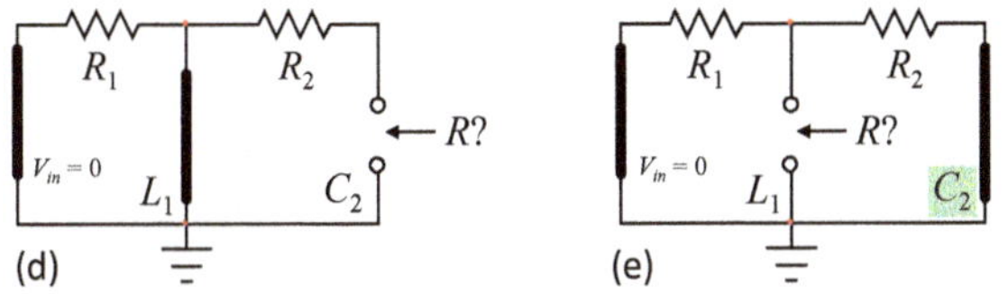

$$D(s) = 1 + \frac{s}{\omega_0 Q} + \left(\frac{s}{\omega_0}\right)^2$$

$$Q = \frac{\sqrt{b_2}}{b_1} = \frac{\sqrt{\frac{L_1}{R_1 \| R_2}R_2C_2}}{\frac{L_1}{R_1} + C_2R_2}$$

$$\omega_0 = \frac{1}{\sqrt{b_2}} = \frac{1}{\sqrt{\frac{L_1}{R_1 \| R_2}R_2C_2}}$$

Determine the gain H^1 when
L_1 is in high-frequency state
C_2 is in dc state

Determine the gain H^2 when
C_2 is in high-frequency state
L_1 is in dc state

Determine the gain H^{12} when
C_2 is in high-frequency state
L_2 is in high-frequency state

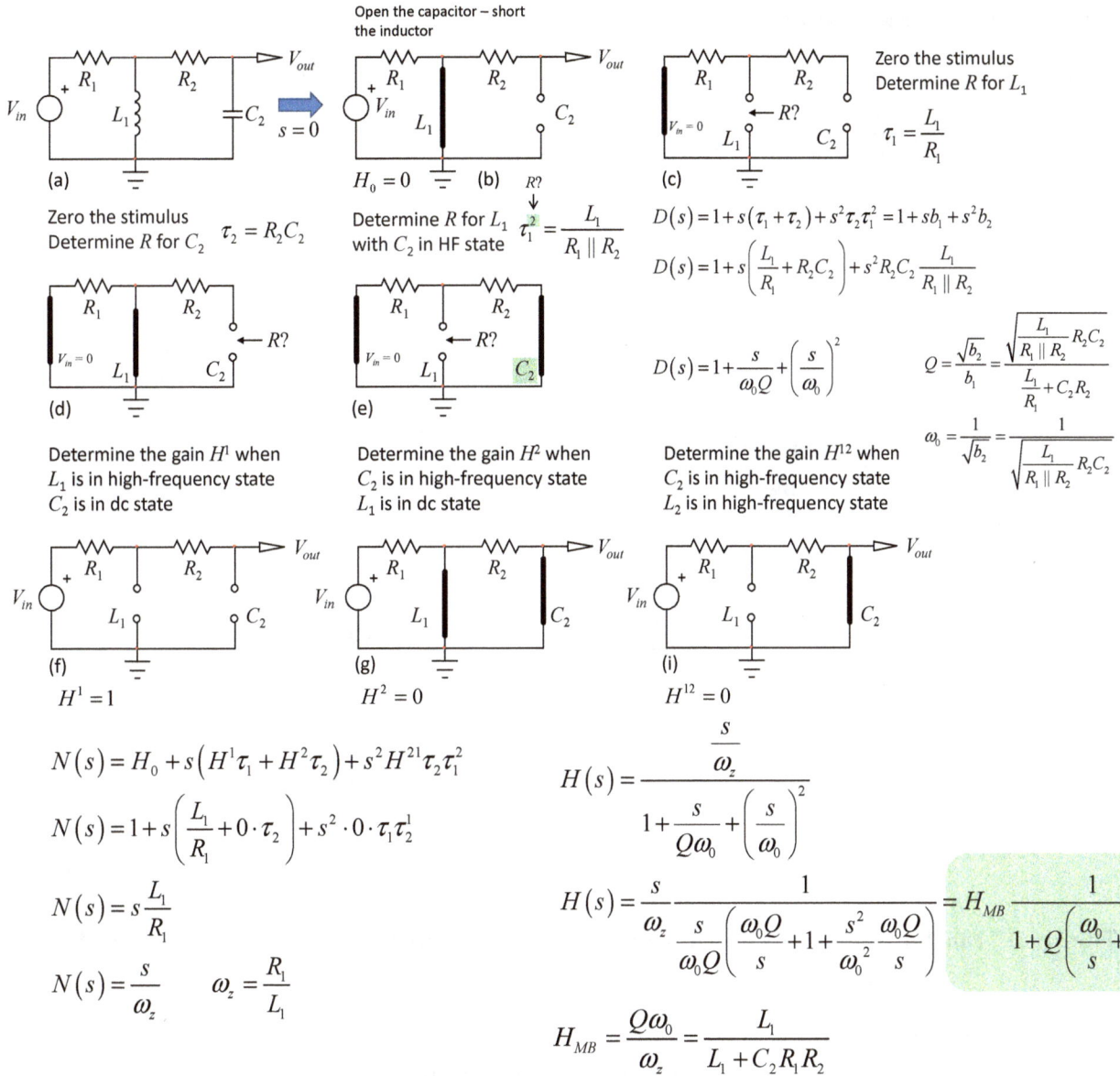

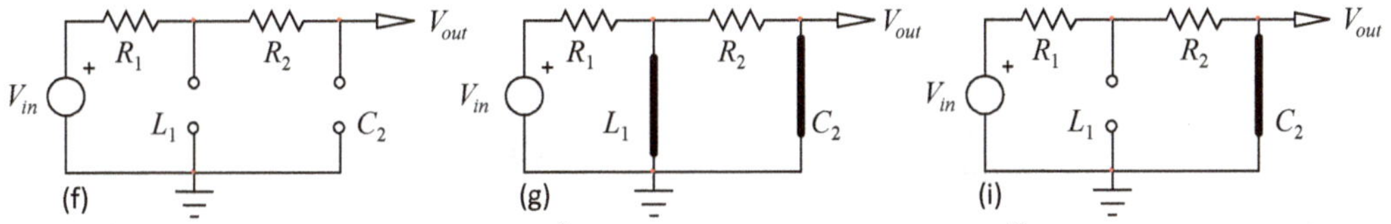

$H^1 = 1 \qquad H^2 = 0 \qquad H^{12} = 0$

$$N(s) = H_0 + s\left(H^1\tau_1 + H^2\tau_2\right) + s^2H^{21}\tau_2\tau_1^2$$

$$N(s) = 1 + s\left(\frac{L_1}{R_1} + 0\cdot\tau_2\right) + s^2\cdot 0\cdot\tau_1\tau_2^1$$

$$N(s) = s\frac{L_1}{R_1}$$

$$N(s) = \frac{s}{\omega_z} \qquad \omega_z = \frac{R_1}{L_1}$$

$$H(s) = \frac{\frac{s}{\omega_z}}{1 + \frac{s}{Q\omega_0} + \left(\frac{s}{\omega_0}\right)^2}$$

$$H(s) = \frac{s}{\omega_z}\frac{1}{\frac{s}{\omega_0 Q}\left(\frac{\omega_0 Q}{s} + 1 + \frac{s^2}{\omega_0^2}\frac{\omega_0 Q}{s}\right)} = H_{MB}\frac{1}{1 + Q\left(\frac{\omega_0}{s} + \frac{s}{\omega_0}\right)}$$

$$H_{MB} = \frac{Q\omega_0}{\omega_z} = \frac{L_1}{L_1 + C_2R_1R_2}$$

The Inductor Clearly Shunts the Path to Ground at Dc, Introducing a Zero at the Origin

Figure 6.44

$R_1 := 100\Omega \quad C_2 := 470nF \quad L_1 := 100\mu H \quad R_2 := 100k\Omega \quad \|(x,y) := \frac{x\,y}{x+y}$

$\tau_1 := \frac{L_1}{R_1} = 1\cdot\mu s \quad \tau_2 := C_2\cdot R_2 = 4.7\times 10^{7}\cdot ns \quad H_0 := 0$

$\tau_{12} := C_2\cdot(R_1+R_2) = 47.047ms \quad \tau_{21} := \frac{L_1}{R_1 \| R_2} = 1.001\cdot\mu s$

$b_1 := \tau_1 + \tau_2 = 4.7\times 10^{4}\cdot\mu s$

$b_2 := \tau_2\cdot\tau_{21} = 4.705\times 10^{4}\cdot\mu s^2 \quad b_{2a} := \tau_1\cdot\tau_{12} = 4.705\times 10^{4}\,\mu s^2$

$D_1(s) := 1 + s\cdot b_1 + s^2\cdot b_2$

$H_1 := 1$

$N_1(s) := s\cdot H_1\cdot\tau_1 \quad \omega_0 := \frac{1}{\sqrt{b_2}} = 4.61\times 10^{3}\,\frac{1}{s} \quad Q := \frac{\sqrt{b_2}}{b_1} = 4.615\times 10^{-3} \quad \omega_z := \frac{R_1}{L_1}$

$H_{10}(s) := \frac{N_1(s)}{D_1(s)} \quad H_5(s) := \frac{\frac{s}{\omega_z}}{1 + \frac{s}{\omega_0\cdot Q} + \left(\frac{s}{\omega_0}\right)^2} \quad H_{mb} := \frac{\omega_0\cdot Q}{\omega_z} = 2.128\times 10^{-5}$

$20\cdot\log(H_{mb}) = -93.442$

$H_6(s) := H_{mb}\cdot\frac{1}{1 + Q\cdot\left(\frac{\omega_0}{s} + \frac{s}{\omega_0}\right)} \quad \frac{L_1}{L_1 + C_2\cdot R_1\cdot R_2} = 2.128\times 10^{-5}$

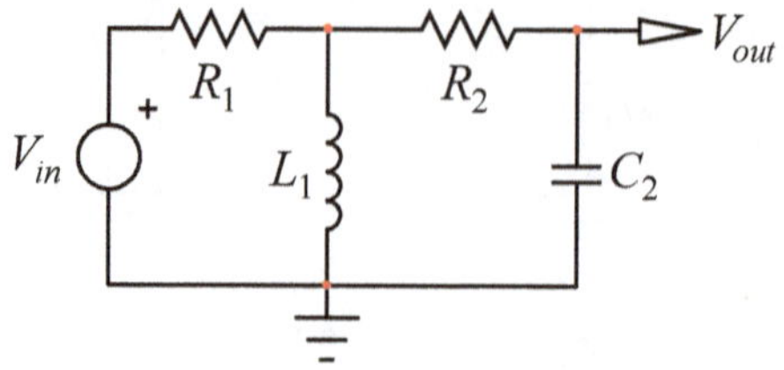

$R_{th}(s) := R_1 \| (s\cdot L_1)$

$H_{ref}(s) := \frac{s\cdot L_1}{R_1 + s\cdot L_1}\cdot\frac{\frac{1}{s\cdot C_2}}{\frac{1}{s\cdot C_2} + R_2 + R_{th}(s)}$

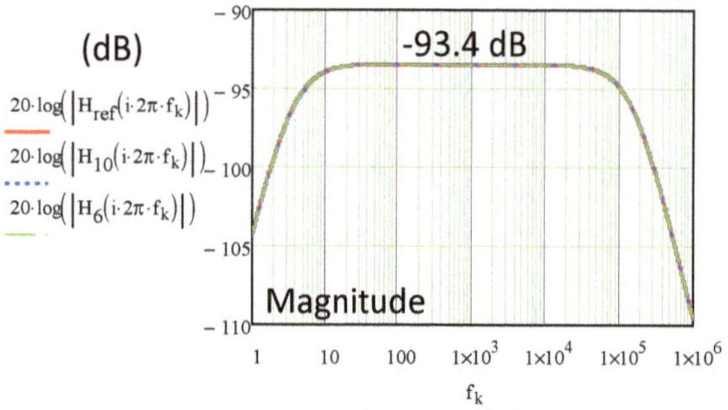

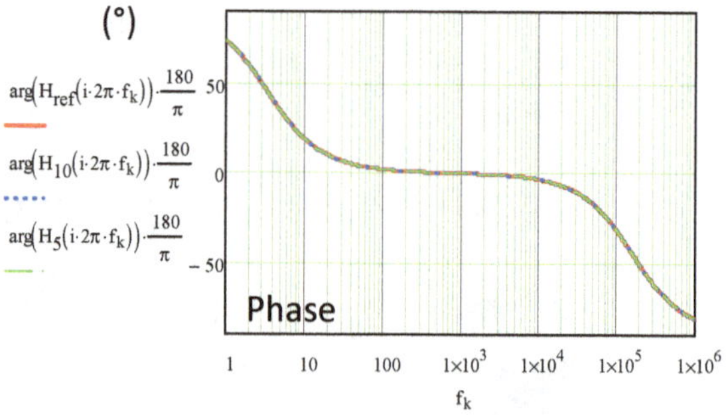

The Final Transfer Function is Reworked to Unveil the Mid-Band Gain as the Leading Term

Figure 6.45

Let's proceed with a simple impedance determination exercise as proposed in Figure 6.46.

A test generator I_T is installed across the connecting terminals. At dc, for $s = 0$, the impedance approaches infinity, but we purposely used the finite value R_{inf} here for further simplifications.

The zero is determined by nulling the response and identifying what impedance combination could bring V_T to 0 V: the impedance made of the series connection of R_1 and C_1 is guilty and brings the zero immediately.

After rearranging the transfer function with an inverted zero, we can check the ac response of this circuit given in Figure 6.47.

This circuit, when associated with an operational transconductance amplifier (OTA), forms a type 2 compensator which is a popular structure in switching converters.

As shown in Figure 6.47, the phase *boost* is maximum at the geometric mean between the pole and the zero and that is usually where you select the crossover frequency. When the pole and zero are coincident, they cancel each other and the ac response represents an integrator.

As you spread them apart—the zero slides towards low frequency while the pole goes up the *x*-axis—the phase curve inflates and peaks to a maximum of 90°.

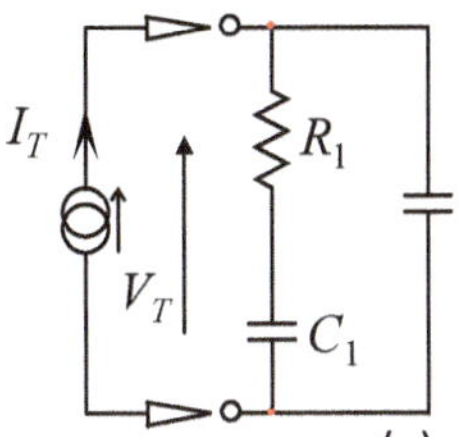

(a)

Set $s = 0$
Open capacitors

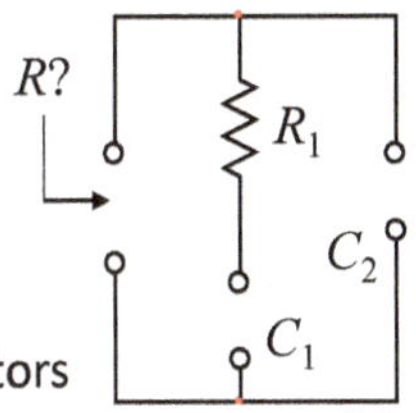

(b)

$R_0 = R_{inf}$

Set stimulus to 0 A and determine the resistance R seen from C_1 and C_2 terminals

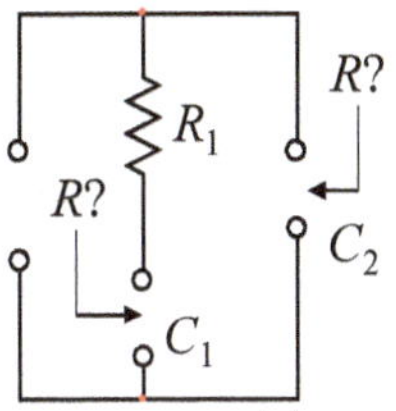

(c)

$$\tau_1 = C_1 R_{\text{inf}}$$

$$\tau_2 = C_2 R_{\text{inf}}$$

$$b_1 = \tau_1 + \tau_2 = R_{\text{inf}}(C_1 + C_2)$$

Determine the resistance R seen from C_2 when C_1 is in high frequency state

$$\tau_2^1 = C_2 R_1$$

$R?$

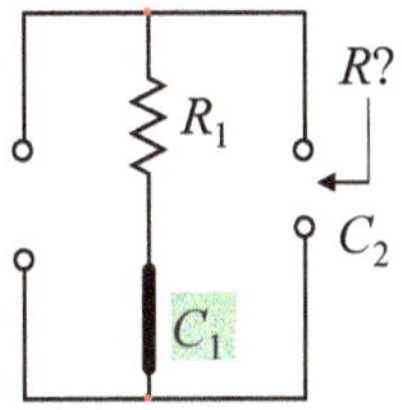

(d)

$$b_2 = \tau_1 \tau_2^1 = R_{\text{inf}} C_1 R_1 C_2$$

$$D(s) = 1 + s b_1 + s^2 b_2$$

$$D(s) = 1 + s R_{\text{inf}}(C_1 + C_2) + s^2 R_{\text{inf}} C_1 R_1 C_2$$

$$D(s) = R_{\text{inf}}\left[\frac{1}{R_{inf}} + s(C_1 + C_2) + s^2 R_1 C_1 C_2\right]$$

What impedance combination could null the response V_T?

$$Z_1(s) = 0$$

$$R_1 + \frac{1}{sC_1} = 0$$

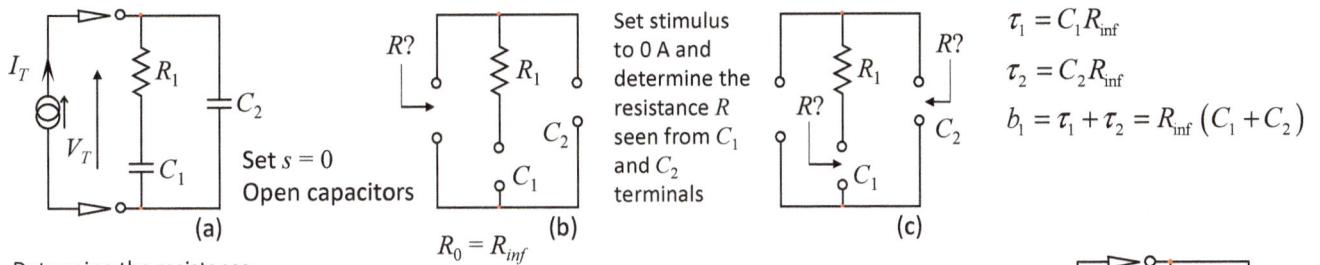

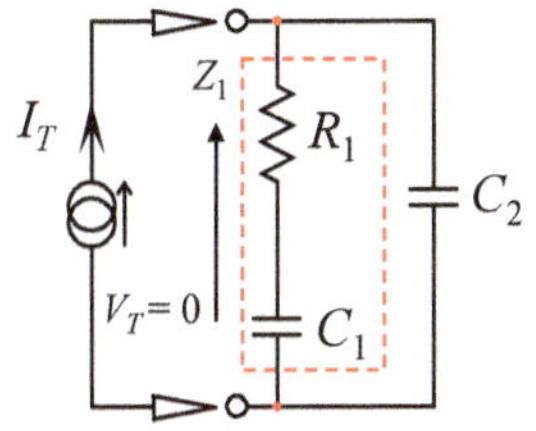

(e)

$$s_z = -\frac{1}{R_1 C_1}$$

$$\omega_z = \frac{1}{R_1 C_1}$$

$$N(s) = 1 + s R_1 C_1$$

$$Z_{in}(s) = R_0 \frac{N(s)}{D(s)} = R_{\text{inf}} \frac{1 + s R_1 C_1}{R_{\text{inf}}\left[\frac{1}{R_{inf}} + s(C_1 + C_2) + s^2 R_1 C_1 C_2\right]}$$

$$Z_{in}(s) = \frac{1 + s R_1 C_1}{s(C_1 + C_2) + s^2 R_1 C_1 C_2} = \frac{s R_1 C_1}{s(C_1 + C_2)} \frac{1 + \frac{1}{s R_1 C_1}}{1 + \frac{s^2 R_1 C_1 C_2}{s(C_1 + C_2)}}$$

$$Z_{in}(s) = \frac{R_1 C_1}{C_1 + C_2} \frac{1 + \frac{1}{s R_1 C_1}}{1 + \frac{s R_1 C_1 C_2}{C_1 + C_2}} = R_{mb} \frac{1 + \frac{\omega_z}{s}}{1 + \frac{s}{\omega_p}}$$

$$\omega_z = \frac{1}{R_1 C_1}$$

$$\omega_p = \frac{C_1 + C_2}{R_1 C_1 C_2}$$

$$R_{mb} = \frac{R_1 C_1}{C_1 + C_2}$$

The Input Impedance of this Simple Circuit is Quickly Determined and Rearranged in a Convenient Form

Figure 6.46

$R_1 := 1500\Omega \quad C_1 := 470nF \quad C_2 := 10nF$

$\|(x,y) := \frac{x \cdot y}{x+y} \quad R_{inf} := 10^{23}\Omega$

$\tau_1 := C_1 \cdot R_{inf} = 4.7\times 10^{16} \cdot s \quad \tau_2 := C_2 \cdot R_{inf} = 1\times 10^{15} \cdot s \quad R_0 := R_{inf}$

$\tau_{12} := C_2 \cdot R_1$

$b_1 := \tau_1 + \tau_2 = 4.8\times 10^{22} \cdot \mu s \quad b_2 := \tau_1 \cdot \tau_{12} = 7.05\times 10^{23} \cdot \mu s^2$

$D_1(s) := 1 + s \cdot b_1 + s^2 \cdot b_2$

$\tau_{1N} := C_1 \cdot R_1 = 7.05\times 10^{-4} \cdot s \quad \tau_{2N} := C_2 \cdot 0 = 0 \cdot s$

$\tau_{12N} := C_2 \cdot 0$

$N_1(s) := 1 + s \cdot R_1 \cdot C_1$

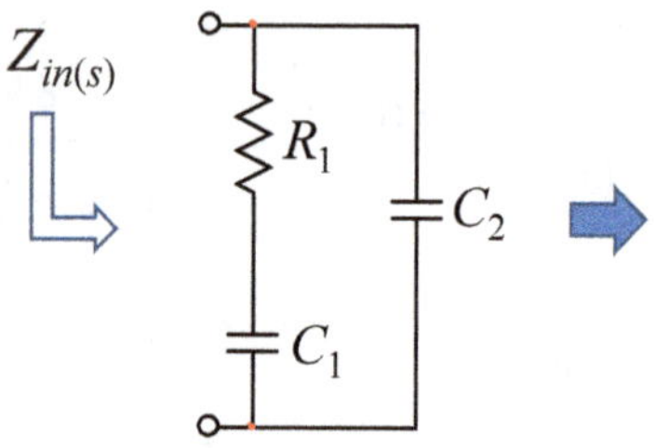

$Z_1(s) := R_1 + \frac{1}{s \cdot C_1}$

$Z_{ref}(s) := Z_1(s) \,\|\, \left(\frac{1}{s \cdot C_2}\right)$

$Z_{10}(s) := R_{inf} \cdot \frac{N_1(s)}{D_1(s)} \quad Z_{30}(s) := R_{inf} \cdot \frac{1 + s \cdot R_1 \cdot C_1}{1 + s \cdot R_{inf} \cdot (C_1 + C_2) + s^2 \cdot C_1 \cdot R_{inf} \cdot C_2 \cdot R_1} \quad Z_{40}(s) := \frac{R_1 \cdot C_1}{C_1 + C_2} \cdot \frac{1 + \frac{1}{s \cdot R_1 \cdot C_1}}{1 + s \cdot \frac{C_1 \cdot C_2}{C_1 + C_2} \cdot R_1}$

$\omega_z := \frac{1}{R_1 \cdot C_1} \quad \omega_p := \frac{C_1 + C_2}{C_1 \cdot C_2 \cdot R_1} \quad R_{mb} := \frac{R_1 \cdot C_1}{C_1 + C_2} \quad Z_{50}(s) := R_{mb} \cdot \frac{1 + \frac{\omega_z}{s}}{1 + \frac{s}{\omega_p}} \quad 20 \cdot \log\left(\frac{R_{mb}}{\Omega}\right) = 63.339 \text{ dBohm} \quad \frac{\sqrt{\omega_p \cdot \omega_z}}{2 \cdot \pi} = 1.564kHz$

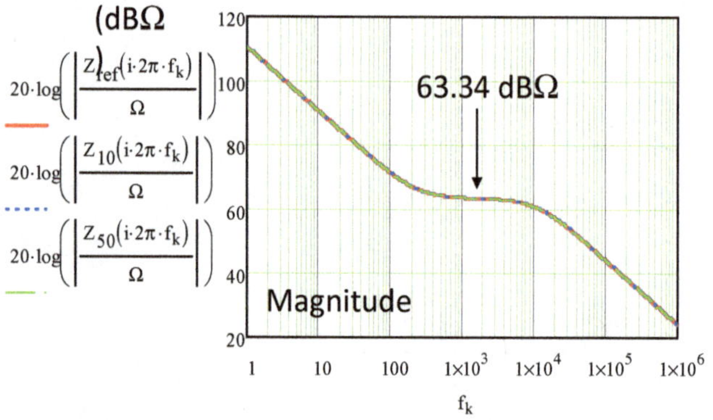

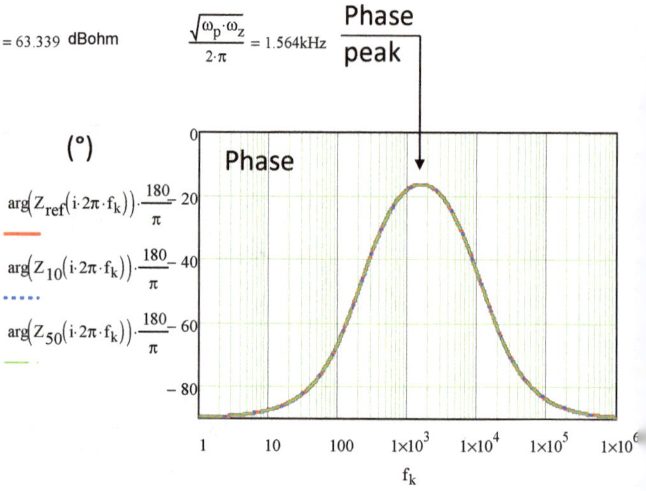

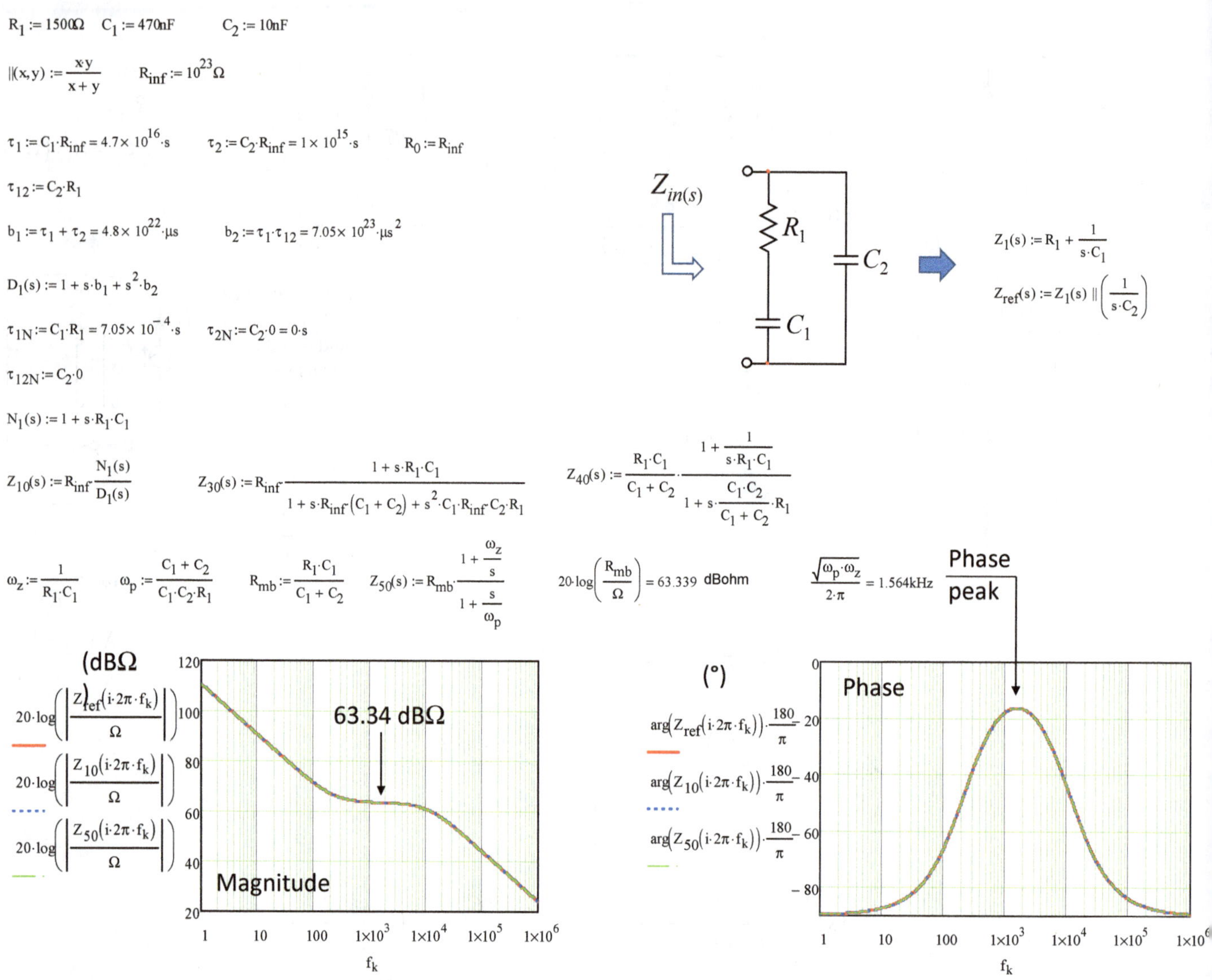

The Inverted Zero Helps Shape the Transfer Function so a Mid-Band Resistance Appears

Figure 6.47

We are back to a simple *LC* network without ohmic losses as drawn in Figure 6.48.

There is no zero in this circuit because if you set L_1 or C_2 in their high-frequency state, the stimulus does not pass through and the response is zero. The ac response is proposed in Figure 6.49. In this sheet, I derived the -3dB cutoff frequency.

To obtain it, you start by replacing s by $j\omega$ in the transfer function and, after expansion, organize the result with ordered real and imaginary parts:

$$\frac{V_{out}(j\omega)}{V_{in}(j\omega)} = \frac{1}{1 + \frac{j\omega}{\omega_0 Q} + \left(\frac{j\omega}{\omega_0}\right)^2} = \frac{1}{1 - \left(\frac{\omega}{\omega_0}\right)^2 + j\frac{\omega}{Q\omega_0}} = \frac{1}{1 - L_1 C_2 \omega^2 + j\frac{L_1}{R_1}\omega} \tag{6.16}$$

Extract the magnitude:

$$\left|\frac{V_{out}(s)}{V_{in}(s)}\right| = \frac{1}{\sqrt{\left(1-L_1C_2\omega^2\right)^2+\left(\frac{L_1}{R_1}\omega\right)^2}} \qquad (6.17)$$

$H_0 = 1$

Set the stimulus to 0 V and determine R seen from L_1's terminals, C_2 is in dc state.

Determine R seen from C_2's terminals, L_1 is in dc state.

$$\tau_1 = \frac{L_1}{R_1}$$

$$\tau_2 = 0\cdot C_2$$

$$b_1 = \tau_1 + \tau_2 = \frac{L_1}{R_1}+0\cdot C_2 = \frac{L_1}{R_1}$$

$$b_2 = \tau_1\tau_2^1 = \frac{L_1}{R_1}C_2R_1$$

Determine the resistance R seen from C_2 when L_1 is in high frequency state

$$D(s)=1+sb_1+s^2b_2=1+s\frac{L_1}{R_1}+s^2\frac{L_1}{R_1}C_2R_1$$

$$D(s)=1+\frac{s}{\omega_0Q}+\left(\frac{s}{\omega_0}\right)^2$$

$$\tau_2^1=C_2R_1$$

$$\omega_0=\frac{1}{\sqrt{b_2}}=\frac{1}{\sqrt{L_1C_2}}$$

$$H(s)=H_0\frac{1}{1+\frac{s}{\omega_0Q}+\left(\frac{s}{\omega_0}\right)^2}=\frac{1}{1+\frac{s}{\omega_0Q}+\left(\frac{s}{\omega_0}\right)^2}$$

$$Q=\frac{\sqrt{b_2}}{b_1}=R_1\sqrt{\frac{C_2}{L_1}}$$

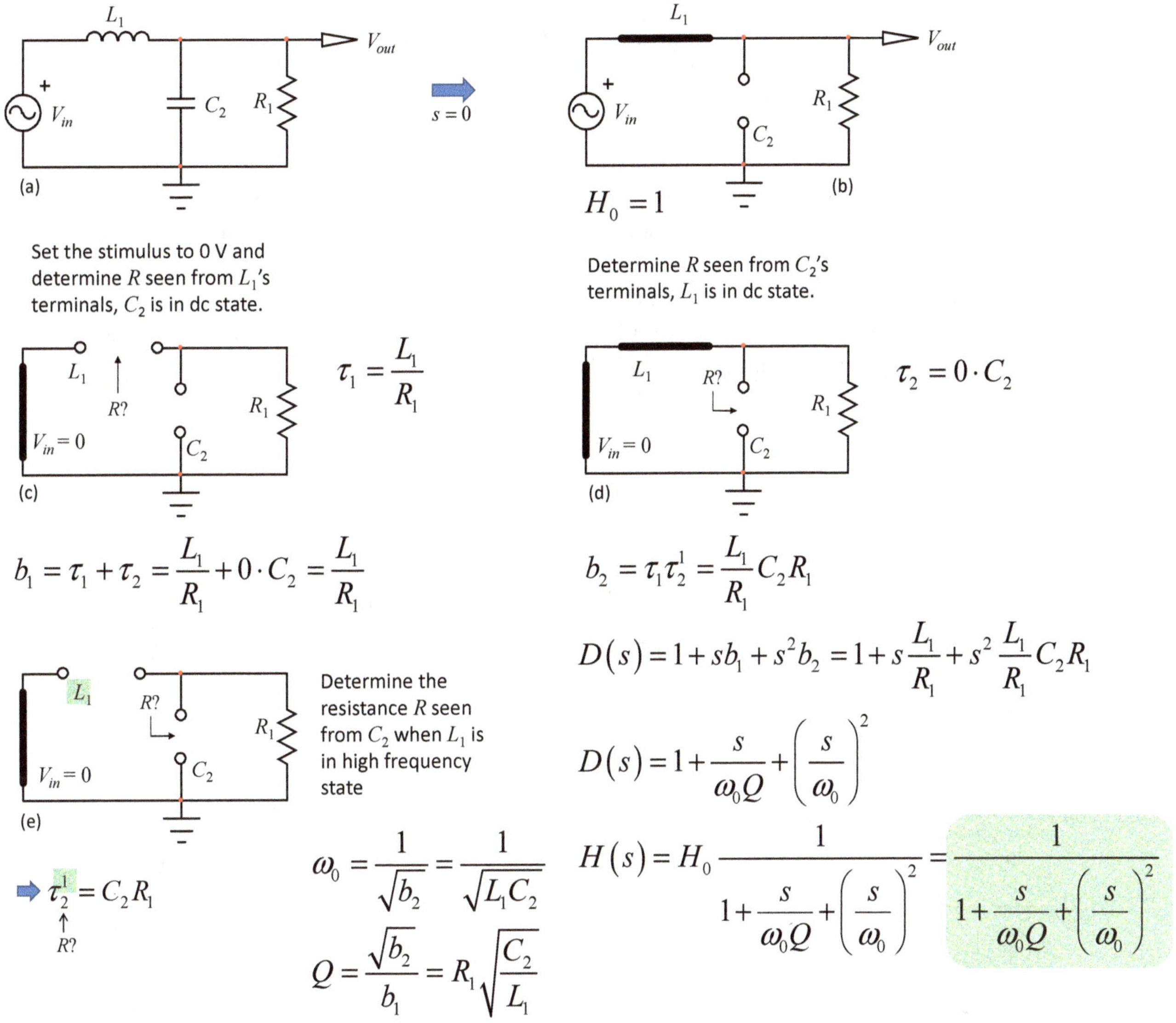

The Transfer Function of this Simple _LC_ Network is Easy to Extract with the FACTs

Figure 6.48

Solve ω_c when this magnitude equals 0.707 or -3 dB:

$$\frac{1}{\sqrt{\left(1 - L_1 C_2 \omega_c^2\right)^2 + \left(\frac{L_1}{R_1}\omega_c\right)^2}} = \frac{1}{\sqrt{2}} \tag{6.18}$$

You obtain:

$$f_c = \frac{1}{2\pi}\sqrt{\frac{\sqrt{8C_2^2 R_1^4 - 4C_2 L_1 R_1^2 + L_1^2} - L_1 + 2C_2 R_1^2}{2C_2^2 L_1 R_1^2}} \tag{6.19}$$

It is now interesting to increase the load resistance and undamp the filter.

The Q peaks to 16 dB in Figure 6.49 and the cutoff frequency increases to 15 kHz with the same natural resonant frequency since neither L or C were touched. If you now drop the loading resistance to 100 Ω, the quality factor collapses to 0.06 and the low-Q approximation can be applied. The transfer function is then rewritten with two cascaded poles whose position depends on Q and w_0.

The original LC filter can now be replaced by two RC filters with an intermediate buffer: the two transfer functions perfectly superimpose in magnitude and phase.

$C_2 := 10nF \qquad L_1 := 25mH \qquad R_1 := 1k\Omega \qquad \|(x,y) := \frac{x\,y}{x+y}$

$H_0 := 1 \qquad 20 \cdot \log(H_0) = 0 \quad dB$

$\tau_1 := \frac{L_1}{R_1} = 25 \cdot \mu s$

$\tau_2 := 0 \cdot C_2 = 0 \cdot \mu s$

$b_1 := \tau_1 + \tau_2 = 25 \cdot \mu s$

$\tau_{12} := C_2 \cdot R_1 = 10 \cdot \mu s$

$b_2 := \tau_1 \cdot \tau_{12} = 250 \cdot \mu s^2$

$H_1(s) := H_0 \cdot \frac{1}{1 + b_1 \cdot s + b_2 \cdot s^2}$

$Q := \frac{\sqrt{b_2}}{b_1} = 0.63246 \quad 20 \cdot \log(Q) = -3.9794 \quad dB \qquad \omega_0 := \frac{1}{\sqrt{b_2}} \qquad f_0 := \frac{\omega_0}{2\cdot\pi} = 10.06584kHz$

$H_3(s) := H_0 \cdot \frac{1}{1 + \frac{s}{\omega_0 \cdot Q} + \left(\frac{s}{\omega_0}\right)^2} \qquad R_1 \cdot \sqrt{\frac{C_2}{L_1}} = 0.63246 \qquad \frac{1}{2\cdot\pi\cdot\sqrt{L_1 \cdot C_2}} = 10.06584kHz$

$f_a := \sqrt{\frac{\sqrt{8\cdot C_2^2 \cdot R_1^4 - 4\cdot C_2 \cdot L_1 \cdot R_1^2 + L_1^2} - L_1 + 2\cdot C_2 \cdot R_1^2}{2\cdot C_2^2 \cdot L_1 \cdot R_1^2}} \cdot \frac{1}{2\cdot\pi} = 8.89433kHz \qquad \omega_a := 2\cdot\pi\cdot f_a$

-3 dB cutoff

$H_{mag}(\omega) := \frac{1}{\sqrt{\left(1 - C_2 \cdot L_1 \cdot \omega^2\right)^2 + \left(\frac{L_1}{R_1}\cdot\omega\right)^2}}$

$H_{mag}(\omega_a) = 0.70711 \quad 20 \cdot \log(H_{mag}(\omega_a)) = -3.0103 \quad dB$

$H_{ref}(s) := \frac{R_1 \| \left(\frac{1}{s \cdot C_2}\right)}{R_1 \| \left(\frac{1}{s \cdot C_2}\right) + s \cdot L_1}$

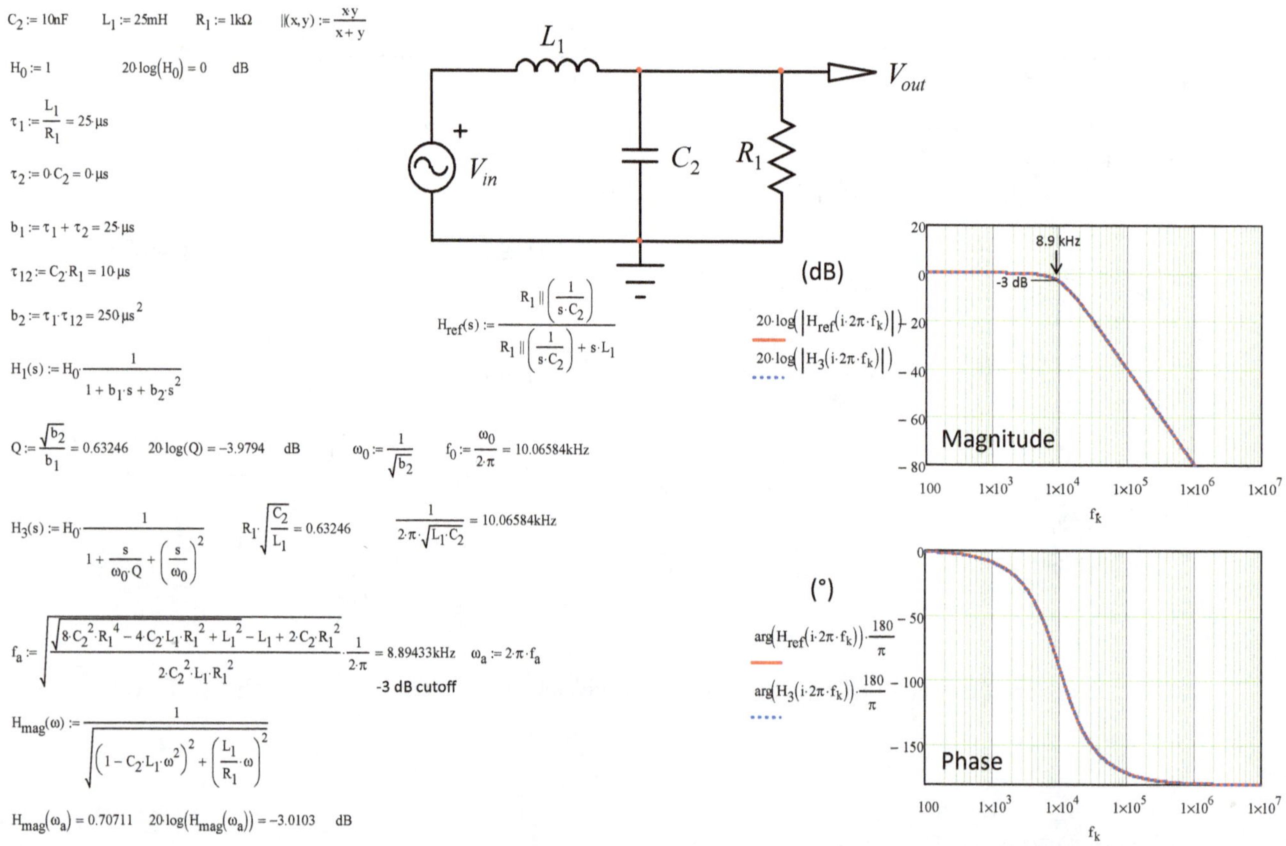

In This Example, the Quality Factor is Moderately Low and There is no Peaking

Figure 6.49

$C_2 := 10\text{nF} \qquad L_1 := 25\text{mH} \qquad R_1 := 10\text{k}\Omega \qquad \|(x,y) := \dfrac{x \cdot y}{x + y}$

$Q := \dfrac{\sqrt{b_2}}{b_1} = 6.32456 \qquad 20 \cdot \log(Q) = 16.0206 \quad \text{dB}$

$f_a := \sqrt{\dfrac{\sqrt{8 \cdot C_2^2 \cdot R_1^4 - 4 \cdot C_2 \cdot L_1 \cdot R_1^2 + L_1^2} - L_1 + 2 \cdot C_2 \cdot R_1^2}{2 \cdot C_2^2 \cdot L_1 \cdot R_1^2}} \cdot \dfrac{1}{2 \cdot \pi} = 15.57086\text{kHz}$

If the loading resistor R_1 now decreases to 100 Ω
Q drops to 0.06 and the low-Q approximation holds

$C_2 := 10\text{nF} \qquad L_1 := 25\text{mH} \qquad R_1 := 0.1\text{k}\Omega \qquad \|(x,y) := \dfrac{x \cdot y}{x + y}$

$Q := \dfrac{\sqrt{b_2}}{b_1} = 0.06325 \qquad 20 \cdot \log(Q) = -23.9794 \quad \text{dB}$

$\omega_{p1} := Q \cdot \omega_0 \qquad \omega_{p2} := \dfrac{\omega_0}{Q} \qquad R_p := 1\text{k}\Omega$

$C_{p1} := \dfrac{1}{R_p \cdot \omega_{p1}} = 250\text{nF} \qquad C_{p2} := \dfrac{1}{R_p \cdot \omega_{p2}} = 1\text{nF}$

$H_{eq}(s) := \dfrac{1}{\left(1 + \dfrac{s}{\omega_{p1}}\right)\cdot\left(1 + \dfrac{s}{\omega_{p2}}\right)}$

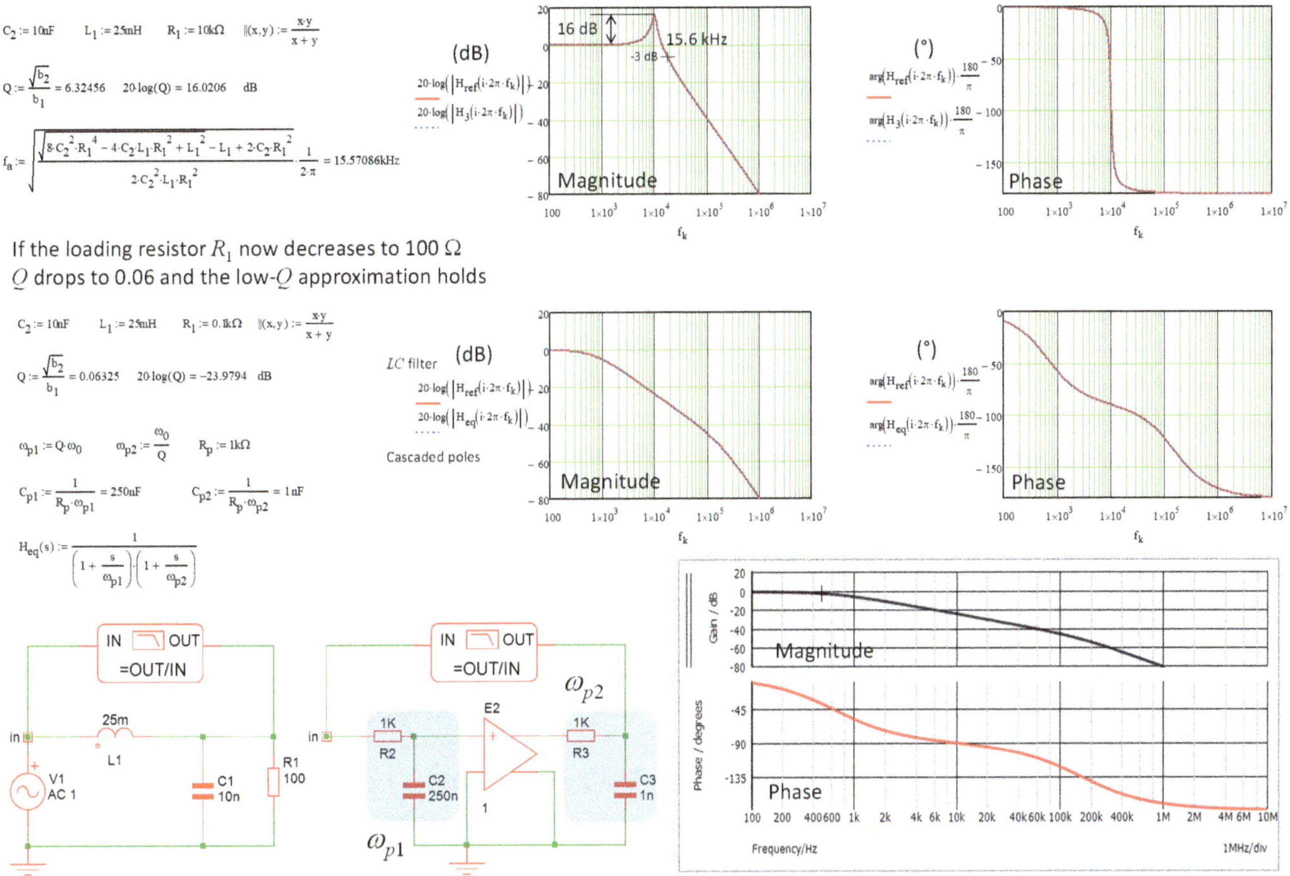

If the Quality Factor is High, a Peak in Magnitude is Observed and the Cutoff Frequency Changes Despite a Similar Natural Resonant Frequency. If Q is very Low, the Low-Q Approximation Holds with Two Separate Poles

Figure 6.50

If a capacitor and an inductor are paralleled, a resonant circuit is created as shown in Figure 6.51.

In this example, a resistive divider drives the resonant network featuring a zero at the origin (L_1 is a short circuit at dc).

The transfer function is rearranged to unveil the peak amplitude as described in Figure 6.52.

The filter is tuned at 1.6 MHz and offers a very selective magnitude response.

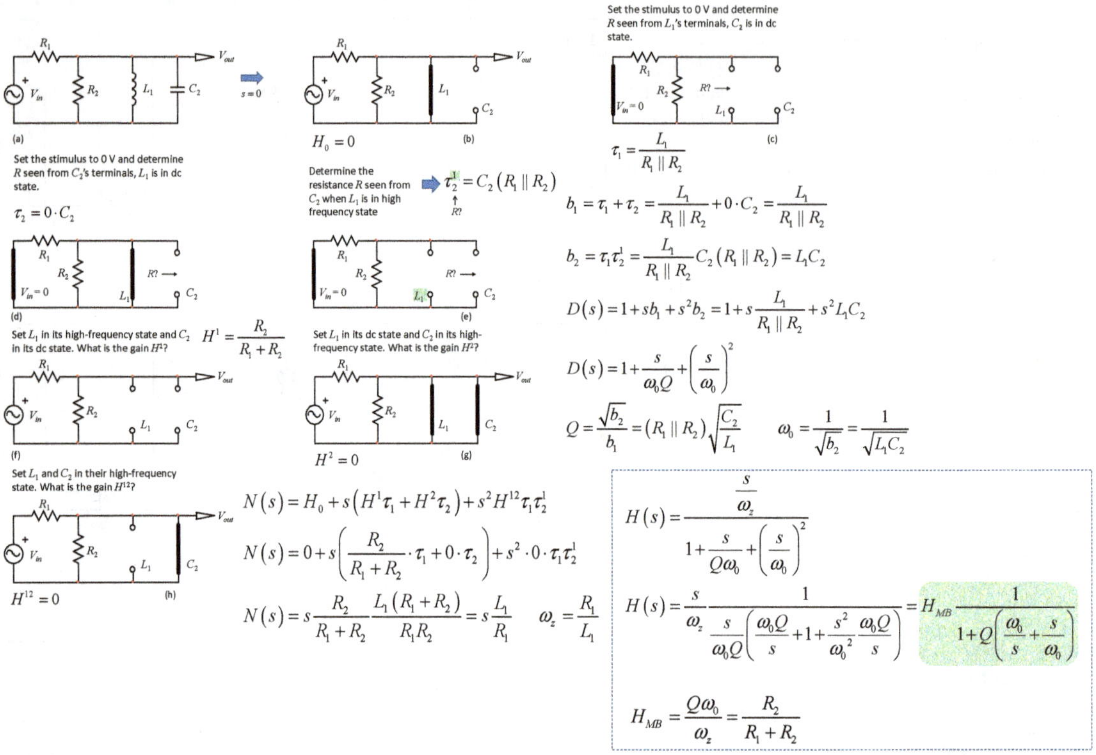
Set the stimulus to 0 V and determine R seen from L1's terminals, C2 is in dc state.
Set the stimulus to 0 V and determine R seen from C2's terminals, L1 is in dc state.
Determine the resistance R seen from C2 when L1 is in high frequency state
Set L1 in its high-frequency state and C2 in its dc state. What is the gain H1?
Set L1 in its dc state and C2 in its high-frequency state. What is the gain H2?
Set L1 and C2 in their high-frequency state. What is the gain H12?

This Resonant Network Peaks at a Value set by the Resistive Divider made of R_1 and R_2

Figure 6.51

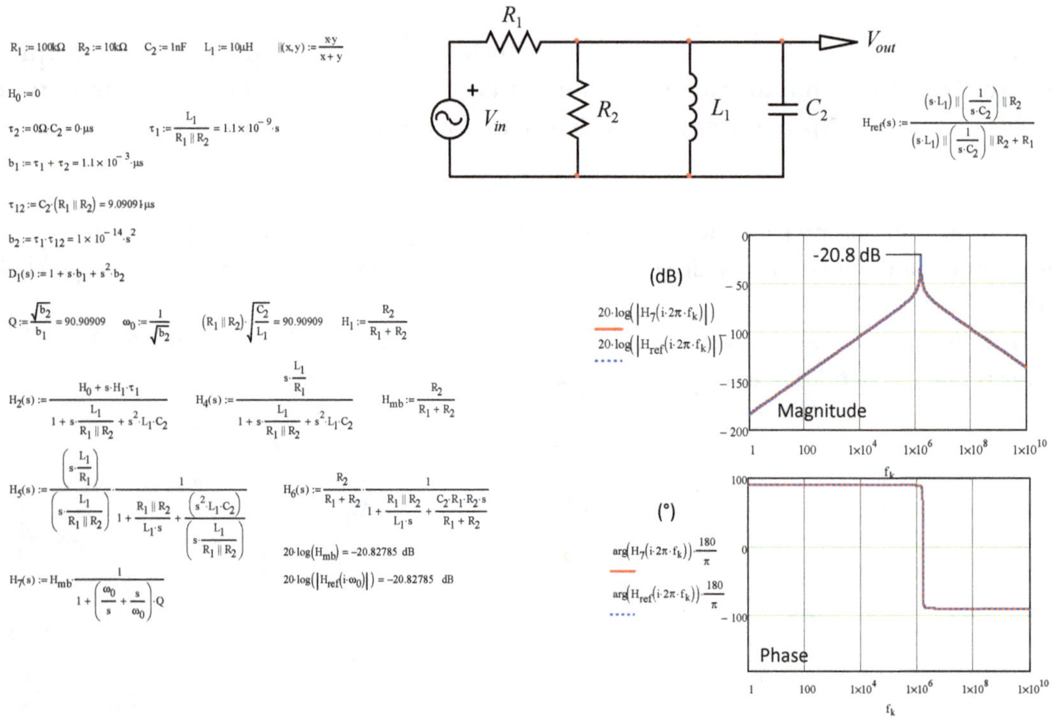
-20.8 dB
(dB)
Magnitude
(°)
Phase

The Quality Factor is Quite High for this Selective Filter

Figure 6.52

We now look at another passive filter, this time with the series-pass connection of the capacitor and the inductor as shown in Figure 6.53.

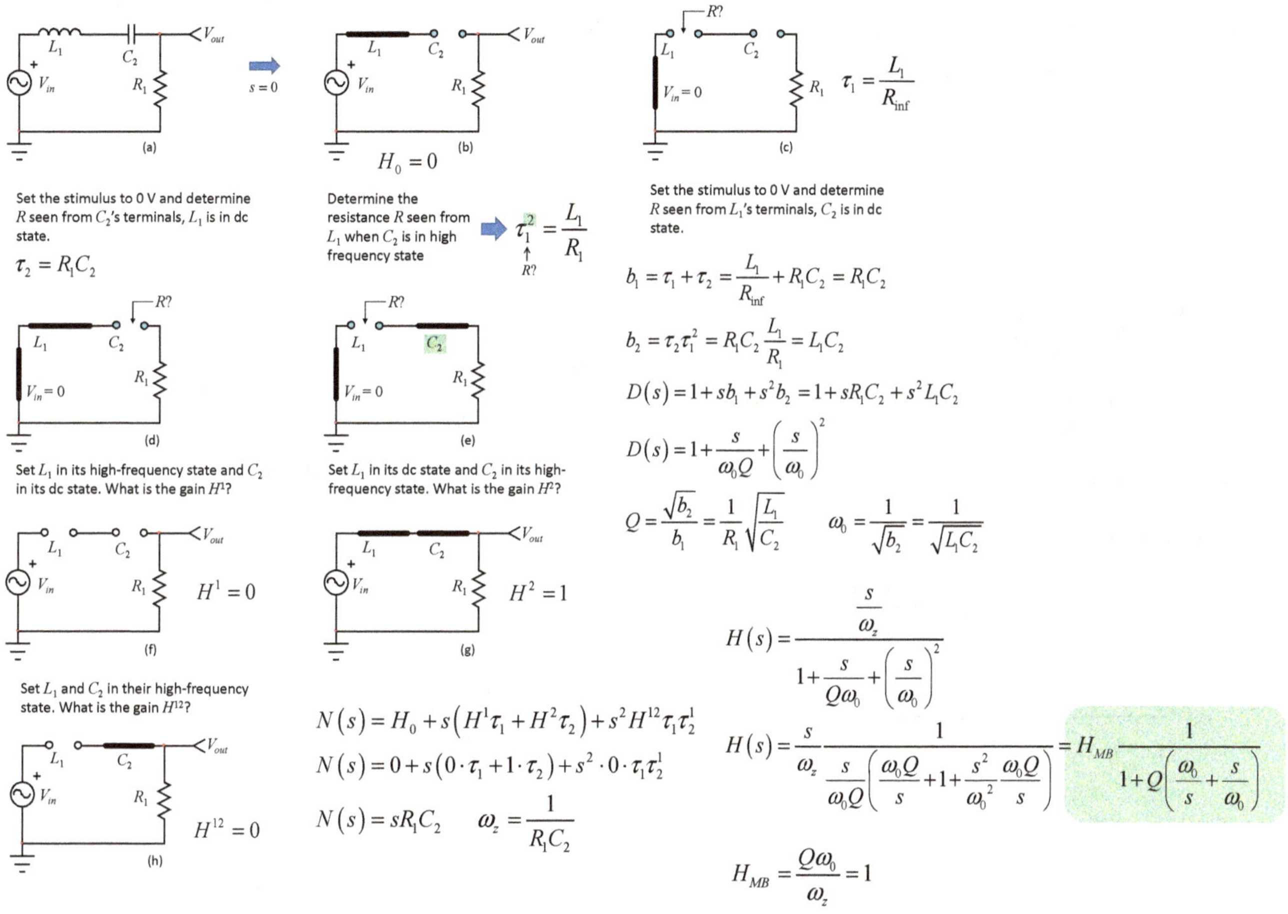
(a)
Set the stimulus to 0 V and determine R seen from C_2's terminals, L_1 is in dc state.
$\tau_2 = R_1C_2$
$s = 0$
(b)
$H_0 = 0$
Determine the resistance R seen from L_1 when C_2 is in high frequency state
$\tau_1^2 = \frac{L_1}{R_1}$
(c)
Set the stimulus to 0 V and determine R seen from L_1's terminals, C_2 is in dc state.
$\tau_1 = \frac{L_1}{R_{inf}}$
$b_1 = \tau_1 + \tau_2 = \frac{L_1}{R_{inf}} + R_1C_2 = R_1C_2$
$b_2 = \tau_2\tau_1^2 = R_1C_2\frac{L_1}{R_1} = L_1C_2$
$D(s) = 1 + sb_1 + s^2b_2 = 1 + sR_1C_2 + s^2L_1C_2$
$D(s) = 1 + \frac{s}{\omega_0 Q} + \left(\frac{s}{\omega_0}\right)^2$
$Q = \frac{\sqrt{b_2}}{b_1} = \frac{1}{R_1}\sqrt{\frac{L_1}{C_2}}$
$\omega_0 = \frac{1}{\sqrt{b_2}} = \frac{1}{\sqrt{L_1C_2}}$
(d)
Set L_1 in its high-frequency state and C_2 in its dc state. What is the gain H^1?
(e)
Set L_1 in its dc state and C_2 in its high-frequency state. What is the gain H^2?
(f)
$H^1 = 0$
(g)
$H^2 = 1$
Set L_1 and C_2 in their high-frequency state. What is the gain H^{12}?
(h)
$H^{12} = 0$
$N(s) = H_0 + s(H^1\tau_1 + H^2\tau_2) + s^2H^{12}\tau_1\tau_2^1$
$N(s) = 0 + s(0\cdot\tau_1 + 1\cdot\tau_2) + s^2\cdot 0\cdot\tau_1\tau_2^1$
$N(s) = sR_1C_2 \quad \omega_z = \frac{1}{R_1C_2}$
$H(s) = \frac{\frac{s}{\omega_z}}{1 + \frac{s}{Q\omega_0} + \left(\frac{s}{\omega_0}\right)^2}$
$H(s) = \frac{s}{\omega_z}\frac{1}{\frac{s}{\omega_0 Q}\left(\frac{\omega_0 Q}{s} + 1 + \frac{s^2}{\omega_0^2}\frac{\omega_0 Q}{s}\right)} = H_{MB}\frac{1}{1 + Q\left(\frac{\omega_0}{s} + \frac{s}{\omega_0}\right)}$
$H_{MB} = \frac{Q\omega_0}{\omega_z} = 1$

The Inductor and the Capacitor are Now Connected in Series with the Source while the Response is Observed across the Resistor

Figure 6.53

There is nothing particularly complicated here.

The ac response presented in Figure 6.54 confirms our analysis.

$R_1 := 50\Omega \quad C_2 := 10nF \quad L_1 := 10\mu H \quad \|(x,y) := \frac{x \cdot y}{x+y} \quad R_{inf} := 10^{23}\Omega$

$H_0 := 0$

$\tau_1 := \frac{L_1}{R_{inf}} = 0 \cdot s \qquad \tau_2 := R_1 \cdot C_2 = 0.5 \cdot \mu s$

$b_1 := \tau_1 + \tau_2 = 0.5 \cdot \mu s$

$\tau_{12} := C_2 \cdot R_{inf} = 1 \times 10^{21} \cdot \mu s \qquad \tau_{21} := \frac{L_1}{R_1} = 0.2 \mu s$

$b_2 := \tau_2 \cdot \tau_{21} = 1 \times 10^{-13} \cdot s^2 \qquad b_{2a} := \tau_1 \cdot \tau_{12} = 1 \times 10^{-13} \cdot s^2$

$D_1(s) := 1 + s \cdot b_1 + s^2 \cdot b_2$

$Q := \frac{\sqrt{b_2}}{b_1} = 0.63246 \qquad \omega_0 := \frac{1}{\sqrt{b_2}} = 3.16228 \times 10^6 \frac{1}{s}$

$\frac{1}{R_1} \cdot \sqrt{\frac{L_1}{C_2}} = 0.63246 \qquad \frac{1}{\sqrt{L_1 \cdot C_2}} = 3.16228 \times 10^6 \frac{1}{s}$

$H_1 := 0 \qquad H_2 := 1 \qquad H_{21} := 0$

$N_1(s) := H_0 + s \cdot (H_1 \cdot \tau_1 + H_2 \cdot \tau_2) + s^2 \cdot H_{21} \cdot \tau_2 \cdot \tau_{21}$

$H_{10}(s) := \frac{N_1(s)}{D_1(s)} \qquad H_{20}(s) := \frac{s \cdot R_1 \cdot C_2}{1 + s \cdot R_1 \cdot C_2 + s^2 \cdot L_1 \cdot C_2} \qquad \omega_z := \frac{1}{R_1 \cdot C_2}$

$H_3(s) := \frac{\frac{s}{\omega_z}}{1 + \frac{s}{\omega_0 \cdot Q} + \left(\frac{s}{\omega_0}\right)^2} \qquad H_{mb} := \frac{\omega_0 \cdot Q}{\omega_z} = 1$

$H_4(s) := H_{mb} \cdot \frac{1}{1 + Q \cdot \left(\frac{s}{\omega_0} + \frac{\omega_0}{s}\right)}$

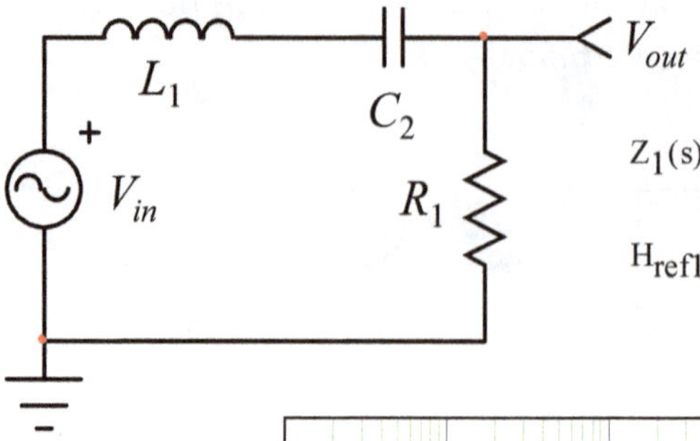

$Z_1(s) := \frac{1}{s \cdot C_2} + s \cdot L_1$

$H_{ref1}(s) := \frac{R_1}{Z_1(s) + R_1}$

(dB)

$20 \cdot \log(|H_{ref1}(i \cdot 2\pi \cdot f_k)|, 10)$

$20 \cdot \log(|H_4(i \cdot 2\pi \cdot f_k)|, 10)$

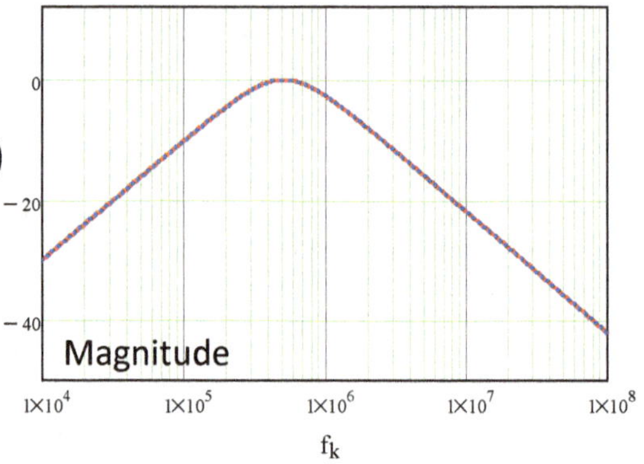

(°)

$\arg(H_{ref1}(i \cdot 2\pi \cdot f_k)) \cdot \frac{180}{\pi}$

$\arg(H_4(i \cdot 2\pi \cdot f_k)) \cdot \frac{180}{\pi}$

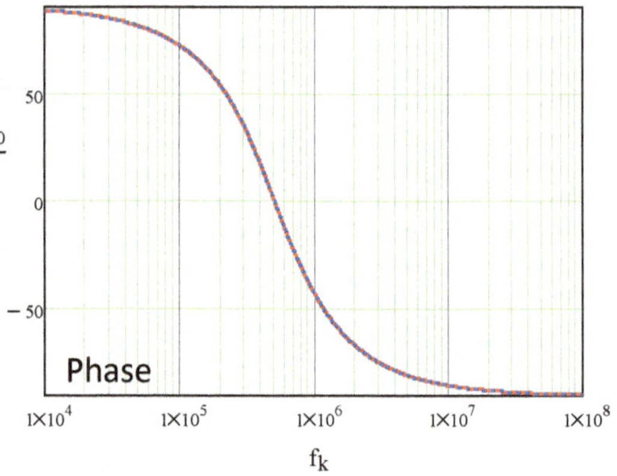

This Passive Filter Offers a Bandpass Ac Response

Figure 6.54

In Figure 6.55, two capacitors are connected in series with resistor R_1 bypassing them while R_2 pulls their junction to ground.

The ac results from Figure 6.56 show the expression obtained by the FACTs versus the brute-force result derived with superposition.

As confirmed by the magnitude difference shown in lower left corner, the transfer functions are rigorously identical.

Phase difference (not shown) has a similar result.

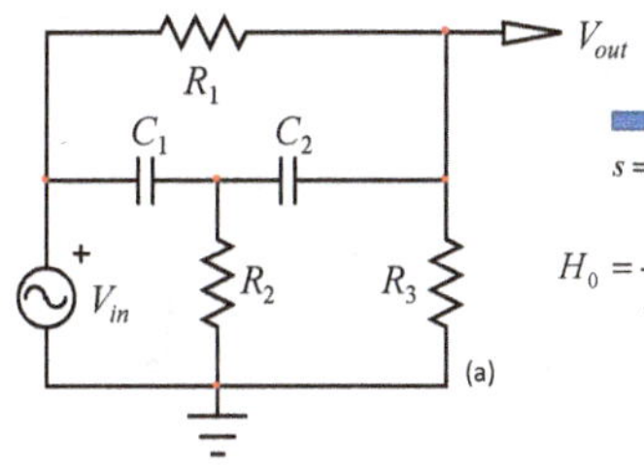

$s = 0$

$$H_0 = \frac{R_3}{R_3 + R_1}$$

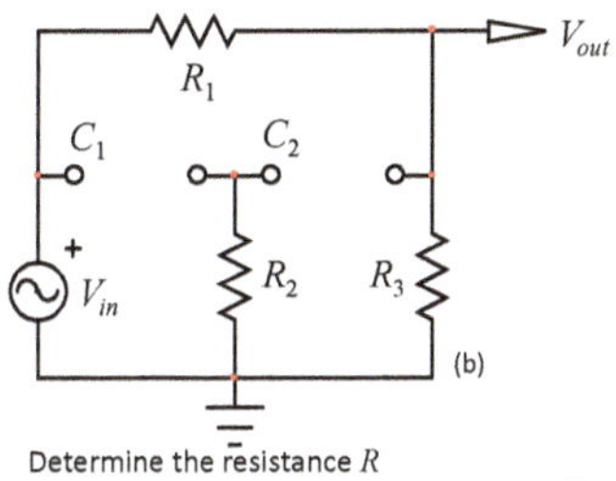

Set the stimulus to 0 V and determine R seen from C_1's terminals while C_2 is in dc state and vice versa:

Determine the resistance R seen from C_2 when C_1 is in high frequency state:

$$\tau_2^1 = C_2\left(R_1 \,||\, R_3\right)$$

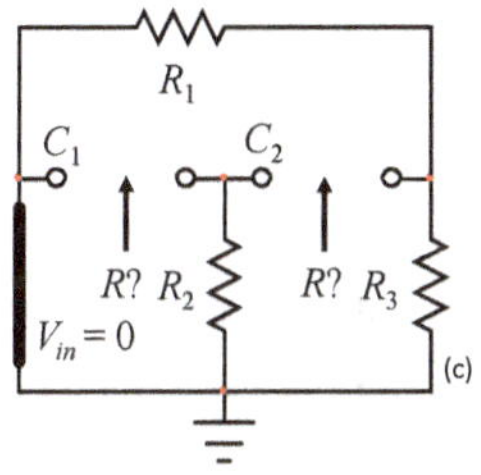

$$\tau_1 = R_2 C_1$$

$$\tau_2 = C_2\left(R_2 + R_1 \,||\, R_3\right)$$

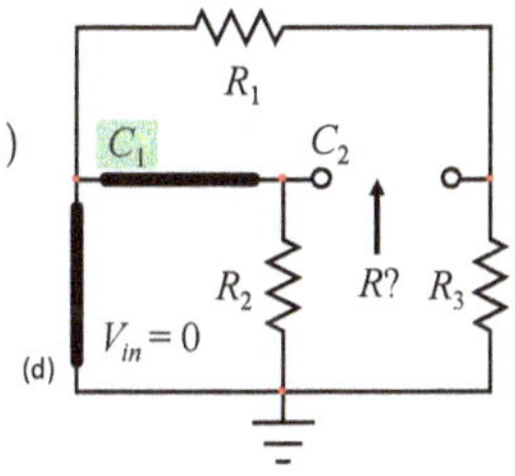

$$b_1 = \tau_1 + \tau_2 = R_2 C_1 + C_2\left(R_2 + R_1 \,||\, R_3\right)$$

$$b_2 = \tau_1 \tau_2^1 = R_2 C_1 C_2\left(R_1 \,||\, R_3\right)$$

$$D(s) = 1 + s b_1 + s^2 b_2 = 1 + s\left[R_2 C_1 + C_2\left(R_2 + R_1 \,||\, R_3\right)\right] + s^2 R_2 C_1 C_2\left(R_1 \,||\, R_3\right)$$

$$D(s) = 1 + \frac{s}{\omega_{0D} Q_D} + \left(\frac{s}{\omega_{0D}}\right)^2$$

$$Q_D = \frac{\sqrt{b_2}}{b_1} = \frac{\sqrt{R_2 C_1 C_2\left(R_1 \,||\, R_3\right)}}{R_2 C_1 + C_2\left(R_2 + R_1 \,||\, R_3\right)} \qquad \omega_{0D} = \frac{1}{\sqrt{b_2}} = \frac{1}{\sqrt{R_2 C_1 C_2\left(R_1 \,||\, R_3\right)}}$$

Set C_1 in its high-frequency state and C_2 in its dc state. What is the gain H^1?

Set C_1 in its dc state and C_2 in its high-frequency state. What is the gain H^2?

Set C_1 and C_2 in their high-frequency state. What is the gain H^{12}?

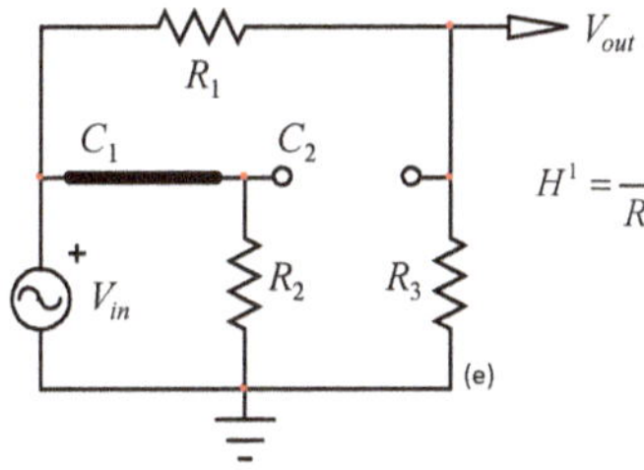

$$H^1 = \frac{R_3}{R_3 + R_1}$$

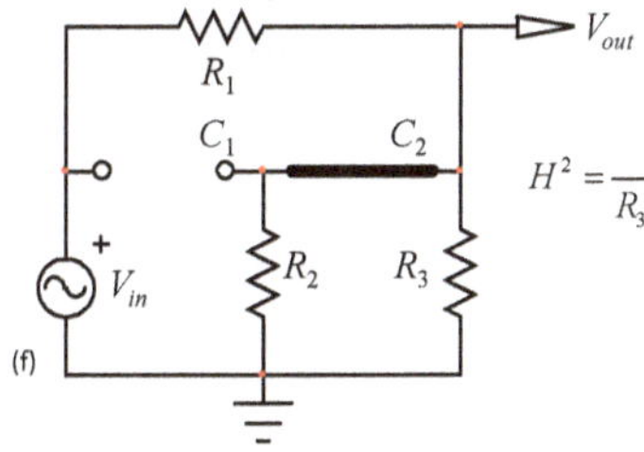

$$H^2 = \frac{R_3 \,||\, R_2}{R_3 \,||\, R_2 + R_1}$$

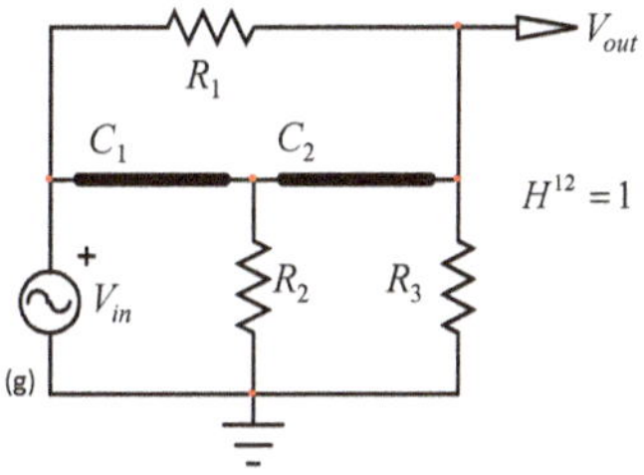

$$H^{12} = 1$$

$$N(s) = H_0 + s\left(H^1 \tau_1 + H^2 \tau_2\right) + s^2 H^{12} \tau_1 \tau_2^1$$

$$N(s) = H_0 + s\left(\frac{R_3}{R_3 + R_1}\tau_1 + \frac{R_3 \,||\, R_2}{R_3 \,||\, R_2 + R_1}\tau_2\right) + s^2 \cdot 1 \cdot \tau_1 \tau_2^1$$

$$N(s) = H_0\left[1 + s\frac{1}{H_0}\left(\frac{R_3}{R_3 + R_1}\tau_1 + \frac{R_3 \,||\, R_2}{R_3 \,||\, R_2 + R_1}\tau_2\right) + s^2 \frac{\tau_1 \tau_2^1}{H_0}\right]$$

$$N(s) = H_0\left[1 + \frac{s}{\omega_{0N} Q_N} + \left(\frac{s}{\omega_{0N}}\right)^2\right]$$

$$Q_N = \frac{\sqrt{a_2}}{a_1} = \frac{\sqrt{C_1 C_2 R_1 R_2}}{R_2\left(C_1 + C_2\right)} \qquad \omega_{0N} = \frac{1}{\sqrt{a_2}} = \frac{1}{\sqrt{C_1 C_2 R_1 R_2}}$$

$$H(s) = \frac{R_3}{R_3 + R_1} \frac{1 + s R_2\left(C_1 + C_2\right) + s^2 C_1 C_2 R_1 R_2}{1 + s\left[R_2 C_1 + C_2\left(R_2 + R_1 \,||\, R_3\right)\right] + s^2 R_2 C_1 C_2\left(R_1 \,||\, R_3\right)}$$

$$H(s) = H_0 \frac{+\frac{s}{\omega_{0N} Q_N} + \left(\frac{s}{\omega_{0N}}\right)^2}{1 + \frac{s}{\omega_{0D} Q_D} + \left(\frac{s}{\omega_{0D}}\right)^2}$$

The Capacitors are Connected in Series with a Pull-Down Resistor at the Junction

Figure 6.55

$R_1 := 1k\Omega$ $R_2 := 1k\Omega$ $R_3 := 1k\Omega$ $C_1 := 22nF$ $C_2 := 22nF$ $\|(x,y) := \dfrac{x\,y}{x+y}$

$H_0 := \dfrac{R_3}{R_1+R_3} = 0.5$ $20\cdot\log(H_0) = -6.021$ dB

$\tau_1 := R_2\cdot C_1 = 22\cdot\mu s$

$\tau_2 := (R_1 \| R_3 + R_2)\cdot C_2 = 33\cdot\mu s$

$b_1 := \tau_1 + \tau_2 = 5.5\times 10^{-5}\cdot s$

$\tau_{12} := C_2\cdot(R_1 \| R_3) = 11\cdot\mu s$

$b_2 := \tau_1\cdot\tau_{12} = 242\cdot\mu s^2$

$D_1(s) := 1 + s\cdot b_1 + s^2\cdot b_2$

$Q_D := \dfrac{\sqrt{b_2}}{b_1} = 0.283$ $\omega_{0D} := \dfrac{1}{\sqrt{b_2}} = 6.428\times 10^4\,\dfrac{1}{s}$

$\dfrac{\sqrt{R_2\cdot C_1\cdot C_2\cdot(R_1 \| R_3)}}{R_2\cdot C_1 + (R_1 \| R_3 + R_2)\cdot C_2} = 0.283$ $\dfrac{1}{\sqrt{R_2\cdot C_1\cdot C_2\cdot(R_1 \| R_3)}} = 6.428\times 10^4\,\dfrac{1}{s}$

$H_1 := \dfrac{R_3}{R_3+R_1} = 0.5$ $H_2 := \dfrac{R_3 \| R_2}{R_1 + R_3 \| R_2} = 0.333$ $H_{12} := 1$

$a_1 := \dfrac{H_1\cdot\tau_1 + H_2\cdot\tau_2}{H_0}$ $a_2 := \dfrac{H_{12}\cdot\tau_1\cdot\tau_{12}}{H_0}$

$N_1(s) := 1 + s\cdot\left(\dfrac{H_1\cdot\tau_1 + H_2\cdot\tau_2}{H_0}\right) + s^2\cdot\left(\dfrac{H_{12}}{H_0}\cdot\tau_1\cdot\tau_{12}\right)$

$Q_N := \dfrac{\sqrt{a_2}}{a_1} = 0.5$ $\omega_{0N} := \dfrac{1}{\sqrt{a_2}} = 4.545\times 10^4\,\dfrac{1}{s}$

$\dfrac{\sqrt{C_1\cdot C_2\cdot R_1\cdot R_2}}{R_2\cdot(C_1 + C_2)} = 0.5$ $\dfrac{1}{\sqrt{C_1\cdot C_2\cdot R_1\cdot R_2}} = 4.545\times 10^4\,\dfrac{1}{s}$

$H_2(s) := H_0\cdot\dfrac{N_1(s)}{D_1(s)}$ $H_5(s) := H_0\cdot\dfrac{1 + \dfrac{s}{\omega_{0N}\cdot Q_N} + \left(\dfrac{s}{\omega_{0N}}\right)^2}{1 + \dfrac{s}{\omega_{0D}\cdot Q_D} + \left(\dfrac{s}{\omega_{0D}}\right)^2}$

$H_4(s) := \dfrac{R_3}{R_1+R_3}\cdot\dfrac{1 + s\cdot[R_2\cdot(C_1+C_2)] + s^2\cdot(C_1\cdot C_2\cdot R_1\cdot R_2)}{1 + s\cdot[R_2\cdot C_1 + (R_1 \| R_3 + R_2)\cdot C_2] + s^2\cdot[R_2\cdot C_1\cdot[C_2\cdot(R_1 \| R_3)]]}$

$V_1(s) := \dfrac{\left[\dfrac{1}{C_2\cdot s} + \left(\dfrac{1}{s\cdot C_1}\right) \| R_2\right] \| R_3}{R_1 + \left[\dfrac{1}{C_2\cdot s} + \left(\dfrac{1}{s\cdot C_1}\right) \| R_2\right] \| R_3}$ Ground C1 (the left side) and determine V1

Superposition for deriving the transfer function

$V_2(s) := \dfrac{R_2}{R_2 + \dfrac{1}{s\cdot C_1}}\cdot\dfrac{R_1 \| R_3}{R_1 \| R_3 + \left[\left(\dfrac{1}{s\cdot C_1}\right) \| R_2\right] + \dfrac{1}{s\cdot C_2}}$ Ground R1 and determine V2

$H_{ref}(s) := V_1(s) + V_2(s)$

Magnitude

$20\cdot\log\left(\left|H_5(i\cdot 2\pi\cdot f_k)\right|\right)$
$20\cdot\log\left(\left|H_4(i\cdot 2\pi\cdot f_k)\right|\right)$
$20\cdot\log\left(\left|H_{ref}(i\cdot 2\pi\cdot f_k)\right|\right)$

Difference between the magnitudes of H_{ref} and H_4

Phase

$\arg\left(H_4(i\cdot 2\pi\cdot f_k)\right)\cdot\dfrac{180}{\pi}$
$\arg\left(H_5(i\cdot 2\pi\cdot f_k)\right)\cdot\dfrac{180}{\pi}$
$\arg\left(H_{ref}(i\cdot 2\pi\cdot f_k)\right)\cdot\dfrac{180}{\pi}$

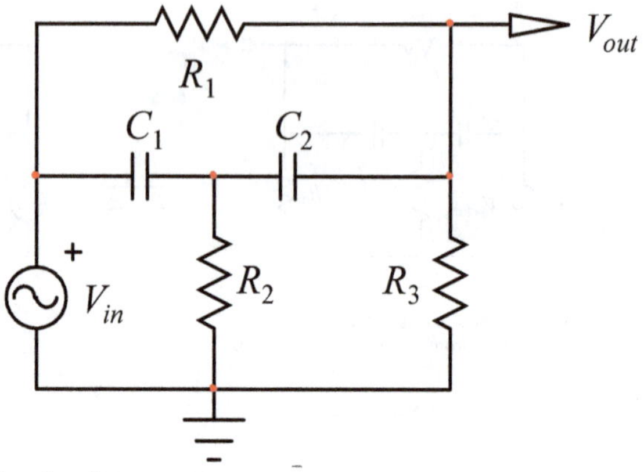

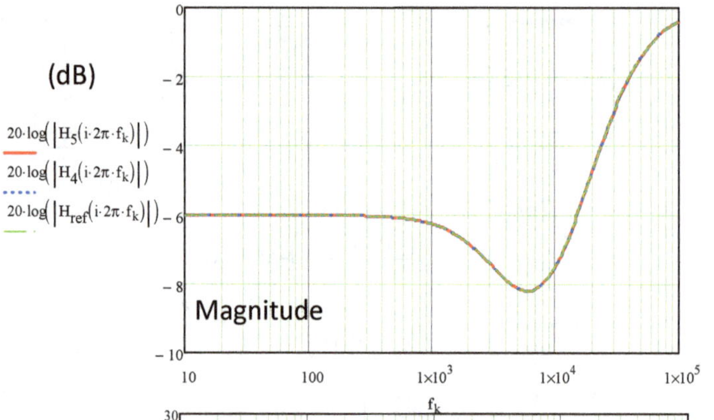

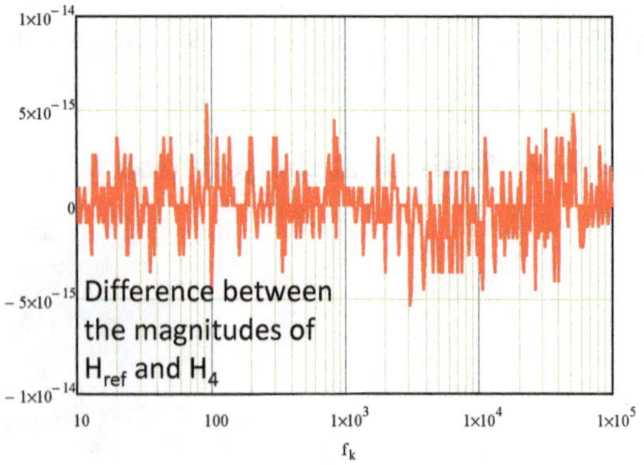

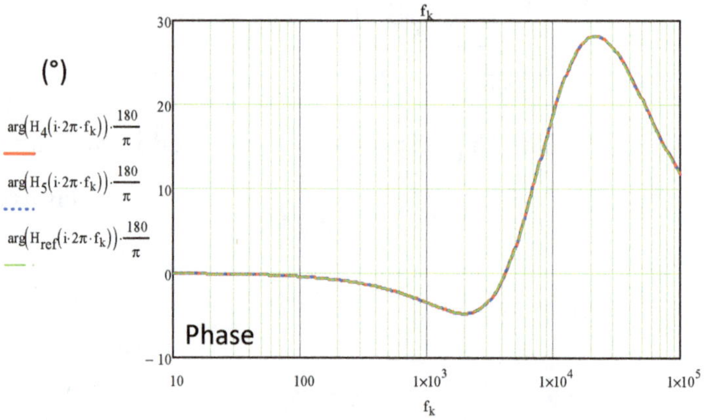

The Response given by the FACTs is Rigorously Similar to that Obtained via Superposition

Figure 6.56

In Figure 6.57, we see two cascaded *RL* networks. Considering the two inductors shunt the main branch to ground, the transfer function is zero at dc and there are two zeroes at the origin. This is confirmed by the high-frequency gains.

The final transfer function is written in a very compact form with inverted poles.

The ac response is given in Figure 6.58.

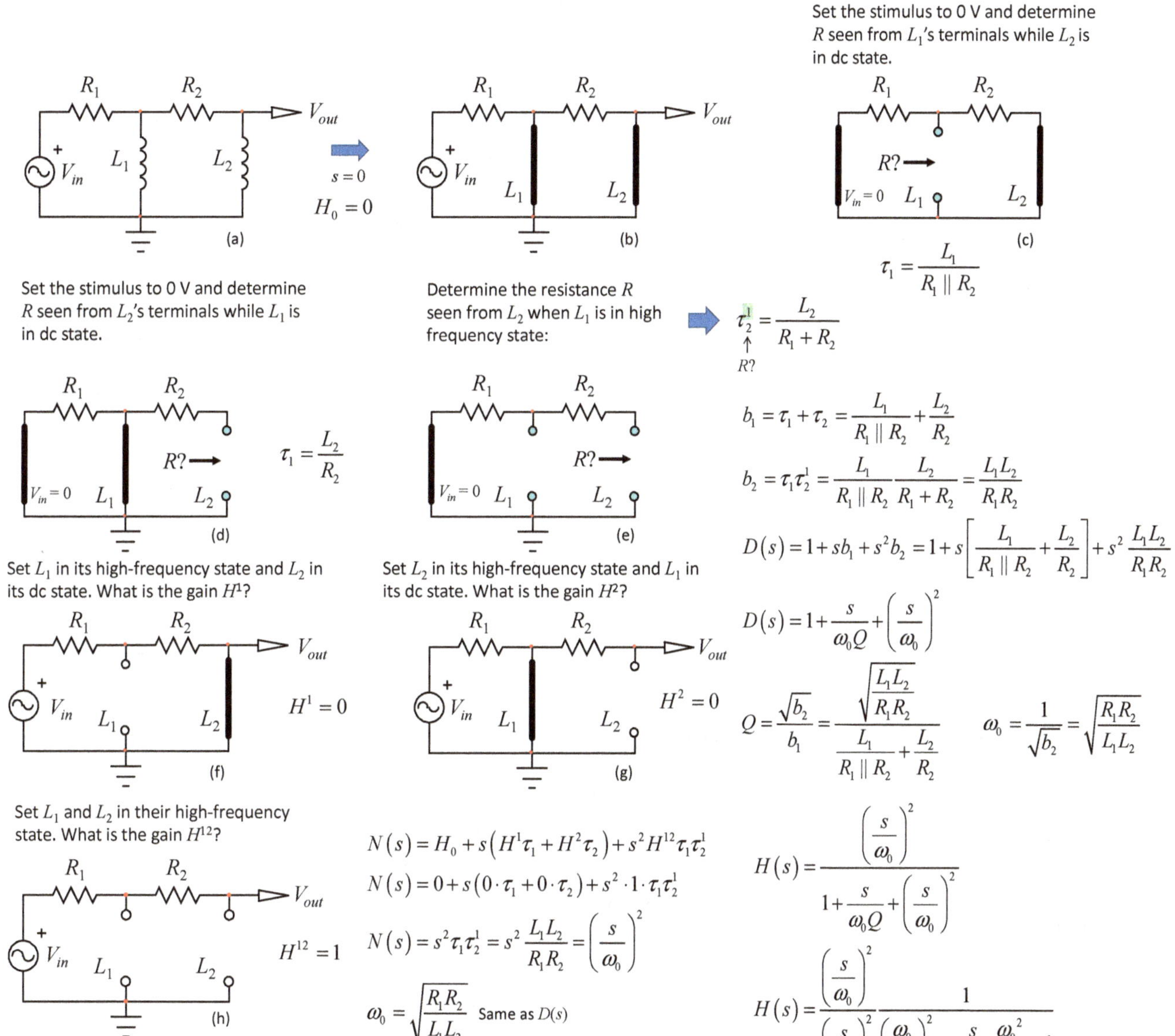
Set the stimulus to 0 V and determine R seen from L1's terminals while L2 is in dc state.
s = 0
H0 = 0
(a)
(b)
(c)
Set the stimulus to 0 V and determine R seen from L2's terminals while L1 is in dc state.
Determine the resistance R seen from L2 when L1 is in high frequency state:
(d)
(e)
Set L1 in its high-frequency state and L2 in its dc state. What is the gain H1?
Set L2 in its high-frequency state and L1 in its dc state. What is the gain H2?
H1 = 0
H2 = 0
(f)
(g)
Set L1 and L2 in their high-frequency state. What is the gain H12?
H12 = 1
(h)
Same as D(s)

$$\tau_1 = \frac{L_1}{R_1 \| R_2}$$

$$\tau_2^1 = \frac{L_2}{R_1 + R_2}$$

$$\tau_2 = \frac{L_2}{R_2}$$

$$b_1 = \tau_1 + \tau_2 = \frac{L_1}{R_1 \| R_2} + \frac{L_2}{R_2}$$

$$b_2 = \tau_1 \tau_2^1 = \frac{L_1}{R_1 \| R_2} \frac{L_2}{R_1 + R_2} = \frac{L_1 L_2}{R_1 R_2}$$

$$D(s) = 1 + sb_1 + s^2 b_2 = 1 + s\left[\frac{L_1}{R_1 \| R_2} + \frac{L_2}{R_2}\right] + s^2 \frac{L_1 L_2}{R_1 R_2}$$

$$D(s) = 1 + \frac{s}{\omega_0 Q} + \left(\frac{s}{\omega_0}\right)^2$$

$$Q = \frac{\sqrt{b_2}}{b_1} = \frac{\sqrt{\frac{L_1 L_2}{R_1 R_2}}}{\frac{L_1}{R_1 \| R_2} + \frac{L_2}{R_2}} \qquad \omega_0 = \frac{1}{\sqrt{b_2}} = \sqrt{\frac{R_1 R_2}{L_1 L_2}}$$

$$N(s) = H_0 + s\left(H^1\tau_1 + H^2\tau_2\right) + s^2 H^{12}\tau_1\tau_2^1$$

$$N(s) = 0 + s\left(0\cdot\tau_1 + 0\cdot\tau_2\right) + s^2\cdot 1\cdot\tau_1\tau_2^1$$

$$N(s) = s^2\tau_1\tau_2^1 = s^2\frac{L_1 L_2}{R_1 R_2} = \left(\frac{s}{\omega_0}\right)^2$$

$$\omega_0 = \sqrt{\frac{R_1 R_2}{L_1 L_2}} \text{ Same as } D(s)$$

$$H(s) = \frac{\left(\frac{s}{\omega_0}\right)^2}{1 + \frac{s}{\omega_0 Q} + \left(\frac{s}{\omega_0}\right)^2}$$

$$H(s) = \frac{\left(\frac{s}{\omega_0}\right)^2}{\left(\frac{s}{\omega_0}\right)^2 \left(\frac{\omega_0}{s}\right)^2} \frac{1}{+\frac{s}{\omega_0 Q}\frac{\omega_0^2}{s^2} + 1}$$

$$H(s) = \frac{1}{1 + \frac{\omega_0}{sQ} + \left(\frac{\omega_0}{s}\right)^2}$$

There is a Double Zero at the Origin in this Cascaded *RL* Network

Figure 6.57

$L_1 := 10\text{mH} \quad L_2 := 100\text{mH} \quad R_1 := 3\text{k}\Omega \quad R_2 := 1\text{k}\Omega \quad \|(x,y) := \frac{x \cdot y}{x+y}$

Transfer function for s=0 $\quad H_0 := 0$

$\tau_1 := \frac{L_1}{R_1 \| R_2} = 13.33333\mu\text{s}$

$\tau_2 := \frac{L_2}{R_2} = 100 \cdot \mu\text{s}$

$b_1 := \tau_1 + \tau_2 = 113.33333\mu\text{s}$

$\tau_{12} := \frac{L_2}{R_1 + R_2}$

$b_2 := \tau_1 \cdot \tau_{12} = 3.33333 \times 10^{-10}\text{s}^2$

$D_1(s) := 1 + s \cdot b_1 + s^2 \cdot b_2$

$Q := \frac{\sqrt{b_2}}{b_1} = 0.16109 \qquad \omega_0 := \frac{1}{\sqrt{b_2}} = 5.47723 \times 10^4 \frac{1}{\text{s}}$

$\frac{R_1 \cdot R_2 \cdot \sqrt{\frac{L_1 \cdot L_2}{R_1 \cdot R_2}}}{L_1 \cdot (R_1 + R_2) + L_2 \cdot R_1} = 0.16109 \qquad \sqrt{\frac{R_1 \cdot R_2}{L_1 \cdot L_2}} = 5.47723 \times 10^4 \frac{1}{\text{s}}$

High-frequency gains H

$H_1 := 0 \quad H_2 := 0 \quad H_{12} := 1$

$\omega_{p1} := \frac{1}{b_1} \qquad f_{p1} := \frac{\omega_{p1}}{2\pi} = 1.40431\text{kHz}$

$\omega_{p2} := \frac{b_1}{b_2} \qquad f_{p2} := \frac{\omega_{p2}}{2\pi} = 54.11268\text{kHz}$

$H_{10}(s) := \frac{H_0 + s \cdot (H_1 \cdot \tau_1 + H_2 \cdot \tau_2) + s^2 \cdot H_{12} \cdot \tau_1 \cdot \tau_{12}}{1 + s \cdot b_1 + s^2 \cdot b_2} \qquad H_{30}(s) := \frac{1}{1 + \frac{\omega_0}{s \cdot Q} + \left(\frac{\omega_0}{s}\right)^2}$

$H_{11}(s) := \frac{\left(s^2 \cdot \frac{L_1 \cdot L_2}{R_1 \cdot R_2}\right)}{\left[1 + s \cdot \left(\frac{L_1}{R_1 \| R_2} + \frac{L_2}{R_2}\right) + s^2 \cdot \frac{L_1 \cdot L_2}{R_1 \cdot R_2}\right]}$

$Z_{th}(s) := R_1 \| (s \cdot L_1)$

$H_{ref}(s) := \frac{s \cdot L_1}{s \cdot L_1 + R_1} \cdot \frac{s \cdot L_2}{s \cdot L_2 + Z_{th}(s) + R_2}$

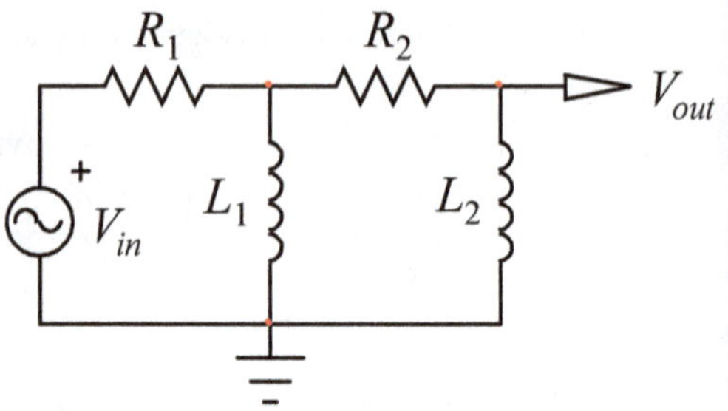

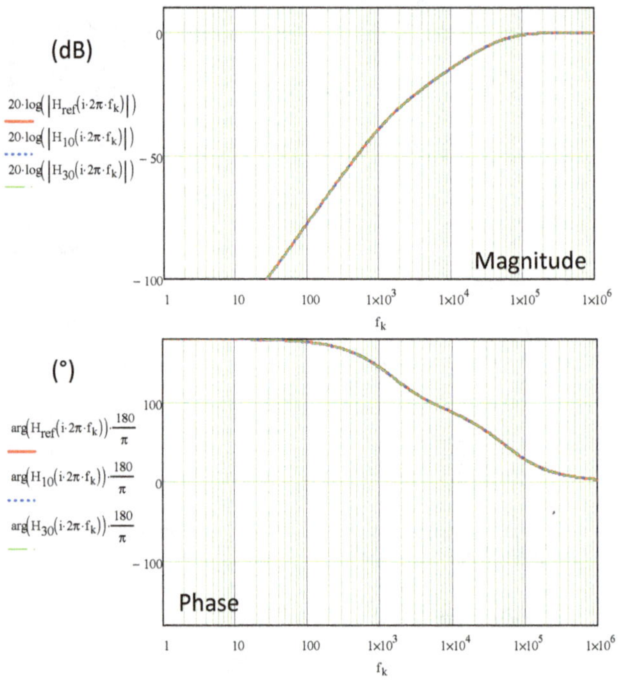

The Compact Notation using an Inverted Pole exactly Matches the Brute-Force Ac Response

Figure 6.58

The circuit from Figure 6.59 associates two capacitors separated by a resistive network.

The dc transfer function is easily obtained by opening all capacitors and then, the excitation is zeroed to determine the time constants. The zero is now unveiled by observing which impedance combination in the circuit could block the stimulus propagation and null the response: if the parallel arrangement of R_1 and C_1 leads to an infinite impedance, then we have a zero.

The final transfer function is obtained and rewritten as two cascaded poles considering the small quality factor.

All ac responses are given in Figure 6.60.

Set the stimulus to 0 V and determine R seen from C_1's terminals while C_2 is in dc state. Then vice versa.

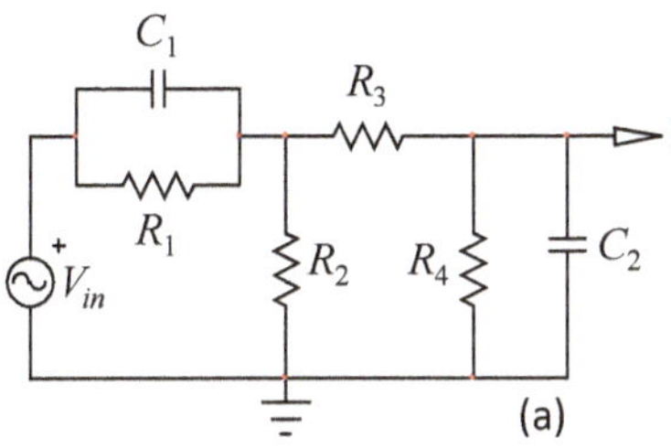

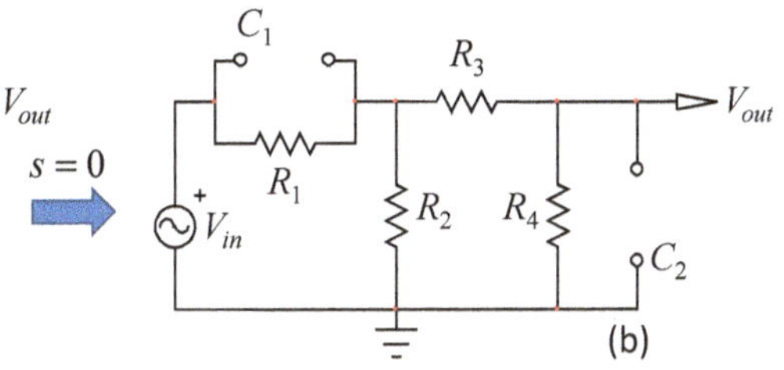

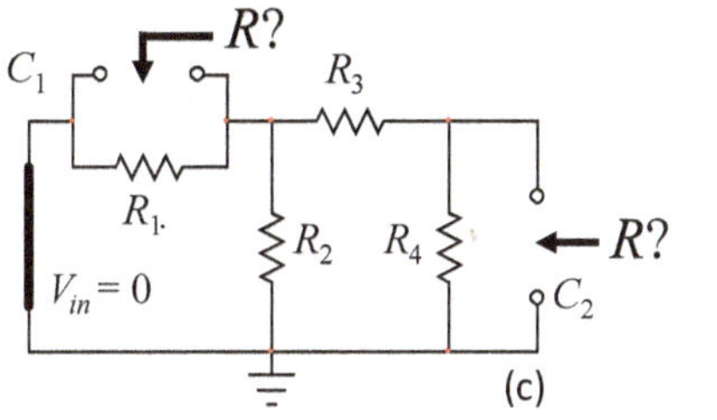

Determine the resistance R seen from C_2 when C_1 is in high frequency state:

$$H_0 = \frac{R_2}{R_2 + R_1}\frac{R_4}{R_4 + R_3 + R_2 \parallel R_1}$$

$$\tau_1 = \left[R_1 \parallel R_2 \parallel (R_3 + R_4)\right]C_1$$

$$\tau_2 = \left[R_4 \parallel (R_3 + R_1 \parallel R_2)\right]C_2$$

$$\tau_2^1 = (R_3 \parallel R_4)C_2$$

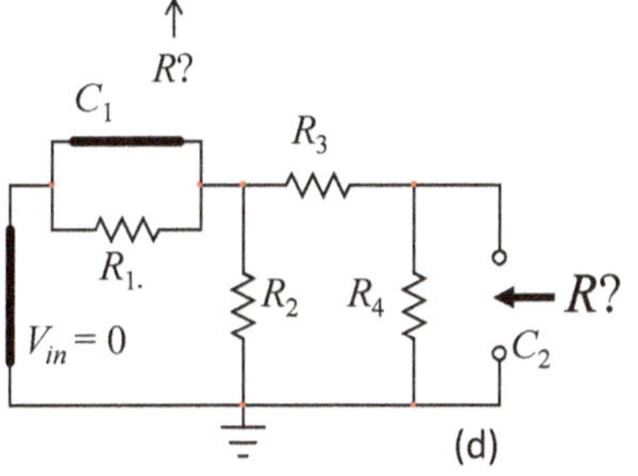

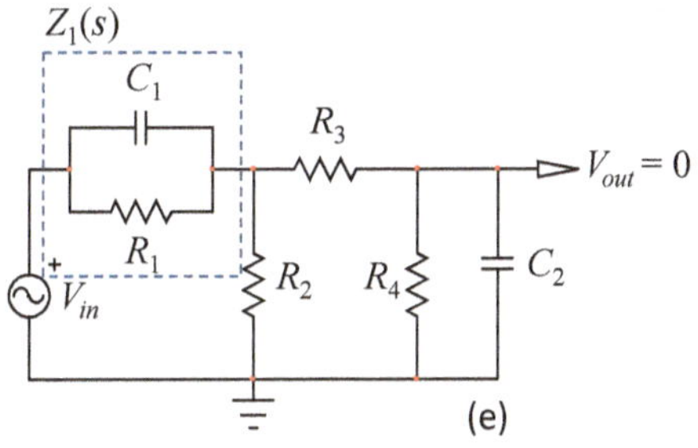

What impedance condition would prevent the stimulus from producing a response?

$$Z_1(s) \to \infty$$

$$R_1 \parallel \frac{1}{sC_1} = \frac{R_1}{1 + sR_1C_1}$$

$$1 + sR_1C_1 = 0 \to s_z = -\frac{1}{R_1C_1}$$

$$\omega_z = \frac{1}{R_1C_1}$$

$$D(s) = 1 + sb_1 + s^2 b_2$$

$$D(s) = 1 + s\left[\left[R_1 \parallel R_2 \parallel (R_3 + R_4)\right]C_1 + \left[R_4 \parallel (R_3 + R_1 \parallel R_2)\right]C_2\right] + s^2\left[R_1 \parallel R_2 \parallel (R_3 + R_4)\right]C_1(R_3 \parallel R_4)C_2$$

$$D(s) = 1 + \frac{s}{\omega_0 Q} + \left(\frac{s}{\omega_0}\right)^2 \qquad Q = \frac{\sqrt{b_2}}{b_1}$$

$$H(s) = \frac{1 + \frac{s}{\omega_z}}{1 + \frac{s}{Q\omega_0} + \left(\frac{s}{\omega_0}\right)^2} \qquad \omega_0 = \frac{1}{\sqrt{b_2}}$$

Low-Q approximation holds:

$$\omega_{p_1} = Q\omega_0 = \frac{1}{b_1} = \frac{1}{\left[R_1 \parallel R_2 \parallel (R_3 + R_4)\right]C_1 + \left[R_4 \parallel (R_3 + R_1 \parallel R_2)\right]C_2}$$

$$\omega_{p_2} = \frac{\omega_0}{Q} = \frac{b_1}{b_2} = \frac{\left[R_1 \parallel R_2 \parallel (R_3 + R_4)\right]C_1 + \left[R_4 \parallel (R_3 + R_1 \parallel R_2)\right]C_2}{\left[R_1 \parallel R_2 \parallel (R_3 + R_4)\right]C_1(R_3 \parallel R_4)C_2}$$

$$H(s) \approx H_0 \frac{1}{\left(1 + \frac{s}{\omega_{p_1}}\right)\left(1 + \frac{s}{\omega_{p_2}}\right)}$$

This Second-Order Network Features a Zero that can be Identified via Inspection

Figure 6.59

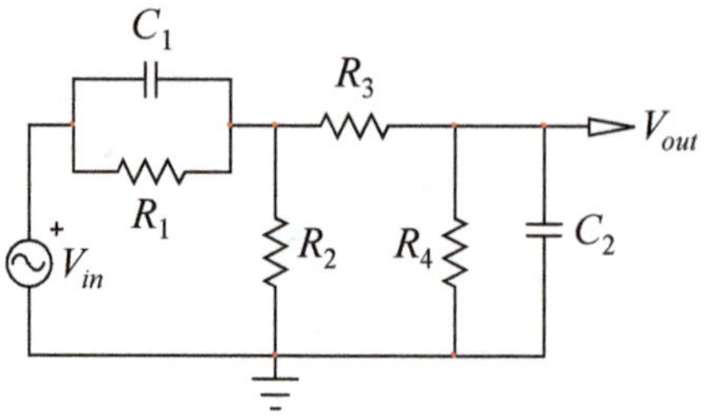

$C_1 := 75nF \quad R_1 := 10k\Omega \quad R_2 := 1000\Omega$

$C_2 := 3nF \quad R_3 := 10k\Omega \quad R_4 := 1k\Omega \quad \|(x,y) := \frac{x \cdot y}{x+y}$

$H_0 := \frac{R_2}{R_1 + R_2} \cdot \frac{R_4}{R_4 + R_3 + R_2 \| R_1} = 7.63359 \times 10^{-3} \quad 20 \cdot \log(H_0) = -42.34543$

$R_{th}(s) := R_1 \| \left(\frac{1}{s \cdot C_1}\right) \| R_2 + R_3$

$H_{ref}(s) := \frac{R_2}{R_2 + \left(\frac{1}{s \cdot C_1}\right) \| R_1} \cdot \frac{R_4 \| \left(\frac{1}{s \cdot C_2}\right)}{R_4 \| \left(\frac{1}{s \cdot C_2}\right) + R_{th}(s)}$

$\tau_1 := [R_1 \| R_2 \| (R_3 + R_4)] \cdot C_1 = 62.97771 \mu s$

$\tau_2 := [R_4 \| (R_3 + R_1 \| R_2)] \cdot C_2 = 2.74809 \mu s$

$b_1 := \tau_1 + \tau_2 = 65.72519 \mu s$

$\tau_{12} := (R_3 \| R_4) \cdot C_2 = 2.72727 \mu s$

$b_2 := \tau_1 \cdot \tau_{12} = 1.71756 \times 10^{-10} s^2$

$D_1(s) := 1 + b_1 \cdot s + b_2 \cdot s^2$

$\omega_z := \frac{1}{C_1 \cdot R_1}$

$H_1(s) := H_0 \cdot \frac{1 + \frac{s}{\omega_z}}{D_1(s)} \quad Q := \frac{\sqrt{b_2}}{b_1} = 0.1994 \quad \omega_0 := \frac{1}{\sqrt{b_2}} = 7.63035 \times 10^4 \frac{1}{s}$

$\omega_{p1} := \omega_0 \cdot Q = 1.52149 \times 10^4 \frac{1}{s} \quad f_{p1} := \frac{\omega_{p1}}{2 \cdot \pi} = 2.42152kHz$

$\omega_{p2} := \frac{\omega_0}{Q} = 3.82667 \times 10^5 \frac{1}{s} \quad f_{p2} := \frac{\omega_{p2}}{2 \cdot \pi} = 60.90329kHz$

$$H_2(s) := H_0 \cdot \frac{1 + \frac{s}{\omega_z}}{\left(1 + \frac{s}{\omega_{p1}}\right) \cdot \left(1 + \frac{s}{\omega_{p2}}\right)}$$

$$H_4(s) := \frac{R_2}{R_1 + R_2} \cdot \frac{R_4}{R_4 + R_3 + R_2 \| R_1} \cdot \frac{1 + s \cdot R_1 \cdot C_1}{1 + s \cdot [[R_1 \| R_2 \| (R_3 + R_4)] \cdot C_1 + [R_4 \| (R_3 + R_1 \| R_2)] \cdot C_2] + s^2 \cdot [[R_1 \| R_2 \| (R_3 + R_4)] \cdot C_1] \cdot [(R_3 \| R_4) \cdot C_2]}$$

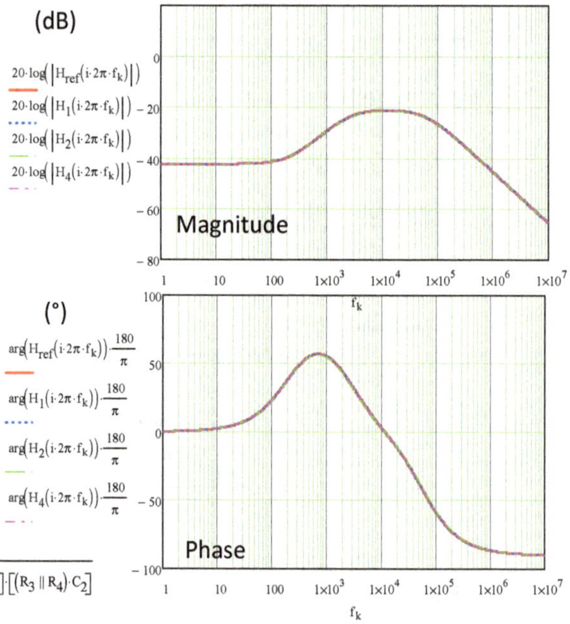

When Involving the Low-*Q* Approximation, the Transfer Function Combines Two Cascaded Poles

Figure 6.60

For this new example, we will go through an input impedance determination as illustrated in Figure 6.61 for the poles determination.

The zeroes are obtained in Figure 6.62 after having nulled the response V_T.

A zeroed response across a current source is similar to replacing this generator by a short circuit.

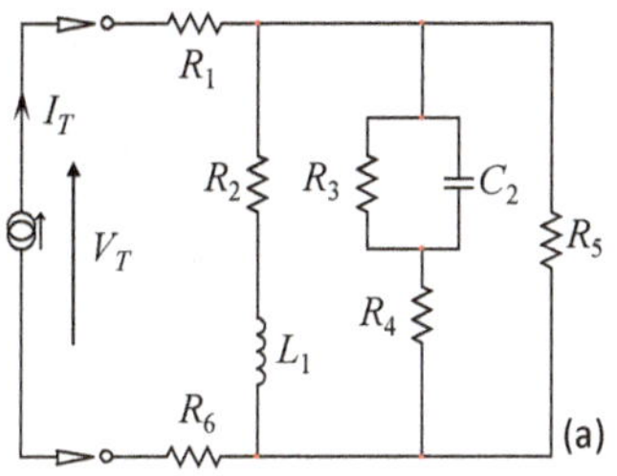

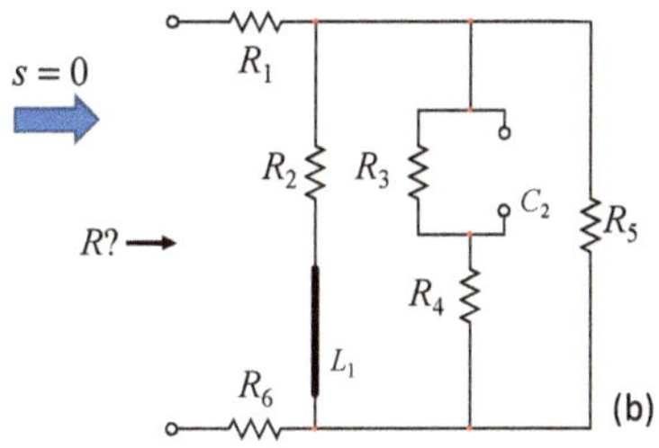

Turn excitation off – $I_T = 0$ A – and determine time constants

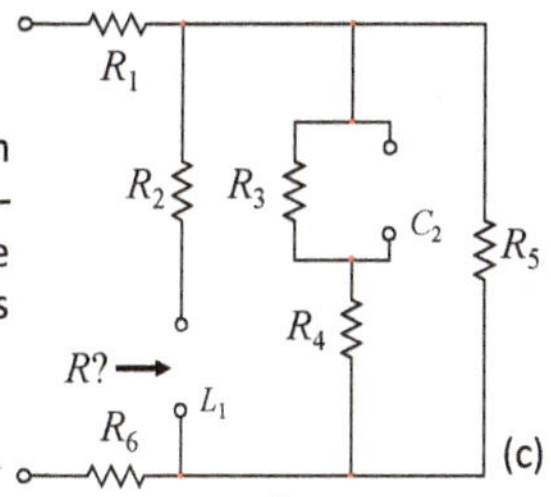

$$R_0 = R_1 + R_6 + R_2 \,||\, R_5 \,||\, (R_3 + R_4)$$

$$\tau_1 = \frac{L_1}{R_2 + (R_3 + R_4) \,||\, R_5}$$

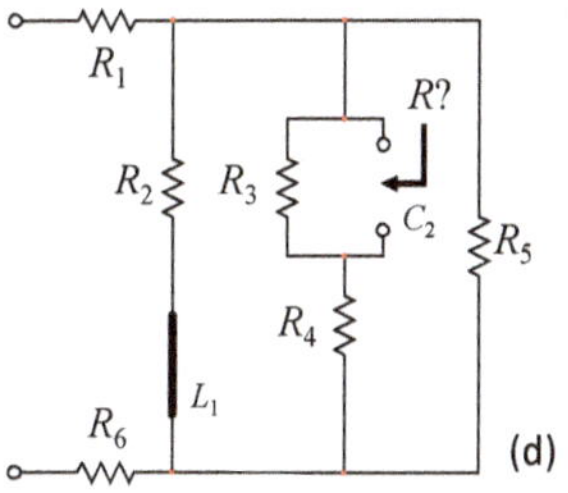

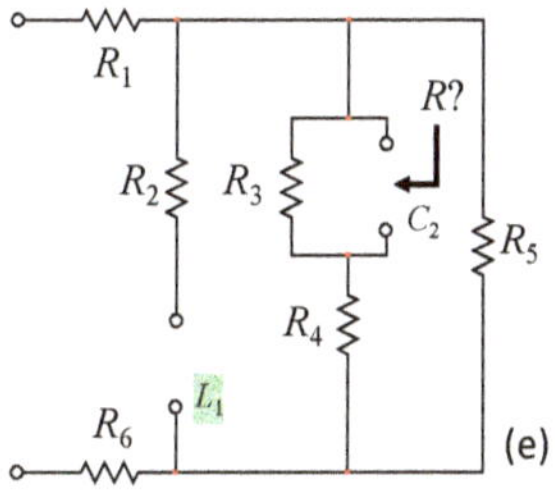

Determine the resistance R seen from C_2 when L_1 is in high frequency state: τ_2^1
↑
R?

$$\tau_2 = C_2\left[R_3 \,||\, (R_4 + R_2 \,||\, R_5)\right] \qquad \tau_2^1 = C_2\left[R_3 \,||\, (R_4 + R_5)\right]$$

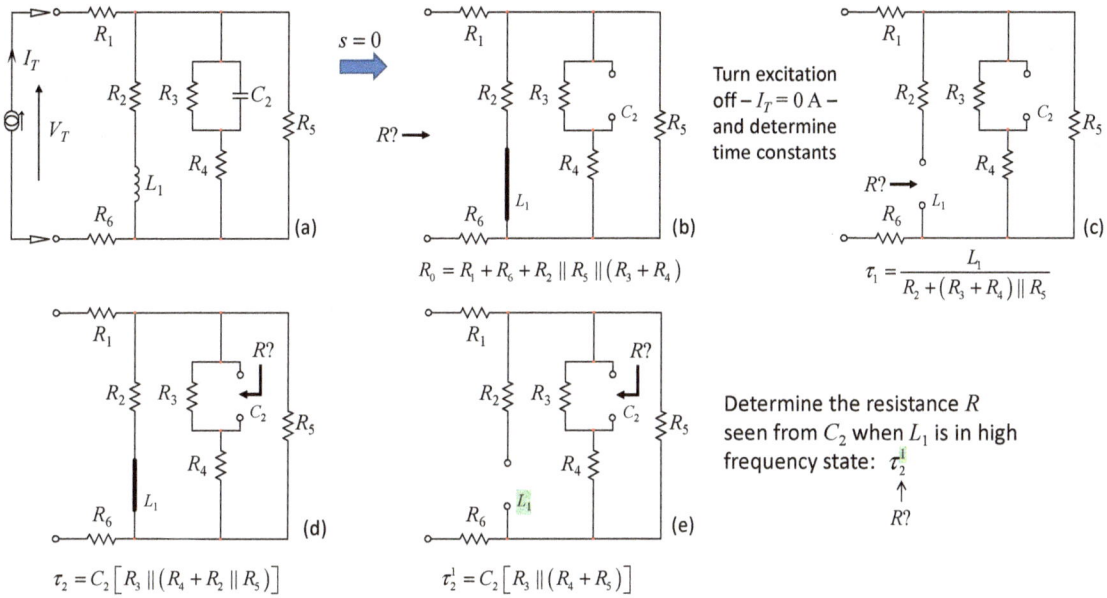

$$D(s) = 1 + sb_1 + s^2 b_2 = 1 + s(\tau_1 + \tau_2) + s^2 \tau_1 \tau_2^1$$

$$D(s) = 1 + s\left[\frac{L_1}{R_2 + (R_3 + R_4) \,||\, R_5} + C_2\left[R_3 \,||\, (R_4 + R_2 \,||\, R_5)\right]\right] + s^2 \frac{L_1}{R_2 + (R_3 + R_4) \,||\, R_5} C_2\left[R_3 \,||\, (R_4 + R_5)\right]$$

$$D(s) = 1 + \frac{s}{\omega_{0D} Q_D} + \left(\frac{s}{\omega_{0D}}\right)^2 \qquad Q_D = \frac{\sqrt{b_2}}{b_1} \qquad \omega_{0D} = \frac{1}{\sqrt{b_2}} \qquad \omega_{p_1} = Q_D \omega_{0D} \qquad \omega_{p_2} = \frac{\omega_{0D}}{Q_D}$$

The Time Constants for the Denominator are Found by Turning the Excitation Off

Figure 6.61

Null the response V_T and determine time constants in this mode.

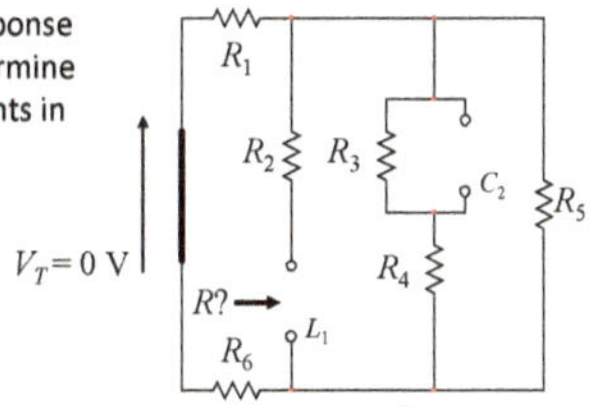

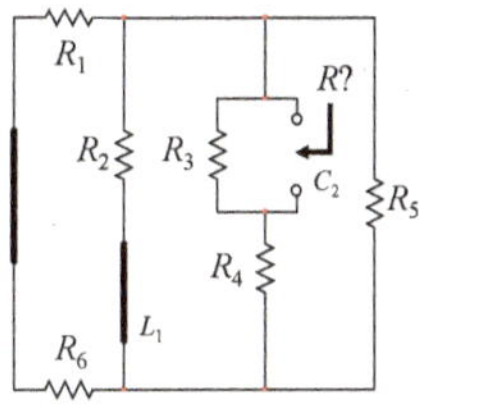

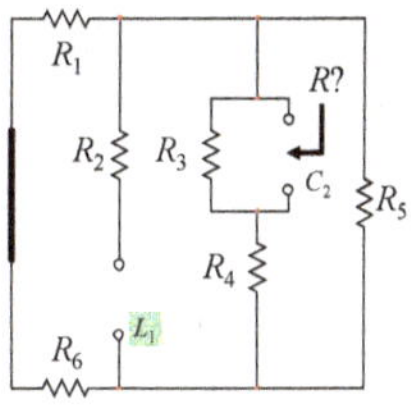

Determine the resistance R seen from C_2 when L_1 is in high frequency state: τ_{2N}^1
↑
R?

$$\tau_{1N} = \frac{L_1}{R_2 + (R_3 + R_4) \,||\, R_5 \,||\, (R_1 + R_6)} \qquad \tau_{2N} = C_2\left[R_3 \,||\, \left[R_4 + R_2 \,||\, R_5 \,||\, (R_1 + R_6)\right]\right] \qquad \tau_{2N}^1 = C_2\left[R_3 \,||\, \left[R_4 + R_5 \,||\, (R_1 + R_6)\right]\right]$$

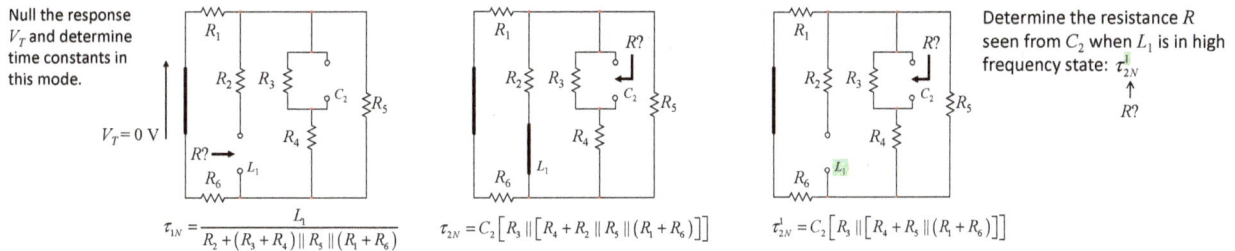

$$N(s) = 1 + sa_1 + s^2 a_2 = 1 + s(\tau_{1N} + \tau_{2N}) + s^2 \tau_{1N} \tau_{2N}^1$$

$$N(s) = 1 + s\left[\frac{L_1}{R_2 + (R_3 + R_4) \,||\, R_5 \,||\, (R_1 + R_6)} + C_2\left[R_3 \,||\, \left[R_4 + R_2 \,||\, R_5 \,||\, (R_1 + R_6)\right]\right]\right] + s^2 \frac{L_1}{R_2 + (R_3 + R_4) \,||\, R_5 \,||\, (R_1 + R_6)} C_2\left[R_3 \,||\, \left[R_4 + R_5 \,||\, (R_1 + R_6)\right]\right]$$

$$N(s) = 1 + \frac{s}{\omega_{0N} Q_N} + \left(\frac{s}{\omega_{0N}}\right)^2 \qquad Q_N = \frac{\sqrt{a_2}}{a_1} \qquad \omega_{0N} = \frac{1}{\sqrt{a_2}} \qquad \omega_{z_1} = Q_N \omega_{0N} \qquad \omega_{z_2} = \frac{\omega_{0N}}{Q_N}$$

$$Z_{in}(s) = R_0 \frac{N(s)}{D(s)} = R_0 \frac{1 + \frac{s}{\omega_{0N} Q_N} + \left(\frac{s}{\omega_{0N}}\right)^2}{1 + \frac{s}{\omega_{0D} Q_D} + \left(\frac{s}{\omega_{0D}}\right)^2} \approx R_0 \frac{\left(1 + \frac{s}{\omega_{z_1}}\right)\left(1 + \frac{s}{\omega_{z_2}}\right)}{\left(1 + \frac{s}{\omega_{p_1}}\right)\left(1 + \frac{s}{\omega_{p_2}}\right)}$$

The Zeroes are Obtained by Nulling Response V_T—similar to Shorting the Current Source

Figure 6.62

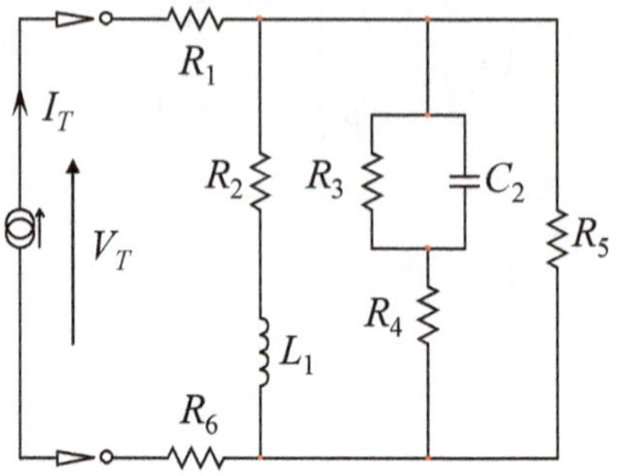

$R_1 := 1k\Omega$ $R_2 := 10k\Omega$ $R_3 := 11k\Omega$ $R_4 := 9k\Omega$

$C_2 := 1000\mu F$ $R_5 := 20k\Omega$ $R_6 := 10\Omega$ $L_1 := 1mH$ $\|(x,y) := \frac{x \cdot y}{x+y}$

Brute force analysis

$Z_1(s) := R_2 + s \cdot L_1$ $Z_2(s) := R_4 + \left(\frac{1}{s \cdot C_2}\right) \| R_3$

$Z_{ref}(s) := R_1 + R_6 + Z_1(s) \| Z_2(s) \| R_5$

$R_0 := R_1 + R_6 + R_2 \| R_5 \| (R_3 + R_4) = 6.01 \times 10^3 \Omega$ $20 \cdot \log\left(\frac{R_0}{\Omega}\right) = 75.577$ dBohms

Current source turned off:

$\tau_1 := \frac{L_1}{R_2 + (R_3 + R_4) \| R_5} = 50 \cdot ns$ $\tau_2 := C_2 \cdot [R_3 \| (R_4 + R_2 \| R_5)] = 6.463s$

$\tau_{12} := C_2 \cdot [R_3 \| (R_4 + R_5)] = 7.975s$

$D_1(s) := 1 + s \cdot (\tau_1 + \tau_2) + s^2 \cdot \tau_1 \cdot \tau_{12}$

$b_1 := \tau_1 + \tau_2 = 6.463 \times 10^6 \cdot \mu s$

$b_2 := \tau_1 \cdot \tau_{12} = 3.988 \times 10^{-7} s^2$

$Q_D := \frac{\sqrt{b_2}}{b_1} = 9.771 \times 10^{-5}$ $\omega_{0D} := \frac{1}{\sqrt{b_2}} = 1.584 \times 10^3 \frac{1}{s}$

$\omega_{p1} := \omega_{0D} \cdot Q_D = 0.155 \frac{1}{s}$ $f_{p1} := \frac{\omega_{p1}}{2 \cdot \pi} = 2.463 \times 10^{-5} \cdot kHz$

$\omega_{p2} := \frac{\omega_{0D}}{Q_D} = 1.621 \times 10^7 \frac{1}{s}$ $f_{p2} := \frac{\omega_{p2}}{2 \cdot \pi} = 2.579 \times 10^3 \cdot kHz$

$Z_{in}(s) := R_0 \cdot \frac{N_1(s)}{D_1(s)}$ $Z_{in2}(s) := R_0 \cdot \frac{1 + \frac{s}{\omega_{0N} \cdot Q_N} + \left(\frac{s}{\omega_{0N}}\right)^2}{1 + \frac{s}{\omega_{0D} \cdot Q_D} + \left(\frac{s}{\omega_{0D}}\right)^2}$

Current source shorted:

$\tau_{1N} := \frac{L_1}{R_2 + (R_3 + R_4) \| R_5 \| (R_1 + R_6)} = 91.597ns$

$\tau_{2N} := C_2 \cdot [R_3 \| [R_4 + R_2 \| R_5 \| (R_1 + R_6)]] = 5.204s$

$\tau_{12N} := C_2 \cdot [R_3 \| [R_4 + R_5 \| (R_1 + R_6)]] = 5.227s$

$a_1 := \tau_{1N} + \tau_{2N} = 5.204 \times 10^6 \cdot \mu s$

$a_2 := \tau_{1N} \cdot \tau_{12N} = 4.788 \times 10^{-7} s^2$

$Q_N := \frac{\sqrt{a_2}}{a_1} = 1.33 \times 10^{-4}$ $\omega_{0N} := \frac{1}{\sqrt{a_2}} = 1.445 \times 10^3 \frac{1}{s}$

$\omega_{z1} := \omega_{0N} \cdot Q_N = 0.192 \frac{1}{s}$ $f_{z1} := \frac{\omega_{z1}}{2 \cdot \pi} = 0.031 \cdot Hz$

$\omega_{z2} := \frac{\omega_{0N}}{Q_N} = 1.087 \times 10^7 \frac{1}{s}$ $f_{z2} := \frac{\omega_{z2}}{2 \cdot \pi} = 1.73 \cdot MHz$

$N_1(s) := 1 + s \cdot (\tau_{1N} + \tau_{2N}) + s^2 \cdot \tau_{1N} \cdot \tau_{12N}$ $Z_{in3}(s) := R_0 \cdot \frac{\left(1 + \frac{s}{\omega_{z1}}\right) \cdot \left(1 + \frac{s}{\omega_{z2}}\right)}{\left(1 + \frac{s}{\omega_{p1}}\right) \cdot \left(1 + \frac{s}{\omega_{p2}}\right)}$

$$Z_{final}(s) := [R_1 + R_6 + R_2 \| R_5 \| (R_3 + R_4)] \cdot \frac{1 + s \cdot \left[\frac{L_1}{R_2 + (R_3 + R_4) \| R_5 \| (R_1 + R_6)} + C_2 \cdot [R_3 \| [R_4 + R_2 \| R_5 \| (R_1 + R_6)]]\right] + s^2 \cdot \frac{L_1}{R_2 + (R_3 + R_4) \| R_5 \| (R_1 + R_6)} \cdot [C_2 \cdot [R_3 \| [R_4 + R_5 \| (R_1 + R_6)]]]}{1 + s \cdot \left[\frac{L_1}{R_2 + (R_3 + R_4) \| R_5} + C_2 \cdot [R_3 \| (R_4 + R_2 \| R_5)]\right] + s^2 \cdot \frac{L_1}{R_2 + (R_3 + R_4) \| R_5} \cdot [C_2 \cdot [R_3 \| (R_4 + R_5)]]}$$

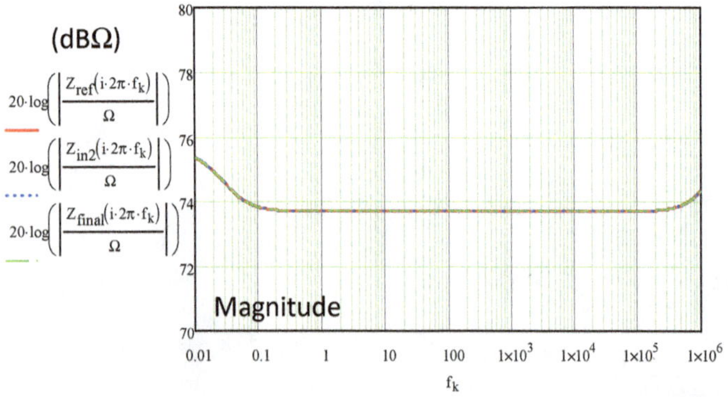

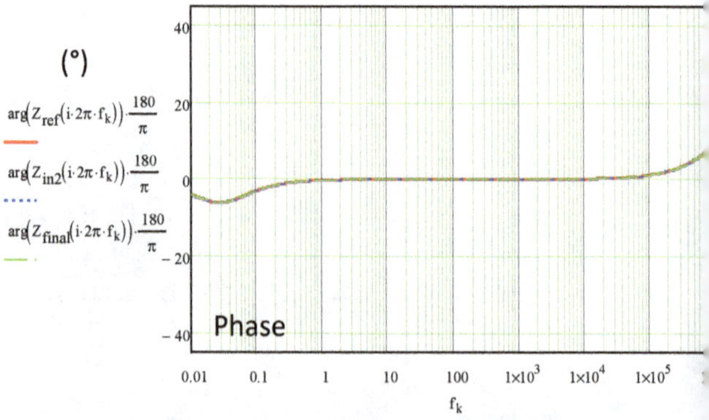

Ac Response from the Brute-Force Expression and that of the FACTs Perfectly Match

Figure 6.63

This last example ends our series of 2nd-order networks and you can exercise your newly acquired skill on them.

6.3 List of Figures and Transfer Functions

For a convenient browsing of the derived transfer functions, below are pictures summarizing the networks studied in this chapter.

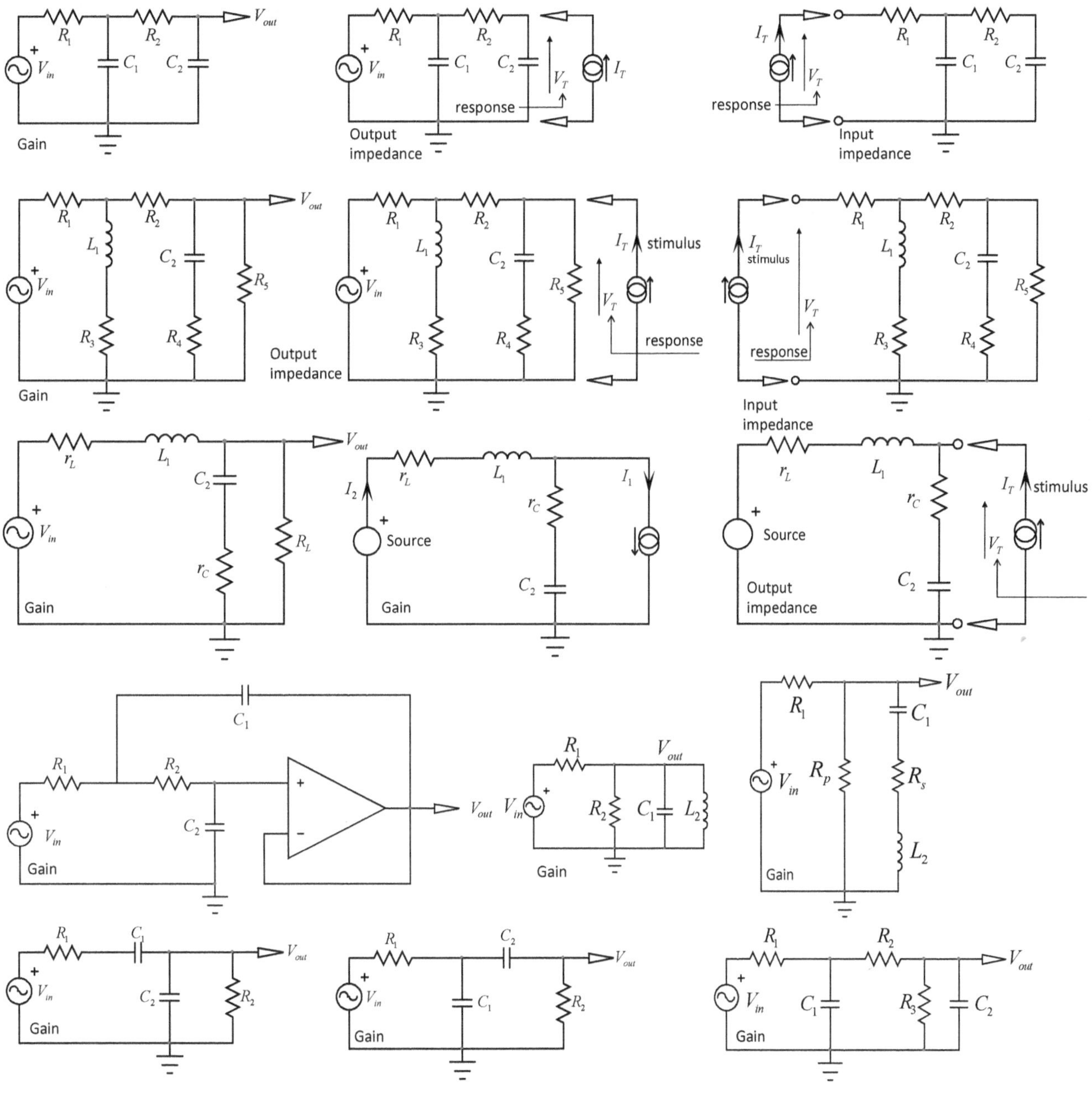

The First Set of 2nd-Order Transfer Functions

Figure 6.64

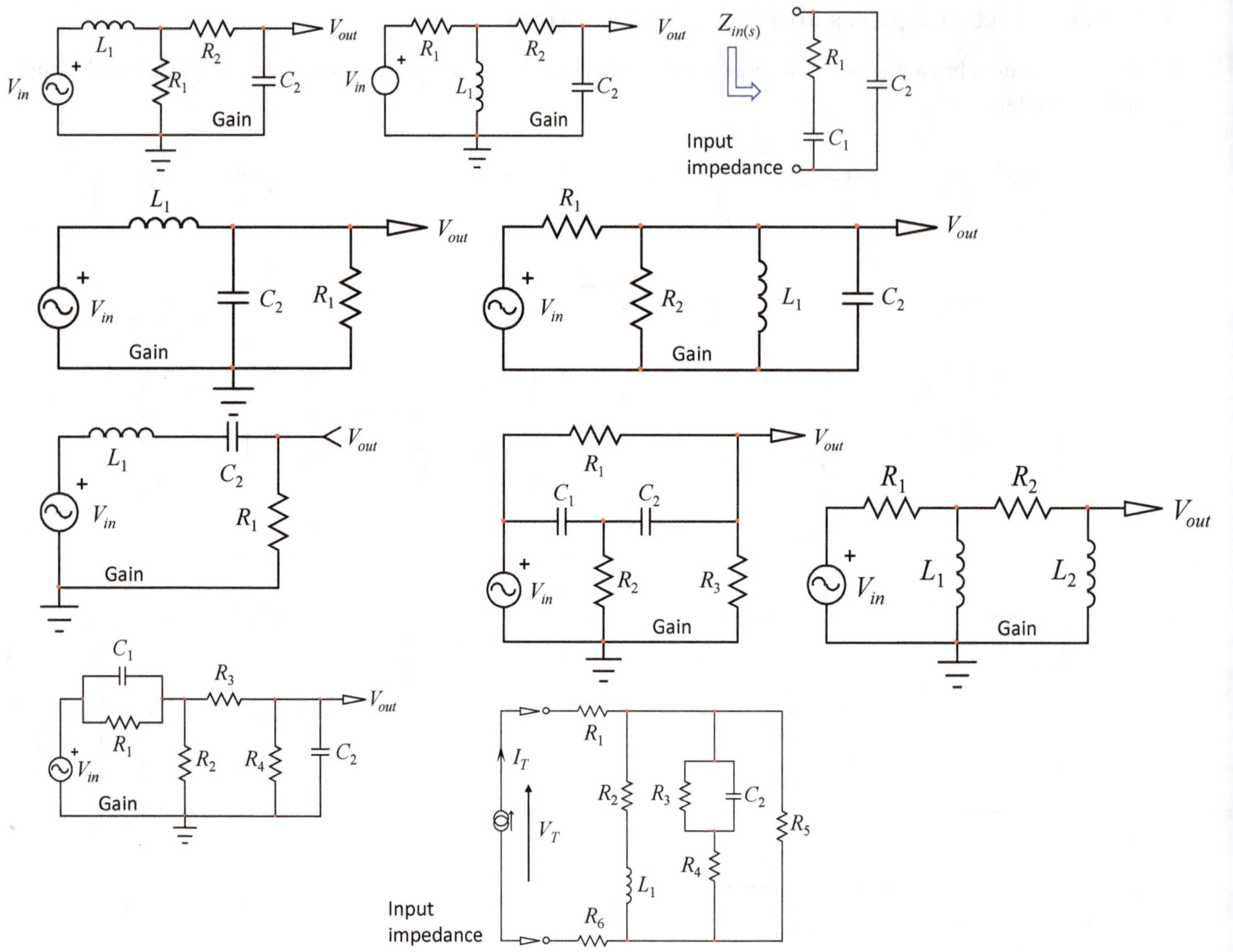

This is the Second Set of 2nd-Order Transfer Functions

Figure 6.65

6.4 References

1. C. Basso, *Input Filter Interactions with Switching Regulators*, Applied Power Electronics Conference, Professional Seminar, Tampa (FL), March 2017

Chapter 7: Third-Order Transfer Functions

IN THIS CHAPTER, I will determine transfer functions of ten networks featuring three energy-storing elements. The principle remains the same: find the time constants with a zeroed excitation to identify the poles, then determine the zeroes by nulling the response or by resorting to the generalized transfer function. We start this chapter with a quick summary of the technique.

7.1 Zeroed Excitation and Null Double Injection for 3rd-Order Systems

The denominator of a third-order transfer function obeys the following expression:

$$D(s)=1+b_1s+b_2s^2+b_3s^3 \tag{7.1}$$

The term b_1 is obtained by summing the time constants associated with each energy-storing element, τ_1, τ_2 and τ_3:

$$b_1 = \tau_1 + \tau_2 + \tau_3 \tag{7.2}$$

During the exercise, the excitation or the stimulus source is zeroed and you determine the resistance R driving the first energy-storing element while the other two are placed in their dc state (an open circuit for a capacitor, a short circuit for an inductor). Then, repeat the process by finding R driving the second element while the first is now placed in its dc state and the third one is still in its dc state. Finally, the first two are back in their dc state and determine the resistance R driving the third one. With the time constants in hand, assemble b_1.

The second term, b_2, is defined by:

$$b_2 = \tau_1\tau_2^1 + \tau_1\tau_3^1 + \tau_2\tau_3^2 \tag{7.3}$$

In this approach, reuse one of the three time constants τ_1, τ_2 or τ_3 then multiply it by another time constant determined as follows:

- τ_2^1 means that energy-storing element labeled 1 is placed in its high-frequency state (a short circuit for a capacitor or an open-circuit for an inductor) and determine the resistance R driving element 2 in this mode. Then assemble the first term in b_2 according to (7.3) by reusing the first time constant τ_1. In this analysis, element 3 is set in its dc state.
- τ_3^1 means that energy-storing element labeled 1 is placed in its high-frequency state and determine the resistance R driving element 3 in this mode. Then assemble the second term in b_2 according to (7.3) by reusing the first time constant τ_1. In this exercise, element 2 remains in its dc state.
- τ_3^2 means that energy-storing element labeled 2 is placed in its high-frequency state and determine the resistance R driving element 3 in this mode. Then assemble b_2 according to (7.3) by reusing the second time constant τ_2. In this step, element 1 is kept in its dc state.

Sometimes, an indeterminacy arises when multiplying time constants or the newly obtained network is difficult to analyze. In that case, redundancy exists. For instance, reshuffle the pairs as follows:

$$\tau_1\tau_2^1 = \tau_2\tau_1^2 \tag{7.4}$$

$$\tau_1\tau_3^1 = \tau_3\tau_1^3 \tag{7.5}$$
$$\tau_2\tau_3^2 = \tau_3\tau_2^3 \tag{7.6}$$

The third term, b_3, is obtained by reusing two previous times constants:

$$b_3 = \tau_1\tau_2^1\tau_3^{12} \tag{7.7}$$

In this expression, reuse the product of the two time constants, τ_1 and τ_2^1, then multiply it by another time constant determined as follows:

- τ_3^{12} means that energy-storing elements labeled 1 and 2 are placed in their high-frequency state and determine the resistance R driving element 3 in this mode. Then assemble the term b_3 according to (7.7) by reusing time constants τ_1 and τ_2^1.

Reshuffling is also an option when an indeterminacy occurs or if the obtained circuit is overcomplicated. For instance, rearrange b_3 as:

$$b_3 = \tau_1\tau_2^1\tau_3^{12} = \tau_2\tau_1^2\tau_3^{21} \tag{7.8}$$

More possibilities on reshuffling exist as detailed in [1].

The numerator of a third-order transfer function will obey the following formula:

$$N(s) = 1 + a_1 s + a_2 s^2 + a_3 s^3 \tag{7.9}$$

The term a_1 is determined by summing the time constants associated with each energy-storing element, τ_{1N}, τ_{2N} and τ_{3N} obtained in a null double injection or NDI:

$$a_1 = \tau_{1N} + \tau_{2N} + \tau_{3N} \tag{7.10}$$

Determine the resistance R offered by the connecting terminals of the first energy-storing element while the other two are placed in their dc state (a capacitor is open while an inductor is replaced by short circuit) and the response is nulled.

Then, repeat the process by finding R driving the second element while the first and the third elements are placed in their dc state with a nulled response.

Finally, the first two are back in their dc state and determine the resistance R driving the third one still with the response nulled. With all the time constants in hand, determine a_1.

This NDI technique requires the installation of a test generator I_T together with the original stimulus brought back in place as documented in the previous chapters.

The second term, a_2, is defined by:

$$a_2 = \tau_{1N}\tau_{2N}^1 + \tau_{1N}\tau_{3N}^1 + \tau_{2N}\tau_{3N}^2 \tag{7.11}$$

In this approach, reuse one of the two time constants already obtained for the term a_1, τ_{1N} or τ_{2N}, and multiply it by another time constant determined as follows:

- τ_{2N}^{1} means that energy-storing element labeled 1 is placed in its high-frequency state (a short circuit for a capacitor or an open-circuit for an inductor) and determine the resistance R driving element 2 while ensuring a null on the output in an NDI configuration. Then assemble a_2 according to (7.11) by reusing the first time constant τ_{1N}. In this analysis, element 3 remains in its dc state.
- τ_{3N}^{1} means that energy-storing element labeled 1 is placed in its high-frequency and determine the resistance R driving element 3 while ensuring a null on the output in an NDI configuration. Then assemble a_2 according to (7.11) by reusing the first time constant τ_{1N}. In this exercise, element 2 is kept in its dc state.
- τ_{3N}^{2} means that energy-storing element labeled 2 is placed in its high-frequency state and determine the resistance R driving element 3 while ensuring a null on the output in an NDI configuration. Then assemble a_2 according to (7.11) by reusing the second time constant τ_{2N}. In this exercise, element 1 is set in its dc state.

The third term, a_3, is obtained by reusing two previous times constants:

$$a_3 = \tau_{1N}\tau_{2N}^{1}\tau_{3N}^{12} \tag{7.12}$$

In this expression, reuse the product of the two time constants τ_{1N} and τ_{2N}^{1} then multiply it by another time constant determined as follows:

- τ_{3N}^{12} means that energy-storing elements labeled 1 and 2 are placed in their high-frequency state and you determine the resistance R driving element 3 while ensuring a null on the output in an NDI configuration. Then form term a_3 according to (7.12) by reusing time constants τ_{1N} and τ_{2N}^{1}.

Of course, when inspection works and the zeroes in the circuit can be visualized—by identifying the impedance conditions bringing an output null—the numerator is immediately expressed and tremendous time is saved.

As well as with the denominator, reshuffling is also possible and follows a scheme similar to that described from (7.4) to (7.6) and (7.8).

Once the numerator and the denominator are identified, write the transfer function after having determined H_0, the gain is obtained for $s = 0$:

$$H(s) = H_0 \frac{N(s)}{D(s)} = H_0 \frac{1 + a_1 s + a_2 s^2 + a_3 s^3}{1 + b_1 s + b_2 s^2 + b_3 s^3} \tag{7.13}$$

The difficulty now lies in factoring this expression in a product of poles and zeroes easily identified.

What matters are the way poles and zeroes are distributed in the frequency domain.

For instance, if poles are well separated, meaning that time constants obey $|b_1| >> \left|\frac{b_2}{b_1}\right| >> \left|\frac{b_3}{b_2}\right|$, the denominator of (7.13) can be advantageously rewritten as:

$$D(s) \approx \left(1+\frac{s}{\omega_{P1}}\right)\left(1+\frac{s}{\omega_{P2}}\right)\left(1+\frac{s}{\omega_{P3}}\right) \quad (7.14)$$

...in which the poles are defined as:

$$\omega_{P1} = \frac{1}{b_1} \quad \omega_{P2} = \frac{b_1}{b_2} \quad \omega_{P3} = \frac{b_2}{b_3} \quad (7.15)$$

If you now realize that one pole dominates the low-frequency response while two poles are grouped at higher frequencies—e.g. like the subharmonic poles in a current-mode switching converter—then the denominator of (7.13) can be rearranged as:

$$D(s) \approx \left(1+\frac{s}{\omega_{P1}}\right)\left(1+\frac{s}{\omega_0 Q}+\left(\frac{s}{\omega_0}\right)^2\right) \quad (7.16)$$

In the above expression, the terms are defined as:

$$\omega_{P1} = \frac{1}{b_1} \quad \omega_0 = \sqrt{\frac{b_1}{b_3}} \quad Q = \frac{\sqrt{b_1 b_3}}{b_2} \quad (7.17)$$

When making these approximations, compare the magnitude and phase plots of the original expression versus its factorization.

Engineering judgement is therefore necessary to decide whether or not the newly factored expression is acceptable or not.

Why is it important to factor these expressions?

Because, ultimately, you will use them to place poles and zeroes by calculating components values. And if you cannot unveil these salient points from a cubic or quadratic expression, then the design becomes extremely complicated. Remember, always work for a *design-oriented analysis* or D-OA.

For those interested in learning more about the proper factorization of these high-order polynomials, Chapter 8 of Ref. [2] brings interesting insights. All the above discussion pertains also to the numerator of course.

When determining the zeroes using an NDI is too complicated or simply because you would like to check your results with an alternative method, you can always resort to the generalized transfer function extended to a 3rd-order network.

The method reuses the time constants already found for the denominator:

$$H(s) = \frac{H_0 + s\left(\tau_1 H^1 + \tau_2 H^2 + \tau_3 H^3\right) + s^2\left(\tau_1\tau_2^1 H^{12} + \tau_1\tau_3^1 H^{13} + \tau_2\tau_3^2 H^{23}\right) + s^3\left(\tau_1\tau_2^1\tau_3^{12} H^{123}\right)}{1 + s\left(\tau_1+\tau_2+\tau_3\right) + s^2\left(\tau_1\tau_2^1 + \tau_1\tau_3^1 + \tau_2\tau_3^2\right) + s^3\tau_1\tau_2^1\tau_3^{12}} \quad (7.18)$$

If H_0 is different than zero, then (7.18) can advantageously be rewritten as:

$$H(s)=H_0\frac{1+s\left(\frac{\tau_1 H^1+\tau_2 H^2+\tau_3 H^3}{H_0}\right)+s^2\left(\frac{\tau_1\tau_2^1 H^{12}+\tau_1\tau_3^1 H^{13}+\tau_2\tau_3^2 H^{23}}{H_0}\right)+s^3\left(\tau_1\tau_2^1\tau_3^{12}\frac{H^{123}}{H_0}\right)}{1+s(\tau_1+\tau_2+\tau_3)+s^2\left(\tau_1\tau_2^1+\tau_1\tau_3^1+\tau_2\tau_3^2\right)+s^3\tau_1\tau_2^1\tau_3^{12}} \tag{7.19}$$

In these expressions, H^1, H^2 and H^3 are high-frequency gains linking the response to the stimulus: bring the stimulus back in the circuit and check when the energy-storing elements are alternately set in their high-frequency state if the response exists. If it does, determine the gain H in this mode. H^1 is obtained when element 1 is set in its high-frequency state while elements 2 and 3 remain in dc state. For H^2, element 2 is placed in its high-frequency state while elements 1 and 3 remain in dc state. H^3 is determined when element 3 is in high-frequency state while element 1 and 2 remain in dc state.

H^{12} indicates that elements 1 and 2 are in high-frequency state (element 3 is in dc), H^{13} means that elements 1 and 3 are in high-frequency state with 2 in its dc state and, finally, H^{23} is determined by having elements 2 and 3 in high frequency while 1 remains in dc state.

H^{123} is a gain obtained with the three energy-storing elements set in their high-frequency state. As previously underlined, the generalized transfer function can sometimes complicate the numerator roots and extra energy may be needed to simplify the final expression. But often it does not, especially when the high-frequency gains are equal to zero or one.

Nevertheless, the results obtained from (7.13) or (7.19) are identical. I often use the generalized transfer function when the reference state obtained for $s = 0$ returns zero (or infinity) and makes the use of (7.13) impractical.

7.2 Circuits with three Energy-Storing Elements

We will start with a classic, where three *RC* networks are cascaded as in Figure 7.1.

Begin with $s = 0$ and determine the dc gain which is unity in absence of load. Then proceed and determine time constants in the different configurations already described. Once assembled, the 3rd-order polynomial can be factored as three distinct poles, offering a better insight into what the transfer function does. When all capacitors and resistors are respectively equal, the transfer function simplifies as shown in the bottom of the figure.

Figure 7.2 confirms the result obtained by FACTs match the reference transfer function obtained after determining the Thévenin resistance at two different locations.

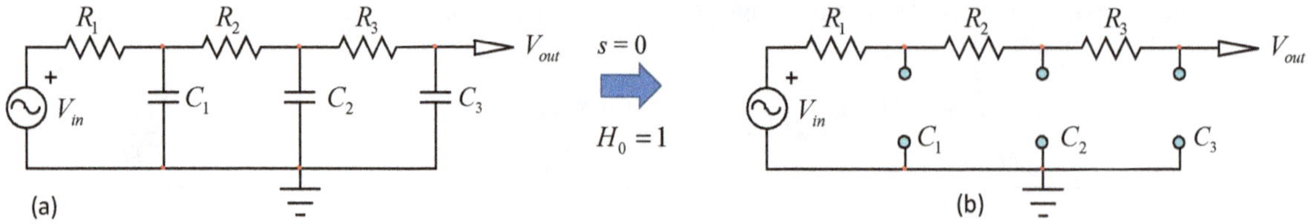

Set the stimulus to 0 V and determine the resistance R driving each capacitor:

$\tau_1 = R_1C_1 \quad \tau_2 = (R_1+R_2)C_2 \quad \tau_3 = (R_1+R_2+R_3)C_3$

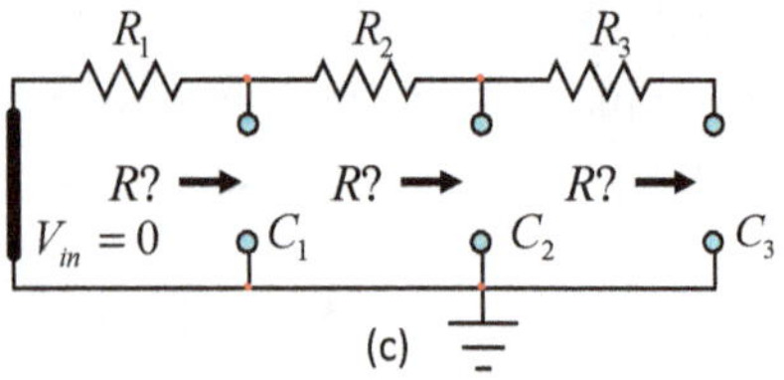

Set capacitor C_1 in its high-frequency state and determine R from C_2's terminals:

$\tau_2^1 = R_2C_2$

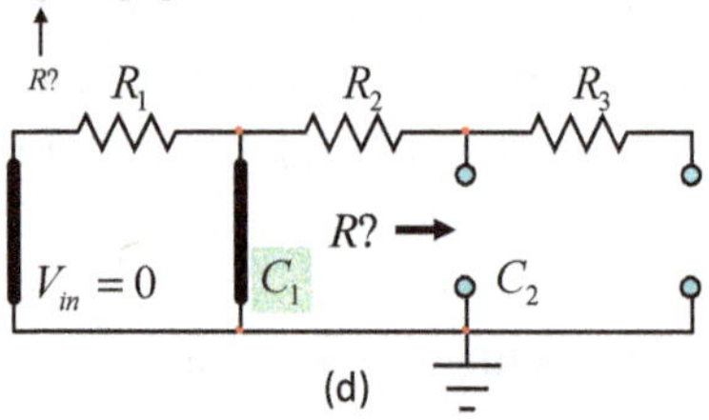

Set capacitor C_1 in its high-frequency state and determine R from C_3's terminals:

$\tau_3^1 = (R_2+R_3)C_3$

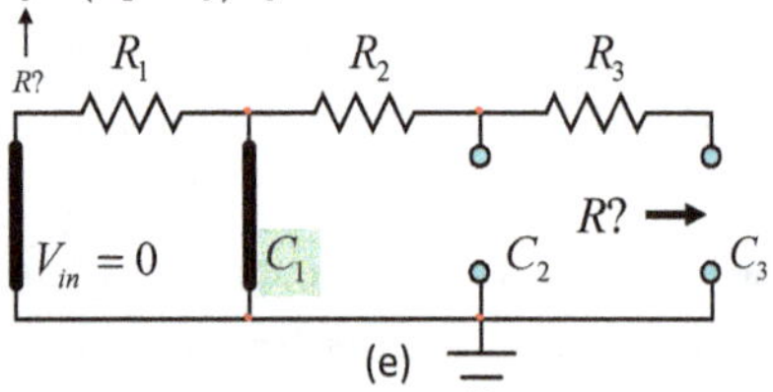

Set capacitor C_2 in its high-frequency state and determine R from C_3's terminals:

$\tau_3^2 = R_3C_3$

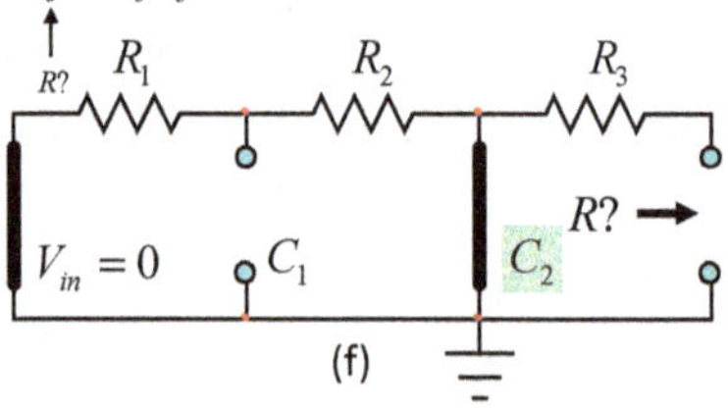

Set capacitor C_1 and C_2 in their high-frequency state and determine R from C_3's terminals:

$\tau_3^{12} = R_3C_3$

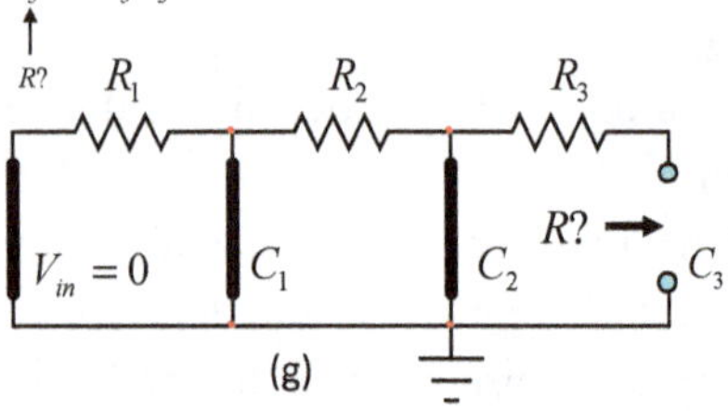

$$b_1 = \tau_1+\tau_2+\tau_3 = R_1C_1+(R_1+R_2)C_2+(R_1+R_2+R_3)C_3$$

$$b_2 = \tau_1\tau_2^1+\tau_1\tau_3^1+\tau_2\tau_3^2 = R_1C_1\left[R_2C_2+C_3(R_2+R_3)\right]+(R_1+R_2)C_2R_3C_3$$

$$b_3 = \tau_1\tau_2^1\tau_3^{12} = R_1C_1R_2C_2R_3C_3$$

$$D(s) = 1+sb_1+s^2b_2+s^3b_3$$

$$H(s) = H_0\frac{1}{D(s)} = \frac{1}{1+sb_1+s^2b_2+s^3b_3}$$

$$H(s) \approx \frac{1}{\left(1+\frac{s}{\omega_{p_1}}\right)\left(1+\frac{s}{\omega_{p_2}}\right)\left(1+\frac{s}{\omega_{p_3}}\right)}$$

$$\omega_{p_1} = \frac{1}{b_1} \quad \omega_{p_2} = \frac{b_1}{b_2} \quad \omega_{p_3} = \frac{b_2}{b_3}$$

If $C_1 = C_2 = C_3 = C$ and $R_1 = R_2 = R_3 = R$

$$H(s) = \frac{1}{1+6RCs+5(RC)^2s^2+(RC)^3s^3}$$

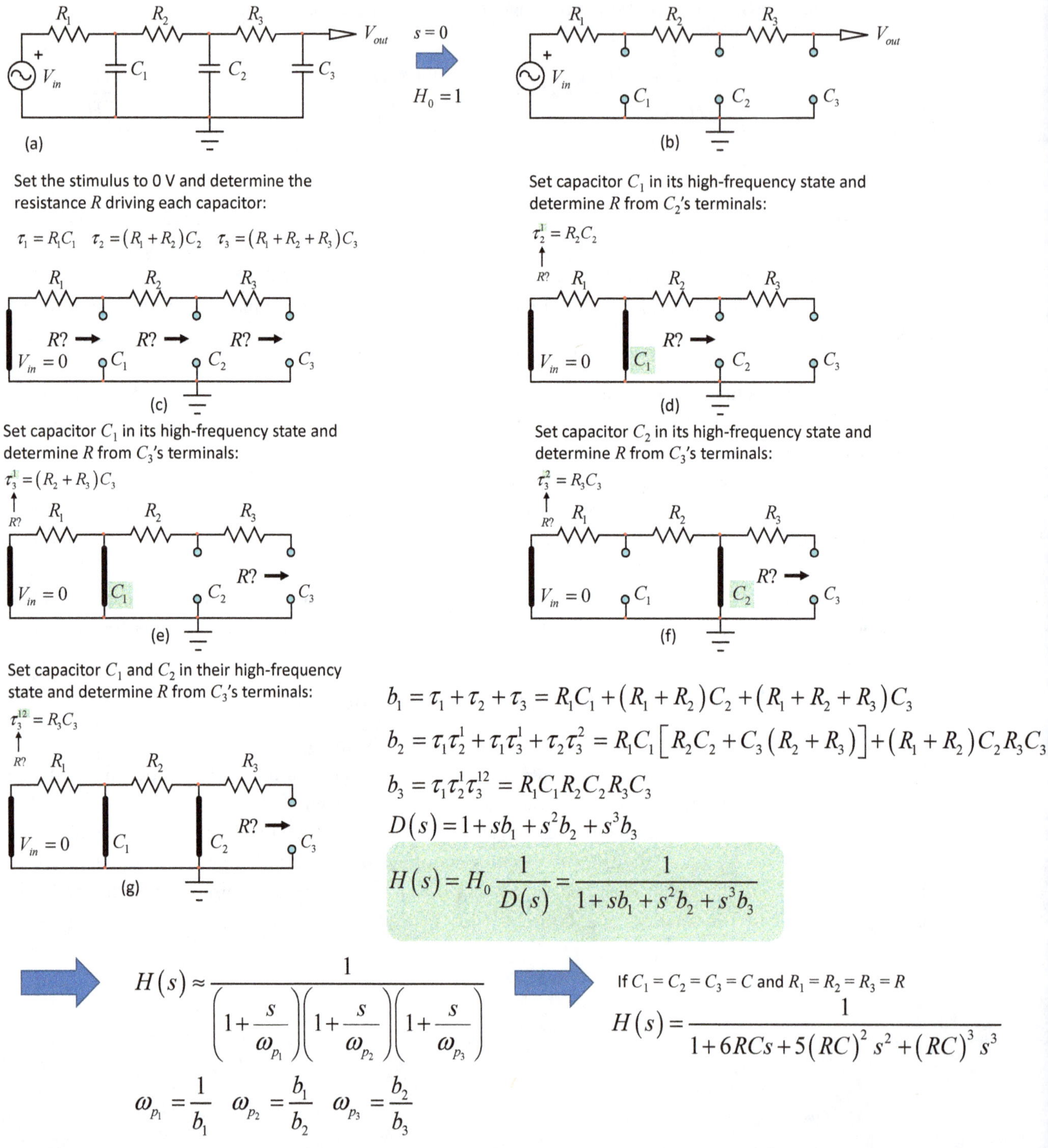

Nothing Complicated Compared to the Second-Order Networks: More Sketches but Principles Remain Similar

Figure 7.1

$R_2 := 1k\Omega \quad C_1 := 100nF \quad R_3 := 1k\Omega$

$R_1 := 1k\Omega \quad C_2 := 100nF \quad C_3 := 100nF$

$H_0 := 1$

$\tau_1 := R_1 \cdot C_1 = 100 \cdot \mu s$

$\tau_2 := (R_1 + R_2) \cdot C_2 = 200 \cdot \mu s$

$\tau_3 := (R_3 + R_2 + R_1) \cdot C_3 = 300 \cdot \mu s$

$b_1 := \tau_1 + \tau_2 + \tau_3 = 600 \cdot \mu s$

$b_{1a} := R_1 \cdot C_1 + (R_1 + R_2) \cdot C_2 + (R_3 + R_2 + R_1) \cdot C_3 = 600 \cdot \mu s$

$\tau_{12} := C_2 \cdot R_2 = 100 \cdot \mu s$

$\tau_{13} := C_3 \cdot (R_3 + R_2) = 200 \cdot \mu s$

$\tau_{23} := C_3 \cdot R_3 = 100 \cdot \mu s$

$b_2 := \tau_1 \cdot \tau_{12} + \tau_1 \cdot \tau_{13} + \tau_2 \cdot \tau_{23} = 5 \times 10^4 \cdot \mu s^2$

$b_{2a} := R_1 \cdot C_1 \cdot [C_2 \cdot R_2 + C_3 \cdot (R_2 + R_3)] + (R_1 + R_2) \cdot C_2 \cdot C_3 \cdot R_3 = 5 \times 10^4 \cdot \mu s^2$

$\tau_{123} := R_3 \cdot C_3 = 100 \cdot \mu s$

$b_3 := \tau_1 \cdot \tau_{12} \cdot \tau_{123} = 1 \times 10^6 \cdot \mu s^3$

$b_{3a} := R_1 \cdot C_1 \cdot (C_2 \cdot R_2) \cdot (R_3 \cdot C_3) = 1 \times 10^6 \cdot \mu s^3 \qquad C_1 \cdot C_2 \cdot C_3 \cdot R_1 \cdot R_2 \cdot R_3 = 1 \times 10^{-12} s^3$

$$H_1(s) := H_0 \cdot \frac{1}{1 + b_1 \cdot s + b_2 \cdot s^2 + b_3 \cdot s^3}$$

$$H_2(s) := H_0 \cdot \frac{1}{(1 + b_1 \cdot s) \cdot \left(1 + s \cdot \frac{b_2}{b_1}\right) \cdot \left(1 + s \cdot \frac{b_3}{b_2}\right)}$$

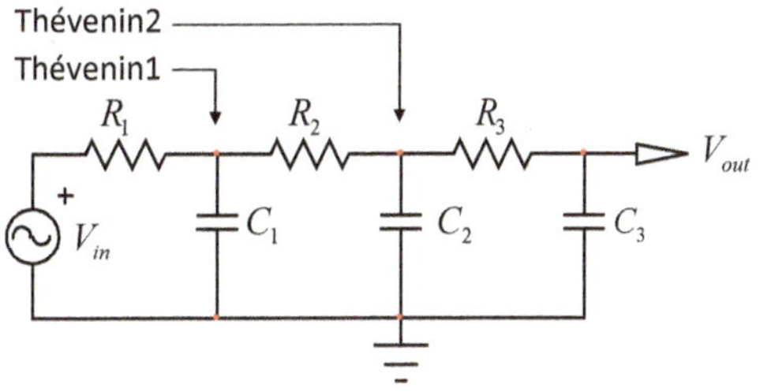

$R_{th1}(s) := \left(\frac{1}{s \cdot C_1}\right) \| R_1 \quad R_{th2}(s) := (R_{th1}(s) + R_2) \| \left(\frac{1}{s \cdot C_2}\right)$ Thévenin resistances

$$H_{ref}(s) := \frac{\frac{1}{s \cdot C_1}}{\frac{1}{s \cdot C_1} + R_1} \cdot \frac{\frac{1}{s \cdot C_2}}{R_{th1}(s) + R_2 + \frac{1}{s \cdot C_2}} \cdot \frac{\frac{1}{s \cdot C_3}}{\frac{1}{s \cdot C_3} + R_{th2}(s) + R_3}$$ Brute-force transfer function

$$H_4(s) := \frac{1}{1 + [R_1 \cdot C_1 + (R_1 + R_2) \cdot C_2 + (R_3 + R_2 + R_1) \cdot C_3] \cdot s + [R_1 \cdot C_1 \cdot [C_2 \cdot R_2 + C_3 \cdot (R_2 + R_3)] + (R_1 + R_2) \cdot C_2 \cdot C_3 \cdot R_3] \cdot s^2 + (C_1 \cdot C_2 \cdot C_3 \cdot R_1 \cdot R_2 \cdot R_3) \cdot s^3}$$

$$H_5(s) := \frac{1}{1 + s \cdot (6 \cdot R \cdot C) + s^2 5 \cdot (R \cdot C)^2 + s^3 \cdot (R \cdot C)^3}$$

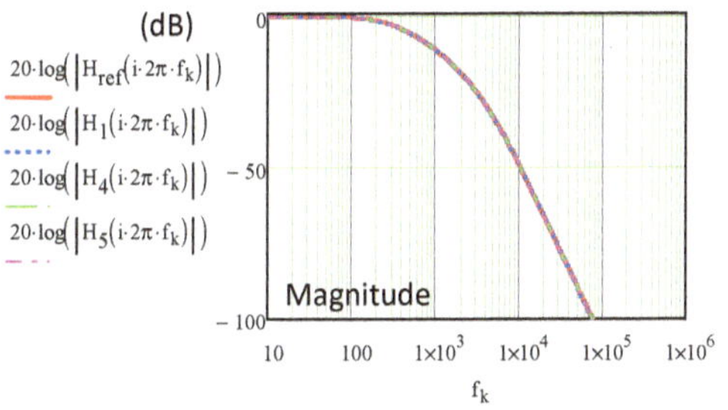

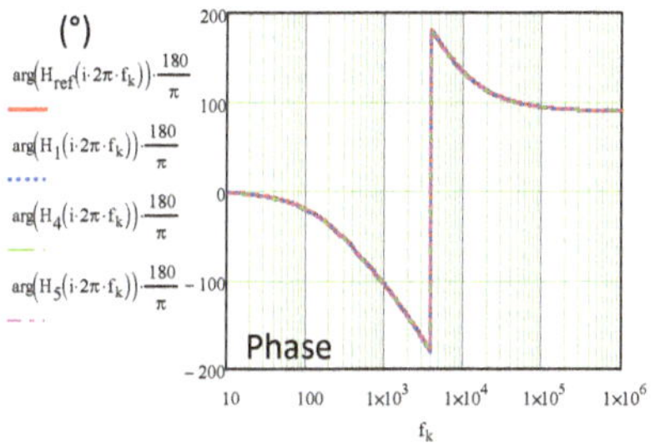

Implement Thévenin Two Times to get Results, but it's Likely to Complicate the Factoring of the Final Equation

Figure 7.2

Now, keep the same circuit, but probe V_{out} across C_2 instead of C_3.

Do we have to restart all over from scratch?

Certainly not because we did not change the structure, just moved the response observation to a different node. And that is where the FACTs really prove to be efficient: because the electrical structure remains similar, the denominator $D(s)$ we determined can be reused and we just need to check for a change in the gain or the presence of zeroes.

As illustrated in Figure 7.3, a single zero brought by R_3C_3 is identified and immediately revealed by inspection.

Finally, if we move the probe one more step to the left and observe the response across C_1, then we can involve the generalized method as described in Figure 7.4.

This is not complicated, just determine high-frequency gains when the capacitors are alternatively replaced by short circuits.

These gains are combined with the denominator time constants to form the numerator as detailed in (7.18). Simplifications occur and the final transfer function comes out quite easily.

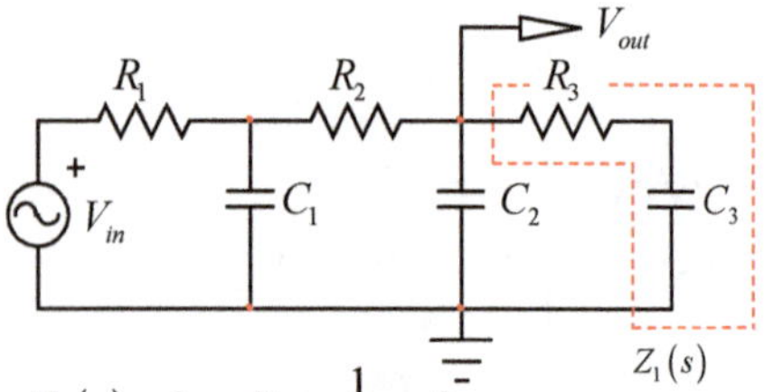

$$Z_1(s)=0 \quad R_3+\frac{1}{sC_3}=0$$

$$\rightarrow s_z=-\frac{1}{R_3C_3} \quad \omega_z=\frac{1}{R_3C_3}$$

$$H(s)\approx\frac{1+\frac{s}{\omega_z}}{\left(1+\frac{s}{\omega_{p_1}}\right)\left(1+\frac{s}{\omega_{p_2}}\right)\left(1+\frac{s}{\omega_{p_3}}\right)}$$

$$\omega_{p_1}=\frac{1}{b_1} \quad \omega_{p_2}=\frac{b_1}{b_2} \quad \omega_{p_3}=\frac{b_2}{b_3} \quad \omega_z=\frac{1}{R_3C_3}$$

If $C_1=C_2=C_3=C$ and $R_1=R_2=R_3=R$

$$H(s)=\frac{1+sRC}{1+6RCs+5(RC)^2s^2+(RC)^3s^3}$$

$$H_1(s):=H_0\cdot\frac{1+s\cdot R_3\cdot C_3}{1+b_1\cdot s+b_2\cdot s^2+b_3\cdot s^3}$$

$$H_4(s):=\frac{1+s\cdot R_3\cdot C_3}{1+\left[R_1\cdot C_1+(R_1+R_2)\cdot C_2+(R_3+R_2+R_1)\cdot C_3\right]\cdot s+\left[R_1\cdot C_1\cdot\left[C_2\cdot R_2+C_3\cdot(R_2+R_3)\right]+(R_1+R_2)\cdot C_2\cdot C_3\cdot R_3\right]\cdot s^2+(C_1\cdot C_2\cdot C_3\cdot R_1\cdot R_2\cdot R_3)\cdot s^3}$$

If R1=R2=R3=R and C1=C2=C3=C then:

$C:=100\text{nF} \quad R:=1\text{k}\Omega$

$$H_5(s):=\frac{1+s\cdot R\cdot C}{1+s\cdot(6\cdot R\cdot C)+s^2\,5\cdot(R\cdot C)^2+s^3\cdot(R\cdot C)^3}$$

$$R_{th1}(s):=\left(\frac{1}{s\cdot C_1}\right)\|R_1 \quad R_{th2}(s):=\left(R_{th1}(s)+R_2\right)\|\left(\frac{1}{s\cdot C_2}\right)$$

$$H_{ref}(s):=\frac{\frac{1}{s\cdot C_1}}{\frac{1}{s\cdot C_1}+R_1}\cdot\frac{\left(\frac{1}{s\cdot C_2}\right)\|\left(R_3+\frac{1}{s\cdot C_3}\right)}{R_{th1}(s)+R_2+\left(\frac{1}{s\cdot C_2}\right)\|\left(R_3+\frac{1}{s\cdot C_3}\right)}$$

Brute-force transfer function

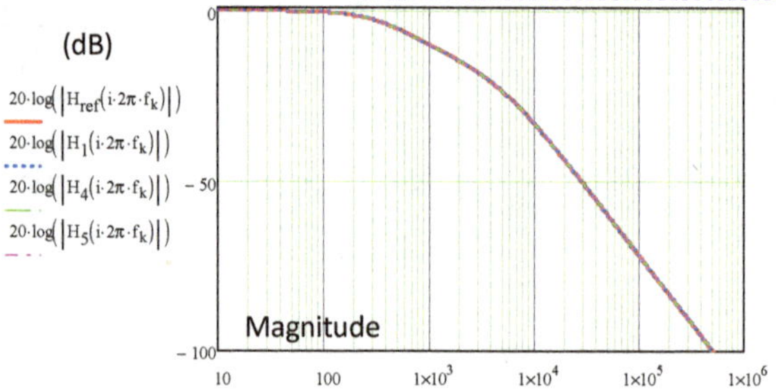

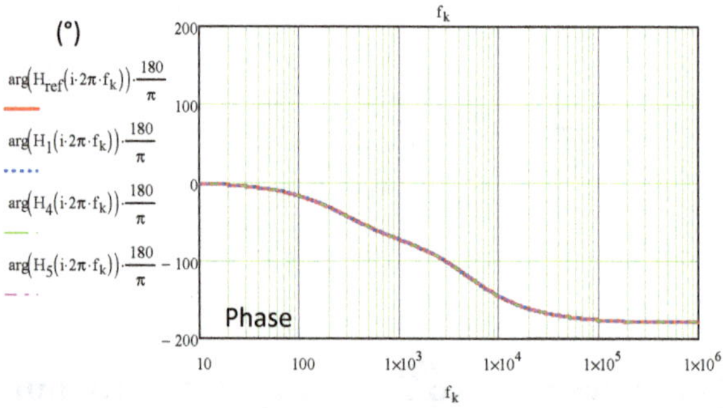

Changing the Position where the Response is Observed does not Affect the Electrical Structure of the Network

Figure 7.3

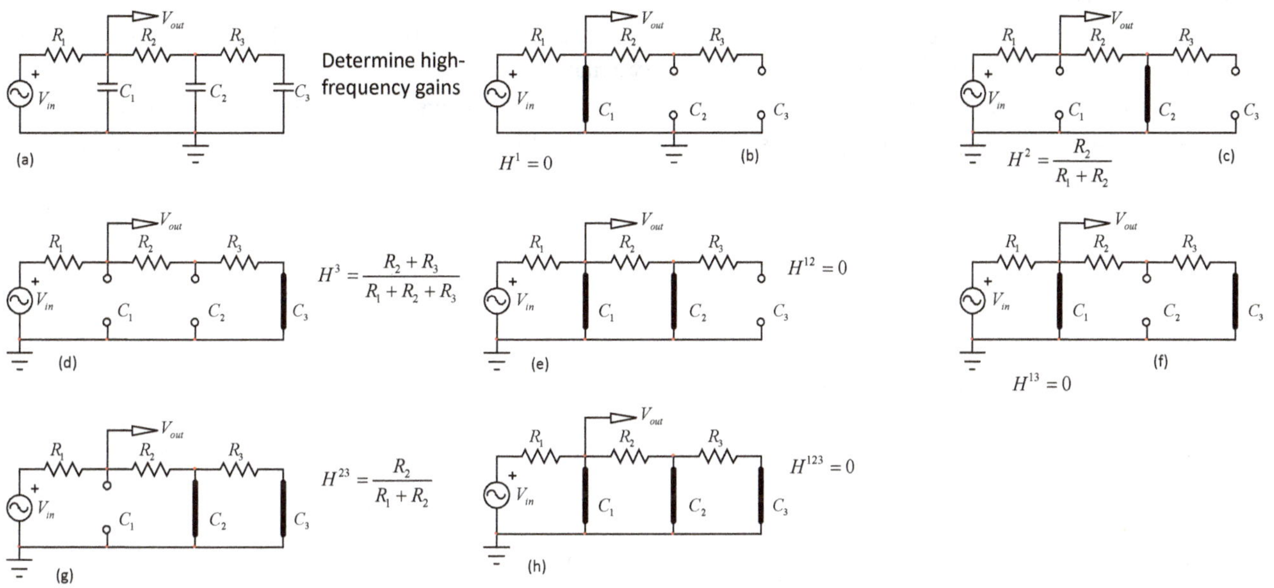

$$N(s) = H_0 + s\left(\tau_1 H^1 + \tau_2 H^2 + \tau_3 H^3\right) + s^2\left(\tau_1\tau_2^1 H^{12} + \tau_1\tau_3^1 H^{13} + \tau_2\tau_3^2 H^{23}\right) + s^3\left(\tau_1\tau_2^1\tau_3^{12} H^{123}\right)$$

$$N(s) = 1 + s\left(\tau_2 H^2 + \tau_3 H^3\right) + s^2\tau_2\tau_3^2 H^{23}$$

$$N(s) = 1 + s\left[(R_1 + R_2)C_2\frac{R_2}{R_1 + R_2} + (R_1 + R_2 + R_3)C_3\frac{R_2 + R_3}{R_1 + R_2 + R_3}\right] + s^2(R_1 + R_2)C_2 R_3 C_3\frac{R_2}{R_1 + R_2}$$

$$N(s) = 1 + s\left[R_2 C_2 + C_3(R_2 + R_3)\right] + s^2 R_2 R_3 C_2 C_3$$

If $C_1 = C_2 = C_3 = C$ and $R_1 = R_2 = R_3 = R$

$$H(s) = \frac{1 + 3RC + s^2(RC)^2}{1 + 6RCs + 5(RC)^2 s^2 + (RC)^3 s^3}$$

$$H_{ref}(s) := \frac{\left(\frac{1}{s \cdot C_1}\right) \| \left[R_2 + \left(\frac{1}{s \cdot C_2}\right) \| \left(R_3 + \frac{1}{s \cdot C_3}\right)\right]}{R_1 + \left(\frac{1}{s \cdot C_1}\right) \| \left[R_2 + \left(\frac{1}{s \cdot C_2}\right) \| \left(R_3 + \frac{1}{s \cdot C_3}\right)\right]}$$

If R1=R2=R3=R and C1=C2=C3=C then:

$C := 100nF \quad R := 1k\Omega$

$$H_5(s) := \frac{1 + s \cdot (3 \cdot R \cdot C) + s^2 \cdot (R \cdot C)^2}{1 + s \cdot (6 \cdot R \cdot C) + s^2 5 \cdot (R \cdot C)^2 + s^3 \cdot (R \cdot C)^3}$$

$$H_4(s) := \frac{1 + s \cdot \left[R_2 \cdot C_2 + C_3 \cdot (R_2 + R_3)\right] + s^2 \cdot (R_2 \cdot R_3 \cdot C_2 \cdot C_3)}{1 + \left[R_1 \cdot C_1 + (R_1 + R_2) \cdot C_2 + (R_3 + R_2 + R_1) \cdot C_3\right] \cdot s + \left[R_1 \cdot C_1 \cdot \left[C_2 \cdot R_2 + C_3 \cdot (R_2 + R_3)\right] + (R_1 + R_2) \cdot C_2 \cdot C_3 \cdot R_3\right] \cdot s^2 + (C_1 \cdot C_2 \cdot C_3 \cdot R_1 \cdot R_2 \cdot R_3) \cdot s^3}$$

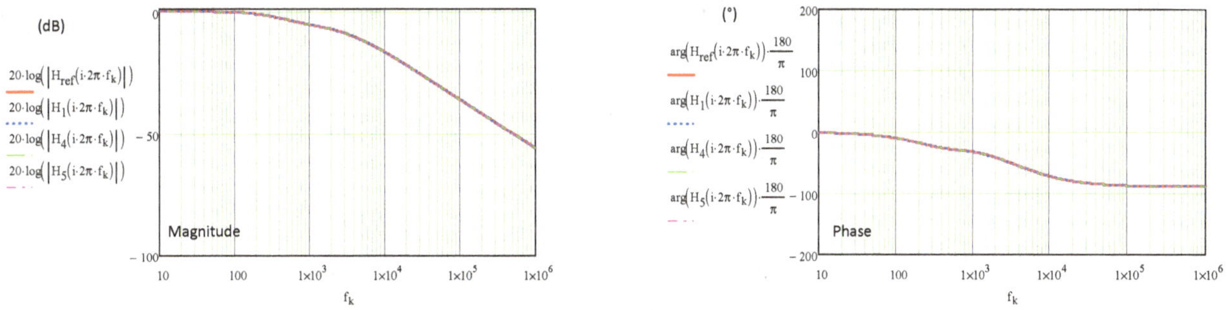

Probing across C_1 Shows the Presence of Two Zeroes that can be Quickly Obtained by Invoking the Generalized Transfer Function

Figure 7.4

An application for this 3rd-order network is the phase-shift oscillator using these cascaded *RC*s. It is shown in Figure 7.5 where the oscillation frequency is determined: express the transfer function in an imaginary format (replace *s* by $j\omega$) and cancel the imaginary part.

Then determine the insertion loss at that frequency and design an amplifier having a gain exactly compensating the attenuation. It is 0.034 or 1/29, easily implemented with an inverting op-amp whose input is buffered by a voltage follower.

As shown in the loop gain plot, the point at which the gain is 1 (0-dB magnitude) with a stimulus returning in phase (0° phase) is located slightly below 4 kHz as calculated.

The simulation on the right side confirms the frequency of oscillations. In a practical realization, this oscillator would start with noise or when the power supply turns on. In SPICE, especially considering perfect elements, the circuit could fail to start.

For that reason, I purposely added an initial condition in one of the capacitors and it does the expected trick.

Note that another example of *CR* phase-shifted oscillator is described later in the chapter.

$$R := 1k\Omega \qquad C := 100nF$$

$$H_{dis}(i\omega) := \frac{1}{\left[1 - 5 \cdot (R \cdot C \cdot \omega)^2\right] + i \cdot \left[6 \cdot \omega \cdot R \cdot C - (\omega \cdot R \cdot C)^3\right]}$$

cancel the imaginary part:

$$6 \cdot \omega \cdot R \cdot C - (\omega \cdot R \cdot C)^3 = 0 \quad \text{for:} \quad \omega_0 := \frac{\sqrt{6}}{C \cdot R}$$

Attenuation at ω_0:

$$Mag_{\omega 0} := \frac{1}{\sqrt{C^6 \cdot R^6 \cdot \omega_0^6 + 13 \cdot C^4 \cdot R^4 \cdot \omega_0^4 + 26 \cdot C^2 \cdot R^2 \cdot \omega_0^2 + 1}} = 0.034 \quad \left(\frac{1}{29}\right)$$

$$20 \cdot \log(Mag_{\omega 0}) = -29.248\ dB$$

$$f_0 := \frac{\omega_0}{2 \cdot \pi} = 3.898\ kHz \qquad \frac{\sqrt{6}}{R \cdot C \cdot 2 \cdot \pi} = 3.898\ kHz$$

amplifier

Loop gain: $T_{OL}(s) = -H(s) \cdot 29$

parameters

R=100k
C=1nF
Ri=100k
Rf=29*Ri

$$f_0 = \frac{1}{\Delta t} = \frac{1}{256.8u} \approx 3.9\ kHz$$

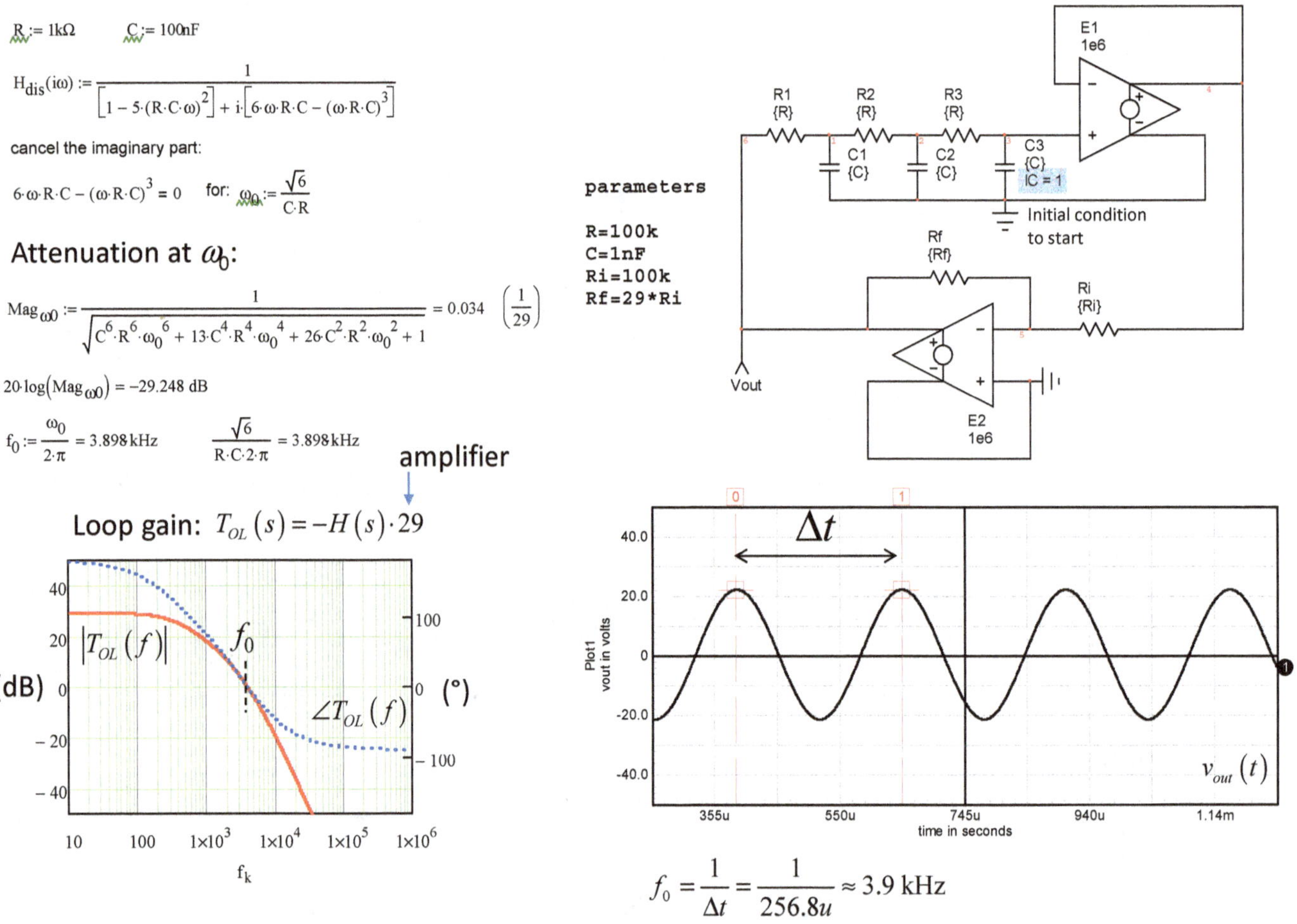

It is Possible to Build a Phase-Shift Oscillator by adding an Amplifier Compensating the Insertion Loss at the Oscillation Frequency

Figure 7.5

In Figure 7.6, we have the small-signal model of a buck converter—without its modulator—made of a *LC* filter loaded by two capacitors. One of them features an equivalent series resistance (ESR) labeled R_2 while the second does not. This is clearly a 3rd-order system and I applied the FACTs to determine its transfer function.

To avoid indeterminacies, I invoked a finite high-value resistance R_{inf} when determining R in some configurations where I had an open circuit. This way, I can run simplifications in the final expression when R_{inf} appears in the numerator and the denominator.

The ac response of the circuit is shown in Figure 7.7 in which I tried to look at the coefficients values to approximate the 3rd-order polynomial as a 1st-order low-pass cascaded with a 2nd-order filter.

With the adopted values, the approximation is confirmed by the superimposed magnitude/phase curves.

$$\tau_1 = \frac{L_1}{R_{\inf}}$$

$$\tau_2 = C_2\left(R_1 + R_2\right)$$

$$\tau_3 = R_1 C_3$$

$$\tau_2^1 = C_2 R_{\inf} \quad \tau_3^1 = C_3 R_{\inf}$$

$$\tau_3^2 = C_3\left(R_1 \parallel R_2\right)$$

$$\tau_3^{12} = R_2 C_3$$

$$Z_1(s) = 0 \quad R_2 + \frac{1}{sC_2} = 0$$

$$\rightarrow s_z = -\frac{1}{R_2 C_2} \quad \omega_z = \frac{1}{R_2 C_2}$$

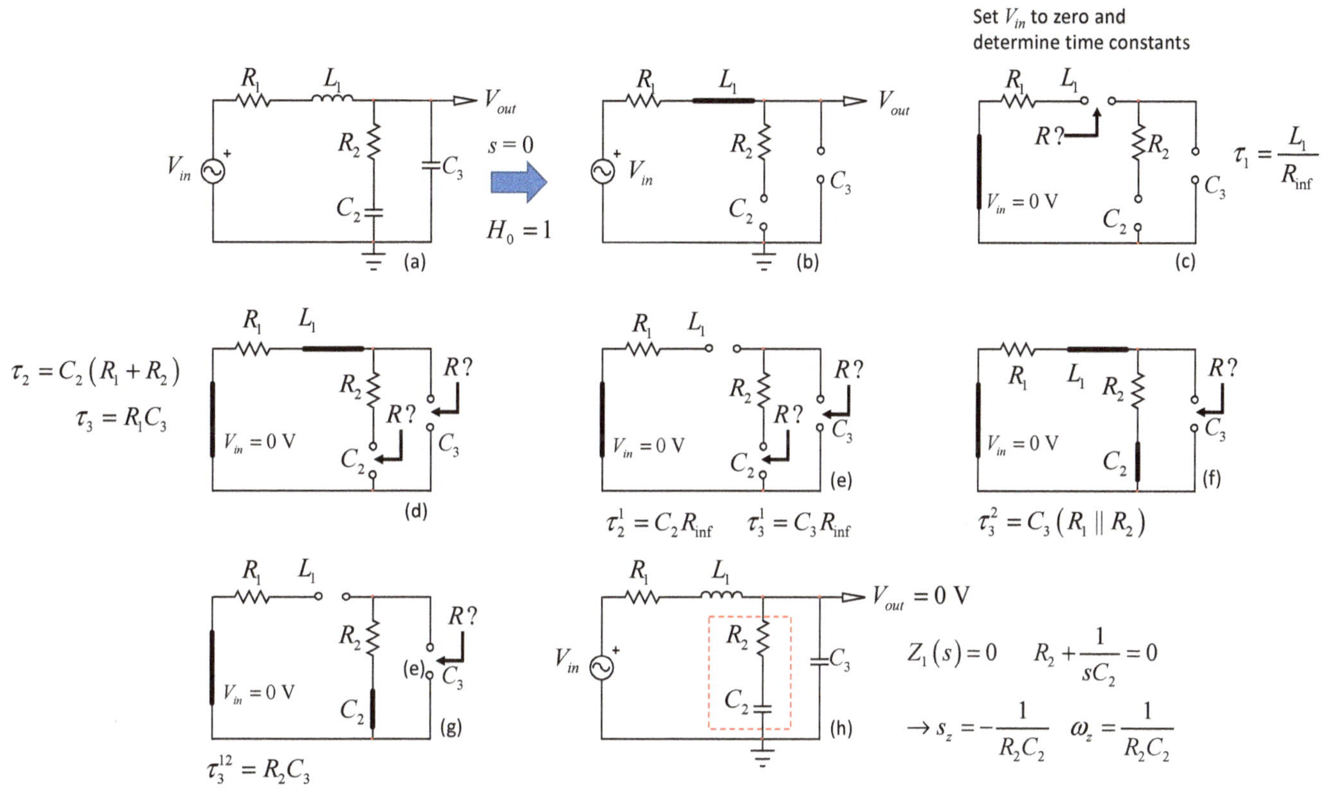

$$b_1 = \tau_1 + \tau_2 + \tau_3 = \frac{L_1}{R_{\inf}} + \left(R_1 + R_2\right)C_2 + R_1 C_3$$

$$b_2 = \tau_1\tau_2^1 + \tau_1\tau_3^1 + \tau_2\tau_3^2 = \frac{L_1}{R_{\inf}}C_2 R_{\inf} + \frac{L_1}{R_{\inf}}C_3 R_{\inf} + C_2 C_3 R_1 R_2$$

$$b_3 = \tau_1\tau_2^1\tau_3^{12} = \frac{L_1}{R_{\inf}}C_2 R_{\inf} R_2 C_3 = L_1 C_2 R_2 C_3$$

$$D(s) = 1 + sb_1 + s^2 b_2 + s^3 b_3$$

$$H(s) = H_0 \frac{N(s)}{D(s)} = \frac{1 + \frac{s}{\omega_z}}{1 + sb_1 + s^2 b_2 + s^3 b_3} \quad \omega_z = \frac{1}{R_2 C_2}$$

On-The-Fly Implementation of the Dummy Finite Resistance $R_{\inf}$ Simplifies the Analysis of this Circuit

Figure 7.6

$R_2 := 0.22\Omega \quad L_1 := 470\mu H \quad C_3 := 10nF$ $\qquad \|(x,y) := \frac{x \cdot y}{x+y}$

$R_1 := 0.1\Omega \quad C_2 := 10nF \quad R_{inf} := 10^{23}\Omega$

$H_0 := 1$

$\tau_1 := \frac{L_1}{R_{inf}} = 0 \cdot \mu s$

$\tau_2 := (R_1 + R_2) \cdot C_2 = 3.2 \times 10^{-3} \cdot \mu s$

$\tau_3 := R_1 \cdot C_3 = 1 \times 10^{-3} \cdot \mu s$

$b_1 := \tau_1 + \tau_2 + \tau_3 = 4.2 \times 10^{-3} \cdot \mu s$

$b_{1a} := (R_1 + R_2) \cdot C_2 + R_1 \cdot C_3 = 4.2 \times 10^{-3} \cdot \mu s$

$\tau_{12} := C_2 \cdot R_{inf} = 1 \times 10^{21} \cdot \mu s$

$\tau_{13} := C_3 \cdot R_{inf} = 1 \times 10^{21} \cdot \mu s$

$\tau_{23} := C_3 \cdot (R_1 \| R_2) = 6.875 \times 10^{-4} \cdot \mu s$

$b_2 := \tau_1 \cdot \tau_{12} + \tau_1 \cdot \tau_{13} + \tau_2 \cdot \tau_{23} = 9.4 \mu s^2$

$b_{2a} := L_1 \cdot C_2 + L_1 \cdot C_3 + (R_1 + R_2) \cdot C_2 \cdot [C_3 \cdot (R_1 \| R_2)] = 9.4 \mu s^2$

$\tau_{123} := R_2 \cdot C_3 = 2.2 \times 10^{-3} \cdot \mu s$

$b_3 := \tau_1 \cdot \tau_{12} \cdot \tau_{123} = 0.01 \cdot \mu s^3$

$b_{3a} := L_1 \cdot C_2 \cdot (R_2 \cdot C_3) = 0.01 \cdot \mu s^3$

$N_1(s) := 1 + s \cdot R_2 \cdot C_2 \qquad \omega_z := \frac{1}{R_2 \cdot C_2}$

$H_1(s) := H_0 \cdot \frac{N_1(s)}{1 + b_1 \cdot s + b_2 \cdot s^2 + b_3 \cdot s^3}$

$H_2(s) := \frac{1 + s \cdot R_2 \cdot C_2}{1 + [(R_1 + R_2) \cdot C_2 + R_1 \cdot C_3] \cdot s + [L_1 \cdot C_2 + L_1 \cdot C_3 + (R_1 + R_2) \cdot C_2 \cdot [C_3 \cdot (R_1 \| R_2)]] \cdot s^2 + [L_1 \cdot C_2 \cdot (R_2 \cdot C_3)] \cdot s^3}$

$H_{ref}(s) := \frac{\left(\frac{1}{s \cdot C_2} + R_2\right) \| \left(\frac{1}{s \cdot C_3}\right)}{\left(\frac{1}{s \cdot C_2} + R_2\right) \| \left(\frac{1}{s \cdot C_3}\right) + R_1 + s \cdot L_1}$

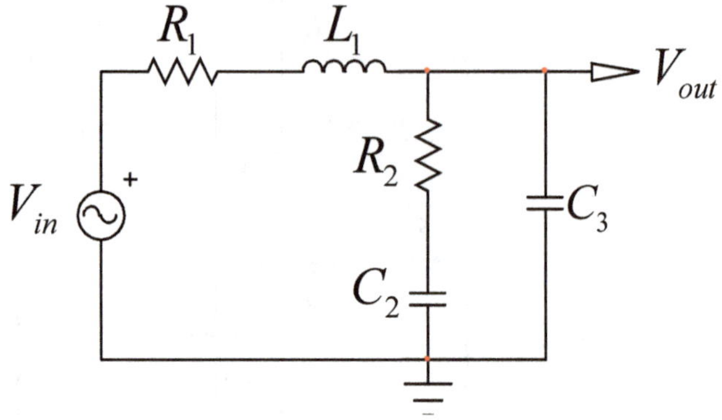

Time constant discussion

$b_1 = 4.2 \cdot ns \qquad \frac{b_2}{b_1} = 2.238 \, ms \qquad \frac{b_3}{b_2} = 1.1 \cdot ns$

$D_3(s) := \left(1 + \frac{s}{\omega_p}\right) \cdot \left[1 + \frac{s}{\omega_0 \cdot Q} + \left(\frac{s}{\omega_0}\right)^2\right] \qquad \omega_p := \frac{b_2}{b_3} \qquad \omega_0 := \frac{1}{\sqrt{b_2}} \qquad Q := \frac{\sqrt{b_2}}{b_1} = 729.986$

$f_0 := \frac{\omega_0}{2 \cdot \pi} = 51.911 \, kHz \qquad f_{0D} := \frac{\omega_0}{2 \cdot \pi} = 51.911 \, kHz$

$H_4(s) := \frac{1 + \frac{s}{\omega_z}}{\left(1 + \frac{s}{\omega_p}\right) \cdot \left[1 + \frac{s}{\omega_0 \cdot Q} + \left(\frac{s}{\omega_0}\right)^2\right]} \qquad H_5(s) := \frac{\left(1 + \frac{s}{\omega_z}\right)}{\left(1 + \frac{b_3}{b_2} \cdot s\right) \cdot \left(1 + s \cdot b_1 + s^2 \cdot b_2\right)}$

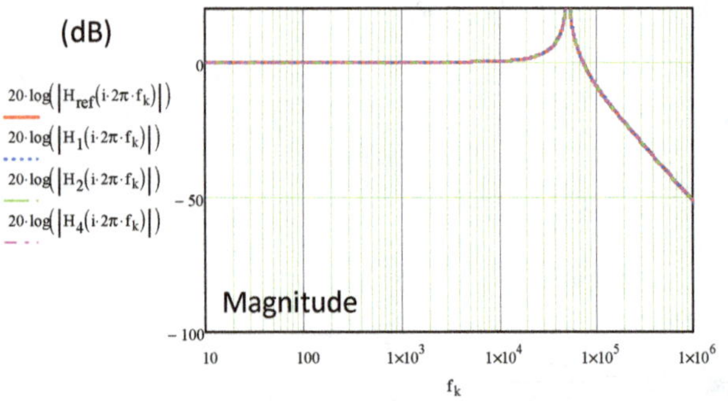

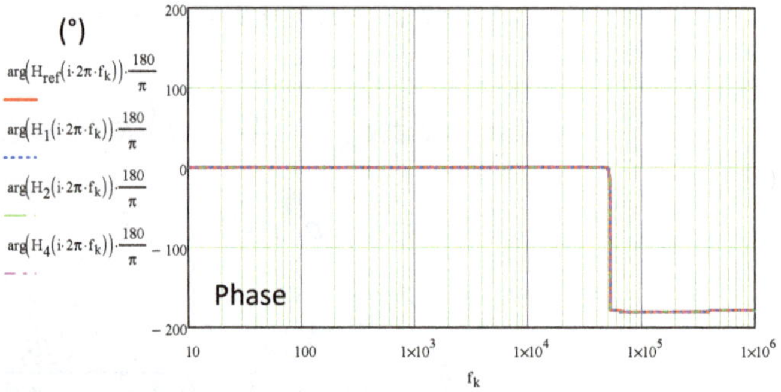

The Approximation of this 3rd-Order Filter by Two Cascaded Filters Gives a Good Result as Confirmed by the Plots

Figure 7.7

Let's now look at Figure 7.8 where three *CR* differentiators are cascaded.

This is a classic often found in phase-shifted oscillators and that is the reason why all resistors and capacitors share similar values. The transfer function is determined the usual way, by first setting the stimulus to 0 V and determining the time constants in this mode.

The rest comes out simply and the denominator appears after a few sketches.

Considering all capacitors in series with the stimulus, there are several zeroes at the origin. To determine the locations of these zeroes, look at the high-frequency gains by alternatively placing each capacitor or a combination of them, in their high-frequency state.

You immediately realize that all these high-frequency gains are equal to zero except when all capacitors are shorted and this is for H^{123}.

I could have avoided drawing all the small sketches but kept them for the sake of the illustration. The numerator, as expected, hosts a triple zero at dc as detailed in Figure 7.9.

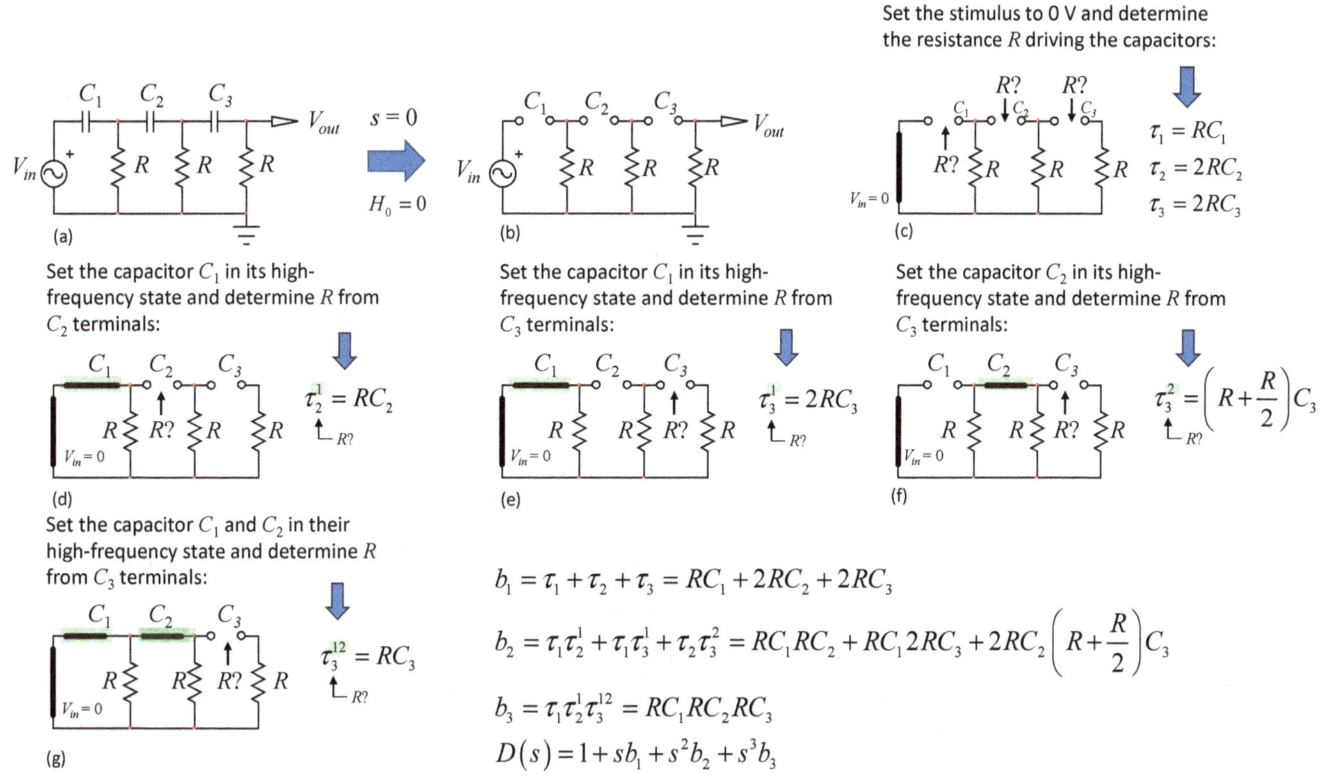
Set the stimulus to 0 V and determine the resistance R driving the capacitors:
(a)
(b)
(c)
s = 0
H_0 = 0
τ_1 = RC_1
τ_2 = 2RC_2
τ_3 = 2RC_3
Set the capacitor C_1 in its high-frequency state and determine R from C_2 terminals:
Set the capacitor C_1 in its high-frequency state and determine R from C_3 terminals:
Set the capacitor C_2 in its high-frequency state and determine R from C_3 terminals:
τ_2^1 = RC_2
τ_3^1 = 2RC_3
τ_3^2 = (R + R/2)C_3
(d)
(e)
(f)
Set the capacitor C_1 and C_2 in their high-frequency state and determine R from C_3 terminals:
τ_3^12 = RC_3
(g)

$$b_1 = \tau_1 + \tau_2 + \tau_3 = RC_1 + 2RC_2 + 2RC_3$$

$$b_2 = \tau_1\tau_2^1 + \tau_1\tau_3^1 + \tau_2\tau_3^2 = RC_1RC_2 + RC_1 2RC_3 + 2RC_2\left(R + \frac{R}{2}\right)C_3$$

$$b_3 = \tau_1\tau_2^1\tau_3^{12} = RC_1RC_2RC_3$$

$$D(s) = 1 + sb_1 + s^2b_2 + s^3b_3$$

$C_1 = C_2 = C_3 = C$

$$b_1 = 5RC$$

$$b_2 = 6(RC)^2$$

$$b_3 = (RC)^3$$

$$D(s) = 1 + s5RC + s^2 6(RC)^2 + s^3(RC)^3$$

Determining the Denominator is Simple even with 3 Energy-Storing Elements

Figure 7.8

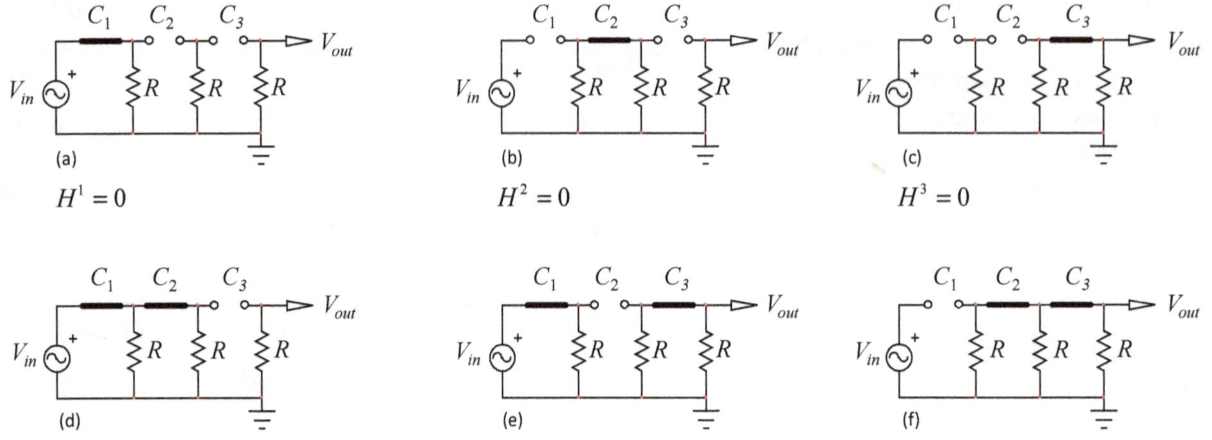

(a) $H^1 = 0$
(b) $H^2 = 0$
(c) $H^3 = 0$
(d) $H^{12} = 0$
(e) $H^{13} = 0$
(f) $H^{23} = 0$

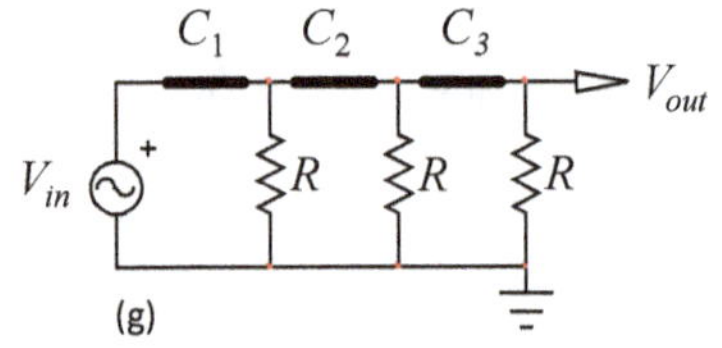

(g) $H^{123} = 1$

$$N(s) = H_0 + s\left(\tau_1 H^1 + \tau_2 H^2 + \tau_3 H^3\right) + s^2\left(\tau_1\tau_2^1 H^{12} + \tau_1\tau_3^1 H^{13} + \tau_2\tau_3^2 H^{23}\right) + s^3\left(\tau_1\tau_2^1\tau_3^{12} H^{123}\right)$$

$$N(s) = 0 + s\left(\tau_1 \cdot 0 + \tau_2 \cdot 0 + \tau_3 \cdot 0\right) + s^2\left(\tau_1\tau_2^1 \cdot 0 + \tau_1\tau_3^1 \cdot 0 + \tau_2\tau_3^2 \cdot 0\right) + s^3\left(\tau_1\tau_2^1\tau_3^{12} \cdot 1\right)$$

$$N(s) = s^3 RC_1 RC_2 RC_3 \xrightarrow[C_1 = C_2 = C_3 = C]{} N(s) = s^3 (RC)^3$$

$$H(s) = \frac{s^3 (RC)^3}{1 + s5RC + s^2 6(RC)^2 + s^3 (RC)^3}$$

Only One High-Frequency Gain is Equal to One while all the Others are Zero

Figure 7.9

The ac response of this transfer function is shown in Figure 7.10.

With the adopted components values, the oscillation frequency is 295 Hz. This type of circuit is often selected as an example in textbooks for implementing a phase-shift oscillator. Each segment contributes a 60° lag which equals -180° in total at the oscillation frequency.

This is shown in Figure 7.11 where the condition for oscillation is determined by finding the root for which the imaginary part cancels.

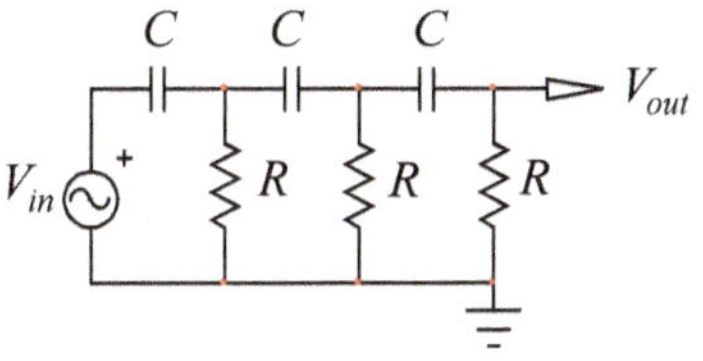

$R_1 := 2.2k\Omega \quad C_1 := 100nF \quad \|(x,y) := \frac{x \cdot y}{x + y}$

$C_3 := 100nF \quad C_2 := 100nF$

$H_0 := 0$

$\tau_1 := R_1 \cdot C_1 = 220\,\mu s$

$\tau_2 := 2R_1 \cdot C_2 = 440\,\mu s$

$\tau_3 := 2R_1 \cdot C_3 = 440\,\mu s$

$b_1 := \tau_1 + \tau_2 + \tau_3 = 1.1 \cdot ms$

$b_{1a} := 5 \cdot R_1 \cdot C_1 = 1.1 \cdot ms$

$\tau_{12} := R_1 \cdot C_2 = 220\,\mu s$

$\tau_{13} := 2 \cdot R_1 \cdot C_3 = 440\,\mu s$

$\tau_{23} := C_3 \cdot (R_1 + R_1 \| R_1) = 330\,\mu s$

$b_2 := \tau_1 \cdot \tau_{12} + \tau_1 \cdot \tau_{13} + \tau_2 \cdot \tau_{23} = 2.904 \times 10^5 \cdot \mu s^2$

$b_{2a} := R_1 \cdot C_1 \cdot (R_1 \cdot C_2) + R_1 \cdot C_1 \cdot (2 \cdot R_1 \cdot C_3) + 2R_1 \cdot C_2 \cdot [C_3 \cdot (R_1 + R_1 \| R_1)] = 2.904 \times 10^5 \cdot \mu s^2$

$b_{2b} := 6 \cdot C_1^2 \cdot R_1^2 = 2.904 \times 10^5 \cdot \mu s^2$

$\tau_{123} := R_1 \cdot C_3 = 220\,\mu s$

$b_3 := \tau_1 \cdot \tau_{12} \cdot \tau_{123} = 1.065 \times 10^7 \cdot \mu s^3$

$b_{3a} := R_1 \cdot C_1 \cdot [R_1 \cdot C_2 \cdot (R_1 \cdot C_3)] = 1.065 \times 10^7 \cdot \mu s^3$

$b_{3b} := C_1^3 \cdot R_1^3 = 1.065 \times 10^{-11} s^3$

$H_{123} := 1$

$f_{osc} := \frac{1}{\sqrt{6} \cdot C_1 \cdot R_1} \cdot \frac{1}{2 \cdot \pi} = 295.34 Hz$

$H_1(s) := \dfrac{H_0 + s^3 \cdot H_{123} b_3}{1 + b_1 \cdot s + b_2 \cdot s^2 + b_3 \cdot s^3}$

$H_4(s) := \dfrac{s^3 \cdot (C_1^3 \cdot R_1^3)}{1 + s \cdot (5 \cdot R_1 \cdot C_1) + s^2 \cdot (6 \cdot C_1^2 \cdot R_1^2) + s^3 \cdot (C_1^3 \cdot R_1^3)}$

$MagH(\omega) := \dfrac{(C_1^3 \cdot R_1^3 \cdot \omega^3)}{\sqrt{(C_1^3 \cdot R_1^3 \cdot \omega^3 - 5 \cdot C_1 \cdot R_1 \cdot \omega)^2 + (1 - 6 \cdot C_1^2 \cdot R_1^2 \cdot \omega^2)^2}}$

$ArgH(\omega) := -atan2[(C_1^3 \cdot R_1^3 \cdot \omega^3 - 5 \cdot C_1 \cdot R_1 \cdot \omega), (1 - 6 \cdot C_1^2 \cdot R_1^2 \cdot \omega^2)]$

$R_{th1}(s) := \left(\dfrac{1}{s \cdot C_1}\right) \| R_1 \qquad R_{th2}(s) := \left(R_{th1}(s) + \dfrac{1}{s \cdot C_2}\right) \| (R_1)$

$H_{ref}(s) := \dfrac{R_1}{\dfrac{1}{s \cdot C_1} + R_1} \cdot \dfrac{R_1}{R_{th1}(s) + R_1 + \dfrac{1}{s \cdot C_2}} \cdot \dfrac{R_1}{\dfrac{1}{s \cdot C_3} + R_{th2}(s) + R_1}$

(dB)

$20 \cdot \log(|H_{ref}(i \cdot 2\pi \cdot f_k)|) - 50$

$20 \cdot \log(MagH(2\pi \cdot f_k))$

Magnitude

(°)

$arg(H_{ref}(i \cdot 2\pi \cdot f_k)) \cdot \dfrac{180}{\pi}$

$ArgH(2\pi \cdot f_k) \cdot \dfrac{180}{\pi}$

Phase

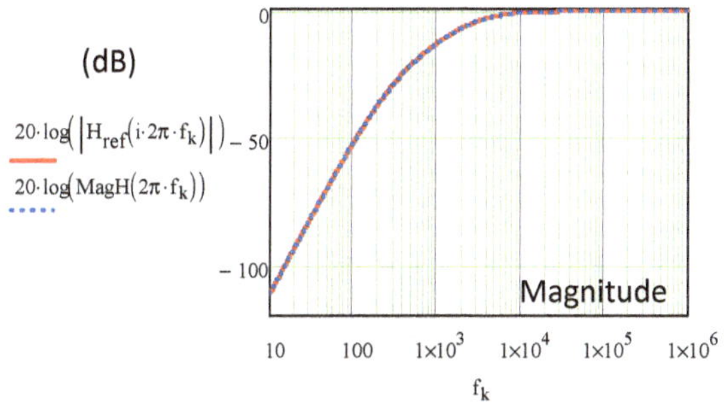

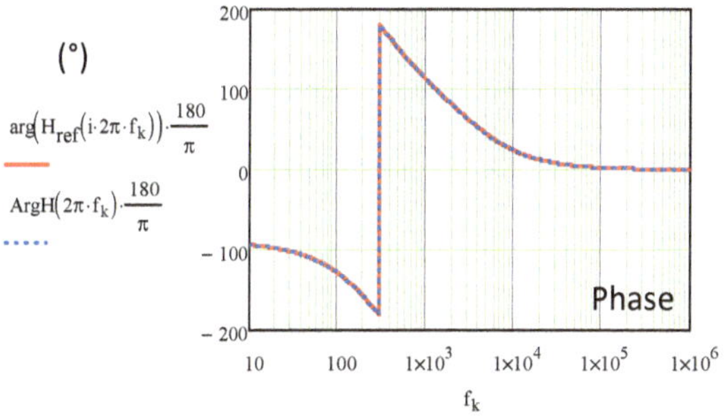

The Transfer Function is Simplified and Rearranged to Reveal a Magnitude and an Argument

Figure 7.10

In a real circuit, a power up sequence or noise would start the oscillator, but in SPICE, use an initial condition in one of the capacitors for an equivalent result.

The frequency is measured around 295 Hz, as calculated in the Mathcad sheet.

$$H(s)=\frac{s^3(RC)^3}{1+s5RC+s^2 6(RC)^2+s^3(RC)^3} \xrightarrow[\text{Factor numerator and simplify}]{s=j\omega} H(j\omega)=\frac{R^3C^3\omega^3}{R^3C^3\omega^3-5RC\omega+j\left(1-6C^2R^2\omega^2\right)}$$

$$\left|H(j\omega)\right|=\frac{R^3C^3\omega^3}{\sqrt{\left(R^3C^3\omega^3-5RC\omega\right)^2+\left(1-6C^2R^2\omega^2\right)^2}}$$

$$\angle H(j\omega)=-\tan^{-1}\left(\frac{1-6C^2R^2\omega^2}{R^3C^3\omega^3-5RC\omega}\right)$$

Cancel the imaginary part:

$$1-6C^2R^2\omega^2=0$$
$$\omega_0=\frac{1}{\sqrt{6}RC}$$

What is the attenuation at ω_0?

$$\left|H\left(\frac{1}{\sqrt{6}RC}\right)\right|=\frac{R^3C^3\left(\frac{1}{\sqrt{6}RC}\right)^3}{\sqrt{\left(R^3C^3\left(\frac{1}{\sqrt{6}RC}\right)^3-5RC\left(\frac{1}{\sqrt{6}RC}\right)\right)^2+\left(1-6C^2R^2\left(\frac{1}{\sqrt{6}RC}\right)^2\right)^2}}=\frac{1}{29}$$

The op-amp must compensate this attenuation by a gain of 29

What is the phase at ω_0?

$$\angle H\left(\frac{1}{\sqrt{6}RC}\right)=-180°$$ Each *RC* segment lags by 60°

We can now plot the loop gain which is: $T_{OL}(s)=H(s)\cdot(-29)$

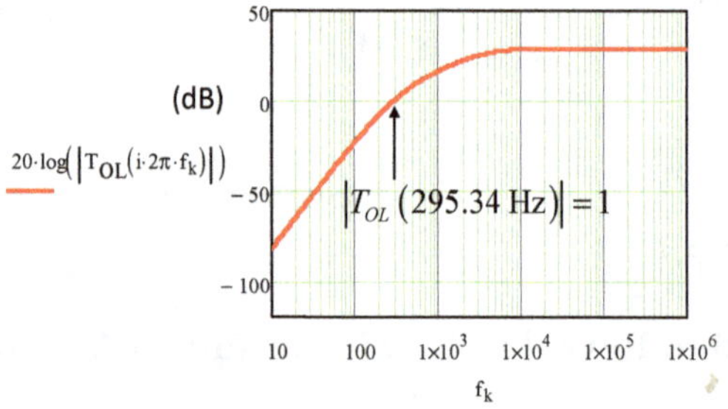

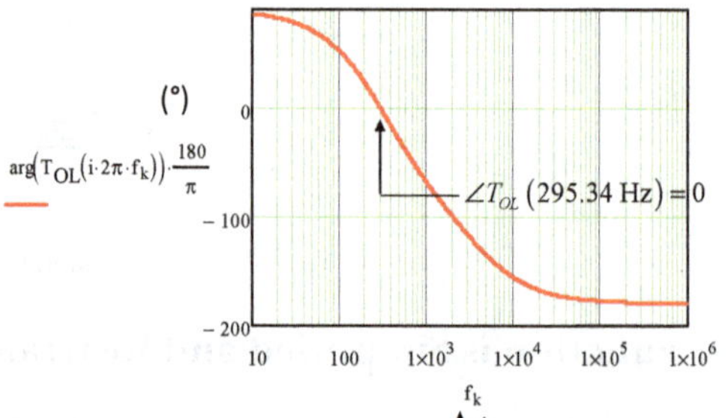

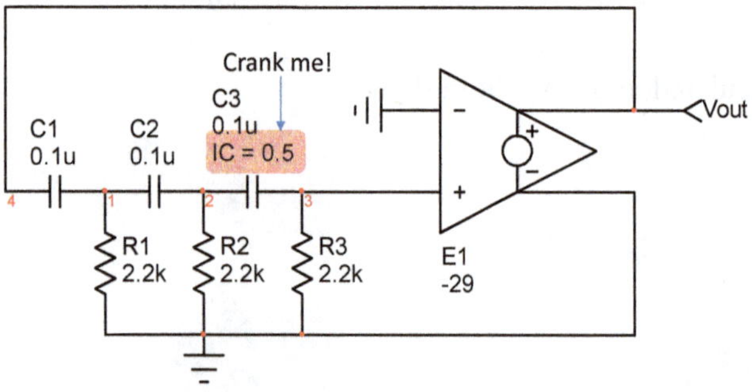

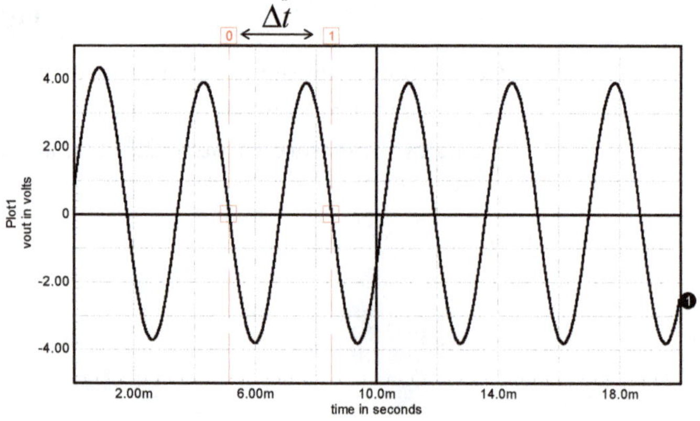

Compensate the loss by a gain of 29

$$f_0=\frac{1}{\Delta t}=\frac{1}{3.384\text{ ms}}=295.5\text{ Hz}$$

The Circuit is Implemented in an Oscillator featuring an Op-Amp and Confirms the Oscillation Frequency of 295 Hz

Figure 7.11

Figure 7.12 illustrates an *LC* filter with two output capacitors affected by different capacitances and equivalent series resistances (ESR): they can't be placed in parallel for reducing the order of the filter, so it is a 3rd-oder structure. It could be that of a voltage-mode-controlled buck converter, for instance. Deriving the transfer function is not difficult and, by now, you should already have spotted the two zeroes brought by the series connection of the capacitor and its ESR.

Set the stimulus to 0 V and determine the resistance R driving the inductor:

Set the stimulus to 0 V and determine the resistance R driving each capacitor:

$$\tau_1 = \frac{L_1}{r_L + R_L}$$

$$\tau_2 = C_2\left(r_{C_2} + r_L \| R_L\right)$$

$$\tau_3 = C_3\left(r_{C_3} + r_L \| R_L\right)$$

Set the inductor L in its high-frequency state and determine R from capacitors 2 and 3 terminals:

Set the capacitor C_2 in its high-frequency state and determine R from capacitors 3 terminals:

$$\tau_2^1 = C_2\left(r_{C_2} + R_L\right)$$

$$\tau_3^1 = C_3\left(r_{C_3} + R_L\right)$$

$$\tau_3^2 = C_3\left(r_{C_3} + r_L \| r_{C_2} \| R_L\right)$$

Set the inductor L_1 and capacitor C_2 in their high-frequency state and determine R from capacitor 3 terminals:

What impedance combination brings an output null when the excitation is back in the circuit?

$$\tau_3^{12} = C_3\left(r_{C_3} + r_{C_2} \| R_L\right)$$

$$Z_2(s) = r_{C_2} + \frac{1}{sC_2} = 0$$

$$Z_3(s) = r_{C_3} + \frac{1}{sC_3} = 0$$

$$\omega_{z_2} = \frac{1}{r_{C_2}C_2} \quad \omega_{z_3} = \frac{1}{r_{C_3}C_3}$$

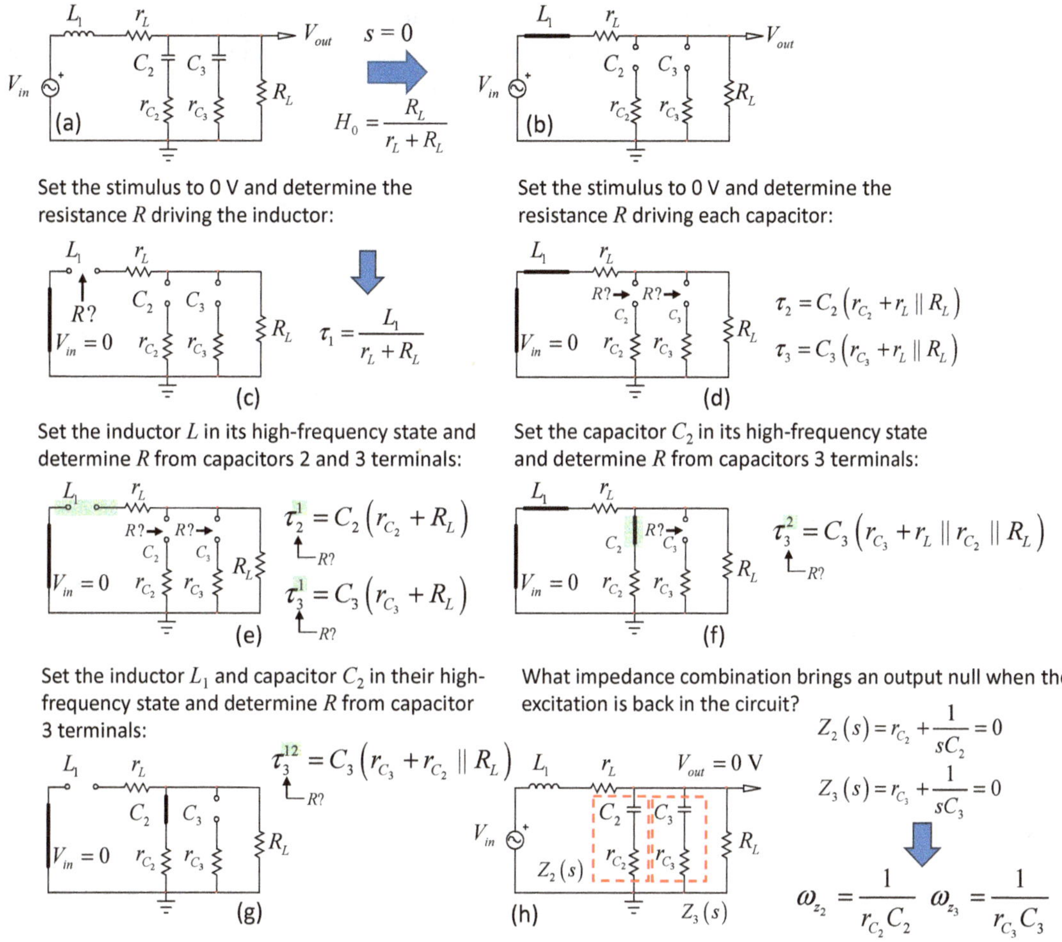

$$b_1 = \tau_1 + \tau_2 + \tau_3 = \frac{L_1}{r_L + R_L} + C_2\left(r_{C_2} + r_L \| R_L\right) + C_3\left(r_{C_3} + r_L \| R_L\right)$$

$$b_2 = \tau_1\tau_2^1 + \tau_1\tau_3^1 + \tau_2\tau_3^2$$

$$b_2 = \frac{L_1}{r_L + R_L}\left[C_2\left(r_{C_2} + R_L\right) + C_3\left(r_{C_3} + R_L\right)\right] + C_2\left(r_{C_2} + r_L \| R_L\right)C_3\left(r_{C_3} + r_L \| r_{C_2} \| R_L\right)$$

$$b_3 = \tau_1\tau_2^1\tau_3^{12} = \frac{L_1}{r_L + R_L}C_2\left(r_{C_2} + R_L\right)C_3\left(r_{C_3} + r_{C_2} \| R_L\right)$$

$$D(s) = 1 + sb_1 + s^2b_2 + s^3b_3 \quad N(s) = \left(1 + sr_{C_2}C_2\right)\left(1 + sr_{C_3}C_3\right)$$

$$H(s) = H_0\frac{N(s)}{D(s)} = \frac{\left(1 + \dfrac{s}{\omega_{z_2}}\right)\left(1 + \dfrac{s}{\omega_{z_3}}\right)}{1 + sb_1 + s^2b_2 + s^3b_3}$$

There are Three Energy-Storing Elements in this Filter as the Two Output Capacitors have Different Capacitances and ESR

Figure 7.12

The ac response is shown in Figure 7.13 and I factored the 3rd-order denominator as a low-pass filter followed by a second-order filter.

Implement this approach by comparing the various time constants and see if one dominates at low or high frequency. Comparing the magnitude and phase responses of the full expression and its approximate version tells you if the simplification holds. It is fairly good in this particular example.

$\|(x,y) := \frac{x \cdot y}{x+y}$

$r_L := 0.1\Omega \quad R_L := 2\Omega \quad r_{C2} := 0.05\Omega \quad r_{C3} := 0.015\Omega \quad C_2 := 10\mu F \quad C_3 := 470\mu F \quad L_1 := 22\mu H$

$H_0 := \frac{R_L}{R_L + r_L} = 0.952 \qquad 20 \cdot \log(H_0) = -0.424$ dB dc gain

$\tau_1 := \frac{L_1}{r_L + R_L} = 10.476 \mu s$

$\tau_2 := (r_{C2} + r_L \| R_L) \cdot C_2 = 1.452 \mu s$

$\tau_3 := (r_{C3} + r_L \| R_L) \cdot C_3 = 51.812 \mu s$

$b_1 := \tau_1 + \tau_2 + \tau_3 = 63.74 \mu s$

$\tau_{12} := C_2 \cdot (r_{C2} + R_L) = 20.5 \mu s$

$\tau_{13} := C_3 \cdot (r_{C3} + R_L) = 947.05 \mu s$

$\tau_{23} := (r_{C3} + r_L \| R_L \| r_{C2}) \cdot C_3 = 22.46 \mu s$

$b_2 := \tau_1 \cdot \tau_{12} + \tau_1 \cdot \tau_{13} + \tau_2 \cdot \tau_{23} = 1.017 \times 10^4 \cdot \mu s^2$

$\tau_{123} := (r_{C3} + r_{C2} \| R_L) \cdot C_3 = 29.977 \mu s$

$b_3 := \tau_1 \cdot \tau_{12} \cdot \tau_{123} = 6.438 \times 10^3 \cdot \mu s^3$

$\omega_{z1} := \frac{1}{r_{C2} \cdot C_2} \qquad f_{z1} := \frac{\omega_{z1}}{2 \cdot \pi} = 318.31 kHz$

$\omega_{z2} := \frac{1}{r_{C3} \cdot C_3} \qquad f_{z2} := \frac{\omega_{z2}}{2 \cdot \pi} = 22.575 kHz$

$H_1(s) := H_0 \cdot \frac{\left(1 + \frac{s}{\omega_{z1}}\right) \cdot \left(1 + \frac{s}{\omega_{z2}}\right)}{1 + b_1 \cdot s + b_2 \cdot s^2 + b_3 \cdot s^3}$

$Q_D := \frac{\sqrt{b_2}}{b_1} = 1.582 \qquad \omega_{0D} := \frac{1}{\sqrt{b_2}}$

$H_5(s) := H_0 \cdot \frac{\left(1 + \frac{s}{\omega_{z1}}\right) \cdot \left(1 + \frac{s}{\omega_{z2}}\right)}{\left[1 + \frac{s}{\omega_{0D} \cdot Q_D} + \left(\frac{s}{\omega_{0D}}\right)^2\right] \cdot \left(1 + s \cdot \frac{b_3}{b_2}\right)}$

Time constants test:

$b_1 = 63.74 \mu s$

$\frac{b_2}{b_1} = 159.535 \mu s$

$\frac{b_3}{b_2} = 0.633 \mu s \qquad \frac{b_2}{b_3} \cdot \frac{1}{2 \cdot \pi} = 251.391 kHz$

Dominates high-frequency response

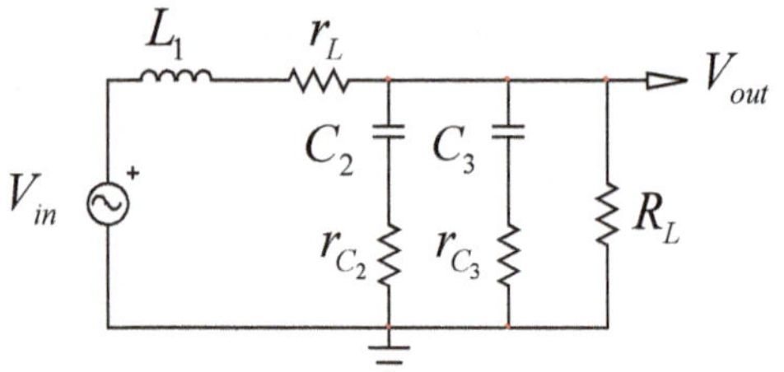

$Z_1(s) := \left(r_{C2} + \frac{1}{s \cdot C_2}\right) \| \left(r_{C3} + \frac{1}{s \cdot C_3}\right) \| R_L$

$H_{ref}(s) := \frac{Z_1(s)}{r_L + s \cdot L_1 + Z_1(s)}$

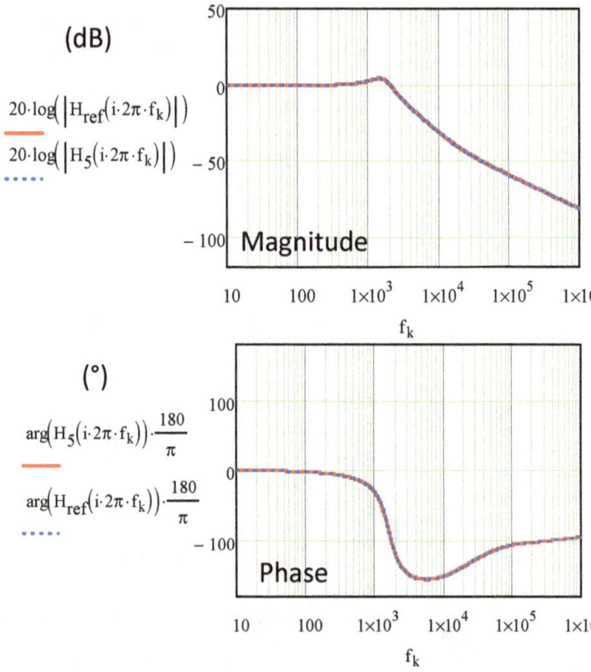

The 3rd-Order Polynomial has been Restructured into Two Cascaded Filters: a Low-Pass One Followed by a 2nd-Order Structure

Figure 7.13

In Figure 7.14, an *RC* network bridges the input to the output. It does not overly complicate the analysis and inspection immediately reveals two zeroes: when C_1 and C_2 are shorted, the stimulus can still propagate to form a response, indicating there are zeroes associated with these capacitors. It is not the case when C_3 is shorted as the response disappears in this case. In this exercise, the resistors all share a similar value R allowing simplifications in the expressions. Figure 7.15 shows the high-frequency gains and only two drawings would have sufficed as most of the configurations return zero. If we consider the poles and zeroes well separated, it is possible to rearrange the transfer function as cascaded low- and high-pass filters, naturally paving the way for a design-oriented expression. This is what is done in Figure 7.16 and all curves, brute-force, FACTs with its approximate version match each other very well.

(a)

$$H_0 = \frac{R}{3R} = \frac{1}{3}$$

(b)

Set the stimulus to 0 V and determine the resistance R driving the capacitors:

Set the capacitor C_1 in its high-frequency state and determine R from capacitor C_2 terminals:

(c)

$$\tau_1 = \left[R + (2R) \| R\right]C_1$$
$$\tau_2 = \left[(2R) \| R\right]C_2$$
$$\tau_3 = \left[(2R) \| R\right]C_3$$

(d)

$$\tau_2^1 = \left[R \| (R + R \| R)\right]C_2$$

Set the capacitor C_1 in its high-frequency state and determine R from C_3 terminals:

Set the capacitor C_2 in its high-frequency state and determine R from C_3 terminals:

(e)

$$\tau_3^1 = \left[R \| \left[(2R) \| R\right]\right]C_3$$

(f)

$$\tau_3^2 = (R \| R)C_3$$

Set capacitors C_1 and C_2 in their high-frequency state and determine R from C_3 terminals:

(g)

$$\tau_3^{12} = (R \| R \| R)C_3$$

(h)

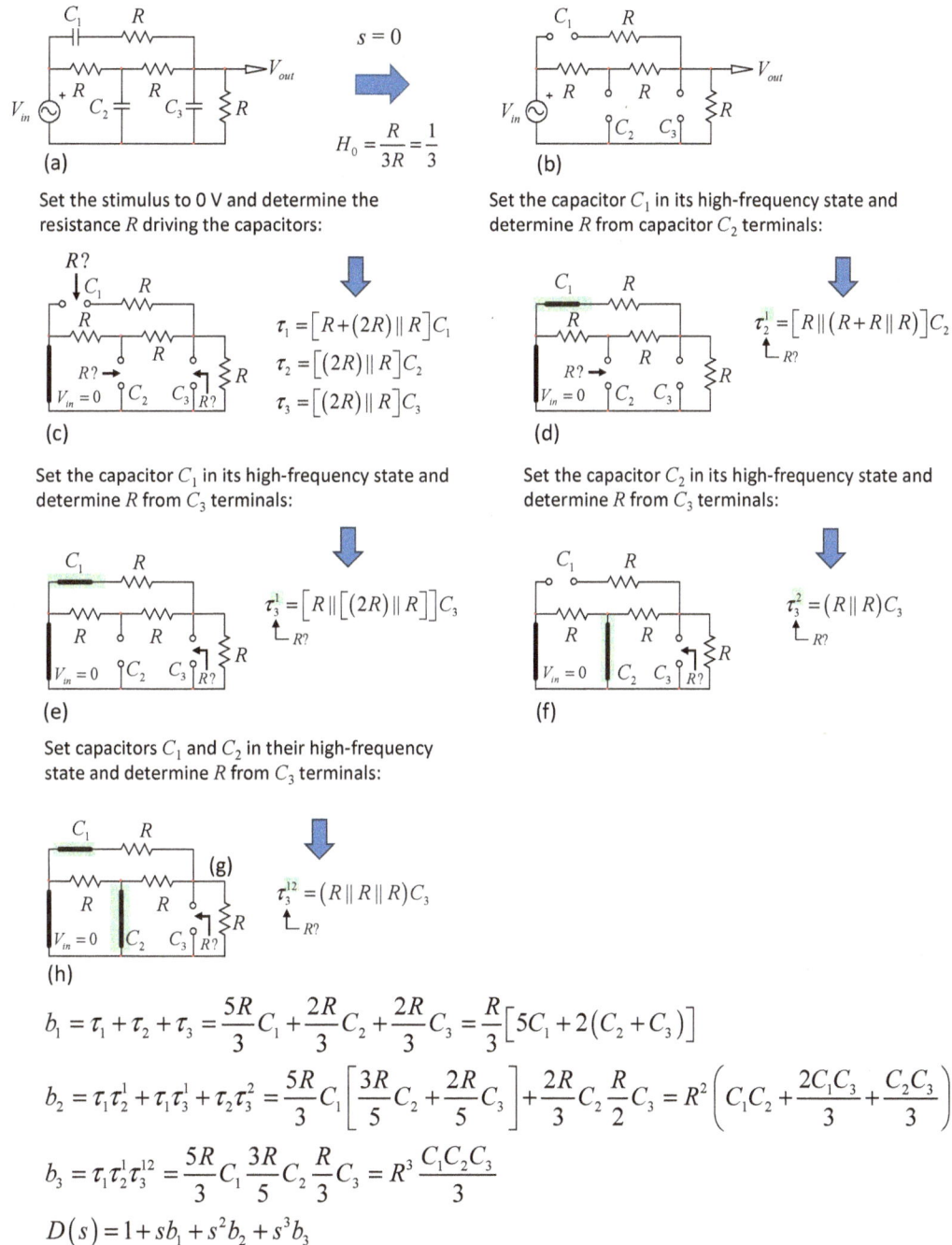

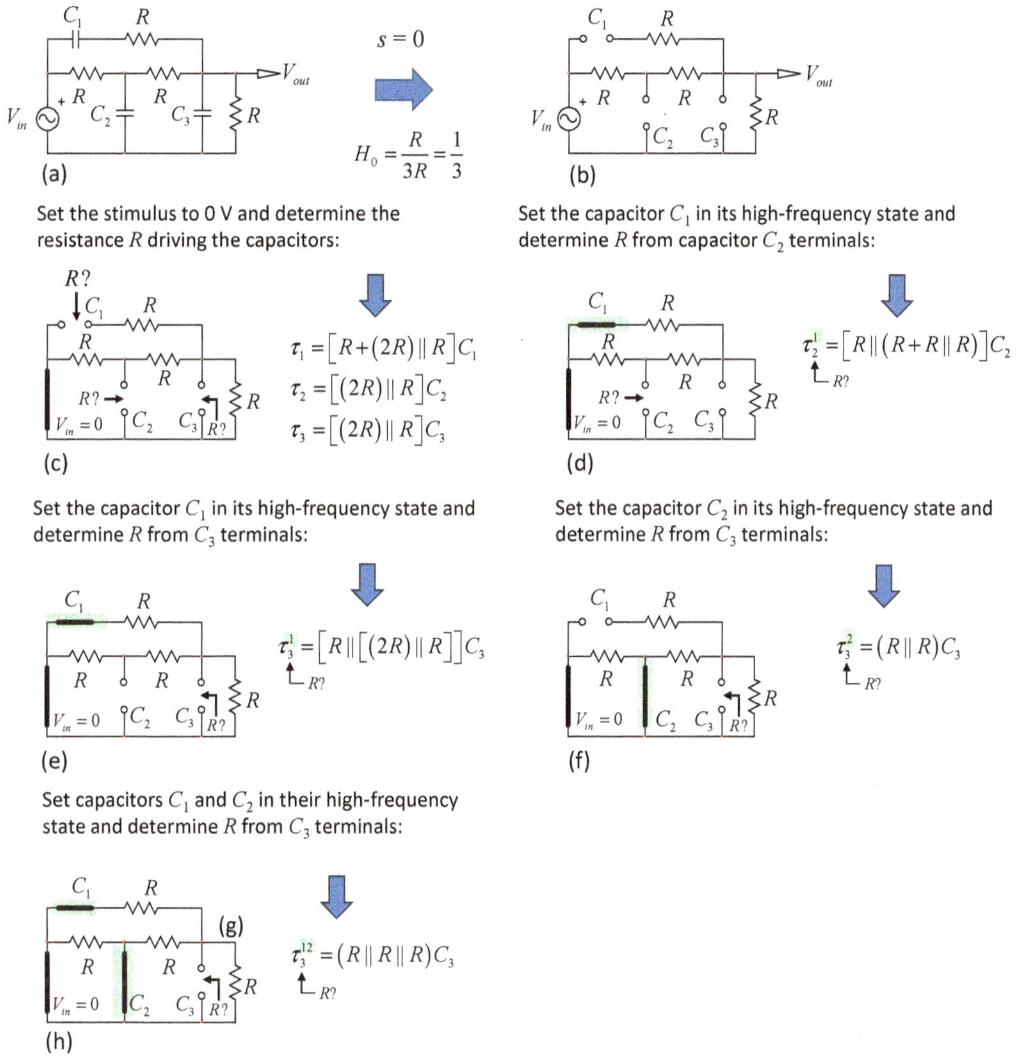

$$b_1 = \tau_1 + \tau_2 + \tau_3 = \frac{5R}{3}C_1 + \frac{2R}{3}C_2 + \frac{2R}{3}C_3 = \frac{R}{3}\left[5C_1 + 2(C_2 + C_3)\right]$$

$$b_2 = \tau_1\tau_2^1 + \tau_1\tau_3^1 + \tau_2\tau_3^2 = \frac{5R}{3}C_1\left[\frac{3R}{5}C_2 + \frac{2R}{5}C_3\right] + \frac{2R}{3}C_2\frac{R}{2}C_3 = R^2\left(C_1C_2 + \frac{2C_1C_3}{3} + \frac{C_2C_3}{3}\right)$$

$$b_3 = \tau_1\tau_2^1\tau_3^{12} = \frac{5R}{3}C_1\frac{3R}{5}C_2\frac{R}{3}C_3 = R^3\frac{C_1C_2C_3}{3}$$

$$D(s) = 1 + sb_1 + s^2b_2 + s^3b_3$$

This Third-Order Circuit hosts Two Zeroes involving C_1 and C_2

Figure 7.14

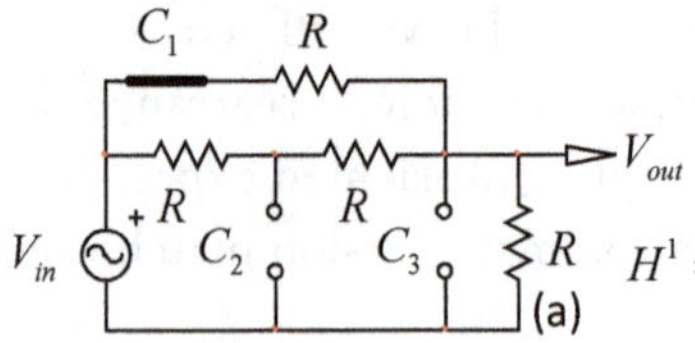

$$H^1 = \frac{R}{R + R\|(2R)} = \frac{3}{5}$$

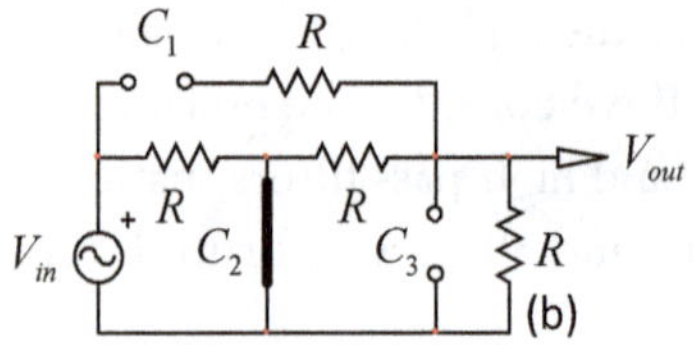

$H^2 = 0$

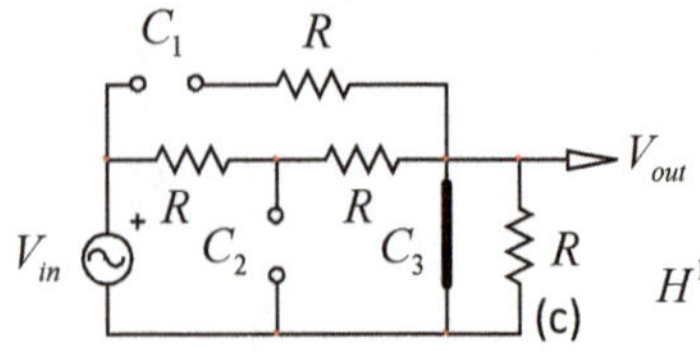

$H^3 = 0$

$$H^{12} = \frac{R\|R}{R\|R + R} = \frac{1}{3}$$

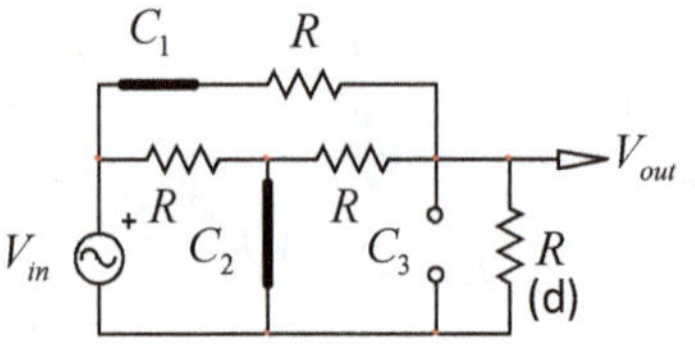

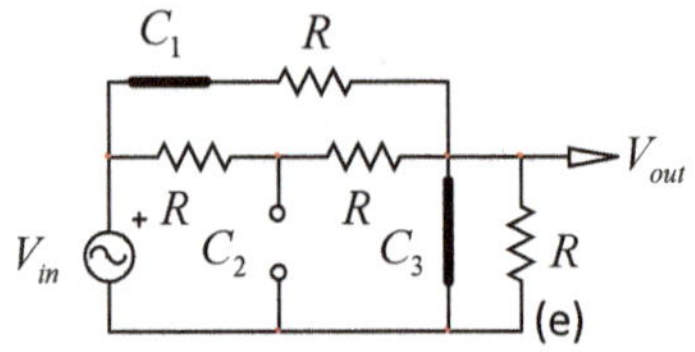

$H^{13} = 0$

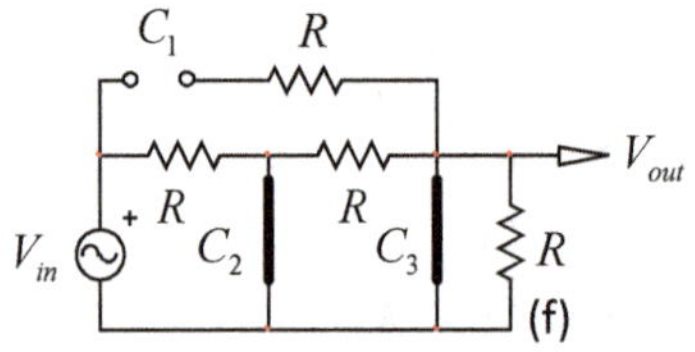

$H^{23} = 0$

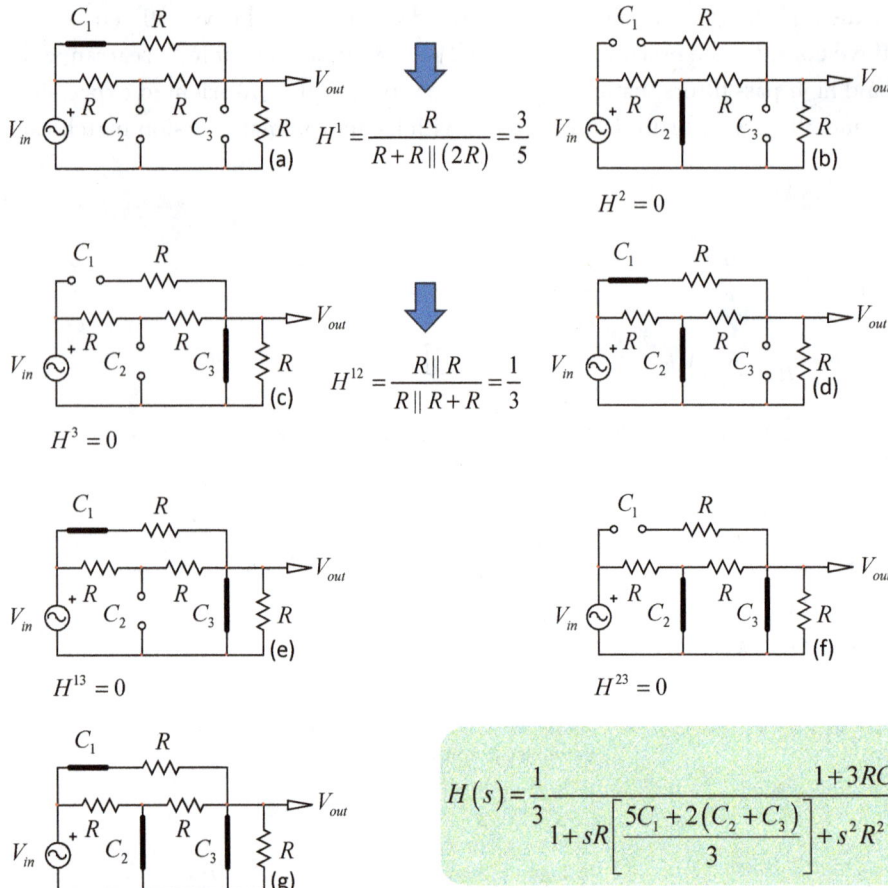

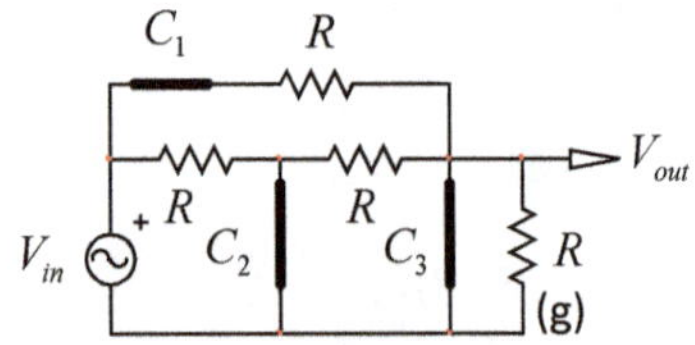

$H^{123} = 0$

$$H(s) = \frac{1}{3}\frac{1 + 3RC_1 s + s^2 C_1 C_2 R^2}{1 + sR\left[\dfrac{5C_1 + 2(C_2 + C_3)}{3}\right] + s^2 R^2\left(\dfrac{3C_1C_2 + 2C_1C_3 + C_2C_3}{3}\right) + s^3 R^3 \dfrac{C_1C_2C_3}{3}}$$

$$N(s) = H_0 + s\left(\tau_1 H^1 + \tau_2 H^2 + \tau_3 H^3\right) + s^2\left(\tau_1\tau_2^1 H^{12} + \tau_1\tau_3^1 H^{13} + \tau_2\tau_3^2 H^{23}\right) + s^3\left(\tau_1\tau_2^1\tau_3^{12} H^{123}\right)$$

$$N(s) = \frac{1}{3} + s\left(\tau_1 \cdot \frac{3}{5} + \tau_2 \cdot 0 + \tau_3 \cdot 0\right) + s^2\left(\tau_1\tau_2^1 \cdot \frac{1}{3} + \tau_1\tau_3^1 \cdot 0 + \tau_2\tau_3^2 \cdot 0\right) + s^3\left(\tau_1\tau_2^1\tau_3^{12} \cdot 0\right)$$

$$N(s) = \frac{1}{3}\left(1 + s3RC_1 + s^2 C_1 C_2 R^2\right)$$

If the poles and zeroes are well separated, it is possible to approximate the transfer function to:

$$H(s) \approx \frac{1}{3}\frac{\left(1 + \dfrac{s}{\omega_{z_1}}\right)\left(1 + \dfrac{s}{\omega_{z_2}}\right)}{\left(1 + \dfrac{s}{\omega_{p_1}}\right)\left(1 + \dfrac{s}{\omega_{p_2}}\right)\left(1 + \dfrac{s}{\omega_{p_3}}\right)}$$

$$\omega_{p_1} = \frac{1}{b_1} \quad \omega_{p_2} = \frac{b_1}{b_2} \quad \omega_{p_3} = \frac{b_2}{b_3}$$

$$\omega_{z_1} = \frac{1}{3RC_1} \quad \omega_{z_2} = \frac{3}{RC_2}$$

Determining the High-Frequency Gains eases the Process of Finding the Zeroes

Figure 7.15

$R_1 := 2.2k\Omega \quad C_1 := 100nF \quad C_3 := 100nF \quad C_2 := 10nF \quad \|(x,y) := \dfrac{xy}{x+y}$

$H_0 := \dfrac{R_1}{3 \cdot R_1} = 0.33333 \quad 20 \cdot \log(H_0) = -9.54243 \quad dB$

$\tau_1 := \left[R_1 + (2R_1) \parallel R_1\right] \cdot C_1 = 366.66667\mu s$
$\qquad \dfrac{5 \cdot R_1}{3} \cdot C_1 = 366.66667\mu s$

$\tau_2 := \left[(2R_1) \parallel R_1\right] \cdot C_2 = 14.66667\mu s$
$\qquad \dfrac{2 \cdot R_1}{3} \cdot C_2 = 14.66667\mu s$

$\tau_3 := \left[(2R_1) \parallel R_1\right] \cdot C_3 = 146.66667\mu s$
$\qquad \dfrac{2 \cdot R_1}{3} \cdot C_3 = 146.66667\mu s$

$b_1 := \tau_1 + \tau_2 + \tau_3 = 0.528\,ms$
$\qquad \dfrac{R_1}{3}\left[5 \cdot C_1 + 2 \cdot (C_2 + C_3)\right] = 0.528ms$

$\tau_{12} := \left[R_1 \parallel (R_1 + R_1 \parallel R_1)\right] \cdot C_2 = 13.2\,\mu s$
$\qquad \dfrac{3 \cdot R_1}{5} \cdot C_2 = 13.2\mu s$

$\tau_{13} := \left[R_1 \parallel \left[(2R_1) \parallel R_1\right]\right] \cdot C_3 = 88\,\mu s$
$\qquad \dfrac{2 \cdot R_1}{5} \cdot C_3 = 88\mu s$

$\tau_{23} := C_3 \cdot (R_1 \parallel R_1) = 110\,\mu s$
$\qquad \dfrac{R_1 \cdot C_3}{2} = 110\mu s$

$b_2 := \tau_1 \tau_{12} + \tau_1 \tau_{13} + \tau_2 \tau_{23} = 3.872 \times 10^4\,\mu s^2$
$\qquad \left(C_1 C_2 + \dfrac{2 \cdot C_1 \cdot C_3}{3} + \dfrac{C_2 \cdot C_3}{3}\right) \cdot R_1^2 = 3.872 \times 10^4\,\mu s^2$

$\tau_{123} := \left[R_1 \parallel (R_1 \parallel R_1)\right] \cdot C_3 = 73.33333\mu s$
$\qquad \dfrac{R_1}{3} \cdot C_3 = 73.33333\mu s$

$b_3 := \tau_1 \tau_{12} \tau_{123} = 3.54933 \times 10^5\,\mu s^3$
$\qquad \dfrac{C_1 \cdot C_2 \cdot C_3}{3} \cdot R_1^3 = 3.54933 \times 10^5\,\mu s^3$

$H_1 := \dfrac{R_1}{R_1 + R_1 \parallel (2 \cdot R_1)} = 0.6 \qquad H_{12} := \dfrac{R_1 \parallel R_1}{R_1 \parallel R_1 + R_1} = 0.33333$

$H_{10}(s) := \dfrac{H_0 + s \cdot H_1 \cdot \tau_1 + s^2 \cdot H_{12} \tau_1 \tau_{12}}{1 + b_1 \cdot s + b_2 \cdot s^2 + b_3 \cdot s^3}$

$H_{11}(s) := H_0 \cdot \dfrac{1 + s \cdot (3 \cdot C_1 \cdot R_1) + s^2 \cdot \left(C_1 \cdot C_2 \cdot R_1^2\right)}{1 + \dfrac{R_1 \cdot (5 \cdot C_1 + 2 \cdot C_2 + 2 \cdot C_3)}{3} \cdot s + \dfrac{R_1^2 \cdot (3 \cdot C_1 \cdot C_2 + 2 \cdot C_1 \cdot C_3 + C_2 \cdot C_3)}{3} \cdot s^2 + \dfrac{C_1 \cdot C_2 \cdot C_3 \cdot R_1^3}{3} \cdot s^3}$

$Q_N := \dfrac{\sqrt{C_1 \cdot C_2 \cdot R_1^2}}{3 \cdot C_1 \cdot R_1} = 0.10541 \qquad \omega_{0N} := \dfrac{1}{\sqrt{C_1 \cdot C_2 \cdot R_1^2}}$

$\omega_{p1} := \dfrac{1}{b_1} \qquad f_{p1} := \dfrac{\omega_{p1}}{2\pi} = 301.42982 Hz$

$\omega_{z1} := \dfrac{1}{3 \cdot C_1 \cdot R_1} \qquad \omega_{z2} := \dfrac{3}{C_2 \cdot R_1} \qquad \omega_{p2} := \dfrac{b_1}{b_2} \qquad f_{p2} := \dfrac{\omega_{p2}}{2\pi} = 2.17029 kHz$

$b_1 = 528\mu s \qquad \dfrac{b_2}{b_1} = 73.33333\mu s \qquad \dfrac{b_3}{b_2} = 9.16667\mu s \qquad \omega_{p3} := \dfrac{b_2}{b_3} \qquad f_{p3} := \dfrac{\omega_{p3}}{2\pi} = 17.36236 kHz$

$H_{20}(s) := H_0 \cdot \dfrac{\left(1 + \dfrac{s}{\omega_{z1}}\right)\left(1 + \dfrac{s}{\omega_{z2}}\right)}{\left(1 + \dfrac{s}{\omega_{p1}}\right) \cdot \left(1 + \dfrac{s}{\omega_{p2}}\right) \cdot \left(1 + \dfrac{s}{\omega_{p3}}\right)}$

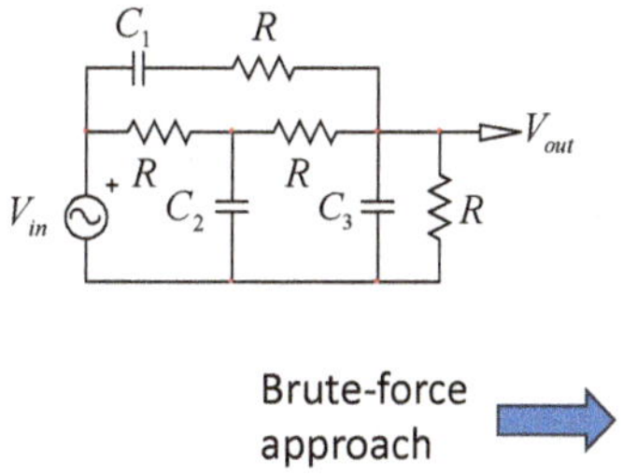

Brute-force approach →

$Z_1(s) := (R_1) \parallel \left(\dfrac{1}{s \cdot C_3}\right) \parallel \left[R_1 + R_1 \parallel \left(\dfrac{1}{s \cdot C_2}\right)\right]$

$V_1(s) := \dfrac{Z_1(s)}{Z_1(s) + R_1 + \dfrac{1}{s \cdot C_1}}$

$Z_2(s) := \left(\dfrac{1}{s \cdot C_3}\right) \parallel \left(\dfrac{1}{s \cdot C_1} + R_1\right) \parallel R_1$

$Z_3(s) := R_1 \parallel \left(\dfrac{1}{s \cdot C_2}\right)$

$V_2(s) := \dfrac{Z_2(s)}{Z_2(s) + R_1 + Z_3(s)} \cdot \dfrac{\dfrac{1}{s \cdot C_2}}{\dfrac{1}{s \cdot C_2} + R_1}$

$H_{ref}(s) := V_1(s) + V_2(s)$

(dB)

$20 \cdot \log\left(\left|H_{11}(i \cdot 2\pi \cdot f_k)\right|\right)$ —

$20 \cdot \log\left(\left|H_{ref}(i \cdot 2\pi \cdot f_k)\right|\right)$ ·····

$20 \cdot \log\left(\left|H_{20}(i \cdot 2\pi \cdot f_k)\right|\right)$ —

(°)

$\arg\left(H_{10}(i \cdot 2\pi \cdot f_k)\right) \cdot \dfrac{180}{\pi}$ —

$\arg\left(H_{ref}(i \cdot 2\pi \cdot f_k)\right) \cdot \dfrac{180}{\pi}$ ·····

$\arg\left(H_{20}(i \cdot 2\pi \cdot f_k)\right) \cdot \dfrac{180}{\pi}$ —

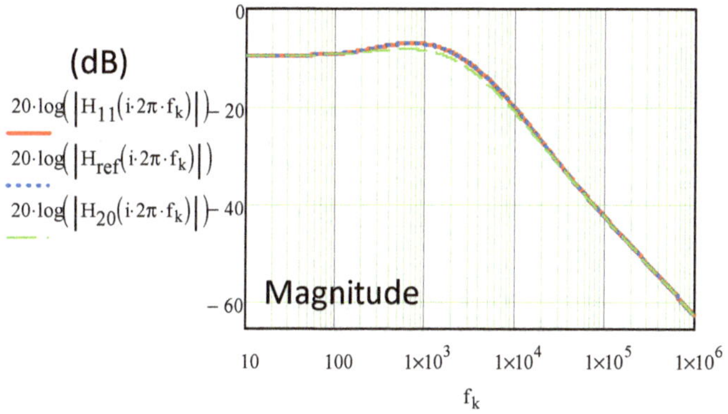

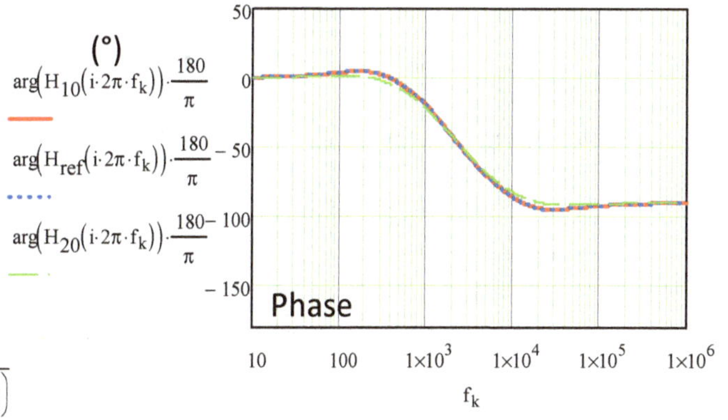

Superposition is used to Obtain the Reference Plot which Perfectly Matches the H_{11} Expression

Figure 7.16

In Figure 7.17, I show a transimpedance transfer function linking a response in volts, V_{out}, to a stimulus in amperes, I_{in}.

The current source is affected by an output resistance R_{th} and it could be that of an operational transconductance amplifier or OTA.

The dc gain R_0 is R_{th} when $s = 0$. Then, turning the stimulus off for the determination of the time constants implies that the current source disappears from the following sketches. It is now easy by inspection to determine the resistance R driving each capacitor in the various configurations.

With the time constants identified, the denominator is assembled.

The zero is obtained by identifying how the response could disappear for a certain impedance combination involving any of the capacitors.

C_1, alone, contributes a zero: when the impedance involving the series connection of R_1 and C_1 becomes a transformed short, the response disappears.

You could also test each capacitor by setting them in their high-frequency state.

It is only when C_1 is replaced by a short circuit, that the stimulus produces a response and the corresponding gain H^1 is non-zero. Any other combinations involving C_2 and C_3 lead to a zero gain.

The final result is given in Figure 7.18. Provided the poles are well separated, it becomes possible to approximate the transfer function and obtain a simpler form.

Figure 7.19 shows the ac response of the derived transfer functions and they match very well with the one obtained using the brute-force approach.

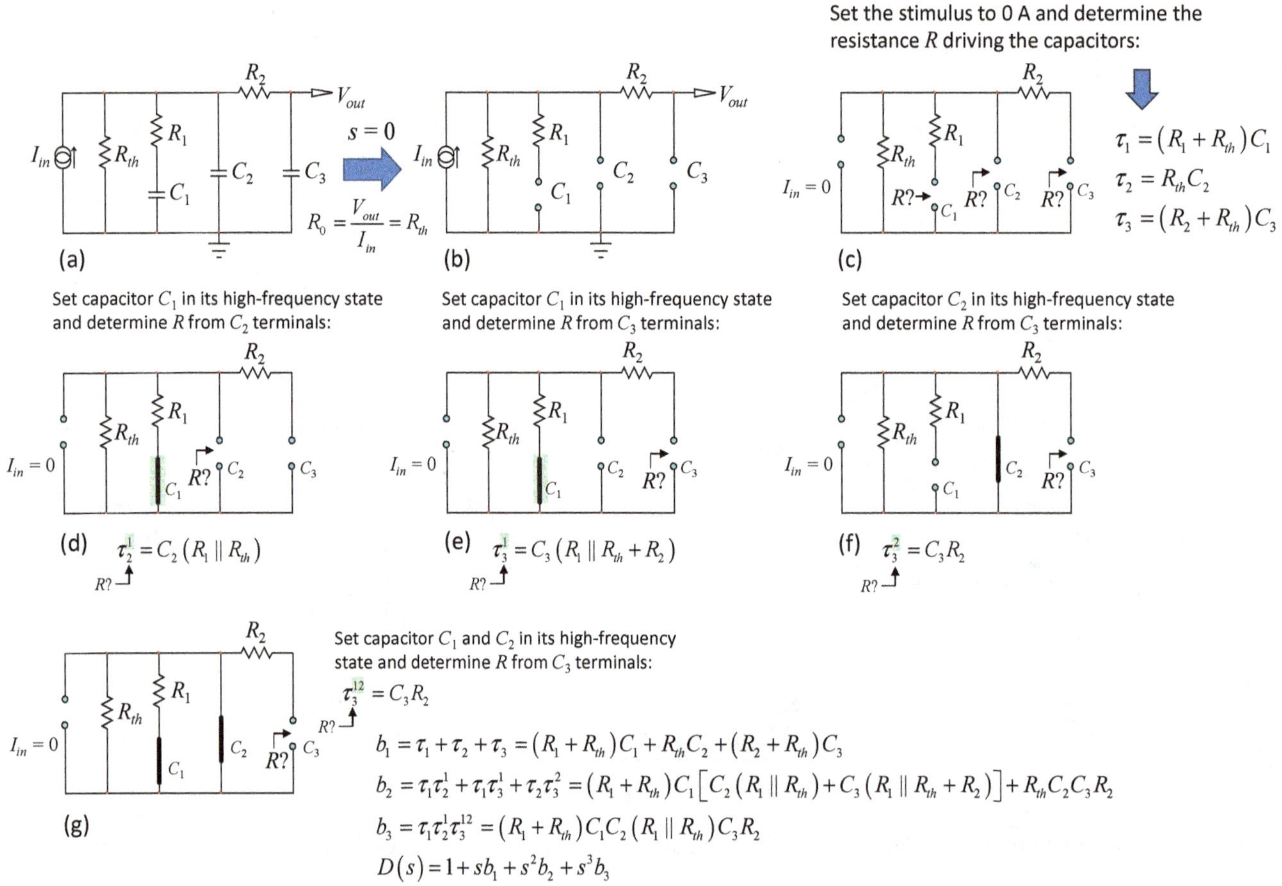

Set the stimulus to 0 A and determine the resistance R driving the capacitors:
(a)
(b)
(c)
s = 0
$R_0 = \frac{V_{out}}{I_{in}} = R_{th}$
$\tau_1 = (R_1 + R_{th})C_1$
$\tau_2 = R_{th}C_2$
$\tau_3 = (R_2 + R_{th})C_3$
Set capacitor C_1 in its high-frequency state and determine R from C_2 terminals:
Set capacitor C_1 in its high-frequency state and determine R from C_3 terminals:
Set capacitor C_2 in its high-frequency state and determine R from C_3 terminals:
(d) $\tau_2^1 = C_2(R_1 \| R_{th})$
(e) $\tau_3^1 = C_3(R_1 \| R_{th} + R_2)$
(f) $\tau_3^2 = C_3R_2$
Set capacitor C_1 and C_2 in its high-frequency state and determine R from C_3 terminals:
$\tau_3^{12} = C_3R_2$
(g)
$b_1 = \tau_1 + \tau_2 + \tau_3 = (R_1 + R_{th})C_1 + R_{th}C_2 + (R_2 + R_{th})C_3$
$b_2 = \tau_1\tau_2^1 + \tau_1\tau_3^1 + \tau_2\tau_3^2 = (R_1 + R_{th})C_1[C_2(R_1 \| R_{th}) + C_3(R_1 \| R_{th} + R_2)] + R_{th}C_2C_3R_2$
$b_3 = \tau_1\tau_2^1\tau_3^{12} = (R_1 + R_{th})C_1C_2(R_1 \| R_{th})C_3R_2$
$D(s) = 1 + sb_1 + s^2b_2 + s^3b_3$

This Filter Receives a Current Stimulus that is Transformed into a Voltage Response

Figure 7.17

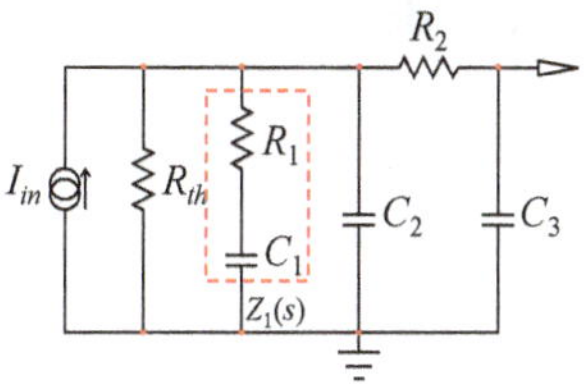

The response is nulled if $Z_1(s) = 0$

$$Z_1(s) = R_1 + \frac{1}{sC_1} = 0 \quad \to s_z = -\frac{1}{R_1C_1}$$

$$\omega_z = \frac{1}{R_1C_1} \qquad N(s) = 1 + \frac{s}{\omega_z}$$

$$Z(s) = \frac{V_{out}(s)}{I_{in}(s)} = R_0\frac{N(s)}{D(s)} = R_0\frac{1+\frac{s}{\omega_z}}{1+sb_1+s^2b_2+s^3b_3}$$

$$Z(s) = R_{th}\frac{1+sR_1C_1}{1+s\left[(R_1+R_{th})C_1+R_{th}C_2+(R_2+R_{th})C_3\right]+s^2\left[(R_1+R_{th})C_1\left[C_2(R_1 \parallel R_{th})+C_3(R_1 \parallel R_{th}+R_2)\right]+R_{th}C_2C_3R_2\right]+s^3\left[(R_1+R_{th})C_1C_2(R_1 \parallel R_{th})C_3R_2\right]}$$

If the roots are well separated, then we can approximate the transfer function:

$$Z(s) \approx R_0\frac{1+\frac{s}{\omega_z}}{\left(1+\frac{s}{\omega_{p_1}}\right)\left(1+\frac{s}{\omega_{p_2}}\right)\left(1+\frac{s}{\omega_{p_3}}\right)} \qquad \omega_{p_1} = \frac{1}{b_1} \quad \omega_{p_2} = \frac{b_1}{b_2} \quad \omega_{p_3} = \frac{b_2}{b_3}$$

The Transfer Function is Complex, but Approximation is possible because the Poles are Spread

Figure 7.18

$R_2 := 1k\Omega \qquad C_1 := 100nF \qquad R_{th} := 4.7k\Omega \qquad \parallel(x,y) := \frac{x \cdot y}{x+y}$

$R_1 := 2.2k\Omega \qquad C_2 := 220nF \qquad C_3 := 150nF$

$R_0 := R_{th} = 4.7 \cdot k\Omega$

$\tau_1 := (R_1 + R_{th}) \cdot C_1 = 690 \cdot \mu s$

$\tau_2 := R_{th} \cdot C_2 = 1.034 ms$

$\tau_3 := (R_{th} + R_2) \cdot C_3 = 855 \cdot \mu s$

$b_1 := \tau_1 + \tau_2 + \tau_3 = 2.579 \times 10^3 \cdot \mu s$

$\tau_{12} := C_2 \cdot (R_1 \parallel R_{th}) = 329.681 \mu s$

$\tau_{13} := C_3 \cdot (R_1 \parallel R_{th} + R_2) = 374.783 \mu s$

$\tau_{23} := C_3 \cdot R_2 = 150 \mu s$

$b_2 := \tau_1 \cdot \tau_{12} + \tau_1 \cdot \tau_{13} + \tau_2 \cdot \tau_{23} = 6.412 \times 10^5 \cdot \mu s^2$

$\tau_{123} := R_2 \cdot C_3 = 150 \mu s$

$b_3 := \tau_1 \cdot \tau_{12} \cdot \tau_{123} = 3.412 \times 10^7 \cdot \mu s^3$

$\omega_{z1} := \frac{1}{R_1 \cdot C_1}$

$$Z_1(s) := R_0 \cdot \frac{1+\frac{s}{\omega_{z1}}}{1+b_1 \cdot s+b_2 \cdot s^2+b_3 \cdot s^3}$$

$$Z_2(s) := R_0 \cdot \frac{1+\frac{s}{\omega_{z1}}}{(1+b_1 \cdot s) \cdot \left(1+s \cdot \frac{b_2}{b_1}\right) \cdot \left(1+s \cdot \frac{b_3}{b_2}\right)}$$

Approximate expression

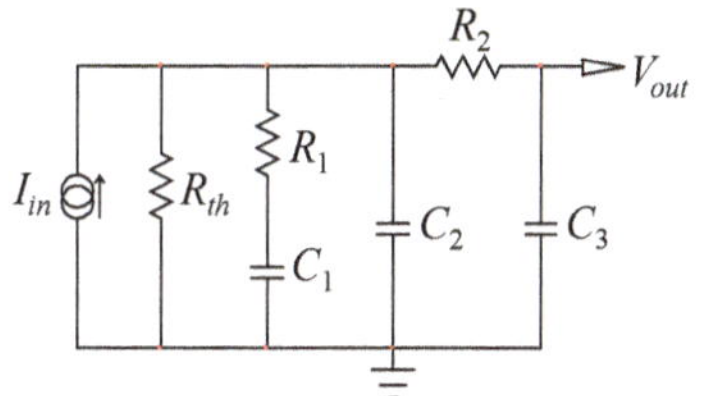

$$Z_{ref}(s) := \left[R_{th} \parallel \left(R_1 + \frac{1}{s \cdot C_1}\right) \parallel \left(\frac{1}{s \cdot C_2}\right) \parallel \left(R_2 + \frac{1}{s \cdot C_3}\right)\right] \cdot \frac{1}{1+s \cdot R_2 \cdot C_3}$$

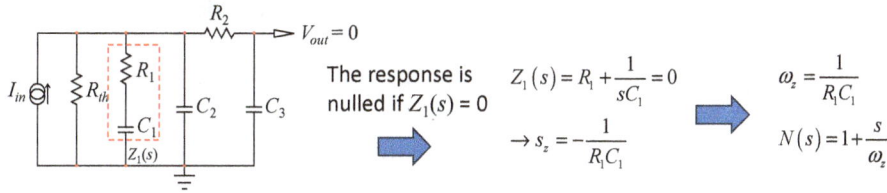

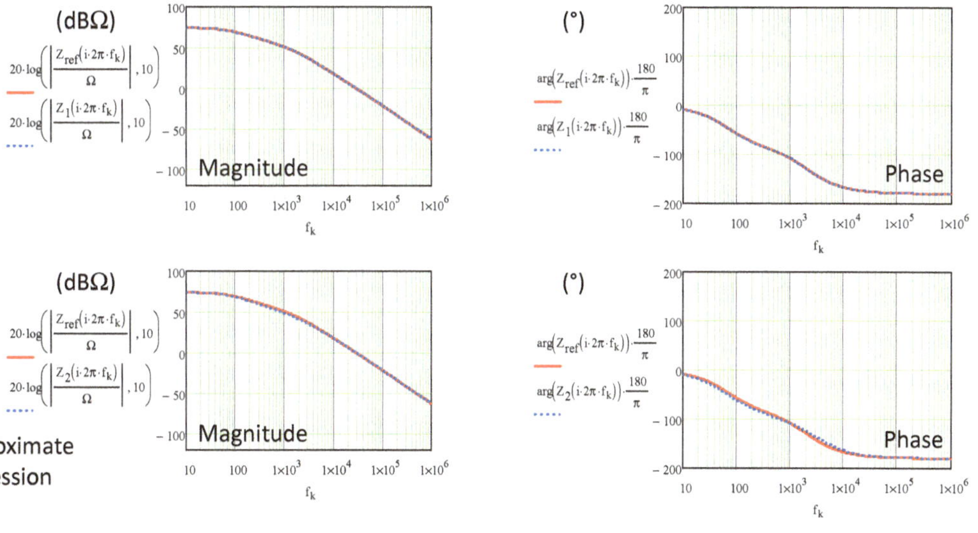

The Transfer Functions—the Complete and Approximate Expressions—are Compared with Expressions obtained through Brute-Force and they Match Well

Figure 7.19

A popular notch filter appears in Figure 7.20. It features three capacitors and looks intimidating at first sight. Fortunately, the FACTs lead to the answer without writing a single line of algebra, just inspection. Start with the dc gain, equal to 1, when all capacitors are open. Then proceed as documented in the previous illustrations. The denominator comes out and it is of 3rd-order as expected.

The zeroes could be obtained by NDI but I prefer, for this example, to resort to the generalized transfer function and its high-frequency gains. As shown in Figure 7.21, the configurations are easy to inspect, quickly leading to the numerator expression.

A practical realization of this filter implies some symmetry in the choice of components like $R_1 = R_2 = 2R$ and $R_3 = R$ then $C_1 = 2C$ and $C_2 = C_3 = C$. Substituting these combinations in the final expression gives a compact transfer function, exhibiting a second-order polynomial in the numerator and denominator. The quality factor is evaluated to 0.25.

The ac response is proposed in Figure 7.22 and confirms the presence of a notch filter tuned around 80 kHz with the adopted values. The reference transfer function is obtained by applying superposition and Thévenin: needless to say, the result would be intractable without a solver to rearrange the coefficients. The response of our transfer function and the reference perfectly match, confirming our analysis.

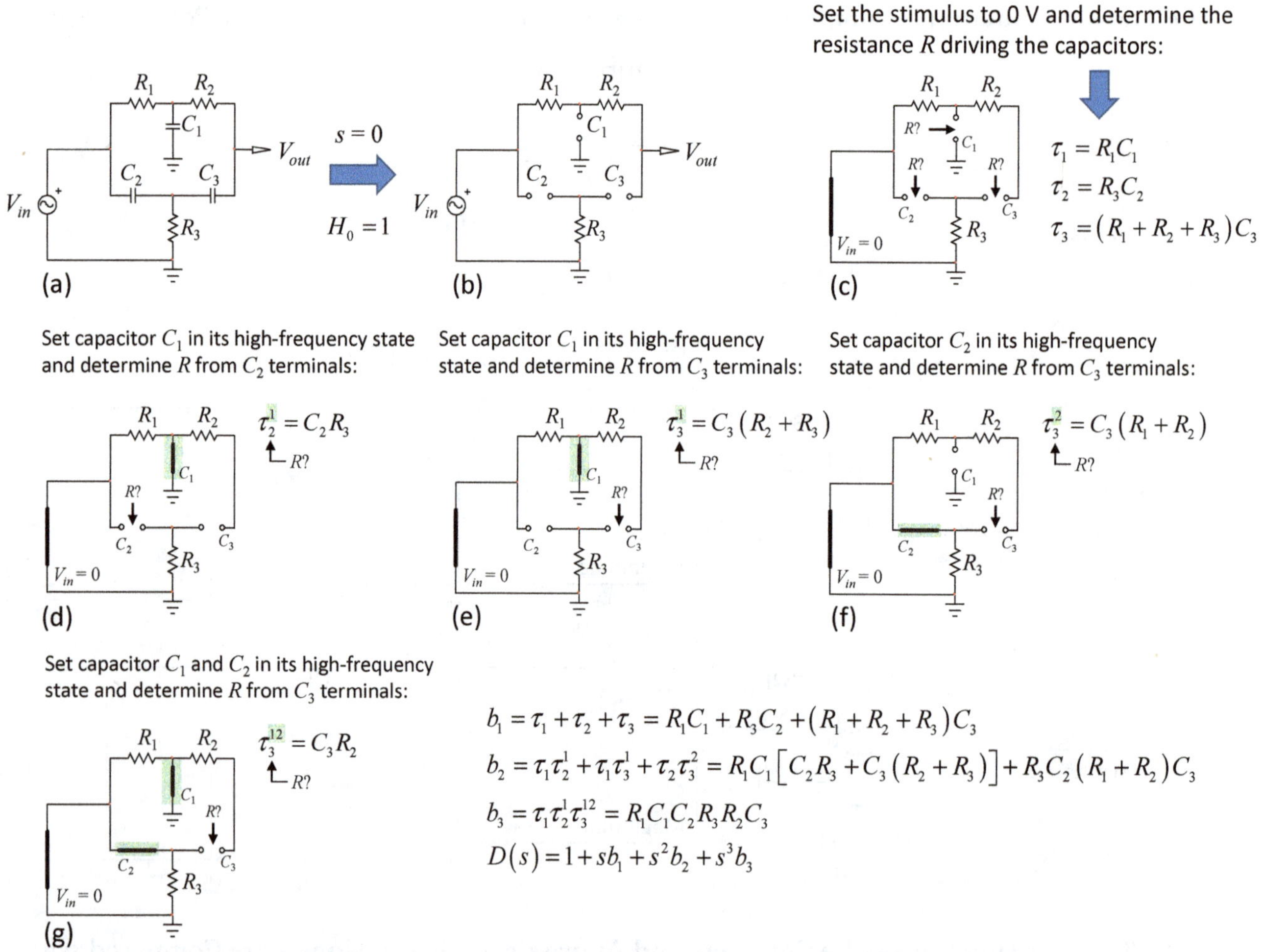
Set the stimulus to 0 V and determine the resistance R driving the capacitors:
s = 0
H0 = 1
Vin
Vout
(a)
(b)
(c)
Vin = 0
R?
τ1 = R1C1
τ2 = R3C2
τ3 = (R1 + R2 + R3)C3
Set capacitor C1 in its high-frequency state and determine R from C2 terminals:
Set capacitor C1 in its high-frequency state and determine R from C3 terminals:
Set capacitor C2 in its high-frequency state and determine R from C3 terminals:
τ2^1 = C2R3
τ3^1 = C3(R2 + R3)
τ3^2 = C3(R1 + R2)
(d)
(e)
(f)
Set capacitor C1 and C2 in its high-frequency state and determine R from C3 terminals:
τ3^12 = C3R2
(g)

$$b_1 = \tau_1 + \tau_2 + \tau_3 = R_1C_1 + R_3C_2 + (R_1 + R_2 + R_3)C_3$$

$$b_2 = \tau_1\tau_2^1 + \tau_1\tau_3^1 + \tau_2\tau_3^2 = R_1C_1\left[C_2R_3 + C_3(R_2 + R_3)\right] + R_3C_2(R_1 + R_2)C_3$$

$$b_3 = \tau_1\tau_2^1\tau_3^{12} = R_1C_1C_2R_3R_2C_3$$

$$D(s) = 1 + sb_1 + s^2b_2 + s^3b_3$$

This Transfer Function can be quickly Determined by Inspection

Figure 7.20

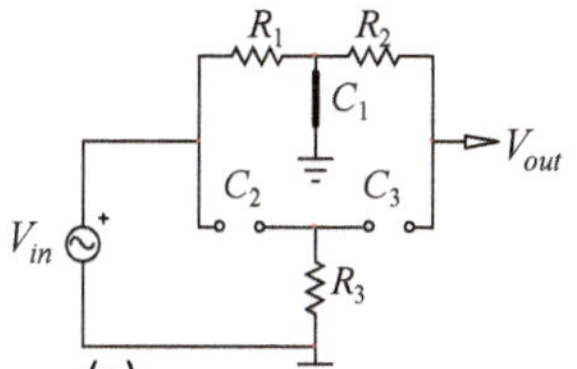

(a)

$H^1 = 0$

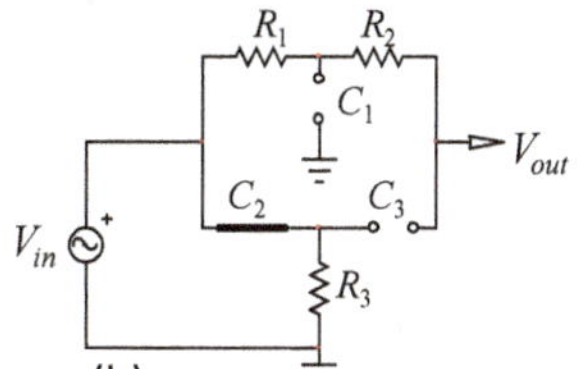

(b)

$H^2 = 1$

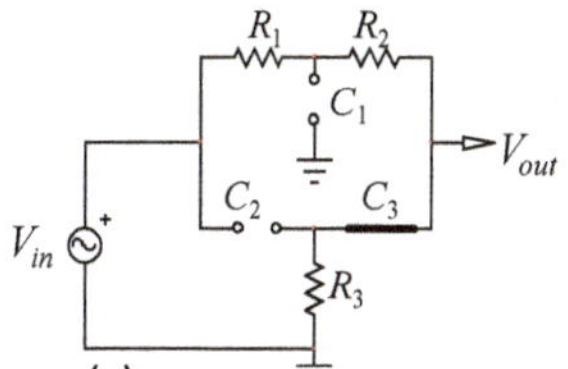

(c)

$$H^3 = \frac{R_3}{R_1 + R_2 + R_3}$$

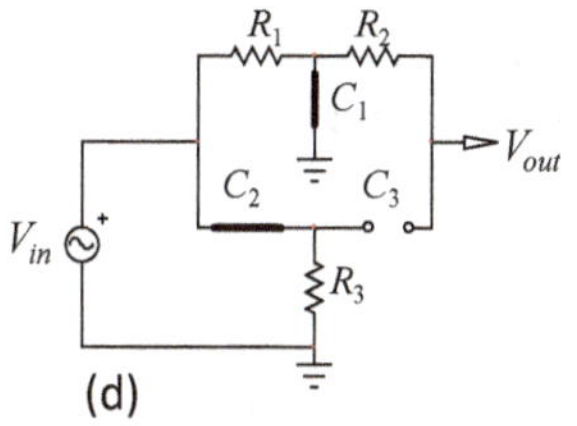

(d)

$H^{12} = 0$

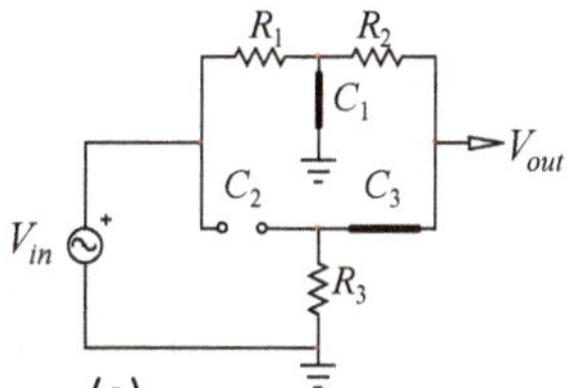

(e)

$H^{13} = 0$

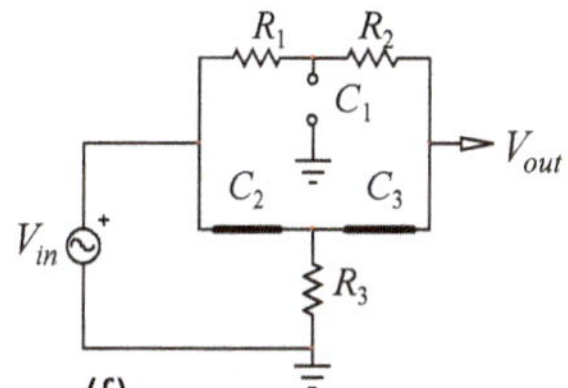

(f)

$H^{23} = 1$

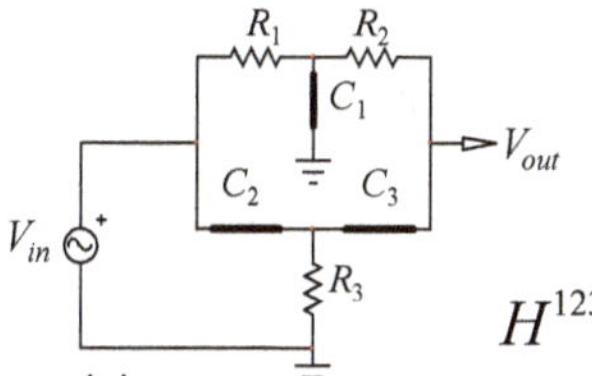

(g)

$H^{123} = 1$

$$N(s) = H_0 + s\left(\tau_1 H^1 + \tau_2 H^2 + \tau_3 H^3\right) + s^2\left(\tau_1\tau_2^1 H^{12} + \tau_1\tau_3^1 H^{13} + \tau_2\tau_3^2 H^{23}\right) + s^3\left(\tau_1\tau_2^1\tau_3^{12} H^{123}\right)$$

$$N(s) = 1 + s\left(\tau_1 \cdot 0 + \tau_2 \cdot 1 + \tau_3 \cdot \frac{R_3}{R_1 + R_2 + R_3}\right) + s^2\left(\tau_1\tau_2^1 \cdot 0 + \tau_1\tau_3^1 \cdot 0 + \tau_2\tau_3^2 \cdot 1\right) + s^3\left(\tau_1\tau_2^1\tau_3^{12} \cdot 1\right)$$

$$N(s) = 1 + sR_3(C_2 + C_3) + s^2 C_2 C_3 R_3 (R_1 + R_2) + s^3 C_1 C_2 C_3 R_1 R_2 R_3$$

$$H(s) = \frac{1 + sR_3(C_2 + C_3) + s^2 C_2 C_3 R_3 (R_1 + R_2) + s^3 C_1 C_2 C_3 R_1 R_2 R_3}{1 + s\left[R_1 C_1 + R_3 C_2 + C_3(R_1 + R_2 + R_3)\right] + s^2\left[R_1 C_1\left[C_2 R_3 + C_3(R_2 + R_3)\right] + R_3 C_2 (R_1 + R_2) C_3\right] + s^3 R_1 C_1 C_2 R_2 R_3 C_3}$$

Now substitute $R_1 R_2 R_3 C_1 C_2$ and C_3 with the below values and rearrange:

$R_1 = 2R \quad R_2 = 2R \quad R_3 = R \quad C_1 = 2C \quad C_2 = C \quad C_3 = C$

$$H(s) = \frac{1 + 4C^2R^2s^2}{1 + s8RC + s^2 4C^2R^2} = \frac{1 + \left(\frac{s}{\omega_0}\right)^2}{1 + \frac{s}{\omega_0 Q} + \left(\frac{s}{\omega_0}\right)^2} \quad \omega_0 = \frac{1}{2RC} \quad Q = \frac{2RC}{8RC} = 0.25$$

In this Example, Resorting to High-Frequency Gains Leads to the Numerator

Figure 7.21

$$\|(x,y) := \frac{x \cdot y}{x + y} \qquad R_{inf} := 10^{20}\,\Omega \qquad R := 1k\Omega \qquad C := 1nF$$

$$R_1 := 2 \cdot R \qquad R_2 := 2 \cdot R \qquad R_3 := R \qquad C_1 := 2 \cdot C \qquad C_2 := C \qquad C_3 := C$$

$$H_0 := 1$$

$$\tau_1 := R_1 \cdot C_1 = 4\,\mu s \qquad \tau_2 := R_3 \cdot C_2 = 1 \times 10^3 \cdot ns \qquad \tau_3 := C_3 \cdot (R_1 + R_2 + R_3) = 5\,\mu s$$

$$b_1 := \tau_1 + \tau_2 + \tau_3 = 1 \times 10^4 \cdot ns$$

$$\tau_{12} := C_2 \cdot R_3 = 1\,\mu s \qquad \tau_{13} := C_3 \cdot (R_3 + R_2) = 3\,\mu s \qquad \tau_{23} := C_3 \cdot (R_1 + R_2) = 4\,\mu s$$

$$b_2 := \tau_1 \cdot \tau_{12} + \tau_1 \cdot \tau_{13} + \tau_2 \cdot \tau_{23} = 2 \times 10^{-11}\,s^2$$

$$\tau_{123} := C_3 \cdot R_2 = 2\,\mu s$$

$$b_3 := \tau_1 \cdot \tau_{12} \cdot \tau_{123} = 8\,\mu s^3$$

$$D_1(s) := 1 + b_1 \cdot s + b_2 \cdot s^2 + b_3 \cdot s^3$$

$$H_1 := 0 \quad H_2 := 1 \quad H_3 := \frac{R_3}{R_1 + R_2 + R_3} \quad H_{12} := 0 \quad H_{13} := 0 \quad H_{23} := 1 \quad H_{123} := 1$$

$$H_{10a}(s) := \frac{H_0 + s \cdot (H_1 \cdot \tau_1 + H_2 \cdot \tau_2 + H_3 \cdot \tau_3) + s^2 \cdot (\tau_1 \cdot \tau_{12} \cdot H_{12} + \tau_1 \cdot \tau_{13} \cdot H_{13} + \tau_2 \cdot \tau_{23} \cdot H_{23}) + s^3 \cdot \tau_1 \cdot \tau_{12} \cdot \tau_{123} \cdot H_{123}}{1 + b_1 \cdot s + b_2 \cdot s^2 + b_3 \cdot s^3}$$

$$H_{10}(s) := H_0 \cdot \frac{1 + s \cdot (\tau_2 + H_3 \cdot \tau_3) + s^2 \cdot (\tau_2 \cdot \tau_{23}) + s^3 \cdot \tau_1 \cdot \tau_{12} \cdot \tau_{123}}{1 + b_1 \cdot s + b_2 \cdot s^2 + b_3 \cdot s^3}$$

$$H_{20}(s) := \frac{1 + R_3 \cdot (C_2 + C_3) \cdot s + s^2 \cdot [C_2 \cdot C_3 \cdot R_3 \cdot (R_1 + R_2)] + s^3 \cdot (C_1 \cdot C_2 \cdot C_3 \cdot R_1 \cdot R_2 \cdot R_3)}{1 + s \cdot [R_1 \cdot C_1 + R_3 \cdot C_2 + C_3 \cdot (R_1 + R_2 + R_3)] + s^2 \cdot [R_1 \cdot C_1 \cdot (C_2 \cdot R_3) + R_1 \cdot C_1 \cdot [C_3 \cdot (R_3 + R_2)] + R_3 \cdot C_2 \cdot [C_3 \cdot (R_1 + R_2)]] + s^3 \cdot (C_1 \cdot C_2 \cdot C_3 \cdot R_1 \cdot R_2 \cdot R_3)}$$

Now substitute the R1, R2, R3, C1, C2 and C3 by their respective value and simplify:

$$H_{30}(s) := \frac{4 \cdot C^2 \cdot R^2 \cdot s^2 + 1}{4 \cdot C^2 \cdot R^2 \cdot s^2 + 8 \cdot C \cdot R \cdot s + 1} \qquad \omega_0 := \frac{1}{2 \cdot R \cdot C} \qquad f_0 := \frac{\omega_0}{2\pi} = 79.57747 kHz \qquad Q := \frac{2 \cdot R \cdot C}{8 \cdot R \cdot C} = 0.25$$

$$H_{40}(s) := \frac{1 + \left(\dfrac{s}{\omega_0}\right)^2}{1 + \dfrac{s}{\omega_0 \cdot Q} + \left(\dfrac{s}{\omega_0}\right)^2}$$

$$R_{th1}(s) := R_1 \,\|\, \left(\frac{1}{s \cdot C_1}\right) \qquad R_{th2}(s) := R_3 \,\|\, \left(\frac{1}{s \cdot C_2}\right)$$

$$k_{th1}(s) := \frac{\dfrac{1}{s \cdot C_1}}{\dfrac{1}{s \cdot C_1} + R_1} \qquad k_{th2}(s) := \frac{R_3}{R_3 + \dfrac{1}{s \cdot C_2}}$$

$$H_{ref}(s) := k_{th1}(s) \cdot \frac{\dfrac{1}{s \cdot C_3} + R_{th2}(s)}{\dfrac{1}{s \cdot C_3} + R_{th2}(s) + R_{th1}(s) + R_2} + k_{th2}(s) \cdot \frac{R_{th1}(s) + R_2}{R_{th1}(s) + R_2 + R_{th2}(s) + \dfrac{1}{s \cdot C_3}}$$

Reference transfer function obtained using Thévenin and superposition.

$$20 \cdot \log\left(\left|H_{10}(i \cdot 2\pi \cdot f_k)\right|, 10\right)$$

$$20 \cdot \log\left(\left|H_{ref}(i \cdot 2\pi \cdot f_k)\right|, 10\right)$$

Magnitude

$$\arg\left(H_{10}(i \cdot 2\pi \cdot f_k)\right) \cdot \frac{180}{\pi}$$

$$\arg\left(H_{ref}(i \cdot 2\pi \cdot f_k)\right) \cdot \frac{180}{\pi}$$

Phase

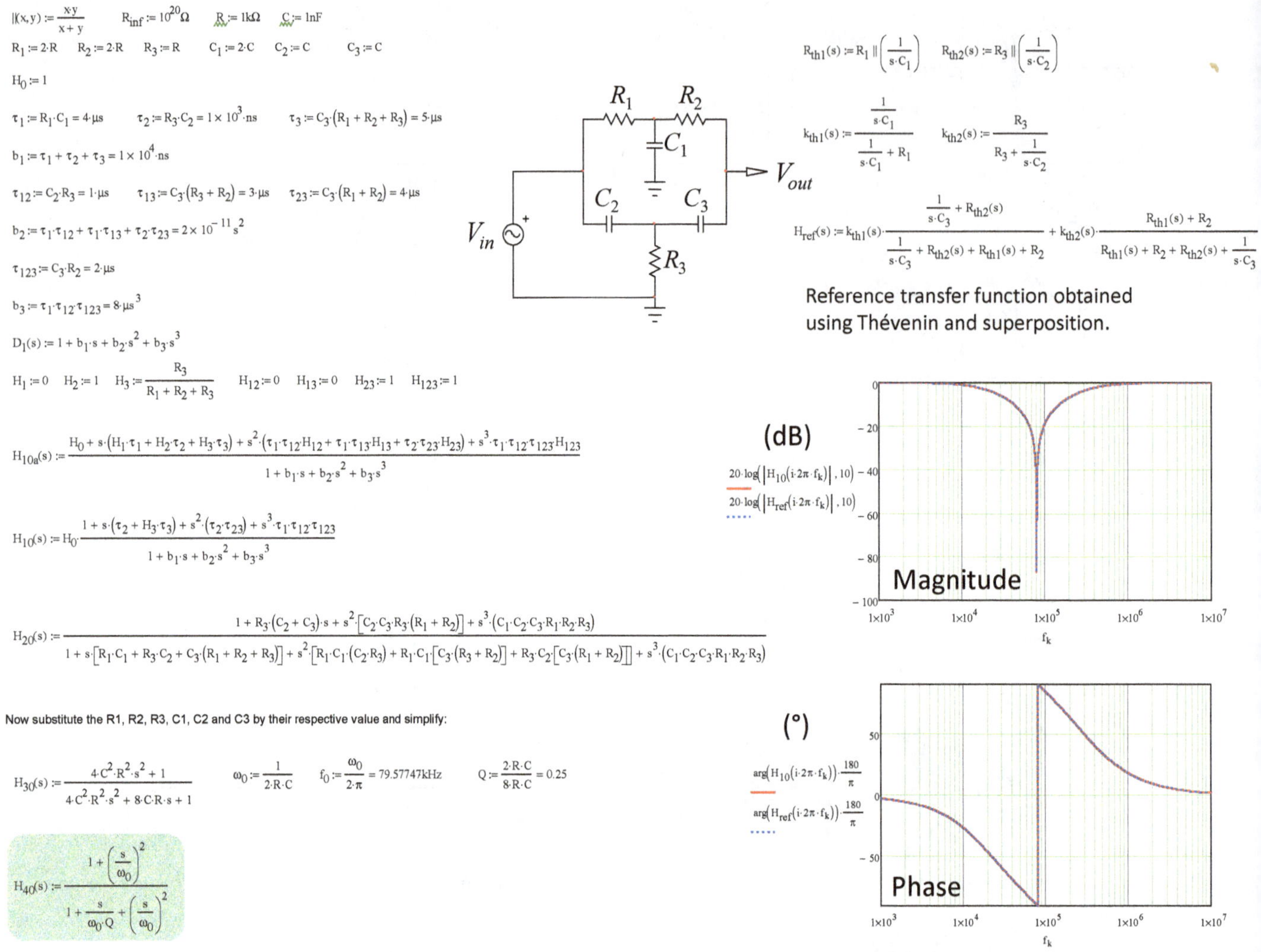

The Ac Response Confirms the Presence of a Notch Tuned below 80 kHz in this Design

Figure 7.22

The next example appears in Figure 7.23 and associates two inductors with a capacitor.

We will determine the impedance Z offered by this network from its connecting terminals. Determine an impedance by connecting a test generator I_T—the stimulus—to the connecting terminals and derive the response V_T you obtain across the current source.

Starting with dc, the inductors are replaced by a short circuit and the capacitor is opened: inspection is immediate to find that R_3 is the dc resistance.

Carry on with the various combinations to find all the time constants and form the denominator.

The numerator, in an impedance determination is obtained by nulling the response V_T which is similar to replacing the generator by a short circuit.

The complete transfer function is assembled after a few simple sketches and shown in Figure 7.24.

Plotting the ac response of the various expressions derived in Figure 7.25 confirms the approach is correct.

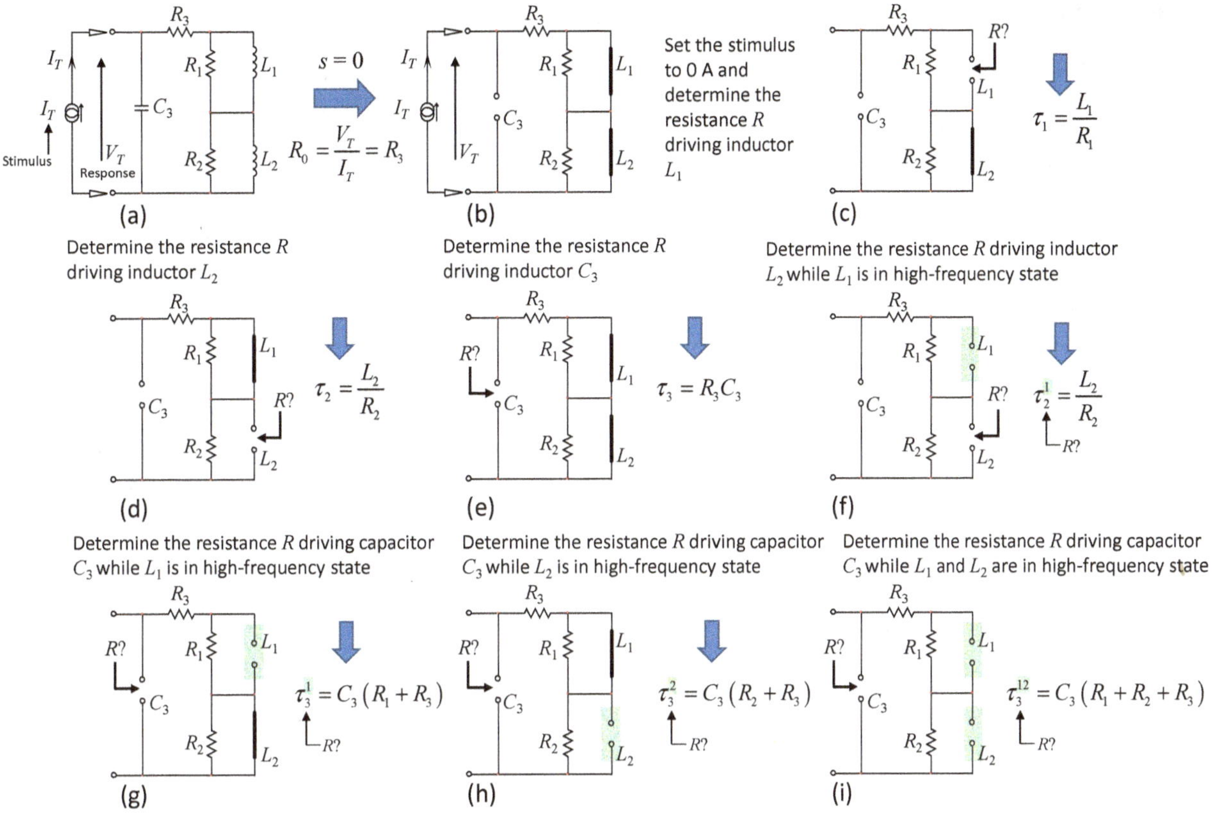
Stimulus
Response
(a)
s = 0
R₀ = V_T / I_T = R₃
(b)
Set the stimulus to 0 A and determine the resistance R driving inductor L₁
(c)
τ₁ = L₁/R₁
Determine the resistance R driving inductor L₂
(d)
τ₂ = L₂/R₂
Determine the resistance R driving inductor C₃
(e)
τ₃ = R₃C₃
Determine the resistance R driving inductor L₂ while L₁ is in high-frequency state
(f)
τ₂¹ = L₂/R₂
Determine the resistance R driving capacitor C₃ while L₁ is in high-frequency state
(g)
τ₃¹ = C₃(R₁ + R₃)
Determine the resistance R driving capacitor C₃ while L₂ is in high-frequency state
(h)
τ₃² = C₃(R₂ + R₃)
Determine the resistance R driving capacitor C₃ while L₁ and L₂ are in high-frequency state
(i)
τ₃¹² = C₃(R₁ + R₂ + R₃)
R?

$$b_1 = \tau_1 + \tau_2 + \tau_3 = \frac{L_1}{R_1} + \frac{L_2}{R_2} + R_3 C_3$$

$$b_2 = \tau_1\tau_2^1 + \tau_1\tau_3^1 + \tau_2\tau_3^2 = \frac{L_1}{R_1}\left[\frac{L_2}{R_2} + C_3(R_1 + R_3)\right] + \frac{L_2}{R_2}(R_2 + R_3)C_3$$

$$b_3 = \tau_1\tau_2^1\tau_3^{12} = \frac{L_1}{R_1}\frac{L_2}{R_2}C_3(R_1 + R_2 + R_3)$$

$$D(s) = 1 + sb_1 + s^2 b_2 + s^3 b_3 = 1 + s\left[\frac{L_1}{R_1} + \frac{L_2}{R_2} + R_3 C_3\right] + s^2\left[\frac{L_1}{R_1}\left[\frac{L_2}{R_2} + C_3(R_1 + R_3)\right] + \frac{L_2}{R_2}(R_2 + R_3)C_3\right] + s^3\left[\frac{L_1}{R_1}\frac{L_2}{R_2}C_3(R_1 + R_2 + R_3)\right]$$

Installing a Test Generator I_T develops a Response V_T Across the Network to Determine Impedance—Fortunately, FACTs Lead Straight to the Result without any Maths

Figure 7.23

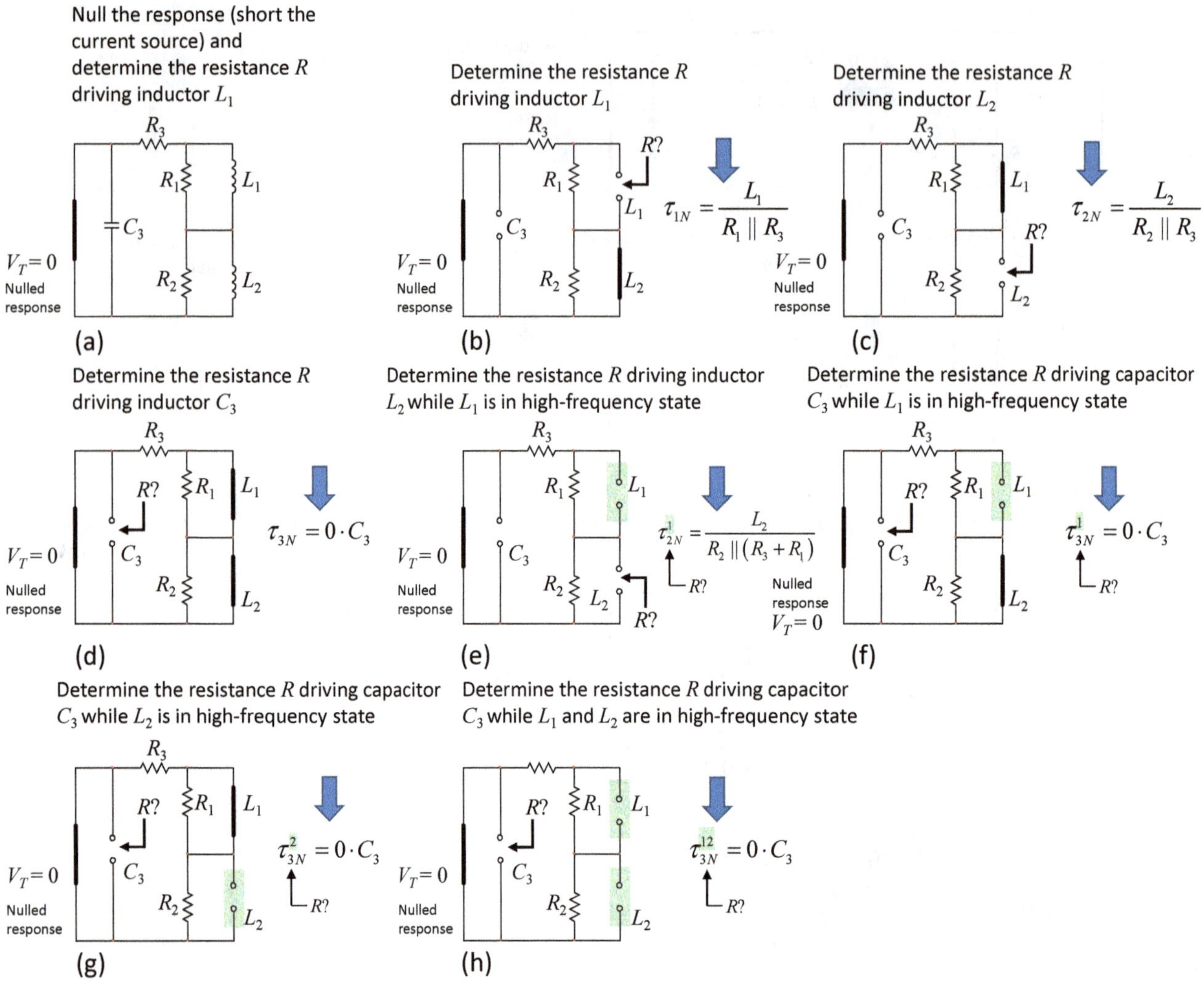

Null the response (short the current source) and determine the resistance R driving inductor L1
Determine the resistance R driving inductor L1
Determine the resistance R driving inductor L2
(a)
(b)
(c)
Determine the resistance R driving inductor C3
Determine the resistance R driving inductor L2 while L1 is in high-frequency state
Determine the resistance R driving capacitor C3 while L1 is in high-frequency state
(d)
(e)
(f)
Determine the resistance R driving capacitor C3 while L2 is in high-frequency state
Determine the resistance R driving capacitor C3 while L1 and L2 are in high-frequency state
(g)
(h)
VT = 0
Nulled response
R?

$$a_1 = \tau_{1N} + \tau_{2N} + \tau_{3N} = \frac{L_1}{R_1 \| R_3} + \frac{L_2}{R_2 \| R_3} + 0 \cdot C_3$$

$$a_2 = \tau_{1N}\tau_{2N}^1 + \tau_{1N}\tau_{3N}^1 + \tau_{2N}\tau_{3N}^2 = \frac{L_1}{R_1 \| R_3}\left[\frac{L_2}{R_2 \| (R_3 + R_1)} + 0 \cdot C_3\right] + \frac{L_2}{R_2 \| R_3} 0 \cdot C_3$$

$$a_3 = \tau_{1N}\tau_{2N}^1\tau_{3N}^{12} = \frac{L_1}{R_1 \| R_3}\frac{L_2}{R_2 \| (R_3 + R_1)} 0 \cdot C_3 = 0$$

$$N(s) = 1 + sa_1 + s^2 a_2 + s^3 a_3 = 1 + s\left[\frac{L_1}{R_1 \| R_3} + \frac{L_2}{R_2 \| R_3}\right] + s^2 \frac{L_1}{R_1 \| R_3}\frac{L_2}{R_2 \| (R_3 + R_1)}$$

$$Z_{in}(s) = R_0 \frac{N(s)}{D(s)} = R_3 \frac{1 + s\left[\frac{L_1}{R_1 \| R_3} + \frac{L_2}{R_2 \| R_3}\right] + s^2 \frac{L_1}{R_1 \| R_3}\frac{L_2}{R_2 \| (R_3 + R_1)}}{1 + s\left[\frac{L_1}{R_1} + \frac{L_2}{R_2} + R_3 C_3\right] + s^2\left[\frac{L_1}{R_1}\left[\frac{L_2}{R_2} + C_3(R_1 + R_3)\right] + \frac{L_2}{R_2}(R_2 + R_3)C_3\right] + s^3\left[\frac{L_1}{R_1}\frac{L_2}{R_2}C_3(R_1 + R_2 + R_3)\right]}$$

The Zeroes are Obtained by Nulling Response V_T—the Same as Shorting the Current Source

Figure 7.24

$$C_3 := 0.1\mu F \qquad \|(x,y) := \frac{x \cdot y}{x + y} \qquad R_2 := 100\Omega \qquad R_3 := 10\Omega$$

$$L_2 := 250\mu H \qquad L_1 := 100\mu H \qquad R_1 := 150\Omega$$

$$R_0 := R_3$$

$$\tau_1 := \frac{L_1}{R_1} = 666.66667ns \qquad \tau_2 := \frac{L_2}{R_2} = 2.5\times 10^{3}\cdot ns \qquad \tau_3 := C_3\cdot R_3 = 1\times 10^{-6}s$$

$$b_1 := \tau_1 + \tau_2 + \tau_3 = 4.16667\times 10^{3}\cdot ns$$

$$b_{1a} := \frac{L_1}{R_1} + \frac{L_2}{R_2} + C_3\cdot R_3 = 4.16667\times 10^{-6}s$$

$$\tau_{12} := \frac{L_2}{R_2} = 2.5\times 10^{-6}s \qquad \tau_{13} := C_3\cdot(R_3 + R_1) = 16\cdot\mu s \qquad \tau_{23} := C_3\cdot(R_3 + R_2) = 1.1\times 10^{-5}s$$

$$b_2 := \tau_1\cdot\tau_{12} + \tau_1\cdot\tau_{13} + \tau_2\cdot\tau_{23} = 3.98333\times 10^{-11}s^2$$

$$b_{2a} := \frac{L_1}{R_1}\cdot\frac{L_2}{R_2} + \frac{L_1}{R_1}\cdot[C_3\cdot(R_3 + R_1)] + \frac{L_2}{R_2}\cdot[C_3\cdot(R_3 + R_2)] = 3.98333\times 10^{-11}s^2$$

$$\tau_{123} := C_3\cdot(R_1 + R_2 + R_3) = 26\cdot\mu s$$

$$b_3 := \tau_1\cdot\tau_{12}\cdot\tau_{123} = 4.33333\times 10^{10}\cdot ns^3 \qquad b_{3a} := \frac{L_1}{R_1}\cdot\frac{L_2}{R_2}\cdot[C_3\cdot(R_1 + R_2 + R_3)] = 4.33333\times 10^{10}\cdot ns^3$$

$$D_1(s) := 1 + b_1\cdot s + b_2\cdot s^2 + b_3\cdot s^3$$

$$\tau_{3N} := 0\cdot C_3 \qquad \tau_{1N} := \frac{L_1}{R_1 \| R_3} \qquad \tau_{2N} := \frac{L_2}{R_2 \| R_3}$$

$$a_1 := \tau_{3N} + \tau_{2N} + \tau_{1N} = 3.81667\times 10^{4}\cdot ns$$

$$\tau_{12N} := \frac{L_2}{R_2 \| (R_3 + R_1)} = 4.0625\times 10^{-6}s \qquad \tau_{13N} := C_3\cdot 0 = 0\cdot\mu s \qquad \tau_{23N} := C_3\cdot 0 = 0$$

$$a_2 := \tau_{1N}\cdot\tau_{12N} + \tau_{1N}\cdot\tau_{13N} + \tau_{2N}\cdot\tau_{23N} = 4.33333\times 10^{-11}s^2$$

$$N_1(s) := 1 + s\cdot a_1 + s^2\cdot a_2 \qquad Z_4(s) := R_0\cdot\frac{N_1(s)}{D_1(s)}$$

$$Z_5(s) := R_3\cdot\frac{1 + s\cdot\left(\frac{L_1}{R_1 \| R_3} + \frac{L_2}{R_2 \| R_3}\right) + s^2\cdot\left[\frac{L_1}{R_1 \| R_3}\cdot\frac{L_2}{R_2 \| (R_3 + R_1)}\right]}{1 + s\cdot\left(\frac{L_1}{R_1} + \frac{L_2}{R_2} + C_3\cdot R_3\right) + s^2\cdot\left[\frac{L_1}{R_1}\cdot\frac{L_2}{R_2} + \frac{L_1}{R_1}\cdot[C_3\cdot(R_3 + R_1)] + \frac{L_2}{R_2}\cdot[C_3\cdot(R_3 + R_2)]\right] + s^3\cdot\left[\frac{L_1}{R_1}\cdot\frac{L_2}{R_2}\cdot[C_3\cdot(R_1 + R_2 + R_3)]\right]}$$

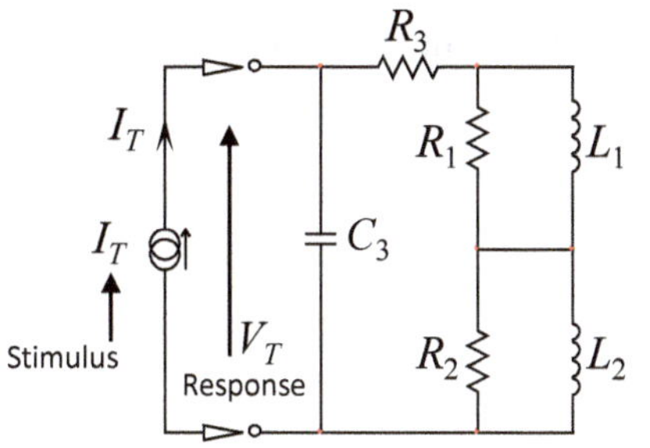

$$Z_1(s) := (s\cdot L_1) \| (R_1)$$

$$Z_2(s) := (s\cdot L_2) \| (R_2)$$

$$Z_{ref}(s) := \left(\frac{1}{s\cdot C_3}\right) \| (Z_1(s) + Z_2(s) + R_3)$$

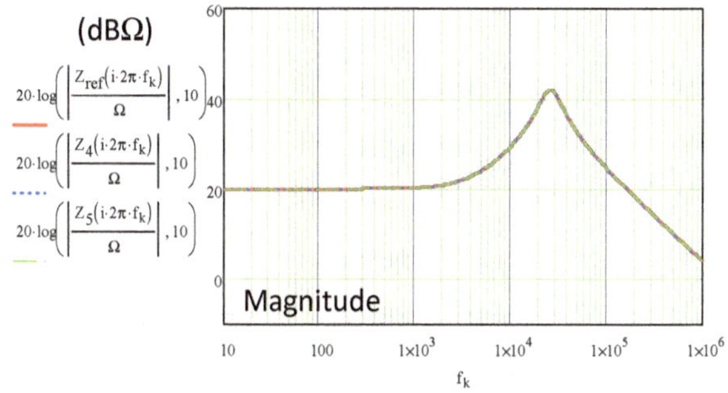

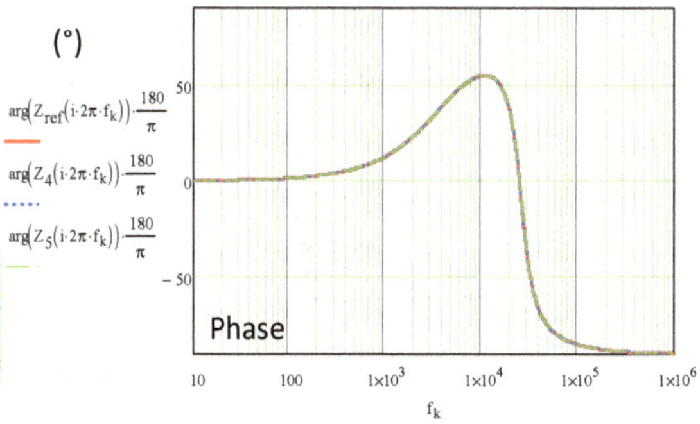

The Magnitude and Phase Responses of this Impedance Confirm the Calculations

Figure 7.25

7.3 List of Figures and Transfer Functions

For a convenient browsing of the derived transfer functions, below are pictures summarizing the networks studied in this chapter.

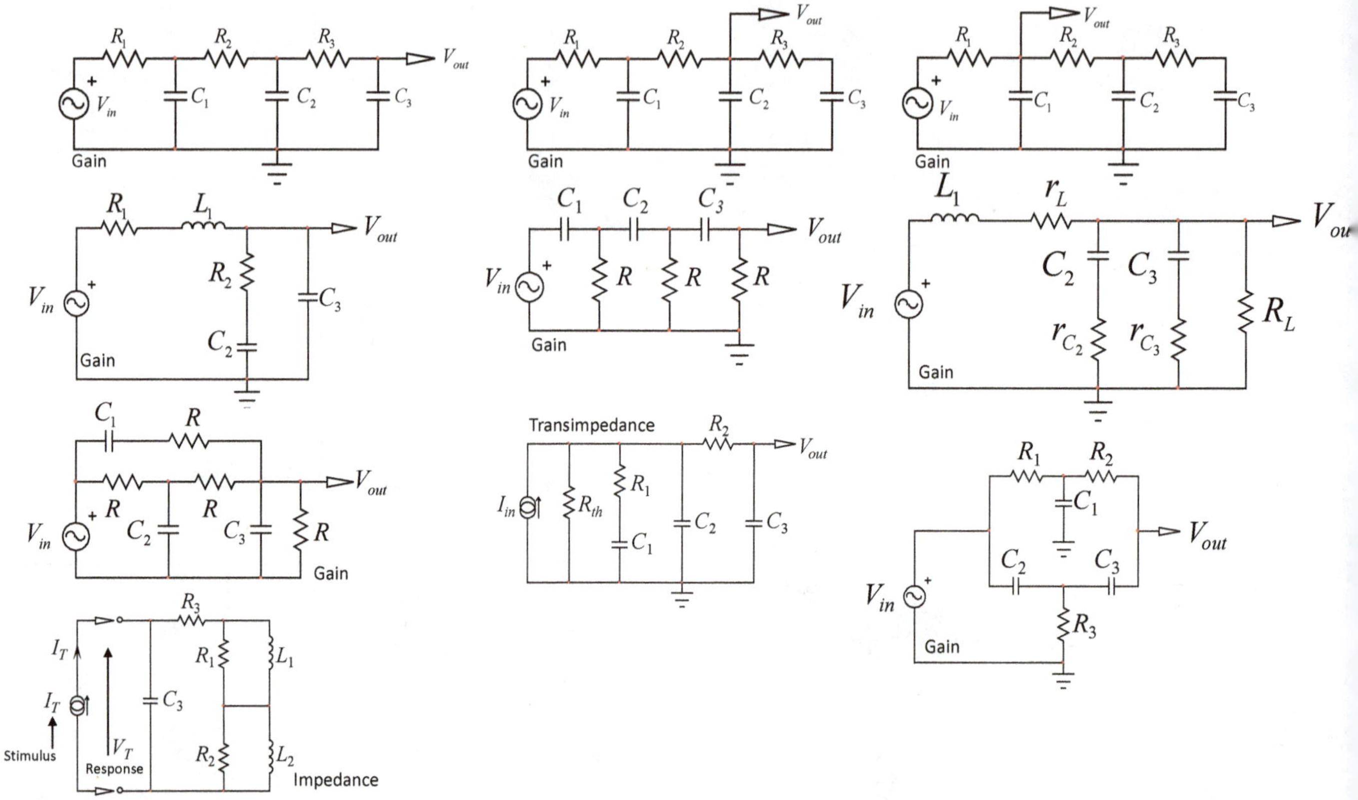

These are the Networks Studied in this Chapter

Figure 7.26

7.4 References

1. C. Basso, *Linear Circuit Transfer Functions – An Introduction to Fast Analytical Techniques*, Wiley, 2016.
2. R. Erickson, D. Maksimović, *Fundamentals of Power Electronics,* Chapter 8 (https://www.ieee.li/pdf/introduction_to_power_electronics/chapter_08.pdf), Springer, 2001.

Appendix A – Illustrating the Process of Determining Poles and Zeroes

THERE IS AN adage saying that a picture is worth a thousand words and it is also true for illustrating the FACTs. Here, I synthesized some parts of the text in pictures you can refer to while acquiring the skill. They depict the flow for determining the poles and the zeroes. Of course, some parts will require reading the text in detail, but it may also offer a different insight from what I wrote.

Determining the pole in a 1st-order circuit

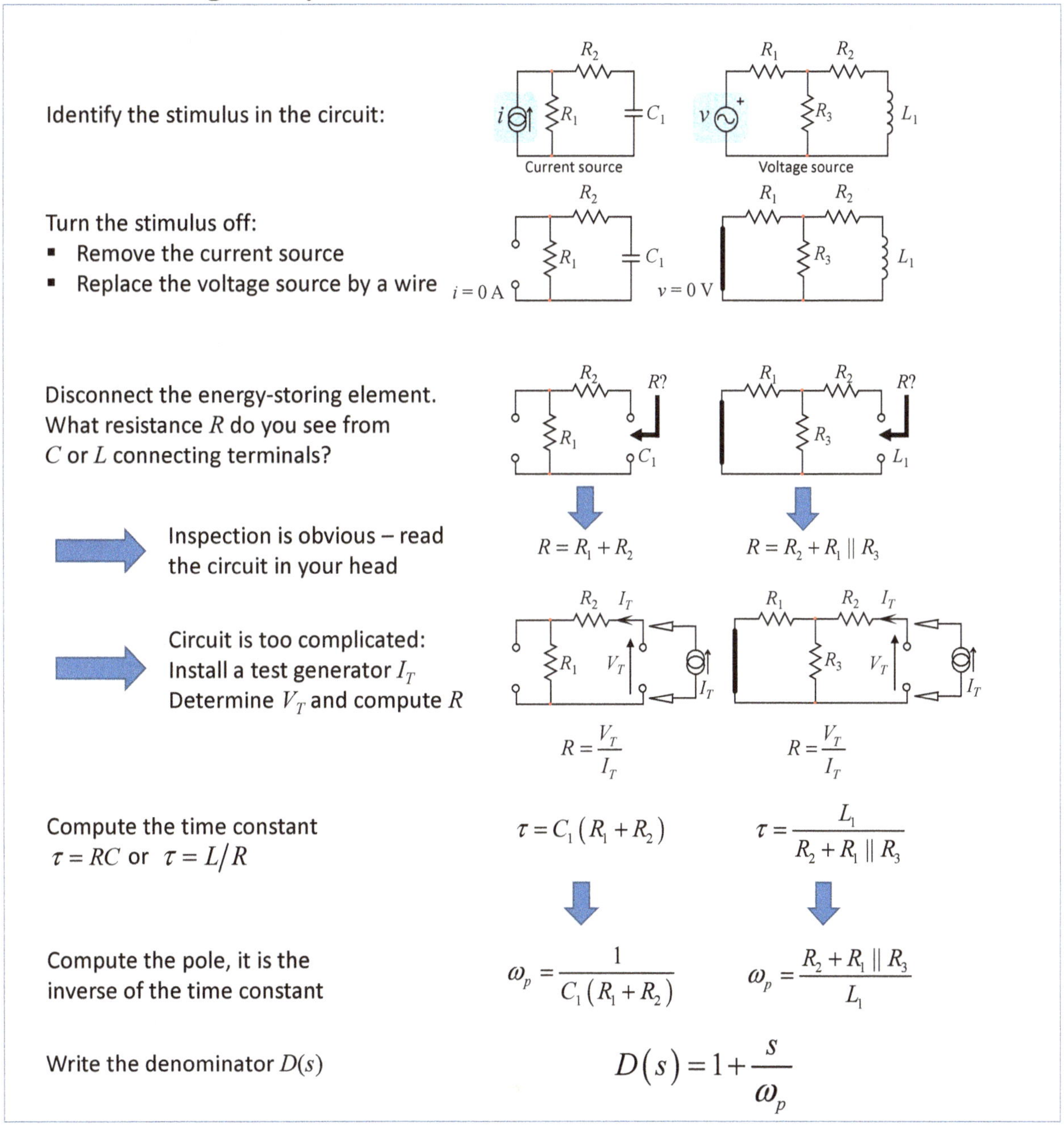

Identify the stimulus in the circuit:
Current source
Voltage source
Turn the stimulus off:
▪ Remove the current source
▪ Replace the voltage source by a wire
$i = 0\ \mathrm{A}$
$v = 0\ \mathrm{V}$
Disconnect the energy-storing element. What resistance R do you see from C or L connecting terminals?
$R?$
Inspection is obvious – read the circuit in your head
$R = R_1 + R_2$
$R = R_2 + R_1 \parallel R_3$
Circuit is too complicated: Install a test generator I_T Determine V_T and compute R
$R = \frac{V_T}{I_T}$
Compute the time constant $\tau = RC$ or $\tau = L/R$
$\tau = C_1(R_1 + R_2)$
$\tau = \frac{L_1}{R_2 + R_1 \parallel R_3}$
Compute the pole, it is the inverse of the time constant
$\omega_p = \frac{1}{C_1(R_1 + R_2)}$
$\omega_p = \frac{R_2 + R_1 \parallel R_3}{L_1}$
Write the denominator $D(s)$
$D(s) = 1 + \frac{s}{\omega_p}$

Determining the poles in a 2nd-order circuit

Count energy-storing elements with *independent* state variables

↓

Assume there are two energy-storing element, L_1 and C_2

↓

The denominator follows the form $D(s) = 1 + b_1 s + b_2 s^2$

↓

Turn the stimulus off and determine time constants for b_1 and b_2

↓ Inspection is working: "look" into the connecting terminals
Circuit is too complicated then use test source I_T

b_1 ⇒

Determine the resistance R_i driving L_1 while C_2 is in dc state (open circuited): $\tau_1 = L_1 / R_i$

Determine the resistance R_j driving C_2 while L_1 is in dc state (short circuited): $\tau_2 = R_j C_2$

Sum the time constants: $b_1 = \tau_1 + \tau_2$

↓

b_2 ⇒

Determine the resistance R_k driving L_1 while C_2 is in high-frequency state (short circuited): $\tau_1^2 = L_1 / R_k$

Determine the resistance R_l driving C_2 while L_1 is high-frequency state (open circuited): $\tau_2^1 = C_2 R_l$

Choose the combination leading to the simplest result: $b_2 = \tau_1 \tau_2^1$ or $b_2 = \tau_2 \tau_1^2$

↓

$$D(s) = 1 + s(\tau_1 + \tau_2) + s^2 \tau_1 \tau_2^1 \iff D(s) = 1 + s(\tau_1 + \tau_2) + s^2 \tau_2 \tau_1^2$$

Determining the zero in a 1st-order circuit

Bring the excitation signal - the stimulus - back in place

↓

Null the response, e.g. $V_{out}(s) = 0\text{ V}$ or $\hat{v}_{out} = 0\text{ V}$

↓

Identify in the *transformed* network, one or several impedances combinations that could block the stimulus propagation: a *transformed* open circuit or a *transformed* short circuit.

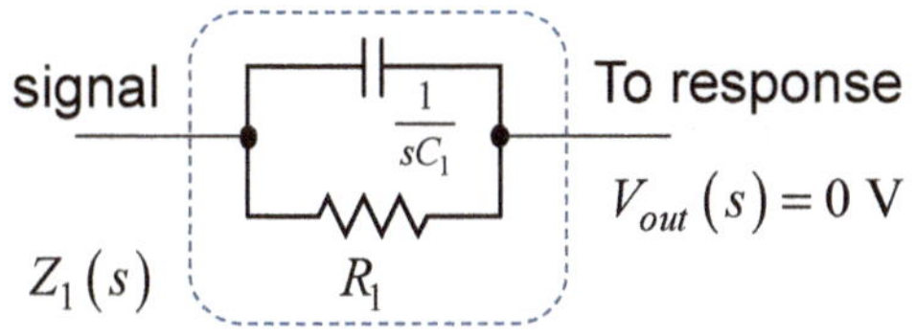
signal
$\frac{1}{sC_1}$
To response
$V_{out}(s) = 0\text{ V}$
$Z_1(s)$
R_1

$Z_1(s_z) \to \infty$

$$N_1(s) = 1 + \frac{s}{\omega_{z_1}}$$

$$\omega_{z_1} = \frac{1}{R_1 C_1}$$

If inspection is not possible, go for a null double injection (NDI)

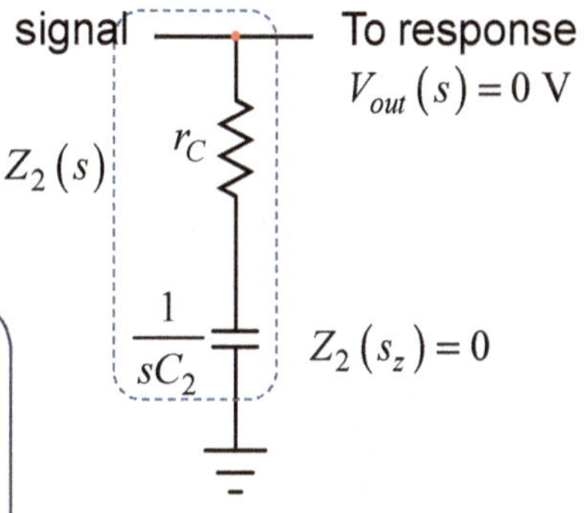
signal
To response
$V_{out}(s) = 0\text{ V}$
$Z_2(s)$
r_C
$\frac{1}{sC_2}$
$Z_2(s_z) = 0$

$$N_2(s) = 1 + \frac{s}{\omega_{z_2}}$$

$$\omega_{z_2} = \frac{1}{r_C C_2}$$

A transformed network simply refers to L and C replaced by their respective impedance values sL and $1/sC$ so that the circuit involves impedances only.

NDI – Creating a null condition

The null double injection or NDI uses the concept of output *null*. It consists of finding a condition in which the network is biased by one fixed stimulus while a second stimulus is tweaked to bring the output voltage to 0 V. See the below picture:

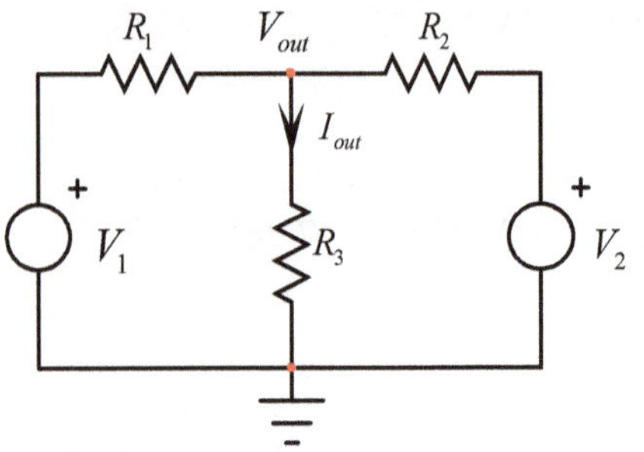

Assuming V_1 is 10 V and all resistors are of equal value, what value will bring node V_{out} to 0 V?

$$V_2 = -V_1$$

A null is different than a short circuit!
Think of it as the virtual ground of an op-amp

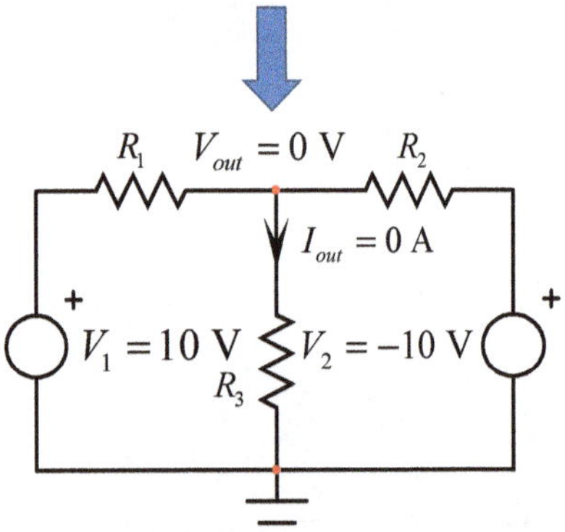

Use SPICE with a bias point analysis to confirm the result

No current circulates in R_3

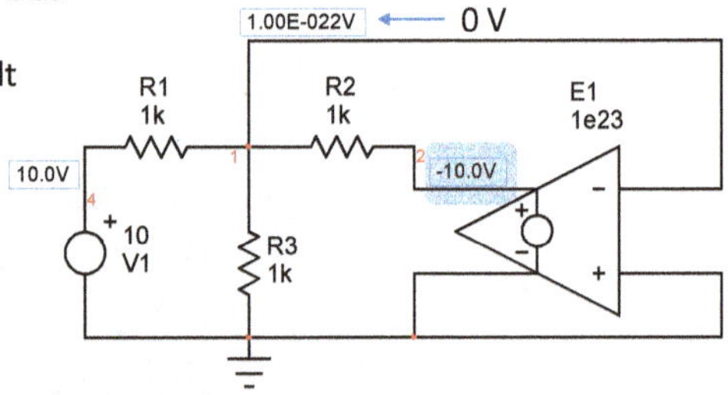

During the FACTs NDI exercises, you will be asked to determine the resistance R "seen'" from one energy-storing element (an L or a C) when the output is nulled. To do so, you will install a test generator I_T injecting a current into the connecting terminals of L or C and adjusted so that it nulls the output. At this moment, the voltage V_T collected across the test generator divided by I_T will be the resistance R you want.

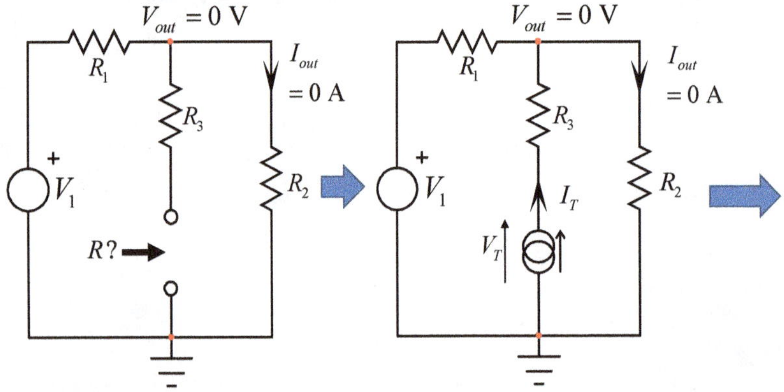

$$R = \frac{V_T}{I_T} = R_3$$

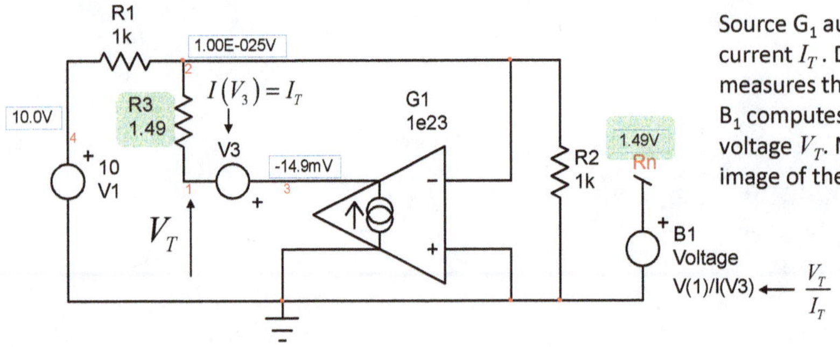

Source G_1 auto-adjusts the test current I_T. Dummy source V_3 measures the injected current I_T and B_1 computes the resistance value with voltage V_T. Node R_n gives a voltage image of the resistance value.

Running a null double injection or NDI

Keep the stimulus in place

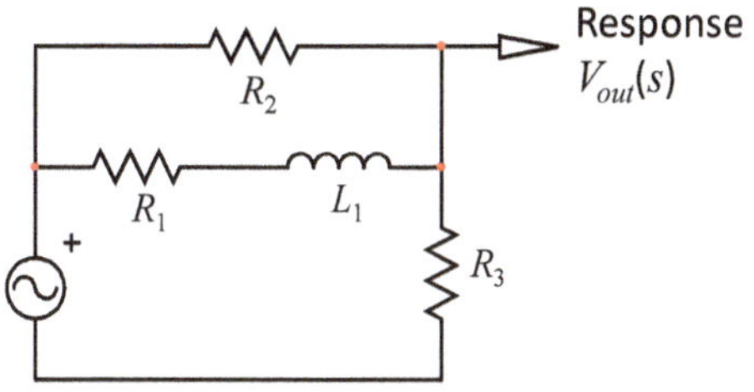
Response
$V_{out}(s)$
R_2
R_1
L_1
R_3
Stimulus

Null the output

"Despite a stimulus in place, the signal cannot reach the output and does not produce a response"

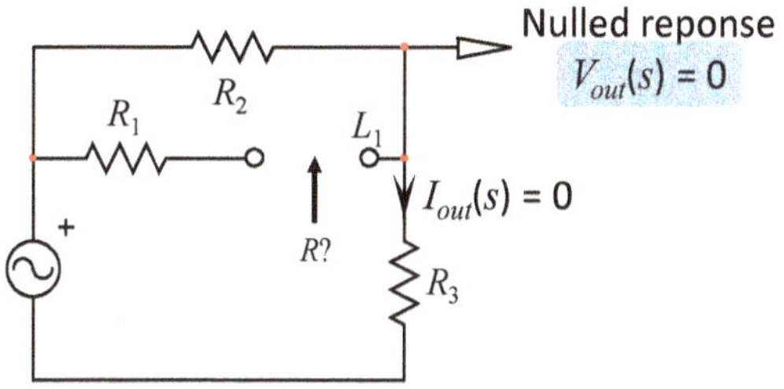
Nulled reponse
$V_{out}(s) = 0$
R_1
R_2
L_1
$I_{out}(s) = 0$
$R?$
R_3
Stimulus

Determine the value of R seen from the inductor (or capacitor) terminals when the output is nulled.

Inspection can work:

"What impedance combination could block the stimulus propagation?"

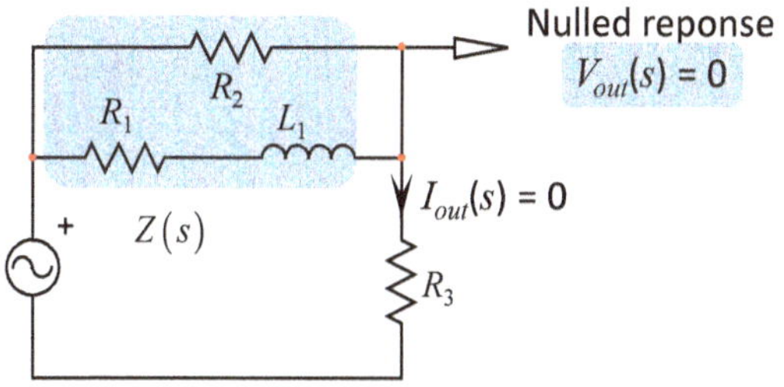
Nulled reponse
$V_{out}(s) = 0$
R_1
R_2
L_1
$I_{out}(s) = 0$
$Z(s)$
R_3
Stimulus

$$Z(s) = R_2 \,||\, (sL_1 + R_1) \to \infty$$

$$Z(s) = R_0 \frac{N(s)}{1 + s\dfrac{L_1}{R_1 + R_2}}$$

Goes infinite if

$$1 + s\frac{L_1}{R_1 + R_2} = 0 \Rightarrow s_z = -\frac{R_1 + R_2}{L_1} \Rightarrow \omega_z = \frac{R_1 + R_2}{L_1}$$

Inspection does not work (circuit too complicated):

Install a test generator I_T across the energy-storing element, this is the second stimulus.

Determine the voltage V_T while the output is nulled and obtain the resistance R in this mode.

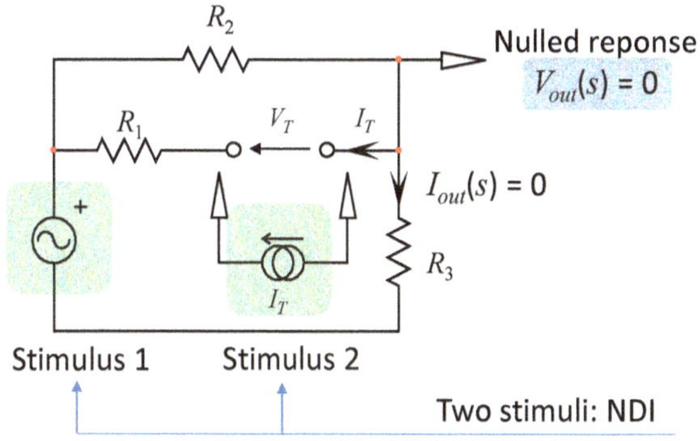
R_2
Nulled reponse
$V_{out}(s) = 0$
R_1
V_T
I_T
$I_{out}(s) = 0$
R_3
I_T
Stimulus 1
Stimulus 2
Two stimuli: NDI

$$R = \frac{V_T}{I_T}$$

$$R = R_1 + R_2 \Rightarrow \tau = \frac{L_1}{R} = \frac{L_1}{R_1 + R_2} \Rightarrow \omega_z = \frac{R_1 + R_2}{L_1}$$

Determining the zeroes in a 2nd-order circuit

Assume there are two energy-storing element, L_1 and C_2

↓

The numerator follows the form $N(s) = 1 + a_1 s + a_2 s^2$

↓

Turn the stimulus on and determine time constants for a_1 and a_2 for a nulled output response

↓ Inspection is working: "look" into the connecting terminals
Circuit too complicated go for NDI and use test source I_T

a_1 ⇒

Determine the resistance R_i driving L_1 while C_2 is in dc state (open circuited): $\tau_{1N} = L_1/R_i$

Determine the resistance R_j driving C_2 while L_1 is in dc state (short circuited): $\tau_{2N} = R_j C_2$

Sum the time constants: $a_1 = \tau_{1N} + \tau_{2N}$

↓

a_2 ⇒

Determine the resistance R_k driving L_1 while C_2 is in high-frequency state (short circuited): $\tau_{1N}^2 = L_1/R_k$

Determine the resistance R_l driving C_2 while L_1 in high-frequency state (open circuited): $\tau_{2N}^1 = C_2 R_l$

Choose the combination leading to the simplest result: : $a_2 = \tau_{1N}\tau_{2N}^1$ or $a_2 = \tau_{2N}\tau_{1N}^2$

⇓

$$N(s) = 1 + s(\tau_{1N} + \tau_{2N}) + s^2(\tau_{1N}\tau_{2N}^1) \Leftrightarrow N(s) = 1 + s(\tau_{1N} + \tau_{2N}) + s^2(\tau_{2N}\tau_{1N}^2)$$

First-order transfer function with the generalized expression

Determine the dc gain ($s = 0$) of the circuit with the stimulus turned on

↓

$H_0 = 10$ Finite gain value

or

$H_0 = 0$ One or several zeroes at the origin

or

$H_0 \to \infty$ Enter a finite high value, simplify later

↓

Determine the time constant of the denominator by turning the stimulus off

↓

$\tau = RC$ or $\tau = \frac{L}{R}$

↓

Bring the stimulus back

Determine the gain H when C or L is set in its high-frequency state

↓ Redraw circuit with C replaced by a wire

Redraw circuit with L removed

$H^1 = 0$ ⇨ There is no zero and $N(s) = H_0$

$H^1 \neq 0$ ⇨ There is a zero and $N(s) = H_0 + sH^1\tau$

↓

$$H(s) = \frac{H_0 + sH^1\tau}{1 + s\tau}$$

↓ $H_0 \neq 0$

$$\Rightarrow \quad H(s) = H_0 \frac{1 + s\frac{H^1}{H_0}\tau}{1 + s\tau} = H_0 \frac{1 + \frac{s}{\omega_z}}{1 + \frac{s}{\omega_p}}$$

$$\omega_z = \frac{H_0}{H^1\tau} \qquad \omega_p = \frac{1}{\tau}$$

Simplify if possible

Index

A

B

C

D

E

F

G

H

I

L

M

N

O

P

Q

R

S

T

U

V

Z

www.ingramcontent.com/pod-product-compliance
Lightning Source LLC
Chambersburg PA
CBHW081057220326
41598CB00038B/7127
9781960405197